D1824766

Folded Pages

LOWER MOTTLED SANDSTONE ON SANDSTONE IN MIDDLE PERMIAN MARL AT GATEFORD SAND-PIT (L 564) PLATE I (*Frontispiece*)

The face of red sandstone below the house is formed of Lower Mottled Sandstone, worked for moulding sand. The red and grey banded sandstone in the nearer face and the grey sandstone in the floor of the pit occur at the top of the Middle Permian Marl and are worked for building sand.

NATURAL ENVIRONMENT RESEARCH COUNCIL

INSTITUTE OF GEOLOGICAL SCIENCES

MEMOIRS OF THE GEOLOGICAL SURVEY OF GREAT BRITAIN
ENGLAND AND WALES

Geology of the Country around East Retford, Worksop and Gainsborough

(*Explanation of One-inch Geological Sheet 101, New Series*)

BY

E. G. Smith, B.Sc., G. H. Rhys, B.Sc.
and R. F. Goossens, B.Sc.

with contributions by
C. G. Godwin, M.Sc. and R. A. Eden, B.Sc., F.R.S.E.

Palaeontology by
M. A. Calver, M.A., Ph.D., J. Pattison, M.Sc., B. Owens, B.Sc., Ph.D.,
G. Warrington, B.Sc., Ph.D., W. H. C. Ramsbottom, M.A., Ph.D.
and H. C. Ivimey-Cook, B.Sc., Ph.D.

Petrography by
R. K. Harrison, M.Sc.

LONDON
HER MAJESTY'S STATIONERY OFFICE
1973

A

ISBN 0 11 880590 8

PREFACE

THIS MEMOIR describes the geology of the district covered by the East Retford (101) New Series Sheet of the One-inch Geological Map of England and Wales. The district was originally surveyed on the scale of one inch to one mile by W. T. Aveline, W. H. Dalton, T. V. Holmes and W. A. E. Ussher, and the results, including revision and additions, were published on Old Series One-inch maps 82 NE, 82 SE and 83 between 1861 and 1897. Accompanying memoirs for Sheets 82 NE and 82 SE were written by W. T. Aveline and published in 1861, with second editions in 1880 and 1879 respectively. The geology of Sheet 83 was described in a memoir by W. A. E. Ussher, A. J. Jukes-Browne and A. Strahan, published in 1888.

Small areas in the west of New Series Sheet 101 were surveyed on the scale of six inches to one mile by Mr. W. N. Edwards and Dr. G. H. Mitchell in 1931 and 1940, but the greater part of the district was mapped on that scale during 1946–7 and 1957–61 by Messrs. E. G. Smith, R. F. Goossens, G. H. Rhys, R. A. Eden and Dr. V. A. Eyles, under the supervision of the last-named, Dr. A. W. Woodland and Mr. D. R. A. Ponsford as successive District Geologists. The new one-inch map was published in 1967.

Most of the present memoir has been written by Mr. E. G. Smith, who has also been responsible for its compilation. Messrs. G. H. Rhys and R. F. Goossens have provided a detailed account of the upper part of the Lower Coal Measures and of the Middle Coal Measures in Chapter IV, and have contributed, along with Messrs. C. G. Godwin and R. A. Eden, to several other parts of the memoir. The animal fossils have been named by Dr. M. A. Calver, assisted by Mr. W. G. E. Graham and Miss D. M. Gregory, (Coal Measures), by Mr. J. Pattison (Permian), Dr. W. H. C. Ramsbottom (Carboniferous Limestone and Millstone Grit Series) and Dr. H. C. Ivimey-Cook (Rhaetic and Jurassic); and palynological accounts have been provided by Dr. B. Owens (Carboniferous) and Dr. G. Warrington (Permo-Trias). Mr. R. K. Harrison, assisted by Mr. K. S. Siddiqui, has supplied mineralogical and petrographical data for several chapters. Dr. P. T. Warren assisted in the preparation of a few diagrams and part of Appendix I. The memoir has been edited by Mr. D. R. A. Ponsford.

Grateful acknowledgment is made to numerous organizations and individuals, in particular to the National Coal Board and the British Petroleum Company, for generous help and for permission to publish certain geological records.

K. C. DUNHAM
Director

Institute of Geological Sciences
Exhibition Road
London, S.W.7
19th October, 1973

v

CONTENTS

ILLUSTRATIONS

TEXT-FIGURES

viii

PLATES

TABLES

LIST OF SIX-INCH MAPS

The following is a list of six-inch geological National Grid sheets included, wholly or in large part, in one-inch map Sheet 101, with the initials of the surveyors and dates of survey. The surveyors were: R. A. Eden, W. Edwards, V. A. Eyles, R. F. Goossens, G. H. Mitchell, G. H. Rhys and E. G. Smith. The sheets marked with an asterisk are printed and published by the Ordnance Survey; copies of the rest are available in dye-line form from the Institute of Geological Sciences, Ring Road Halton, Leeds LS15 8TQ.

SK 57 NE*	Worksop	R.F.G., W.E., R.A.E.	1940–57
SK 58 SE*	Woodsetts, Carlton in Lindrick and Shireoaks..	W.E., R.F.G., R.A.E., E.G.S.	1940–64
58 NE*	Firbeck, Oldcoates, Letwell and Langold	R.F.G.	1946
SK 59 SE*	Tickhill and Sandbeck ..	R.F.G., G.H.M., R.A.E., E.G.S. ..	1931, 1946, 1957–60
SK 67 NW*	Manton and Clumber Park	R.F.G., V.A.E., R.A.E.	1946–59
67 NE*	Elkesley	V.A.E., R.A.E., E.G.S.	1947, 1957–63
SK 68 SW*	Scofton	R.F.G., R.A.E., W.E.	1946–59
68 SE*	Barnby Moor, Sutton and Babworth	R.A.E., V.A.E. ..	1947, 1957
68 NW*	Blyth	R.F.G., E.G.S. ..	1946, 1958
68 NE*	Mattersey, Ranskill, Torworth and Lound ..	E.G.S.	1957–59
SK 69 SW*	Styrrup, Harworth and Bircotes	E.G.S., R.F.G. ..	1946, 1958–60
69 SE*	Bawtry, Scrooby, Scaftworth and Everton	E.G.S.	1957–59
SK 77 NW*	Ordsall, Grove, Eaton, Gamston and Headon ..	V.A.E., R.A.E., E.G.S.	1947–66
77 NE	Treswell, Rampton, Stoke-ham and East Drayton ..	G.H.R.	1957
SK 78 SW	Hayton, Clarborough and East Retford	R.A.E., V.A.E., E.G.S.	1947, 1957–63
78 SE	Sturton le Steeple, North Leverton and South Leverton	G.H.R., R.A.E. ..	1947, 1959–61
78 NW	Wiseton and Clayworth ..	E.G.S., R.A.E. ..	1947, 1957–59
78 NE	Saundby, Bole and North Wheatley	E.G.S., R.A.E. ..	1947, 1957–60
SK 79 SW	Gringley on the Hill ..	E.G.S.	1957–59
SK 79 SE	Walkeringham and Beckingham	E.G.S.	1957
SK 87 NW	Cottam and Laneham ..	G.H.R.	1957–61
SK 88 SW	Knaith and Littleborough	G.H.R.	1957–61
88 NW	Gainsborough (south) and Lea	G.H.R.	1960–61
SK 89 SW	Morton and Gainsborough (north)	E.G.S.	1960

A narrow strip of ground along the western margin of the one-inch map is included in six-inch sheets SK 57 SW, 57 NW, 58 SW, 58 NW and 59 SW, the first three of which are published. A strip of ground along the southern margin is included in six-inch sheets SK 57 SE, 67 SW, 67 SE, 77 SW, 77 SE and 87 SW, which are incomplete and unpublished.

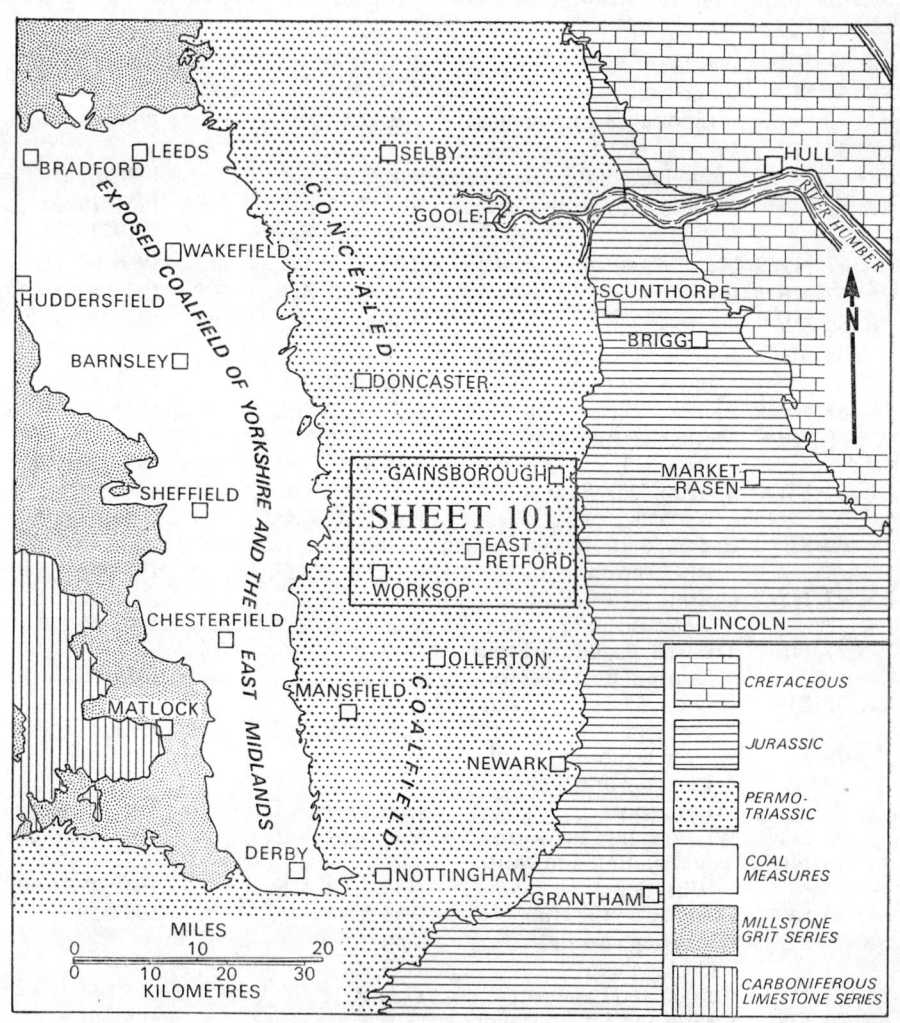

FIG. 1. *Sketch-map showing the location and general geological relations of the East Retford district*

Chapter I

INTRODUCTION

GENERAL AND GEOGRAPHICAL

THIS MEMOIR describes the district[1] covered by the East Retford (101) Sheet of the One-inch Geological Map of England and Wales, a district which lies largely in Nottinghamshire but includes a portion of the West Riding of Yorkshire in the north-west, a corner of Derbyshire in the south-west and parts of Lindsey (Lincolnshire) in the east. Apart from a small area of basal Lias in the north-east, the solid rocks at surface are of Permo-Triassic age (Fig. 1), dipping gently eastwards and not extensively obscured by superficial deposits except along the main rivers. The Permo-Triassic rocks everywhere rest upon Coal Measures, and these contain large reserves of coal which, in the west, have been proved in numerous exploratory boreholes and are being exploited from several collieries (Fig. 2). The district thus forms an important part of the Yorkshire–Nottinghamshire Concealed Coalfield. Beneath the Coal Measures the full succession of the Millstone Grit Series has been proved in a number of deep borings for oil that reach Carboniferous Limestone, the oldest known strata in the district. Commercial quantities of oil in the folded Carboniferous rocks have recently been discovered and tapped at several places. Sites of all the boreholes sunk in search of oil are shown on Figs. 2 and 3.

The district is one of subdued relief, presenting in general an undulating, rural aspect. Only a small area rises above 200 ft O.D.—with 366 ft on the Bunter Pebble Beds at Manor Hills, south of Worksop, as the highest point. The lowest part of the district is an area of fenland in the north-east which is only a few feet above sea level. The district lies almost entirely within the catchment area of the River Trent and, except in the west where the River Ryton and its tributaries run eastwards down the Permian dip-slopes, the drainage is predominantly to the north. The Ryton turns northwards to the east of Worksop and flows over the outcrop of the Bunter Pebble Beds to join the River Idle at Scrooby. The Idle, originating from the confluence of the rivers Poulter and Maun at West Drayton in the south, flows northwards over the Waterstones to East Retford, and thence over the Pebble Beds to Bawtry, receiving no significant tributary other than the Ryton on the way. A few minor becks drain the Keuper Marl outcrop eastwards to the River Trent, which runs along the eastern edge of the district between Dunham Bridge and Walkerith, meandering across a wide flood plain which falls less than 10 ft in 12 miles. The Trent is tidal throughout the district.

The district is predominantly agricultural, and can be broadly divided into three regions corresponding to the principal outcrops. The loams and clays derived from the Permian rocks in the west and the heavy claylands of the

---

[1]Throughout the memoir the word 'district' refers to the area included in the East Retford (101) Sheet.

1

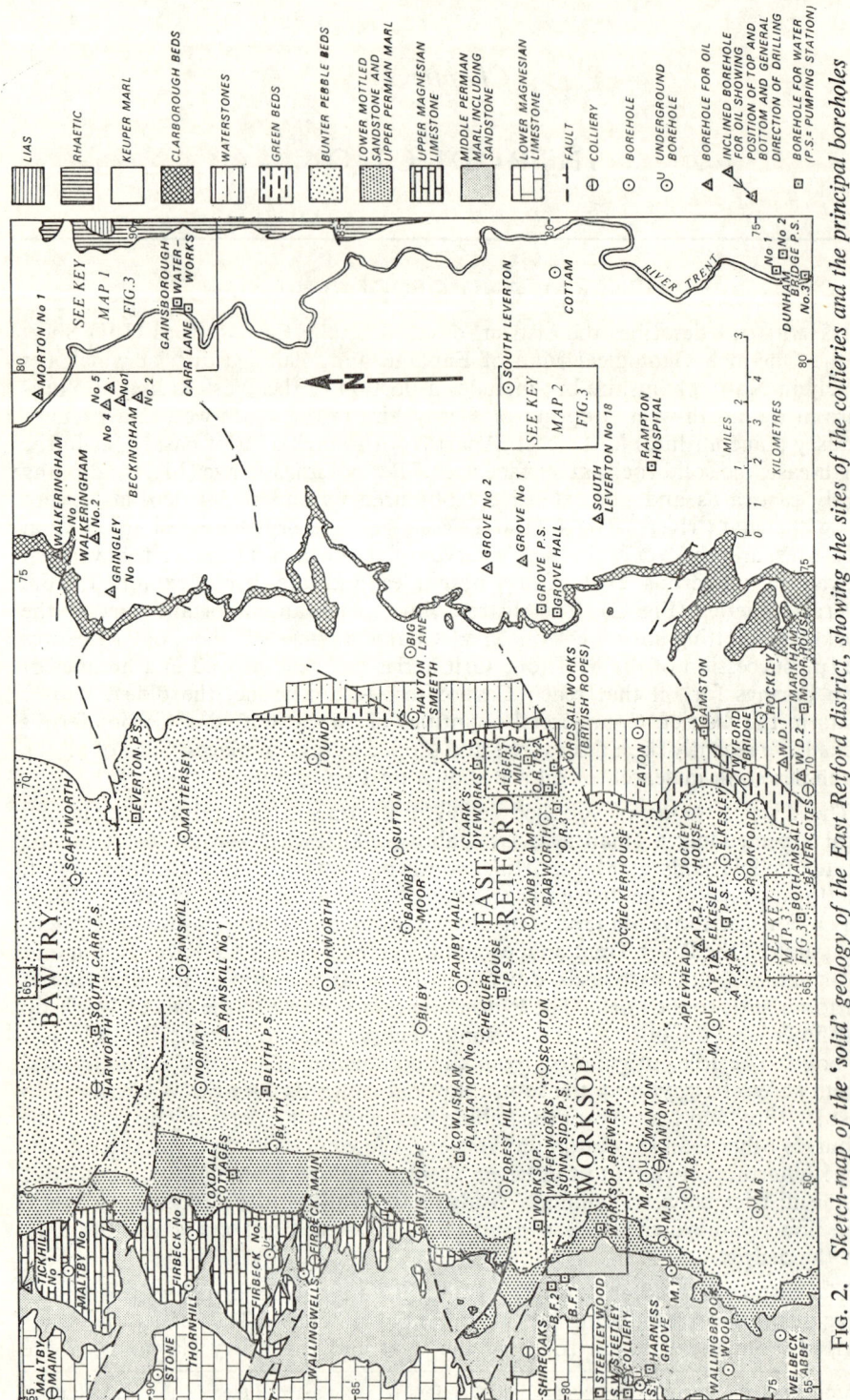

Fig. 2.   Sketch-map of the 'solid' geology of the East Retford district, showing the sites of the collieries and the principal boreholes
Abbreviations of borehole names are listed on the opposite page.

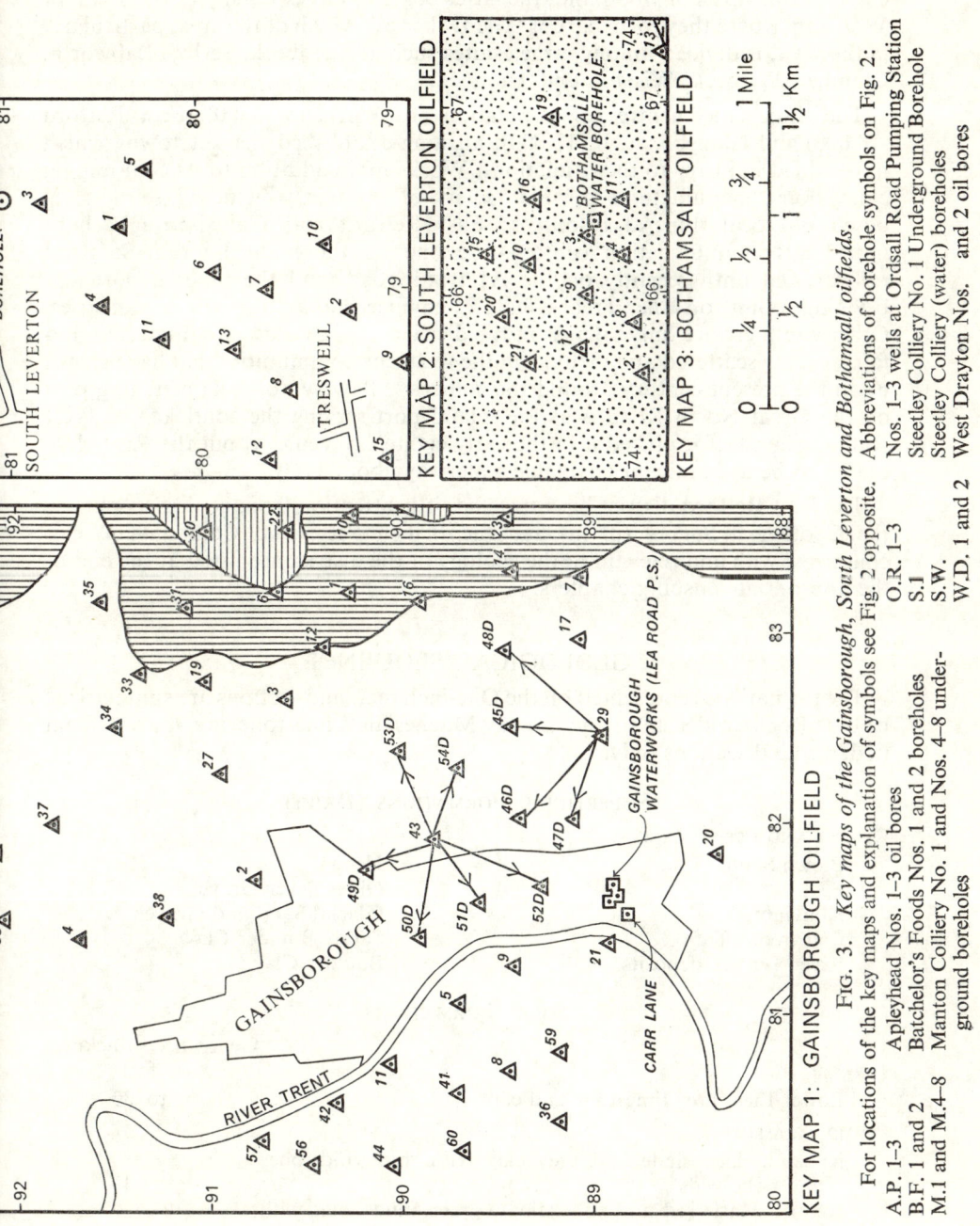

KEY MAP 2: SOUTH LEVERTON OILFIELD

KEY MAP 3: BOTHAMSALL OILFIELD

KEY MAP 1: GAINSBOROUGH OILFIELD

Fig. 3. *Key maps of the Gainsborough, South Leverton and Bothamsall oilfields.*

For locations of the key maps and explanation of symbols see Fig. 2 opposite.   Abbreviations of borehole symbols on Fig. 2:

A.P. 1–3	Apleyhead Nos. 1–3 oil bores	O.R. 1–3	Nos. 1–3 wells at Ordsall Road Pumping Station
B.F. 1 and 2	Batchelor's Foods Nos. 1 and 2 boreholes	S.I	Steetley Colliery No. 1 Underground Borehole
M.1 and M.4–8	Manton Colliery No. 1 and Nos. 4–8 under-	S.W.	Steetley Colliery (water) boreholes
	ground boreholes	W.D. 1 and 2	West Drayton Nos. and 2 oil bores

Keuper Marl in the east are largely devoted to arable farming, while the light and relatively hungry soils of the Pebble Beds support mixed farming and considerable areas of woodland, the latter being found especially to the south of Worksop, where they form part of 'The Dukeries'. Much of the area, particularly in the west, is divided among large estates such as Sandbeck, Serlby, Babworth, Clumber, Welbeck, Wiseton and Thonock.

The three chief centres of population are Worksop (3500), East Retford (18 000) and Gainsborough (17 000), all long-established market towns which have added industry to their functions. Worksop, in addition to its coal-mining connexions, manufactures a wide variety of products, including refractories based on local mineral resources; East Retford and Gainsborough, both important communication centres—the one on the main London–Scotland railway and until recently on the A1 trunk road, and the other a port and bridging point on the Trent—feature engineering among their industries. Otherwise there are only villages in the district, though these are numerous and fairly evenly scattered and include, in the west, the communities that have grown up in the present century around the collieries. Bawtry, for long a staging post on the Great North Road, was the inland port serving the south of the West Riding when Defoe visited it in the early eighteenth century, but the River Idle ceased to be a significant navigable waterway about 1760.

The East Retford district has attracted little attention from geologists outside the Geological Survey (p. iii) and the British Petroleum Company. Works concerned with and relevant to the geology of the district are listed at the end of the appropriate ensuing chapters.

GEOLOGICAL SEQUENCE

The formations represented on the One-inch map and sections are summarized below. Rocks older than the Lower Magnesian Limestone are known from underground sections only.

SUPERFICIAL FORMATIONS (DRIFT)

RECENT AND PLEISTOCENE

Blown Sand	Head
Peat	Older River Gravel
Alluvium	Glacial Sand and Gravel
Calcareous Tufa	Sandy Boulder Clay
River Terrace deposits	Boulder Clay

SOLID FORMATIONS

	Generalized thickness in feet
JURASSIC	
Lower Lias: grey limestone and clay	to 30
PERMO-TRIASSIC	
Rhaetic: dark shale and grey clay with thin sandstones and bone-beds	40
Keuper Marl: red and subordinate green mudstones with skerries and Clarborough Beds, gypsum and anhydrite; thin Tea Green Marl at top	650 to 900
Waterstones: red and brown mudstones, siltstones and sandstones	0 to 180

Generalized thickness
in feet

Green Beds: mainly green and grey mudstones; anhydrite and basal conglomerate locally	0 to 70
Bunter Pebble Beds: red and brown sandstone with pebbles	600 to 1000
Lower Mottled Sandstone: red fine-grained sandstone ..	30 to 150
Upper Permian Marl: red mudstone and sandstone with anhydrite and gypsum	0 to 190
Upper Magnesian Limestone: thinly bedded dolomite and limestone	0 to 130
Middle Permian Marl: red mudstone and red and grey sandstone; gypsum and anhydrite bands	80 to 180
Lower Magnesian Limestone: dolomite and limestone ..	80 to 350
Lower Permian Marl: grey argillaceous carbonate-siltstones with dolomite and limestone bands; Marl Slate 0 to 30 ft at base	0 to 180
Basal Permian Sands and Breccia: grey sand, conglomerate and breccia	0 to 90

Unconformity

CARBONIFEROUS

Coal Measures (Westphalian)	
Upper Coal Measures: grey and primary red mudstones, sandstones and siltstones with thin coals	to 1100
Middle Coal Measures: grey mudstones, siltstones and sandstones with marine bands and numerous coals, several of which are workable	800 to 1900
Lower Coal Measures: grey mudstones, siltstones and sandstones; marine bands in lower part; numerous coals, several of which are workable, in upper part ..	800 to 1600
Millstone Grit Series (Namurian): grey and dark mudstones with sandstones, siltstones and marine bands; thin limestone bands in lower part; local volcanic rocks in upper part	750 to 1800+
Carboniferous Limestone Series (Dinantian): grey and dark limestones with dark shales at top	to 300

INTRUSIVE IGNEOUS ROCKS

Dolerite in Millstone Grit Series

GEOLOGICAL HISTORY

The uppermost beds of the Carboniferous Limestone Series consist of limestone and shales laid down in the sea that covered the greater part of England at this time. The succeeding Millstone Grit Series, or Namurian, consists largely of mudstones and shales with beds of sandstone, particularly towards the top. Much of the lower part of these deposits was probably laid down under marine conditions and, because subsidence of the sea floor was more marked in the north of the district, in the area of the so-called Gainsborough Trough, they are several times thicker there than in the south. The Millstone Grit deposits are progressively less marine upwards, and in later Namurian times were probably laid down for the most part in fresh or brackish water. The sandstones provide evidence of deltaic conditions invading the district from the north, and periodically the whole area was flooded by the sea, which left behind thin but stratigraphically important bands of marine shale. Towards the end of Namurian times there were outbursts of volcanic activity in the south of the district, and these are represented in the succession by tuffs, agglomerates and lavas.

The upper Namurian sediments have a clear cyclic pattern resulting from the rhythmic subsidence of the inshore shelf area on which they were deposited. The Coal Measures, laid down conformably on top of them under similar conditions, show the same repetitive sequence of deposits, but, in general, marine incursions were less frequent, there is a smaller proportion of sandstone —and that generally of finer grain—and there is a marked increase in the number and importance of seatearths and coals. The Coal Measures succession is comparable to that found elsewhere in the Pennine coalfields, and clearly was accumulated in the same broad basin of deposition. In the East Retford district the sequence thickens towards the west, where more than 4500 ft of strata were laid down. The bulk of the Coal Measures are grey, but primary red beds make their appearance in the Upper Coal Measures. These red beds are transitional deposits between the grey cyclic measures and the rocks of Etruria Marl facies known farther south, and foreshadow the uplift which was to bring Carboniferous sedimentation to an end.

The Hercynian earth-movements which set in towards the end of Coal Measures times uplifted the whole region and imposed moderate folding and faulting on the Carboniferous rocks. There followed a long period of terrestrial conditions, lasting through Lower Permian times, during which considerable thicknesses of Carboniferous strata were removed by denudation.

When the sea returned to the district in Upper Permian times it was to an arid, peneplained area, thinly and irregularly covered by Basal Permian Sands and Breccia. This sea, part of the north European Zechstein Sea, apparently had only restricted connexion with the oceans of the time, for its deposits are characterized by a restricted and peculiar fauna, and by chemical precipitates and considerable thicknesses of red measures. Limestone, dolomite and gypsum, together with mudstone, were laid down in a series of evaporite cycles, and near the margin of the sea sands and sporadic thin breccias were deposited. The south-western part of the East Retford district is typical of this marginal development, and it is here that facies changes, a much-discussed feature of the Permo-Triassic rocks of the East Midlands, are most marked.

The Zechstein deposits, over 800 ft thick in the north-east of the district, are succeeded by the Triassic rocks, the lower part of which consists of a thick arenaceous formation, the Bunter Pebble Beds. These beds are evidently continental in origin, and were presumably derived from the south, as the false bedding suggests and because the pebbles become smaller and rarer northwards. There followed a period of uplift and erosion, which was succeeded by another invasion of the sea. The Triassic sediments deposited in this sea consist mainly of Keuper Marl, a red mudstone formation with evaporites, but in the lower part of the sequence in the south of the district they include the Waterstones, which contain a high proportion of sandstone and indicate littoral conditions. There is a slight stratigraphical break between the Keuper Marl and the overlying Rhaetic rocks. The latter consist mainly of dark shales and grey marls, which herald a change to a fully marine environment in ensuing Lower Lias times, though there is evidence in the Upper Rhaetic of a local reversion to Keuper Marl conditions of deposition. The only Jurassic rocks preserved in the district are a few feet of basal Liassic limestones and clays in the extreme north-east. The rest of the Jurassic, together with the Cretaceous and any early Tertiary deposits, has been completely removed by denudation after a period of uplifting movements in mid-Tertiary times. These movements were responsible for the gentle easterly tilt exhibited by the post-Carboniferous rocks, and for limited faulting—mainly adjustments along pre-existing fractures—and minor folding.

The next recorded episode in the geological history of the district is the appearance of an ice-sheet in Pleistocene times. This pre-Weichselian glaciation left behind tills and outwash deposits, which evidently were once extensive, but which have been reduced to relics by subsequent erosion. In Weichselian times the district was apparently ice-free, and from then until the present day, gravels and other alluvia have been accumulating on the lower ground. The recent deposition of calcareous tufa, peat and blown sand completes the geological column.

E.G.S.

Chapter II

CARBONIFEROUS LIMESTONE SERIES
(DINANTIAN)

GENERAL ACCOUNT

ROCKS of this Series, the oldest known in the district, have been proved in four oil bores—Bothamsall No. 3, Apleyhead No. 1, Grove No. 1 and South Leverton No. 1. They consist of the uppermost 300 ft or so of limestones and shales (Fig. 4) of Upper *Posidonia* (P_2) and *Dibunophyllum* (D) ages. Lower parts of the Series are unknown and can only be speculated upon in the light of sections outside the district, particularly that provided by Eakring No. 146 Well, 9 miles to the south, where nearly 3000 ft of limestones and other rocks of C_1 and higher zones rest on about 1700 ft of conglomeratic sandstones, which may be partly or possibly wholly of Devonian age, and which in turn rest on supposed Cambrian rocks (Lees and Taitt 1946, pp. 280–3 and fig. 10; Edwards 1967, pp. 9 and 14–23).

Details of the Carboniferous Limestone in the above-listed oil bores in the East Retford district are known chiefly from descriptions of chippings by geologists of the British Petroleum Co. and from interpretations of the geophysical logs. Only a small amount of coring was undertaken, and in each case this was well down in the succession. In these circumstances difficulty arises in defining the upper boundary of the Carboniferous Limestone Series. In the country farther south Edwards (1967, pp. 14, 32), following the practice of the British Petroleum Co.'s geologists, has taken the boundary at the junction between strata consisting predominantly of limestones and predominantly of shales, although it was realised that the true junction between the Dinantian and Namurian as defined by the goniatite faunas lies in many cases within the shales at a higher level. In the present district it is believed that, by comparing the gamma-ray logs of the bores with those of sections known in detail elsewhere, a more precise delimitation can be attempted. Boreholes in the Ashover area of Derbyshire yielded a detailed faunal sequence and show that there the shales of Lower *Eumorphoceras* (E_1) age are much more radioactive than the underlying and overlying strata, giving a characteristic high on the gamma-ray profile (Ramsbottom and others 1962, p. 119; Cosgrove in *ibid.*, p. 164, pl. viii). A similar high is a feature of the gamma-ray logs of the strata at or near the limestone/shale junction in the East Retford oil bores, and by analogy the top of the Carboniferous Limestone is taken at the same place on the logs as, for example, at Highoredish Borehole (Fig. 4). It should be noted, however, that between Highoredish and the Bothamsall–Apleyhead bores lies Calow No. 1 Oil Bore, where cores were taken in the P_2 and E_1 strata (Smith and others 1967, pp. 37, 65) and where gamma-ray logging revealed a coincidence of the highest part of the profile with the upper part of the P_2 Zone, but there was no marked radioactivity peak.

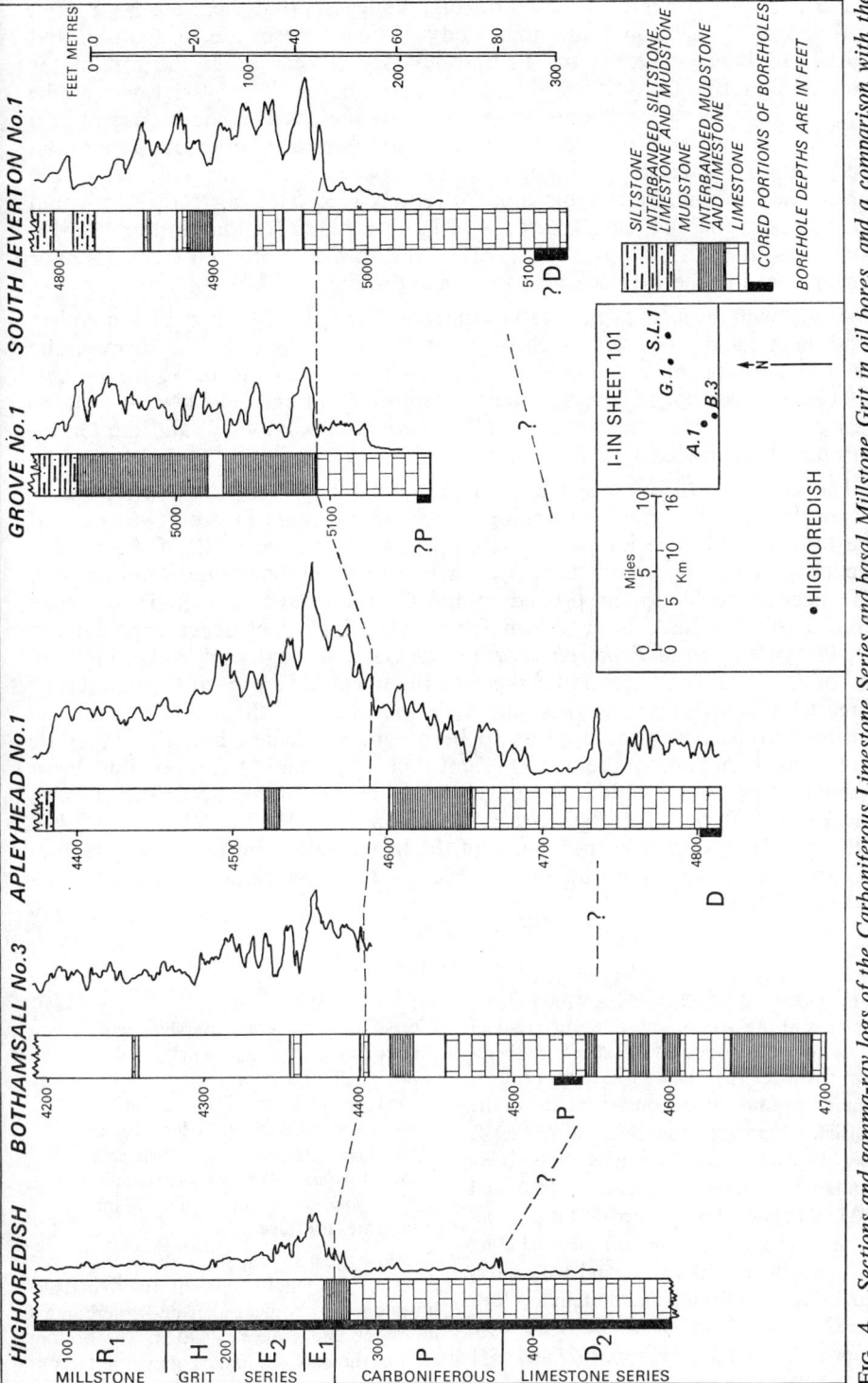

FIG. 4. Sections and gamma-ray logs of the Carboniferous Limestone Series and basal Millstone Grit in oil bores, and a comparison with the succession in the Ashover area of Derbyshire

The tentative correlation shown on Fig. 4 suggests that there are 50 to 65 ft of shales of P_2 age above the main body of limestone in the Bothamsall and Apleyhead bores, whereas the P_2/E_1 boundary is near or at the top of the limestone in the Grove and South Leverton bores. This variation can be accounted for by facies change or, possibly, by the eastward development of a non-sequence at the top of the Carboniferous Limestone Series. In Bothamsall No. 3 Bore P_2 fossils have been found between 4526 and 4541 ft: if the top of the Carboniferous Limestone Series has been fixed correctly this would indicate that, there at least, the rocks of P_2 age are thicker than in the Ashover boreholes, though still, possibly, within the thickness limits of the Cawdor Group rocks in the Matlock area (Smith and others 1967, p. 8).

Apart from the shales at the top of the sequence in the Bothamsall and Apley-head bores and at intervals throughout the succession in the former, the Dinantian rocks proved consist of pale, with subordinate dark, limestones. They are evidently of 'massif' facies, deposited on the 'block' between the basinal areas of the Widmerpool Gulf (Falcon and Kent 1960) and the Gains-borough Trough (Kent 1966).

The Gainsborough Trough crosses the northern part of the district and is responsible for the great thickening which occurs there in the lower part of the Millstone Grit Series (Kent 1966, pp. 334–7; and see p. 12 of the present account). It is inferred to have been active in Dinantian times (Kent 1967). The deepest rocks so far proved in the Gainsborough Trough in the East Retford district have been shown from palynological evidence (pp. 27 and 28; Plate II) to be of *Eumorphoceras* age at Gainsborough and Walkeringham, but oil bores at Trumfleet and Askern to the north have proved Carboniferous Limestone below a thick Millstone Grit sequence. In these latter bores the Carboniferous Limestone appears to be of 'massif' facies, but, since they lie on the northern flank of the trough (Kent 1966, figs. 3 and 5), it is possible that a basinal facies, comparable to the development in the Widmerpool Gulf (Falcon and Kent 1960, p. 20; *A. Rep. Inst. geol. Sci. for 1966*, 1967, p. 71 and *for 1967*, 1968, pp. 81–2) occurs in the middle of the trough. Ramsbottom (1969, p. 766) has speculated upon the possible presence of reefs associated with the Gains-borough Trough.   E.G.S.

DETAILS

In *Bothamsall No. 3 Oil Bore*, 4700 ft deep, the base of the Millstone Grit Series is taken at about 4405 ft (see Fig. 4). This bore therefore provides the thickest section of Carboniferous Limestone—almost 300 ft—in the district. Chippings show that the top 50 ft consist of dark grey calcareous shale, interbanded with limestone between 4420 and 4435 ft. Below this, down to 4550 ft, is limestone, largely pale grey or buff, but including dark grey and black types, and with interbanded shale in the top and bottom few feet. Cores, of which about 70 per cent were recovered, were taken between 4525 and 4541 ft. Details of these are as follows:

	ft
Limestone, pale grey, massive, with stylolites and a few small vughs; rare shell fragments..	$1\frac{1}{2}$
Limestone, dark grey, with pale bands and bands of dark shaly limestone; stylolites; *Leiopteria sp.*, *Limipecten sp.*, *Martinia?*, *Productus concinnus*, *Neoglyphioceras sp.* This fauna indicates a P_2 age	$14\frac{1}{2}$

Below 4550 ft chippings show shale and limestone to about 4605 ft, pale massive limestone to about 4640 ft and interbanded limestone and shale to 4690 ft. The bottom 10 ft of the bore are in dark grey argillaceous limestone.

In *Apleyhead No. 1 Oil Bore* some 223 ft of strata are tentatively assigned to the Carboniferous Limestone Series (Fig. 4). Chippings indicate that the top 10 ft or so consist of dark calcareous shale, and the next 50 ft, down to 4653 ft, of dark shale interbanded with dark grey limestone. The rest of the section is limestone, largely pale grey to grey, but with dark bands in the lower part; crinoid debris and calcite veins are recorded. The upper part of the limestone contains bands of dark shale, probably down to 4690 ft according to the gamma-ray log (Fig. 4), and shale bands also occur at the bottom. The lowermost 13 ft, to a depth of 4805½ ft, were cored, but only 5½ ft of cores were recovered. Details are as follows:

	ft	in
Limestone, grey, thinly bedded, with bands and partings of calcareous shale; *Gigantoproductus sp.*, *Spirifer sp.* of *S. bisulcatus* group; smooth spiriferoid fragments in bottom 6 in .. .. ..·	2	7
Shale, hard, dark grey, with thin limestone bands; crinoid debris ..	0	8
Limestone, grey to dark grey ..	2	3

The fossils are not diagnostic, but are consistent with a *Dibunophyllum* age for this part of the Limestone.

In *Grove No. 1 Oil Bore* about 65 ft of strata are assigned to the Carboniferous Limestone Series (Fig. 4). These consist of limestone, described from chippings as grey or brownish grey crystalline limestone with a relict oolitic and calcarenitic texture, containing, in the upper part, argillaceous partings. Coring was carried out from 5154 ft to the bottom of the bore at 5161 ft, the basal 1 ft 5 in of core being lost. Details are as follows:

	ft	in
Limestone, grey to brownish grey; massive in upper part, unevenly but generally thinly bedded below; dolomitized bed at 5156 ft; crinoid debris, costate productoid fragment and smooth spiriferoid fragment below 5156 and 5157½ ft	5	7

In *South Leverton No. 1 Oil Bore* the Millstone Grit–Carboniferous Limestone boundary is taken at about 4965 ft, near the top of a succession of limestones which extend to the bottom of the bore at 5125 ft. These limestones, as described from chippings, are generally buff-coloured though locally pale grey or white, are mostly crystalline and show partial dolomitization near the top. Cores from 5105 to 5125 ft are described as follows:

	ft
Limestone, pale brownish grey, coarsely crystalline and massive; *Antiquatonia sp., Krotovia?, Schizophoria sp., Spirifer sp.*, smooth spiriferoids indet. *Core broken, and lost between 5109 and 5112 ft and below 5119 ft.* The fossils do not indicate an age other than fairly high Viséan	20

E.G.S.

REFERENCES

COSGROVE, M. E. *in* RAMSBOTTOM, W. H. C., RHYS, G. H. and SMITH, E. G. 1962. *q.v.*

EDWARDS, W. 1967. Geology of the country around Ollerton. 2nd edit. *Mem. geol. Surv. Gt Br.*

FALCON, N. L. and KENT, P. E. 1960. Geological results of petroleum exploration in Britain 1945–1957. *Mem. geol. Soc. Lond.*, No. 2.

KENT, P. E. 1966. The structure of the concealed Carboniferous rocks of northeastern England. *Proc. Yorks. geol. Soc.*, **35**, 323–52.

—— 1967. A contour map of the sub-Carboniferous surface in the north-east Midlands. *Proc. Yorks. geol. Soc.*, **36**, 127–33.

LEES, G. M. and TAITT, A. H. 1946. The geological results of the search for oilfields in Great Britain. *Q. Jnl geol. Soc. Lond.*, **101** for 1945, 255–317.

RAMSBOTTOM, W. H. C. 1969. Reef distribution in the British Lower Carboniferous. *Nature, Lond.*, **222**, 765–6.

—— RHYS, G. H. and SMITH, E. G. 1962. Boreholes in the Carboniferous rocks of the Ashover district, Derbyshire. *Bull. geol. Surv. Gt Br.*, No. 19, 75–168.

SMITH, E. G., RHYS, G. H. and EDEN, R. A. 1967. Geology of the country around Chesterfield, Matlock and Mansfield. *Mem. geol. Surv. Gt Br.*

Chapter III

MILLSTONE GRIT SERIES (NAMURIAN)

INTRODUCTION

THE MILLSTONE GRIT SERIES or Namurian (the terms are synonymous) has been proved in 73 deep boreholes for oil in the East Retford district, but only four of these—at South Leverton, Grove, Apleyhead and Bothamsall—have reached its base. These four bores show thicknesses of between about 750 and 1100 ft, comparable to those recorded in the northern and western parts of the Ollerton (113) district to the south (Edwards 1967, pp. 32–3, pl. iv). Although largely uncored, they show that the upper part of the succession has a clear cyclic pattern and contains thick sandstones, while the lower part consists essentially of shales, with bands of limestone which are increasingly common towards the base. In general, therefore, this succession is not unlike that known at outcrop in north Derbyshire (Smith and others 1967, p. 60), though it is on average substantially thinner and there is no evidence as to whether or not the complete zonal sequence is present.

In terms of 'massif' or 'block' sedimentation on the one hand and 'basin', 'gulf' or 'trough' sedimentation on the other, the sections of the oil bores at South Leverton, Grove, Apleyhead and Bothamsall are of the former type, while those in the northern part of the district are of the latter, lying within the Gainsborough Trough (Kent 1966, pp. 333–9). Here the Millstone Grit Series is much thicker than over the 'block' areas to the south (Fig. 5): nearly 1800 ft of strata have been proved without reaching the base in Walkeringham No. 1 Oil Bore, and there are 1750 ft in a faulted, and thereby shortened, sequence in Gainsborough No. 2 Oil Bore, where the lowest beds proved are of Arnsbergian (E_2) age (Table 1). The Gainsborough Trough was thus a region of relatively rapid subsidence, and formed an offshoot from the main basin of deposition to the west (Kent 1966, fig. 4). There is no evidence that deeper water conditions prevailed within the trough, though it is possible that some of the sandy and silty beds below the Kinderscout Grit, in e.g. Tickhill No. 1 Oil Bore (Plate II), are turbidites.

One of the most striking features of the Millstone Grit succession in the area of the Gainsborough Trough is the thickness of the Kinderscout Grit. This mass of medium- to coarse-grained sandstone, almost 300 ft thick in Walkeringham No. 1 Oil Bore, is absent or represented by only a relatively thin development of finer grained sandstone in the bores situated on the 'block' (Fig. 5 and Plate II). The feldspar content of the Kinderscout Grit, like that of the higher sandstones, indicates a northerly derivation (Gilligan 1920), and the bulk of the sediment may have been carried into the East Retford district from the north-west—along the axis of the trough from the main basin of accumulation.

The downwarping movements that produced the Gainsborough Trough became much less marked after the end of Kinderscoutian times, so that the higher Namurian rocks show little variation throughout the East Retford district (Plate II).

TABLE 1. *Classification of the Millstone Grit Series*

Namurian Divisions (Heerlen 1937)	Stages	Zones recognized in East Retford District	Stratigraphical Groups
C	Yeadonian (G₁)	G. cumbriense G. cancellatum	Rough Rock Group
B	Marsdenian (R₂)	R. superbilingue R. bilingue R. gracile	Middle Grit Group
	Kinderscoutian (R₁)	R. reticulatum at top	Kinderscout Grit Group
A	Alportian (H₂)		and underlying beds
	Chokierian (H₁)		
	Arnsbergian (E₂)		
	Pendleian (E₁)		

Diagnostic goniatites are recorded from only three horizons in the Millstone Grit Series of the district—the *Reticuloceras reticulatum*, *Gastrioceras cancellatum* and *G. cumbriense* marine bands. The first and last of these marine bands have yielded goniatites at only one locality each. The *G. cancellatum* Marine Band is goniatite-bearing in five oil bores, and it can be identified in several others, where it is of *Lingula* facies or is partly represented by crinoidal limestone. Several other post-Alportian (R and G₁) marine bands (Table 1) are of *Lingula* facies locally, and there is evidence in places of limestone. The replacement of goniatites by *Lingula* and the presence of limestone with benthonic fossils is regarded as indicative of approach to a shore-line (Ramsbottom 1969, pp. 224–5). The marine bands above the *R. reticulatum* horizon are all unrepresented in places, and it may well be that some, those of the *R. bilingue* Zone in particular, are absent throughout the greater part of the district.

Very little coring has taken place in the oil bores below the *R. reticulatum* horizon and no diagnostic macrofossils have been recovered. This, combined

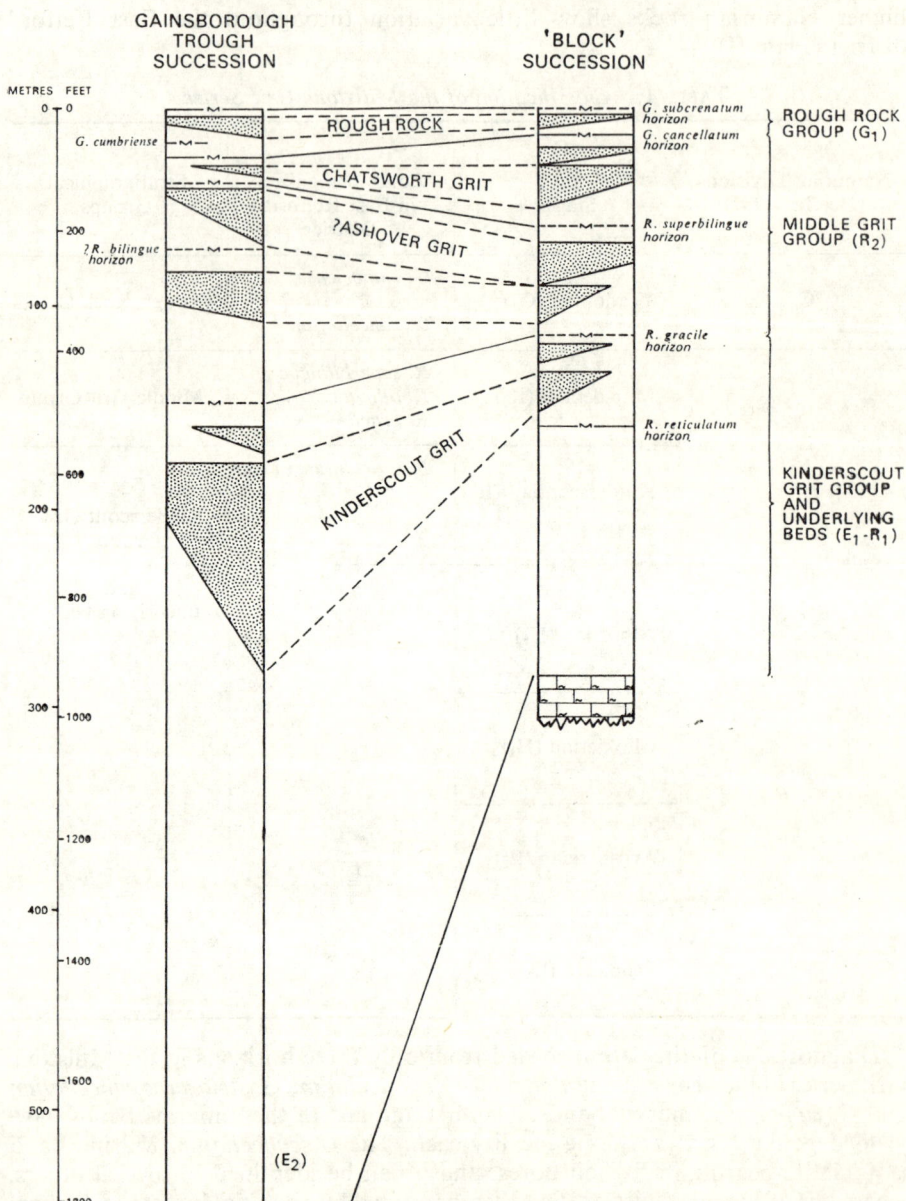

FIG. 5. *Comparative generalized sections of the Millstone Grit Series in the Gains-borough Trough and on the 'Block' to the south*

with incomplete coring and the failure of goniatite faunas at many horizons in higher strata, leads to difficulties in correlating and classifying the Millstone Grit Series of the district. Brunstrom (1963) and Taylor and Howitt (1965) have suggested a general correlation for the top and bottom of the Series and for the *R. gracile* horizon in some of the oil bores, and Downing and Howitt (1969) have attempted a more detailed and comprehensive correlation based on gamma-ray logs. The largely tentative correlations shown on Plate II are based on comparisons of known details of the succession, taking into account bores outside the district boundaries, supplemented by geophysical and palynological information.

The presence of rocks of Pendleian (E_1) age (Table 1) is inferred from the gamma-ray profiles of the rocks overlying the Carboniferous Limestone at South Leverton, Grove, Apleyhead and Bothamsall. These bear a striking resemblance (Fig. 4) to the gamma-ray profiles of the highly radioactive strata proved in boreholes near Ashover, Derbyshire (Ramsbottom and others 1962, p. 119; Cosgrove in *ibid.*, p. 164, pl. viii). Palynological determinations show that the lowest rocks in Walkeringham No. 1 Oil Bore are of Arnsbergian (E_2) age, but there is no proof of the presence of strata of Chokierian (H_1) and Alportian (H_2) ages. Rocks of Kinderscoutian (R_1) age are shown to be present by the occurrence of *R.* cf. *reticulatum* in South Leverton No. 1 Oil Bore and by palynological determinations from Gainsborough No. 2 Oil Bore. The Marsdenian (R_2) age of part of the succession can be confidently inferred from correlations with goniatite-bearing sequences beyond the district boundaries at Corringham and Bothamsall. That the uppermost part of the succession is of Yeadonian (G_1) age is shown by the presence of the *G. cancellatum* Marine Band and, locally of the *G. cumbriense* Marine Band some distance below the *G. subcrenatum* horizon.

Volcanic rocks of R_2 and possibly later age occur in the oil bores at Bothamsall and West Drayton, and there are traces of pyroclastic material elsewhere at a similar horizon. The only other igneous rock proved in the Millstone Grit Series is a dolerite intruded into E_2 shales at Gainsborough No. 2 Oil Bore.

GENERAL STRATIGRAPHY

Kinderscout Grit Group and underlying beds (E_1 to R_1). These strata, extending from the base of the Millstone Grit Series to the base of the *R. gracile* Marine Band horizon, are up to 600 ft thick on the 'block' in the south of the district. In the Gainsborough Trough to the north an incomplete succession of more than 1300 ft has been proved. Here the lowest beds reached in one of the deepest oil wells are shown by their miospore content to be of Arnsbergian (E_2) age.

The strata are predominantly mudstones, with bands of limestone occurring towards the base in the 'block' succession, and with sandstone near the top. The only important sandstone is the Kinderscout Grit, medium- to coarse-grained and up to about 300 ft thick in the Gainsborough Trough. Except in the Grove oil bores, this grit is poorly developed or even absent on the 'block'.

The only diagnostic goniatites recovered from these strata are from South Leverton No. 1 Oil Bore, where *R.* cf. *reticulatum* was found about 25 ft below the 65 ft of siltstones with thin sandstones which there represent the Kinderscout Grit.

At Gainsborough there is a bed of dolerite, over 100 ft thick and presumably a sill, in the E_2 shales near the bottom of No. 2 Bore.

Middle Grit Group (R_2). These beds, lying between the base of the *R. gracile* Marine Band and the base of the *G. cancellatum* Marine Band, are about 250 to 500 ft thick. The variations in thickness appear to be random, with no significant thickening over the area of the Gainsborough Trough. The beds consist of a cyclic sequence of mudstones and sandstones with local seatearths and thin coals. No goniatites have been found in the East Retford district, but the position of the *R. gracile* Marine Band can be inferred in some oil bores by comparison with sections outside the district, and the *R. superbilingue* Marine Band can be identified in the Bothamsall and Apleyhead bores and tentatively recognized at Gainsborough. A calcareous facies of the latter marine band is present in the south, with a crinoid/brachiopod fauna at Bothamsall, and there is some evidence that limestone also occurs at the *R. gracile* horizon in the north-west of the district. The *R. bilingue* Marine Band is evidently present in Gainsborough No. 1 Oil Bore, but has yielded only *Lingula*, *Orbiculoidea* and fragmentary bivalves.

One or more beds of sandstone, in places with substantial thicknesses of siltstone, occur almost everywhere between the horizons of the *R. gracile* and *R. bilingue* marine bands, but lack of knowledge about the marine bands of the *R. bilingue* Zone makes their correlation with sandstones in other areas uncertain. The name "?Ashover Grit" has been given to the prominent, medium- to coarse-grained sandstone, up to about 100 ft thick, which lies between the assumed horizons of the *R. bilingue* and *R. superbilingue* marine bands in the north-west of the district. This sandstone is apparently well developed also at South Leverton and Grove (see below), but is thin in the north-east. The Chatsworth Grit consists of up to 70 ft of fine- to coarse-grained sandstone in the southern oil bores, and is probably represented by one of a number of thin beds of sandstone elsewhere in the district.

In the South Leverton and Grove oil bores nearly all the strata of the Middle Grit Group consist of sandstone. Evidently this is partly ?Ashover Grit and partly Chatsworth Grit and also includes lower and higher sandstones.

In the extreme south of the district much of the Middle Grit Group is represented by tuffs, agglomerates and lavas originating from a volcano in the Bothamsall area. In the west of the Bothamsall Oilfield, volcanic rocks, totalling almost 140 ft in No. 3 Bore, occupy most of the interval between the *R. gracile* and *R. superbilingue* marine bands; in the east of this oilfield and in West Drayton No. 2 Bore (see below) they extend up to the top of the Middle Grit Group.

Rough Rock Group (G₁). These beds, extending from the base of the *G. cancellatum* Marine Band to the base of the *G. subcrenatum* Marine Band, which marks the top of the Millstone Grit Series, are only about 30 to 100 ft thick. The *G. cancellatum* Marine Band contains diagnostic goniatites in a number of oil bores, and at Morton No. 1 Bore there are, in addition, crinoids and brachiopods, preserved in limestone. The marine band is a distinctive horizon in several other sections with crinoids, calcareous brachiopods and sponge spicules.

The *G. cumbriense* Marine Band occurs in Tickhill No. 1 Oil Bore in the extreme north-west of the district, and is apparently represented by a *Lingula* band in a few other sections.

The Rough Rock, varying from fine-grained sandstone with siltstone to coarse-grained sandstone, ranges in thickness from a few feet in the north-west to about 50 ft in the north-east. Its thickness cannot be determined in the south, where it is continuous with the overlying Crawshaw Sandstone of the Coal Measures. At West Drayton No. 2 Oil Bore the Rough Rock rests directly on volcanic rocks.

The Pot Clay Coal has been recorded in a few sections, but is absent in others.

E.G.S.

DETAILS OF STRATIGRAPHY

KINDERSCOUT GRIT GROUP AND UNDERLYING BEDS

The Kinderscout Grit Group and underlying beds are about 450 to about 600 ft thick in South Leverton No. 1, Grove No. 1, Apleyhead No. 1 and Bothamsall No. 3 oil bores, where the base of the Millstone Grit Series, although not precisely recognized (p. 8), has been penetrated. The *Reticuloceras gracile* Marine Band, the base of which defines the top of the Kinderscout Grit Group, has not been proved within the district, but its approximate position in the above bores can be inferred from provings in Bothamsall No. 1 Oil Bore to the south and in Corringham Nos. 1 and 7 oil bores to the east (pp. 18–9 and Plate II). The South Leverton, Grove, Apleyhead and Bothamsall bores provide a 'block'-type sequence, but the bores in the northern part of the district lie within the area of the Gainsborough Trough, which had its maximum known effects in Kinderscoutian and earlier times (p. 12), and here the Kinderscout Grit Group and underlying beds are much thicker and have not been bottomed. The thickness at Walkeringham No. 1 Oil Bore is probably more than 1300 ft, and at Gainsborough No. 2 Oil Bore the lowest level reached is at least 1250 ft, and probably considerably more because of inferred faulting, below the *R. gracile* horizon. The deepest rocks from both these bores have yielded miospore assemblages indicating a *Eumorphoceras* (E) age, more precisely Upper *Eumorphoceras* (Arnsbergian or E_2) at Gainsborough (p. 27).

Few cores have been taken in the Kinderscout Grit Group and underlying beds, and little can be done to subdivide them. Apart from the Kinderscout Grit, of which there is a thick development in some bores (see below) the rocks, as shown by chipping samples and geophysical records, are predominantly mudstones and shales. Siltstones and minor sandstones also occur, particularly in the upper part. The basal shales at South Leverton, Grove, Apleyhead and Bothamsall are commonly pyritic and partly calcareous. They contain limestone bands, which are numerous at South Leverton and Grove, where they occur for up to about 100 and about 250 ft above the Carboniferous Limestones respectively, but which are inconspicuous at Apleyhead and Bothamsall.

Chippings indicate a few feet of impure limestone with crinoid debris at about 4605 ft in South Leverton No. 1 Oil Bore some 30 ft below the occurrence of *Reticuloceras* cf. *reticulatum* (see below), and thin limestones are also recorded some 40 and 225 ft below the Kinderscout Grit in Ranskill No. 1 and Walkeringham No. 1 oil bores respectively (Plate II).

The only goniatites recorded are from South Leverton No. 1 Oil Bore: *R.* cf. *reticulatum* in 6 in of dark silty micaceous mudstone at $4575\frac{1}{2}$ ft, and indeterminate ghosts and fragments in similar mudstone between 4644 ft 7 in and 4646 ft 5 in. The latter mudstone is part of a thicker band which also contains *Lingula mytilloides*, a conodont assemblage and palaeoniscid scales. *L. mytilloides* also occurs between $4558\frac{3}{4}$ ft and 4563 ft 7 in, not far above the *R.* cf. *reticulatum* horizon, in this bore, and is here associated with *Serpuloides sp.*

The only other macrofossils recorded in the district are from Ranskill No. 1 and Walkeringham No. 1 oil bores. At Ranskill fish debris was found between 5662 ft 2 in and 5667 ft 11 in, near the bottom of the bore. At Walkeringham dark, largely canky mudstones, from the bottom 50 ft of the bore and proved by miospore evidence (p. 28) to be of E age, yielded *Modiolus sp.* at two horizons (a 4-in band at 6347 ft 4 in and a 9-in band at $6358\frac{3}{4}$ ft) and also contained scattered fish fragments including palaeoniscid scales. Details of miospores, principally from Walkeringham No. 1, Morton No. 1, Gainsborough No. 2 and Grove No. 1 oil bores, are given on pp. 26–8.

For details of the dolerite, which is more than 100 ft thick and occurs in E_2 shales near the bottom of Gainsborough No. 2 Oil Bore, see pp. 28–9.

Kinderscout Grit. This name is given to the locally thick sandstone which occurs, in one or more beds, at depths ranging between 300 and 600 ft below the *Gastrioceras cancellatum* Marine Band horizon. What is apparently the same sandstone lies below the *R. gracile* Marine Band in Corringham Nos. 1 and 7 oil bores, near Gainsborough but outside the district (Plate II), and equivalent siltstones with thin sandstones are underlain in South Leverton No. 1 Oil Bore by a marine band which has yielded *R.* cf. *reticulatum* (see above). In addition, miospore assemblages obtained from argillaceous measures just above and just below the sandstone in Gainsborough No. 2 Oil Bore show that the sandstone is of R_1 age. It may be, however, from evidence provided by the Corringham bores that this sandstone is at the horizon of the Lower Kinderscout Grit of outcrop areas to the west. In Corringham No. 1 Oil Bore, where the *R. gracile* Marine Band occurs at 5230 ft, the sandstone is overlain by dark to grey mudstone with *L. mytilloides* and *Orbiculoidea nitida* between 5298 and $5303\frac{1}{2}$ ft. The latter horizon may well be that of the Butterly Marine Band (Bromehead and others 1933, pp. 15, 145) between the Lower and Upper Kinderscout Grits.

The Kinderscout Grit is thickest in the area of the Gainsborough Trough (Plate II). Indeed, except at Grove, where at this horizon there are about 70 ft of sandstone, which is medium-grained in No. 1 Bore and medium- to coarse-grained in No. 2 Bore, the Kinderscout Grit is poorly represented or absent in oil bores outside the trough. This is also the case in the numerous oil bores to the south of the district (Edwards 1967, pp. 36–52) and in Torksey No. 1 Oil Bore to the east.

The thickest proved section of Kinderscout Grit is in Walkeringham No. 1 Oil Bore, where there are almost 300 ft of mainly medium- to coarse-grained sandstone. Cores taken in the upper part of the sandstone (Plate II) show that it is very coarse in parts. Over 100 ft of Kinderscout Grit, divided by beds of shale, have been proved in Beckingham No. 1 and Morton No. 1 oil bores, and in the deep bores at Gainsborough. Again much of it is medium- to coarse-grained, but finer sandstone occurs at the top and bottom of the Grit in most sections. In Tickhill No. 1 Bore the Kinderscout Grit consists of nearly 200 ft of fine- to coarse-grained sandstone with thin bands of shale. The Grit is relatively thin in Ranskill No. 1 Bore, consisting of 35 ft of mostly medium-grained sandstone with a thin bed of fine- to medium-grained sandstone 24 ft below.

Above the Kinderscout Grit of this district there are several thin beds of sandstone and siltstone which may be at the horizon of the Upper Kinderscout Grit of western outcrop areas (see above). The sandstone is chiefly fine- to medium-grained, but one bed at Tickhill No. 1, Ranskill No. 1 and Walkeringham No. 1 bores, which attains a thickness of nearly 40 ft at the first locality, is partly coarse-grained.

MIDDLE GRIT GROUP

These beds are 350 to 400 ft thick in the Bothamsall Oilfield and probably have the following approximate thicknesses in other oil bores: Tickhill 350 ft, Ranskill and Walkeringham 400 ft, Morton 300 ft, Beckingham 325 ft, Gainsborough 450 ft, South Leverton and Grove 250 ft, Apleyhead 500 ft (but possibly only 300 ft). The *G. cancellatum* Marine Band, the base of which marks the top of the group, occurs in several bores and its horizon can be identified with some confidence in others (pp. 23–4), but at the base of the group the position of the *R. gracile* Marine Band is largely inferred (see below). Downing and Howitt (1969, figs. 5 and 6) arrived at thicknesses different from the above for the Middle Grit Group in some bores, particularly in the Gainsborough area, chiefly because they selected a different position for the *R. gracile* horizon.

No goniatites are recorded from the Middle

Grit Group, but few cores have been taken in the lower part. Marine bands of the *R. bilingue* Zone appear to be present in Gainsborough No. 1 Oil Bore (p. 20), but there is no evidence that they occur anywhere in the district outside this oilfield. The *R. superbilingue* Marine Band, recorded in Bothamsall No. 1 Oil Bore to the south (Edwards 1967, p. 37), is, where its horizon has been cored within the district, either absent or not of goniatite facies.

Reticuloceras gracile Marine Band. This marine band has been proved just beyond the district boundary in Bothamsall No. 1 Oil Bore to the south (Edwards 1967, p. 37) and in four of the Corringham oil bores to the east (Plate II). At Corringham (5217 ft 10 in to 5230 ft in No. 1 Bore, 5274 ft 10 in to 5276 ft 4 inches in No. 2 Bore, 5223 ft to 5233 ft 7 inches in No. 6 Bore, and 5083 ft 7 in to 5088 ft $6\frac{1}{2}$ inches in No. 7 Bore) it has yielded, in addition to *R. gracile: Lingula mytilloides, Caneyella sp., Anthracoceratites sp.*, mollusc spat and fish debris.

With the probable exceptions of Grove No. 2 and South Leverton No. 1 oil bores (see below), no cores have been taken at this horizon in the East Retford district, and its assumed position depends chiefly upon lithological correlations northwards from Bothamsall and westwards from Corringham, assisted in some instances by the interpretation of gamma-ray logs. Its position is in doubt at Apleyhead No. 1 Oil Bore, where the ash at 3950 ft (see p. 31) has been tentatively correlated with the base of the volcanic sequence, some distance above the *R. gracile* Band, in Bothamsall No. 1 Oil Bore (Plate II). If, however, the ash band at about 3802 ft in Apleyhead No. 1 Bore (p. 31) is correlated with the base of the Bothamsall volcanic sequence, the *R. gracile* horizon would be much higher at Apleyhead, giving a thickness for the Middle Grit Group more compatible with those at Grove and South Leverton.

In Grove No. 2 Bore the marine band at 4631¾ ft probably represents the *R. gracile* horizon. Details of cores taken between 4600 and 4632 ft are as follows:

	ft
Sandstone, soft light grey, coarse-grained, micaceous (*4 ft of core lost*)	9½
Core lost	3
Siltstone with sandstone bands and laminae; pyrite nodules; plants ..	12
Mudstone, dark silty, with bands of ironstone and calcareous siltstone; crinoid columnals, *Belinurus* cf. *belullus* and fish debris	7¼
Mudstone, grey silty	0¼

In South Leverton No. 1 Oil Bore the horizon of the marine band is probably between 4438 and 4467 ft, but cores of dark silty mudstone with much pyrite taken between these depths showed only vague traces of organic remains.

In Tickhill No. 1 Oil Bore a thin limestone band at about 4895 ft, evidently corresponding to a band of calcareous siltstone at about 4865 ft in Ranskill No. 1 Oil Bore, is probably at or near the *R. gracile* horizon. It may therefore be that this marine band, like the *R. superbilingue* (p. 21) and *Gastrioceras cancellatum* (p. 23) marine bands, is developed as a calcareous facies locally.

The interval between the assumed position of the *R. gracile* Marine Band and the horizon of the *R. superbilingue* Marine Band (see below) is largely occupied by volcanic rocks at Bothamsall (pp. 29–30), and by medium- to coarse-grained and pebbly sandstone at Grove and South Leverton.

In both Grove oil bores a coal is recorded in the lower half of the sandstone, but the substantial seam thickness reported is in doubt. In No. 1 Bore the coal is at the top of an argillaceous parting thought to be at about the *R. bilingue* horizon (see below), which contains further traces of coal. Cores of the basal part of the sandstone in No. 2 Bore are described above.

Cores of the top 68½ ft of the sandstone were obtained in South Leverton No. 1 Oil Bore. They showed coarse-grained, false-bedded sandstone except in the top few feet which were fine- to medium-grained and more thinly bedded. In South Leverton No. 2 Bore the sandstone at this horizon is continuous with the overlying ?Ashover Grit.

Elsewhere in the district there are several beds of sandstone in this interval. They can be considered in two groups, above and below the presumed horizon of the *R. bilingue* Marine Band. Identification of the latter horizon depends upon correlation with Gainsborough No. 1 Oil Bore which was cored through much of these measures. Details of cores taken in the Middle Grit Group in this bore are as follows:

	Thickness		Depth	
	ft	in	ft	in
G. cancellatum Marine Band (see p. 24) ..		to	4690	0
Mudstone, brown micaceous	0	4	4690	4
Sandstone with mudstone partings	11	1	4701	5
Mudstone, dark grey; scattered *Lingula mytilloides*	7	5	4708	10
Mudstone, dark grey silty, with sandstone partings	4	2	4713	0
Siltstone with sandstone partings	3	3	4716	3
Sandstone with bands of grey micaceous mudstone	2	9	4719	0

	Thickness ft in	Depth ft in
Mudstone, dark grey silty micaceous; *Serpuloides sp.*, *L. mytilloides* and fish debris (?*R. superbilingue* Marine Band)	4 3	4723 3
Sandstone	11 2	4734 5
Mudstone, dark grey, very silty at top; 1-in ironstone band; fish debris at base ..	4 2	4738 7
Mudstone, dark grey; *L. mytilloides* and fish debris	10 5	4749 0
Mudstone, grey silty, with plant debris ..	0 6	4749 6
Coal	0 1	4749 7
Seatearth, silty mudstone	0 2	4749 9
Core lost	6 3	4756 0
Sandstone with siltstone partings	1 6	4757 6
Core lost	2 6	4760 0
Mudstone, grey silty micaceous; 4-in sandstone band; plant debris	4 11	4764 11
Mudstone, dark grey; *L. mytilloides* and fish debris	4 1	4769 0
Sandstone with siltstone partings	25 4	4794 3
Siltstone, grey micaceous	2 9	4797 0
No core	13 0	4810 0
Mudstone, grey silty; abundant *L. mytilloides*	6 3	4816 3
Siltstone, grey; plant debris	5 3	4821 6
Mudstone, grey silty; *L. mytilloides* and palaeoniscid scales; *Orbiculoidea nitida*, pectinoid and fragments of nuculoids between 4843 and 4846 ft (?*R. bilingue* Marine Band)	28 10	4850 4
Siltstone, grey micaceous, with thin ironstone bands	17 8	4868 0
Mudstone, grey silty, with 11-in ironstone band at 4885½ ft; *L. mytilloides* and fish debris between 4870 ft and 4875 ft 4 in ..	20 0	4888 0
Siltstone, grey micaceous, with thin sandy bands; *L. mytilloides* at 4904 ft 5 in	22 0	4910 0
Siltstone, grey micaceous, with sandstone bands	4 4	4914 4
No core	14 8	4929 0
Sandstone	31 0	4960 0

The sandstone at 4960 ft in this bore is, according to chipping samples, underlain by siltstone with sandstone bands. Elsewhere in the district one or more beds of sandstone occur at this general horizon below the assumed position of the *R. bilingue* Marine Band, and there are substantial thicknesses of siltstone at Tickhill, Beckingham and Apleyhead. These sandstones cannot be correlated with named sandstones at outcrop to the west and north-west of the district (see e.g. Ramsbottom 1966, pl. 4) because their relative positions with respect to the marine bands of the *R. bilingue* Zone are not known. The mudstones and siltstones between 4794¼ ft and 4914 ft 4 inches in Gainsborough No. 1 Bore are apparently not represented in the South Leverton bores. Elsewhere few cores have been taken in these beds, which may attain a thickness of about 130 ft. Chipping samples show them to consist largely of silty mudstone, with siltstone and thin bands of sandstone in some of the thicker sections.

Two thin coals are recorded close below the ?Ashover Grit in Beckingham No. 1 Oil Bore, and, in a similar position, there is evidence of coal in a few of the bores at Gainsborough, and of seatearth at Ranskill and Morton. At a lower horizon traces of coal or ganister are recorded in some of the Gainsborough bores.

?Ashover Grit. One or more of the beds of sandstone between the presumed *R. bilingue* and *R. superbilingue* horizons in Gainsborough No. 1 Oil Bore (see above), but probably only the lowest, evidently represents the Ashover Grit, which in the Chesterfield district occurs between the *R. bilingue* late mut. and *R. superbilingue* marine bands (Smith and others 1967, pp. 61–3 and 70–1). The lowest or main bed of sandstone in this section (at 4794¼ ft) is correlated, via the Morton and Beckingham bores, with a prominent sandstone in the Walkeringham,

Ranskill and Tickhill bores, 95 ft thick at the last locality (Plate II). Apparently it is also the equivalent of the upper part of the thick sandstone development above the presumed *R. bilingue* horizon at South Leverton and Grove (p. 19). The ?Ashover Grit is medium- to coarse-grained in nearly all the provings in the district, and is pebbly in many sections.

There are two *Lingula* bands close below the presumed *R. superbilingue* Marine Band in Gainsborough Nos. 1 and 2 oil bores, which may be compared to the two *Lingula* bands between the Ashover Grit and the *R. superbilingue* Marine Band in north Derbyshire (Smith and others 1967, p. 68).

Reticuloceras superbilingue Marine Band. No goniatites have been found at this horizon within the district, but *R. cf. superbilingue* is recorded in Bothamsall No. 1 Oil Bore (Plate II and Edwards 1967, p. 37) just beyond the southern boundary. A comparison of the section of this bore with that of Bothamsall No. 4 Bore clearly identifies the *R. superbilingue* Marine Band at 3602¾ ft in the latter. Here it consists of 1½ ft of dark grey calcareous mudstone on 1¾ ft of grey limestone, and has yielded abundant *Serpuloides sp.* in addition to crinoid columnals, *Schizophoria sp.* and fish remains. Both lithologically and faunally it thus resembles the development of the band in No. 1 Bore (Edwards *idem*). In Bothamsall No. 3 Bore the marine band is evidently represented by 3 in of dark grey silty mudstone with *Serpuloides sp.* and fish debris at 3584¼ ft. In Apleyhead No. 1 Oil Bore (3770 ft 9½ in) it apparently consists of 2½ ft of dark grey mudstone with *L. mytilloides* and palaeoniscid scales, on 8½ in of calcitic sandstone with mudstone partings which have yielded an indeterminate coiled shell.   E.G.S.

A specimen of the calcitic sandstone from 3770 ft 8 inches in Apleyhead No. 1 Bore was examined in thin section (E 35419)[1]. This is a hard, graded sandstone with clastic grains of quartz, sparse potash feldspar and rock particles, scattered in a coarsely crystalline calcite cement which marginally replaces the clastics. The sparse rock particles include felsitic lava, chert, quartzite and silty mudstone. Slivers of brown, isotropic material may be organic phosphate. Selvedges of an indeterminate brown mineral occur around scattered quartz grains.   R.K.H.

Outside the Bothamsall–Apleyhead area the *R. superbilingue* horizon can be tentatively recognized only in Gainsborough Nos. 1 and 2 bores at 4723¼ ft (see p. 20) and 4593½ ft respectively, and in Tickhill No. 1 Bore at 4555 ft 5 in. Details of cores in Gainsborough No. 2 Bore between 4573 ft 11 in and 4593½ ft are as follows:

	ft	in
Mudstone, dark grey; *L. mytilloides*, *Orbiculoidea sp.* and fish including *Elonichthys sp.* and *Rhadinichthys sp.*	6	4
Mudstone, dark grey, slightly silty, with sandstone partings	0	9
Mudstone, dark grey, silty at top and bottom; *Planolites ophthalmoides*, *L. mytilloides*, *Curvirimula?* and fish debris (*2 ft of core missing*)	12	6

In Tickhill No. 1 Bore the 9 ft 11 in of dark grey silty mudstone at 4555 ft 5 in contain *Serpuloides sp.*, *L. mytilloides* and fish debris.

The *R. superbilingue* Marine Band is absent at South Leverton. In No. 2 Bore the ?Ashover Grit is continuous with the Chatsworth Grit, and in No. 1 Bore less than 10 ft of unfossiliferous mudstone and sandstone, all cored, separate the two Grits.

The interval between the supposed *R. superbilingue* Marine Band and the *G. cancellatum* Marine Band is only 11 ft 10 inches in Tickhill No. 1 Oil Bore and is unlikely to exceed 40 ft in any of the northern oil bores. Sandstone is present everywhere in the north in one or, locally, two thin beds. In Gainsborough No. 1 Oil Bore *Lingula* occurs between the two sandstones (p. 19), which may well represent the Chatsworth Grit and Redmires Flags (see below). In the south of the district the strata between the *R. superbilingue* and *G. cancellatum* marine bands are much thicker, with a maximum of

---

[1]Numbers prefixed 'E' refer to thin sections in the English Sliced Rock Collection of the Institute of Geological Sciences.

C

250 ft in Apleyhead No. 1 Oil Bore. In the east of the Bothamsall Oilfield and in West Drayton No. 2 Bore they include volcanic rocks (see pp. 29–30).

Chatsworth Grit. This name is given to the prominent sandstone, up to 70 ft thick, in the lower part of the measures between the *R. superbilingue* and *G. cancellatum* marine bands in the Grove, Apleyhead and Botham-sall oil bores. It may be that this sandstone is not the precise equivalent of the Chatsworth Grit of the type area (Smith and others 1967, pp. 68–9). The horizon of the Baslow Coal, which there defines the top of the Chatsworth Grit, cannot be identified in the East Retford district, and it may be that the *Lingula* band that overlies the so-called Chatsworth Grit at Apleyhead and Bothamsall (see below) is the equivalent of the *Lingula*-bearing measures above the 'Brown Edge Flags' in Derbyshire (*ibid.* pp. 69 and 74–5).

The Chatsworth Grit is thick at South Leverton, but it is apparently continuous with higher sandstones in No. 1 Bore and with both higher and lower sandstones in No. 2 Bore. It is probably present as a thin sandstone in most of the northern oil bores, but it cannot be positively identified (see above).

In the South Leverton, Grove and Botham-sall bores the Chatsworth Grit is coarse-grained or pebbly at the base and generally becomes finer grained upwards. In Apleyhead No. 1 Oil Bore the lower part may be represented by 32½ ft of siltstones, calcareous in part, which are separated by 24 ft of silty mudstone from an upper, 42-ft leaf of fine- to medium-grained sandstone (Plate II).

Lingula mytilloides occurs in dark grey mudstone close above the Chatsworth Grit in Apleyhead No. 1 (in a 3-in bed at 3646¾ ft), Bothamsall No. 2 (3597 ft 4 in to 3599 ft), Bothamsall No. 3 (3505 ft 11 in to 3506 ft 1 in) and Bothamsall No. 4 (3529 ft 7 in to 3530 ft 11 in) oil bores. In Apleyhead No. 1 and Bothamsall No. 2 bores there are associated palaeoniscid scales. This *Lingula* band may be the same as that recorded at 4708 ft 10 inches in Gainsborough No. 1 Bore (see p. 19).

Fourteen inches above the *Lingula* band in Bothamsall No. 4 Bore specimens of an anthraconaioid *Carbonicola* were collected from 5 in of dark grey mudstone. This occurrence may be compared to the mussel-bands found between the Chatsworth Grit and Redmires Flags in districts to the west, in for instance the Rod Moor boreholes (Pulfrey 1934, pp. 260–2; Eden and others 1957, p. 20) and Calow No. 1 Borehole (Smith and others 1967, pp. 74–5).

The strata between the Chatsworth Grit and the *G. cancellatum* Marine Band contain one or more beds of fine- to medium-grained sandstone at Grove, Apleyhead and Botham-sall. That between 4407 and 4435 ft in Grove No. 1 Bore is probably the Redmires Flags, but the correlation of this bed with sandstones in the other oil bores is uncertain.

In Grove No. 2, Apleyhead No. 1 and Bothamsall Nos. 2 and 4 oil bores there is a *Lingula* band 13 to 29 ft below the base of the *G. cancellatum* Marine Band. This is represented by a fish-band 30½ ft below the marine band in Bothamsall No. 3 Bore. At Bothamsall No. 2 Bore only, there is a further occurrence of *Lingula* 10 to 12 ft lower, here associated with *Serpuloides sp.* and fish fragments.

In Morton No. 1, Apleyhead Nos. 1 and 2 and Bothamsall Nos. 2 and 3 oil bores there is a *Lingula* band close below the *G. cancellatum* Marine Band. The 1½ to 5 ft of intervening measures contain sandstone except at Morton, and, except in the Apley-head bores, they also include seatearth. This *Lingula* band is described with the *G. cancellatum* Marine Band (p. 23) because, in districts to the west, it may pass laterally into goniatite-bearing beds and become part of the marine band.

ROUGH ROCK GROUP

The strata of this group have been com-pletely cored in several oil bores, and range in thickness from about 30 ft to a little over 100 ft. The thickest proved section, and the only one in which the *G. cancellatum, G.* *cumbriense* and *G. subcrenatum* marine bands all contain diagnostic goniatites is that of Tickhill No. 1 Bore. Details of the log are as follows:

	Thickness		Depth	
	ft	in	ft	in
G. subcrenatum Marine Band (for details see p. 48)			4428	3
ROUGH ROCK Sandstone,fine-grained, with dark micaceous laminae; roots in top 1¾ ft	7	2	4435	5
Mudstone, grey silty; sporadic plant debris ..	25	7	4461	0
Mudstone, dark grey, slightly silty	6	0	4467	0
G. cumbriense Marine Band Mudstone, dark grey, silty and 'canky'; Serpuloides sp., L. mytilloides, Dunbarella?, Palaeoneilo cf. mansoni, G. crenulatum and G. cumbriense ..	4	0	4471	0
Mudstone, grey silty micaceous	2	0	4473	0
Siltstone, micaceous ..	2	9	4475	9
Mudstone, dark grey; Lingula sp.[juv.], P. aff. mansoni and fish debris	1	0	4476	9
Mudstone, grey silty; Lingula sp. and Curvirimula sp.	5	3	4482	0
Mudstone, grey, silty at top, with sporadic thin ironstone bands ..	43	6	4525	6
G. cancellatum Marine Band Mudstone, dark grey, with sandy bands and marine fossils (for details see below) ..	8	2	4533	8

Gastrioceras cancellatum Marine Band.
Diagnostic goniatites have been found at this horizon in Tickhill No. 1, Walkeringham No. 1, Morton No. 1, Apleyhead Nos. 1 and 2 and Bothamsall Nos. 2 and 4 oil bores.

In Tickhill No. 1 Oil Bore the marine band, 8 ft 2 in thick at 4533 ft 8 in and resting on sandstone, consists of dark grey mudstone, which is very silty with thin sandy bands in the lower half. The following fauna has been collected: *Crurithyris sp., Lingula mytilloides, Productus carbonarius, Caneyella multirugata,*

Dunbarella elegans, nuculoid indet., *Polidevcia sp., Agastrioceras carinatum, Anthracoceras sp., Gastrioceras cancellatum.*

In Walkeringham No. 1 Oil Bore *Agastrioceras carinatum* and *G. crencellatum* occur with *L. mytilloides* in dark mudstone between 4690 ft 2 in and 4693 ft 10 in. *L. mytilloides* is also recorded from the overlying 8 ft of mudstones.

In Morton No. 1 Oil Bore coring commenced in the marine band at 4658 ft and proved its base at 4663 ft 3½ in. The topmost 4 ft 10 in consist of dark grey mudstone with ironstone lenses, containing *L. mytilloides, Agastrioceras carinatum, G. crencellatum,* conodonts and fish debris. At the base is a 5½-in bed of compact grey limestone with crinoid debris, abundant brachiopods (*Cruxrithyris sp.* and *Rugosochonetes sp.*), *G.* cf. *crencellatum* and fish fragments. Below the limestone and separated from it by 1 ft 6½ in of dark grey silty mudstone-seatearth, are 1 ft 11 in of dark grey mudstone, silty at the top and containing rootlets which die out downwards. This mudstone has yielded *Serpuloides sp., L. mytilloides* and fish fragments. The Morton section is noteworthy for the occurrence of crinoidal limestone with abundant brachiopods in the goniatite-bearing marine band, and indicates proximity of the shore-line at this period (see p. 13 and Ramsbottom 1969, pp. 224–5).

At Apleyhead the marine band is about 14 ft thick in No. 1 Oil Bore (but see below). There are 11 ft 8 in of dark grey mudstone with *A. carinatum* and *G.* cf. *crencellatum* at 3558¾ ft. The mudstone is silty and micaceous in part and, as at Tickhill, contains partings of sandstone towards the base. This rests on 26 in of interbanded mudstone and sandstone, which have not yielded fossils, and these in turn rest on fine calcarenitic sandstone with crinoid debris and a doubtful goniatite, which passes down at about 3561½ ft into unfossiliferous sandstone. At 3563 ft 2 in this sandstone rests on dark grey to black shale with *L. mytilloides* and fish in the top 4 in. The limestone band is comparable to that occurring at the base of the marine band in Morton No. 1 Oil Bore, and the *Lingula*-bearing shales are reminiscent of the marine mudstone below the main band in that bore (see above). E.G.S.

A thin section (E 35416) of the calcarenitic sandstone from 3561 ft in Apleyhead No. 1

Oil Bore shows roughly bedded, fine, sandy clastics cemented by calcite, with dark shaly intercalations. Resistates include poorly sorted quartz and feldspars (0·03 to 1·2 mm; averaging 0·1 mm) with rare mica flakes and rock particles in addition to the fossil fragments. The latter include brown to yellow particles probably composed of phosphate. A pale yellowish brown mineral, possibly chlorite, rims clastic grains. Minor constituents include zircon, specks of opaque ore, pyrite and leucoxene. The darker shaly material contains abundant, oriented shell fragments, small proportions of clastic grains, very fine dark carbonate, opaque carbonaceous films, chlorite and granular opaque ore.   R.K.H.

In Apleyhead No. 2 Oil Bore *L. mytilloides*, *Caneyella multirugata*, *Dunbarella sp.*, *A. carinatum*, *Anthracoceras sp.* and *G.* cf. *crencellatum* occur in 6 ft 1 in of dark grey mudstone at 3595 ft 1 in. Below is a 6-in limestone band in which no fossils were found and which is separated from 2 ft of sandstone by 4 in of grey silty micaceous mudstone. Immediately below the sandstone are 15 in of dark grey shale with abundant *Lingula* in the top few inches.

In Bothamsall No. 2 Oil Bore *L. mytilloides*, *A.* cf. *carinatum*, *G. crencellatum* and palaeoniscid scales occur in mudstone between 3490½ and 3493 ft. *Lingula* continues to 3487 ft 2 in, and also occurs at 3497 ft 11 inches in a grey siltstone which, at 3496 ft 7 in, underlies 3 ft 5 in of sandstone with rootlets at the top, which in turn is separated from the goniatite-bearing strata by 2 in of unfossiliferous mudstone.

In Bothamsall No. 4 Oil Bore 2 ft 1 in of dark grey mudstone with *Lingula sp.* [juv.], *?A. carinatum* and *G.* cf. *crencellatum* are recorded at 3434 ft. *Planolites ophthalmoides* occurs in a 1-ft core of paler and slightly silty mudstone recovered from the 4 ft of strata above the band, but the core between 3434 ft and a ganister at about 3439 ft was almost entirely lost during drilling.

Goniatites have also been recorded in Bothamsall No. 3 Oil Bore at what is apparently the *G. cancellatum* horizon. These have crenulate lirae, but positive identification is not possible. They occur at 3413 ft in grey mudstone which, between 3412 ft 2 in and 3413 ft 10 in, also contains *L. mytilloides*, *Orbiculoidea?* and fish debris.

Lingula also occurs in dark mudstone at 3418½ ft close below 1 ft 10 in of sandstone and ganister.

In Beckingham No. 1 Oil Bore the *G. cancellatum* horizon is evidently represented by 4 ft 10½ in of grey to dark grey mudstone with *Serpuloides?*, *Planolites ophthalmoides*, *Lingula sp.* and *Orbiculoidea?* at 4638 ft 10½ in.

In the Gainsborough Oilfield the marine band is a distinctive horizon consisting of mudstone with a hard basal band, variously described as sandstone, ironstone or limestone and containing brachiopods, crinoid debris and sponge spicules. This band was correlated with the Pot Clay Marine Band (p. 48) by Howitt and Brunstrom (1966, p. 555). In Gainsborough No. 1 Oil Bore the marine band, at 4690 ft, consists of 4¾ ft of dark green silty mudstone with *L. mytilloides* and fish debris on 6 in of sandstone with bands of dark grey silty mudstone and containing *Serpuloides sp.*, crinoid columnals, *L. mytilloides* and *O. nitida*. In No. 7 Bore there are 2 ft of dark grey silty mudstone on 10 in of ironstone at 4698 ft 5 in, and the fauna comprises: hexactinellid sponge spicules, *Hyalostelia sp.*, *Serpuloides sp.*, crinoid columnals and fish debris including *Rhabdoderma sp.* There is a similar development in No. 43 Bore where, at 4579 ft 7 in, there is a 1½-in limestone below 4 ft 4½ in of grey to dark grey mudstone, and the fauna comprises: hexactinellid sponge spicules, *Hyalostelia sp.* [spicules 0·6 mm in diameter], *Serpuloides stubblefieldi*, *Paraconularia sp.* and fish debris including *Rhabdoderma?*. In No. 6 Bore the marine band is evidently represented by dark grey mudstone between 4689 and 4692 ft containing *L. mytilloides*, *Orbiculoidea sp.* and fish debris.

The *G. cancellatum* Marine Band is absent in South Leverton No. 1 Oil Bore, but the *Lingula* band at 4207¾ ft in No. 2 Bore is apparently at this horizon. In Grove No. 2 Oil Bore the marine band consists of 1 ft 10 in of dark micaceous siltstone with mudstone laminae, on 2 ft of very hard grey calcareous siltstone with mudstone laminae and sporadic quartz grains, at 4353 ft 10 in. *L. mytilloides* and fish debris occur in the upper part, and the fauna of the hard band comprises abundant *Serpuloides sp.* and crinoid debris with *O. nitida*, *Rugosochonetes sp.* and a smooth spiriferoid.

The beds between the *G. cancellatum* Marine Band horizon and the Rough Rock consist largely of mudstones with some siltstones, but include a 4 ft 11-in bed of sandstone in Walkeringham No. 1 Oil Bore, and, according to the chipping samples, two thin beds of sandstone in Ranskill No. 1 Oil Bore.

Only in Tickhill No. 1 Oil Bore, where it contains diagnostic goniatites, can the *G. cumbriense* Marine Band be positively identified (see p. 23). In Walkeringham No. 1 Oil Bore *Lingula* occurs at three levels between the *G. cancellatum* and *G. subcrenatum* horizons, and the highest of these, in grey silty mudstone between 4631¾ ft and 4635 ft 7 in, overlying the thin sandstone (see above), is taken to be the *G. cumbriense* horizon. The two lower *Lingula* occurrences are in micaceous siltstone at 4661 and 4671½ ft, and together probably represent the marine horizon close below the *G. cumbriense* Marine Band at Tickhill (p. 23). A *Lingula* band above the *G. cancellatum* horizon has also been recorded in several of the Gainsborough oil bores, but it is not possible to say whether this is at the *G. cumbriense* horizon or corresponds to the lower marine band of Tickhill No. 1 Bore.

At South Leverton and Bothamsall the argillaceous beds between the *G. cancellatum* horizon and the Rough Rock are thin, and in West Drayton No. 2 Oil Bore the Rough Rock rests directly on volcanic rocks (p. 29).

Rough Rock. This sandstone is apparently present throughout the district, ranging in thickness from under 10 ft at Tickhill to about 50 ft in parts of the Gainsborough Oilfield. The latter figure may, however, be exceeded in some of the Bothamsall bores, where the sandstone is not a discrete member (see below). In the South Leverton and Gainsborough areas the sandstone here identified as the Rough Rock was classed as Crawshaw Sandstone by Taylor and Howitt (1965, p.

198) and Howitt and Brunstrom (1966, p. 557, fig. 3, pl. 31).

Including ganisteroid sandstone at the top, the Rough Rock is 7 ft 2 in thick in Tickhill No. 1 Oil Bore (see p. 23), and, according to the chippings, consists of about 15 ft of fine-grained micaceous sandstone with some siltstone in Ranskill No. 1 Oil Bore. At Walkeringham No. 1 Oil Bore it is represented by 3 ft 10 in of sandstone with roots, on 19 ft of siltstone with sandstone beds and roots in the top 8 ft. Elsewhere except at Bothamsall it generally consists of fine-grained, occasionally fine- to medium-grained, sandstone which is commonly micaceous and contains carbonaceous laminae and bands and partings of siltstone and silty mudstone. Medium- to coarse-grained sandstone occurs in the lower part of the Rock in two oil bores at South Leverton.

The thick sandstone above the *G. cancellatum* horizon in the Bothamsall Oilfield is thought to be a combined Rough Rock and Crawshaw Sandstone (see p. 49) and it therefore straddles the Millstone Grit Series–Coal Measures boundary (Downing and Howitt 1969, p. 247). Detailed petrographical work is needed to distinguish the junction between the component sandstones. The lower part of the combined rock is very variable, ranging from fine-grained sandstone with siltstone in some bores to coarse-grained sandstone in others.

The Pot Clay Marine Band (p. 48) may rest directly on the ganisteroid top of the Rough Rock, as in Tickhill No. 1 Oil Bore, or be separated from the sandstone by a few feet of silty or argillaceous beds. In South Leverton No. 7 Oil Bore 2½ ft of greyish brown siltstone-seatearth were cored between the marine band and the sandstone. In several oil bores a thin coal, the Pot Clay Coal, is recorded, but no cores have been obtained.

E.G.S

PALYNOLOGY

Samples from seven oil bores—Gainsborough No. 2, Grove No. 1, Morton No. 1, Ranskill No. 1, South Leverton No. 1, Tickhill No. 1 and Walkeringham No. 1— have been examined for spores. Of the 19 samples prepared using standard preparation techniques, 14 have yielded microfloral assemblages. The samples examined were derived from both cores and chippings. In the case of the latter contamination by caving of horizons higher in the well has resulted in diagnostic Westphalian spore and Permo-Triassic pollen types being recorded along with the Namurian assemblages.

In the stratigraphical interpretation of the samples, therefore, care has been taken to omit any record considered to be the result of contamination. Permanent mounts of all productive samples studied are stored in the Palynology Collection of the Institute of Geological Sciences at Leeds; the catalogue numbers of individual samples (e.g. SAL 1250) are given in the text.

Gainsborough No. 2 Oil Bore. Five samples were examined from this bore, and all yielded reasonably well-preserved miospore assemblages.

Depth 5047 to 5050 ft (SAL 1271). The miospore assemblage recovered from this sample was dominated by representatives of the genus *Lycospora*, which made up 73·5 per cent of the total assemblage. Of greater significance, however, is the presence of representatives of *Crassispora kosankei*, which account for 6 per cent of the total population. This species is known to occur in sediments as old as high Viséan, but it is not until the onset of Namurian B times that it becomes a significant member of the miospore population (Neves 1961; Owens and Burgess 1965). Several of the accessory spores present in the assemblage are of stratigraphical value: *Knoxisporites seniradiatus* was originally described from R_1–G_1 sediments in the southern Pennine region (Neves 1961) and has since been recorded from beds of E_2 to early R age in the Stainmore Outlier, Westmorland (Owens and Burgess 1965). *Proprisporites laevigatus* was recorded by Neves (1961) in sediments of E_1–R_2 age in the southern Pennine region, but was found to be restricted to high Namurian A deposits in the Stainmore Outlier. *Rugospora corporata* and *Spelaeotriletes arenaceus* are known respectively from high Namurian A–Namurian C and Namurian A to low Namurian C beds in northern England (Neves and Owens 1966). The presence of *Cirratriradites rarus* is worthy of note. This species has only previously been recorded in Britain by Owens (1963) from Namurian B deposits in the Stainmore Outlier. The evidence available from these accessory spores, together with the presence in significant numbers of *Crassispora kosankei*, suggests an R_1 age for this sample.

Depth 5292 ft (SAL 1272) [Core sample]. As in SAL 1271, the dominant elements in the assemblage were representatives of the genus *Lycospora*, which made up 60 per cent of the total population. *Crassispora kosankei* was less frequent at 2·5 per cent, but may still be of stratigraphical significance. Of the stratigraphically useful accessory spores present in the assemblage the following are of interest: *Knoxisporites dissidius*, described from E_2–G_2 sediments in the southern Pennines by Neves (1961), and subsequently reported by Owens and Burgess (1965) from E_2–R_2 sediments in the Stainmore Outlier; *Krauselisporites sp.*, an undescribed group of spores recorded by Owens (1963) from Namurian A to low Namurian C sediments in the Stainmore Outlier. Representatives of the latter group of spores were also recorded in SAL 1271. *Spelaeotriletes arenaceus* and *Proprisporites laevigatus*, both of which occurred in SAL 1271 (see above), were also recorded from this horizon. The limited evidence available from the accessory spores together with the reduced frequency of *Crassispora kosankei* in the sample suggest an R (Namurian B) age.

Depth 5509 to 5512 ft (SAL 176). The miospore assemblage here was associated with plant and wood debris. Representatives of the genus *Lycospora* were the most frequent spores; no accessory spores of diagnostic significance were recorded.

Depth 6003 to 6006 ft (SAL 175). Numerous representatives of the genus *Lycospora* dominated this assemblage. Several of the accessory spores present are of stratigraphical significance, in particular representatives of the genus *Schulzospora*, which are characteristic components of high Viséan and Namurian A miospore assemblages in many parts of Britain. *Leiotriletes* cf. *tumidus*, *Acanthotriletes castanea*, *Convolutispora* cf. *tesselata*, *Microreticulatisporites* cf. *microreticulatus*, *?Bellispores nitidus*, *Vallatisporites vallatus* and *Spinozonotriletes uncatus* form an association which is characteristic of Namurian A deposits. The total evidence available from the assemblage suggests that an E_2 age is probable for the sample.

Depth 6249 to 6252 ft (SAL 174). Representatives of the genus *Lycospora* were again the most commonly recorded spore type in the assemblage. Amongst the accessory spores present, the following association is of particular significance: *Bellispores nitidus*, *Crassispora kosankei* and *?Remysporites* *magnificus*. This distinctive association has been found to be diagnostic of E_{2b} assemblages in northern England. The only other accessory spore of note is *Rugospora corporata*, which has previously been recorded by Neves and Owens (1966) from high Namurian A to Namurian C deposits in the southern Pennines and the Stainmore Outlier. On the evidence available from the accessory spores an E_2 age is suggested.

Grove No. 1 Oil Bore. Three samples, all from chippings, were examined from this bore and all yielded miospore assemblages.

Depth 4578 to 4581 ft (SAL 1267). The dominant elements in the miospore assemblage were the specimens of *Lycospora*, which made up 86 per cent of the total population. Densospores, particularly *Densosporites anulatus*, made up 7·5 per cent and *Crassispora kosankei* a further 1·5 per cent. Only a few accessory spores of stratigraphical significance were recorded and, of these, only *Lycospora subtriquetra* is worthy of note. This species has previously been recorded in Britain only from Namurian B to Westphalian A sediments in the Stainmore Outlier (Owens 1963). The limited evidence available does not permit a precise age determination to be made but does possibly indicate a Namurian B affinity.

Depth 4650 to 4653 ft (SAL 1269). Representatives of the genus *Lycospora* account for 83 per cent of the miospore population. The number of specimens of *Crassispora kosankei* at 3 per cent of the assemblage is greater than in SAL 1267 (see above). Several of the accessory spores present are of stratigraphical significance: *Ibrahimispores brevispinosus* was described by Neves (1961) from sediments of H to early Westphalian A age in the southern Pennine region and has since been recorded by Owens and Burgess (1965) from beds of E_2–R age in the Stainmore Outlier. *Secarisporites* cf. *remotus* was described from R_2–G_1 beds in the southern Pennines by Neves (1961) and has since been recorded by Owens and Burgess (1965) in post R_1 and lower Westphalian beds in the Stainmore Outlier. *Knoxisporites dissidius* and *Rugospora corporata* (see above) also occur. The evidence available from these accessory spores together with the presence in reasonable numbers of *Crassispora kosankei* suggest an R (Namurian B) age.

Depth 4767 to 4770 ft (SAL 1268). Although a varied miospore assemblage was recovered none of the species recorded is of stratigraphical value.

Morton No. 1 Oil Bore. Two samples, both from chippings, were examined from this bore. In view of their close stratigraphical relationship and the general similarity in composition of their miospore assemblages they are interpreted together.

Depths 5310 to 5314 ft (SAL 178) and 5478 to 5482 ft (SAL 177). The dominant elements in both of the assemblages were the representatives of the genus *Lycospora*. Several of the stratigraphically useful accessory spores recorded were common to both assemblages and the following are considered of greatest significance. *Punctatisporites sinuatus* has been recorded by Neves (1961) from beds of R_2 to basal Coal Measures age in the southern Pennine region and by Owens and Burgess (1965) from upper Namurian A to lower Westphalian B deposits in the Stainmore Outlier. *Secarisporites lobatus* was described from H–G_1 beds in the southern Pennine region and has since been recorded by Owens (1963) from Namurian B to lower Westphalian A sediments in the Stainmore Outlier. *S.* cf. *remotus* (see above) was also present. *Reticulatisporites karadenizensis* was described by Artuz (1957) from Westphalian A coals of the Zonguldak Coalfield, Turkey, and has since been recorded by Owens (1963) in Namurian B to lower Namurian C deposits in the Stainmore Outlier. *Lycospora subtriquetra* and *Spelaeotriletes arenaceus* (see p. 26) are also significant. Whilst the previous records of these accessory spores do not permit a precise age determination to be made, they suggest a Namurian B affinity. Such a conclusion would be in agreement with the presence in reasonable numbers in both assemblages of representatives of *Crassispora kosankei*.

Ranskill No. 1 Oil Bore. One sample of cored material from 5669 ft (SAL 1266) was examined. The organic residue recovered contained abundant wood and plant debris but only a very small number of poorly preserved spores, which were difficult to identify even at generic level.

South Leverton No. 1 Oil Bore. One sample of cored material from 4441 ft (SAL 1272) was examined. The microfloral assemblage was dominated by representatives of the genus *Lycospora*, but also contained significant numbers of *Crassispora kosankei* which suggests an age younger than the base of R_1 for the sample. None of the accessory spores recorded was of stratigraphical value.

Tickhill No. 1 Oil Bore. Four samples, three from chippings and one from core, were examined from the following depths: 4851 to 4853 ft (SAL 1263); 4912 ft, core, (SAL 1264); 5193 to 5196 ft (SAL 1262); 5598 to 5601 ft (SAL 1261). A small number of miospores was recovered from SAL 1263 and SAL 1261, but none of the species recorded has any detailed stratigraphical value. The microfloral residues recovered from SAL 1262 and SAL 1264 were composed entirely of wood and plant debris.

Walkeringham No. 1 Oil Bore. Three samples, two from chippings and one from core, were examined, and all yielded miospore assemblages.

Depth 5294 to 5297 ft (SAL 1265). The dominant elements in the miospore assemblage were the numerous representatives of the genus *Lycospora* which made up 68·5 per cent of the total population. Of greater significance is the presence of *Crassispora kosankei* with a frequency of 7·5 per cent, indicating an age younger than the base of R_1. Of the accessory spores present the following are of stratigraphical value: *Punctatisporites sinuatus* (p. 27); *Secarisporites* cf. *remotus* (p. 27); *Reticulatisporites karadenizensis* and *Cirratriradites rarus*, which previously have only been recorded in Britain by Owens and Burgess (1965) from R_1–G_1 beds in the Stainmore Outlier; *Knoxisporites dissidius*, *Spelaeotriletes*

arenaceus and *Rugospora corporata* (see p. 26). The evidence available from the accessory spores together with the fact that significant numbers of specimens of *Crassispora kosankei* are present in the assemblage suggest a Namurian B age.

Depth 5700 to 5703 ft (SAL 1270). The numerous representatives of the genus *Lycospora* were the most commonly recorded spore types in SAL 1270. Specimens of *Crassispora kosankei* were also common and suggest an age younger than the base of R_1. Only two of the accessory spores present are worthy of note: *Proprisporites laevigatus* and *Spelaeotriletes arenaceus* (p. 26).

Depth 6340 to 6390 ft (SAL 1250) [Core sample]. Representatives of the genus *Lycospora* made up 81 per cent of the assemblage. It is significant that the number of specimens of *Crassispora kosankei* was very small, which suggests a pre-R_1 age. Of the accessory spores present, the following are worthy of note: *Microreticulatisporites punctatus*, a characteristic component of upper Viséan and lower Namurian assemblages in northern England and the Midland Valley of Scotland; *Camptotriletes cristatus*, described by Sullivan and Marshall (1966) from P_2–E_1 beds in the Midland Valley of Scotland and recorded by Neville (1968) from upper Viséan deposits of the Fife Coast; *Dictyotriletes tuberosus*, described by Neves (1961) from E_2 to H beds in the southern Pennine region and recorded by Owens and Burgess (1965) from beds of E_1 to early R_1 age in the Stainmore Outlier. The limited evidence available from these accessory spores also suggests a Namurian A age for the sample.   B.O.

IGNEOUS ROCKS

Igneous rocks were encountered in Gainsborough No. 2 Oil Bore, in the Bothamsall Oilfield and in West Drayton No. 2 Oil Bore. Traces of igneous rocks are also recorded in Apleyhead No. 1 and Tickhill No. 1 oil bores.

Gainsborough Oilfield. No. 2 is the deepest of the bores (6259 ft) in this field, and it encountered about 105 ft of dolerite, presumably in the form of a sill, with base at 6170 ft (Plate II), over 1500 ft below the horizon of the Pot Clay Marine Band. Palynological samples from above and below the dolerite show that it lies within shales of E_2 age (p. 27). The lateral extent of the dolerite is not known, but it is absent in the

few other bores in the district that penetrate beds of E age and also in Corringham No. 1 Oil Bore 5 miles to the east-north-east. An intrusive dolerite at a similar horizon in the Millstone Grit Series is found, however, in Egmanton No. 57 Oil Bore (Edwards 1967, p. 39, fig. 4), about 3 miles to the south of the present district. The presence of a substantial dolerite intrusion at depth in the Gainsborough Trough is reminiscent of the situation in the Widmerpool Gulf, where, in the Duffield Borehole (*A. Rep. Inst. geol. Sci. for 1967*, 1968, p. 82), sills have been proved in the argillaceous Widmerpool Formation of the Carboniferous Limestone Series.   E.G.S.

Thin sections (E 35709–18) of chippings show, throughout the thickness recorded (6066 to 6170 ft), dolerite in variable states of alteration. Other lithologies (excluding chlorite, which almost certainly stems from the altered dolerite) are very sparse and include shale and siltstone, probably representing cavings.

Throughout the chippings sectioned, the dolerite is medium- to coarse-grained with average granularity from about 0·2 to 0·6 mm. Poikilitic plates of coarse, pinkish titanaugite range, however, up to 3 mm across (E 35715) and tabular feldspars attain 0·8 mm in length. The titanaugite encloses laths, which, where unaltered, approach labradorite in composition. Most of the plagioclase, however, is variably albitized, or altered to carbonate or pink clay minerals. There are coarse, subhedral plates and rods of ilmenite, mostly altered to leucoxene. The finer grained, interstitial components include chlorite, carbonate, feldspars and opaque dust. Apatite needles are a conspicuous accessory. One grain of chloritic, very fine-grained, fluxioned basalt was noted (E 35711), though the exact horizon of this is uncertain.   R.K.H.

Bothamsall Oilfield and West Drayton. Igneous rocks have been proved in all the deeper Bothamsall oil bores that lie within the district, and at West Drayton. They consist of tuffs, agglomerates and lavas, possibly with minor intrusions, and, as Edwards (1967, p. 33 and pl. iv) has noted, evidently represent a number of outbursts from a volcano situated in the Bothamsall area—see also Falcon and Kent (1960, p. 25 and fig. 11). In the west the volcanic rocks probably lie between the horizons of *Reticuloceras gracile* and *R. superbilingue:* in the east their base has not been proved, but they extend upwards to the *Gastrioceras cancellatum* horizon (Bothamsall No. 13 Oil Bore) and in West Drayton No. 2 Oil Bore are immediately overlain by the Rough Rock.

The only proved complete sequence is that at Bothamsall No. 3 Oil Bore where the tuff, agglomerate and lava (see below and Harrison *in* Edwards, 1967, pp. 52–7), including two

bands of silty shale near the base, have a total thickness of nearly 140 ft. This compares with 175 ft and over 200 ft of volcanic rocks in Bothamsall Nos. 1 and 5 bores, the latter sequence including an epiclastic intercalation near the top. Bothamsall Nos. 1 and 5 bores lie south of the district boundary, and their volcanic rocks have been described in detail by Harrison (*idem*).

Bothamsall No. 2 Oil Bore proved volcanic rock between 3722 ft and the bottom of the hole at 3738 ft. This is described in the British Petroleum Company's log as "volcanic ash with tuff at base". No. 4 Bore finished drilling at 3625 ft, having penetrated 10 ft into "pale grey tuff", and No. 10 Bore at 3679 ft in a 10-ft band described as "grey sandy pyritic ash" (but see below). In Bothamsall No. 13 Oil Bore the rocks between 3630 ft and the bottom of the hole at 3757 ft consist largely of tuff, but a 20-ft band of pale grey, fine-grained, quartz-sandstone is recorded at 3690 ft resting on about 16 ft of dolerite, which represents either a lava or an intrusion (see below). In Bothamsall No. 15 Oil Bore 20 ft of volcanic rock are recorded at the bottom of the hole (3788 ft). Described as "grey and green, calcareous, gritty tuff", they evidently consist (see below) largely of lava, probably with some tuff.

In West Drayton No. 2 Oil Bore about 75 ft of volcanic rocks, described in the BP Company's log as "black and speckled tuff with much calcite", are sandwiched between two thick sandstones—probably the Chatsworth Grit and Rough Rock cum Crawshaw Sandstone (see pp. 25 and 49)—at 3794 ft. Falcon and Kent (1960, p. 25) noted an ash bed about 5 ft thick in West Drayton No. 1 Oil Bore, about ¼ mile NE of No. 2 Bore. This is presumably part of the "grey spotted siltstone with detrital igneous material and plant debris; partly cemented to hard rock" which Mr. W. Edwards described from the cores between 3882 and 3892 ft (see also below). It is apparently at the same stratigraphical horizon as the tuff in No. 2 Bore.
                                                                      E.G.S.

Bothamsall No. 3 Oil Bore. Chipping samples were taken of the major part of the volcanic succession between 3592 and 3729 ft, cores being available for only the uppermost 16 ft. The latter consist of coarse volcanic breccia succeeded in turn by thin beds of vesicular lava, non-vesicular basic lava and vesicular tuff. This altered lava was provisionally identified as andesite, although a more basic (basaltic) parentage was suggested (Harrison, *idem*), and an overall basaltic affinity now seems more likely as suggested by the less-altered chippings described below.

The chippings (averaging about 4 mm) were sampled over 10-ft intervals, and thin sections (E 35602–15) prepared. In the chippings, lava predominates throughout the succession (with chlorite and carbonate, stemming perhaps from amygdales), though sparse sedimentary lithologies occur, particularly in the lower part. It is not possible to determine whether the lava chippings represent lava flows, or medium- to coarse-grained pyroclastic rocks containing abundant lava fragments. The lithological log of this bore provided by the BP Exploration Co. indicates, however, a predominance of pyroclastic rocks with minor lava flows.

The dominant lava in the chippings is microporphyritic with microphenocrysts of albitized plagioclase and unidentified pseudomorphs up to 1 mm across, set in strongly fluxioned groundmass swarms of plagioclase needles averaging 0·2 by 0·06 mm, and ranging down to 0·09 by 0·01 mm. Though plagioclase lies chiefly in the compositional range of albite to oligoclase, rare labradorite was found in one specimen (E 35614). Interstitial components include dark olive-green to brown chlorite, leucoxene, opaque dust and sporadic pyrite. Amygdaloidal structures are sporadic, and contain zoned, radially-fibrous chlorite, carbonate, chalcedony and, in places, pyrite. Fragments of apparently non-porphyritic lava are similar to the groundmass of the microporphyritic type described. The textures of the lava chippings vary little except towards the base (3714 to 3717 ft) where the feldspar laths become a little coarser, and the interstitial components increasingly turbid and dark.

The highly altered state, particularly of the groundmass, of the lava particles precludes firm identification. The general fluxioned texture, however, and the labradorite noted above suggest albitized basalt, but the relatively high feldspar to (altered) ferromagnesian minerals ratio could indicate albitized andesitic basalt. Differentiation trends which may have occurred cannot, however, be assessed.

Bothamsall No. 10 Oil Bore. Chippings (E 35616–7) were examined from depths of 3668 to 3671 ft and 3677 to 3679 ft. Thin sections show particles of tuffaceous calcareous sandstone consisting of angular, poorly sorted quartz-sand (0·2 mm) with carbonate granules, sporadic pyroclastic microlitic lava and yellowish brown basic glass. Siltstone and mudstone chippings also occur.

Bothamsall No. 13 Oil Bore. Sections (E 35618–26) were examined of chippings taken at approximately 9-ft intervals from 3634 to 3754 ft. In the higher part of the sequence (3634 to 3661 ft: E 35618 and 35620) the chippings range from turbid and opaque pumiceous particles to pumiceous tuff, sandy spicular tuff with coarse carbonate, and sphaerosideritic siltstone. From 3694 to 3706 ft, however, the chippings sampled consist of dolerite, with relatively fresh phenocrystic feldspars averaging 1·8 by 0·5 mm and groundmass feldspars averaging 0·9 by 0·09 mm. They are mainly labradorite, and form a framework filled with chlorite, carbonate and ilmenite-magnetite. There are clusters of finely granular (0·04 mm) anhedral clinopyroxene particles. In the lower part of the sequence (3715 to 3754 ft) the chippings consist of predominantly pumiceous tuff and pumice, with spheroidal amygdales (0·1 mm diameter) containing green chlorite rims and chalcedonic interiors. The base is formed of green to dark brown chlorite. Chloritic pumiceous tuff predominates from 3727 to 3754 ft (E 35624–6).

Bothamsall No. 15 Oil Bore. Chippings from 3778 to 3782 ft (E 35627) consist entirely of olive-green to brown, fluxioned microlitic lava (plagioclase microlites averaging 0·3 by 0·03 mm) with much interstitial yellowish brown chlorite and leucoxene. At 3786 to 3788 ft the chippings (E 35628) include similar lava, with probable tuff and a trace of calcareous siltstone.

West Drayton No. 1 Oil Bore. A specimen (E 19190) of a gritty, coarse sandy litho-clastic rock from 3885 ft was described by Dr. J. Phemister. Angular lithic fragments up to 2 mm across include carbonaceous argillite, altered (argillized) vesicular lava and tuff, and are cemented mainly by calcite which has partly replaced some of the lithic particles.

R.K.H.

Apleyhead No. 1 Oil Bore. Here the BP Company's geologists have recognized traces of volcanic ash in chippings from two horizons. Apparently the higher horizon represents a thin band of light grey feldspathic ash of siltstone grade at about 3802 ft, and the lower horizon represents at least two thin ash bands interbedded with black silty pyritous shales and grey carbonaceous, micaceous sandstone between 3940 and 3950 ft. These deposits are probably the distal equivalents of the thick pyroclastics at Bothamsall.

Tickhill No. 1 Oil Bore. Mr. R. K. Harrison has examined thin sections of the coarser particles from chippings between 4825 and 4896 ft. He records, besides a variety of epiclastic rocks, possible pumice and pumiceous mudstone at 4825 to 4829 ft, pumiceous siltstone at 4831 to 4833 ft and ?tuffaceous calcilutite at 4839 to 4841 ft. This occurrence appears to be at approximately the same stratigraphical horizon as the occurrences at Apleyhead and Bothamsall (Plate II). E.G.S.

REFERENCES

ARTUZ, S. 1957. Die Sporae dispersae der Turkischen Steinkohle von Zonguldak Gebiet. *Rev. de la Faculté des Sciences de l'Université d'Istanbul.* Ser. B, **22**, 240–63.

BROMEHEAD, C. E. N., EDWARDS, W., WRAY, D. A. and STEPHENS, J. V. 1933. The geology of the country around Holmfirth and Glossop. *Mem. geol. Surv. Gt Br.*

BRUNSTROM, R. G. W. 1963. Recently discovered oilfields in Britain. *Wld Petrol. Congr.*, Frankfurt, Section 1, 1–10.

COSGROVE, M. E. *in* RAMSBOTTOM, W. H. C., RHYS, G. H. and SMITH, E. G. 1962. *q.v.*

DOWNING, R. A. and HOWITT, F. 1969. Saline ground-waters in the Carboniferous rocks of the English East Midlands in relation to the geology. *Q. Jnl engng Geol.*, **1**, 241–69.

EDEN, R. A., STEVENSON, I. P. and EDWARDS, W. 1957. Geology of the country around Sheffield. *Mem. geol. Surv. Gt Br.*

EDWARDS, W. 1967. Geology of the country around Ollerton. 2nd edit. *Mem. geol. Surv. Gt Br.*

FALCON, N. L. and KENT, P. E. 1960. Geological results of petroleum exploration in Britain 1945–1957. *Mem. geol. Soc. Lond.*, No. 2.

GILLIGAN, A. 1920. The petrography of the Millstone Grit of Yorkshire. *Q. Jnl geol. Soc. Lond.*, **75**, 251–94.

HARRISON, R. K. *in* EDWARDS, W. 1967. *q.v.*

HOWITT, F. and BRUNSTROM, R. G. W. 1966. The continuation of the East Midlands Coal Measures into Lincolnshire. *Proc. Yorks. geol. Soc.*, **35**, 549–64.

KENT, P. E. 1966. The structure of the concealed Carboniferous rocks of north-eastern England. *Proc. Yorks. geol. Soc.*, **35**, 323–52.

NEVES, R. 1961. Namurian plant spores from the southern Pennines, England. *Palaeontology*, **4**, 247–79.

—— and OWENS, B. 1966. Some Namurian camerate miospores from the English Pennines. *Pollen et Spores*, **8**, 337–60.

NEVILLE, R. S. W. 1968. Ranges of selected spores in the Upper Viséan of the East Fife coast section between St. Monance and Pittenweem. *Pollen et Spores*, **10**, 431–62.

OWENS, B. 1963. A palynological investigation of the Namurian and Westphalian sediments of the Stainmore Outlier, Westmorland. Unpublished Ph.D. Thesis, Univ. of Sheffield.

—— and BURGESS, I. C. 1965. The stratigraphy and palynology of the Upper Carboniferous Outlier of Stainmore, Westmorland. *Bull. geol. Surv. Gt Br.*, No. 23, 17–44.

PULFREY, W. 1934. A boring in the Millstone Grits, Rod Moor, Sheffield. *Proc. Yorks. geol. Soc.*, **22**, 254–64.

RAMSBOTTOM, W. H. C. 1966. A pictorial diagram of the Namurian rocks of the Pennines. *Trans. Leeds geol. Ass.*, **7**, 181–4.

—— 1969. The Namurian of Britain. *C. r. 6th Cong. Int. de Strat. et de Géol. du Carb.*, **1**, 219–32.

—— RHYS, G. H. and SMITH, E. G. 1962. Boreholes in the Carboniferous rocks of the Ashover district, Derbyshire. *Bull. geol. Surv. Gt Br.*, No. 19, 75–168.

SMITH, E. G., RHYS, G. H. and EDEN, R. A. 1967. Geology of the country around Chesterfield, Matlock and Mansfield. *Mem. geol. Surv. Gt Br.*

SULLIVAN, H. J. and MARSHALL, A. E. 1966. Viséan spores from Scotland. *Micropaleontology*, **12**, 265–85.

TAYLOR, F. M. and HOWITT, F. 1965. Field meeting in the U.K. East Midlands oil-fields and associated outcrop areas. *Proc. Geol. Ass.*, **76**, 195–209.

Chapter IV

COAL MEASURES (WESTPHALIAN)

INTRODUCTION

THE COAL MEASURES, present at depth throughout the East Retford district, conformably succeed the Millstone Grit Series and are unconformably overlain by Permian rocks. Exploratory boreholes, colliery shafts and oil bores show that up to 4500 ft of strata, divided into Lower, Middle and Upper divisions, are preserved beneath the sub-Permian unconformity. The Lower Coal Measures, 800 to 1600 ft thick, are everywhere complete, but the Middle Coal Measures, 800 to 1900 ft thick, are incomplete over the Whitwell and Bothamsall anticlines in the south-west and part of the Gainsborough Anticline in the north-east, due to the denudation which preceded the deposition of the Permian rocks. Pre-Permian denudation has also been responsible for the absence of the Upper Coal Measures in the same areas, and has removed an unknown thickness of strata younger than the 1100 ft of this division remaining in the district.

A marked thinning towards the east or south-east is apparent in all three divisions of the Coal Measures (Figs. 7 and 9 and Plate IX). It is taken to indicate an approach to the margin of the Pennine basin of deposition, which is also reflected in the faunal composition of the marine bands (Calver 1968). As Howitt and Brunstrom (1966, pp. 558–61, pl. 31, fig. 4) have demonstrated, there is, superimposed on this general thinning, a local thickening in the area of the Gainsborough Trough (see p. 12). While in Coal Measures times this area continued to subside more rapidly than the areas on each side of it, its effects on sedimentation were insignificant compared to the striking depositional pattern produced by the trough in early Millstone Grit times (Plate II).

The Coal Measures comprise cyclic sequences of sediments consisting of mudstones, siltstones, sandstones, seatearths and coals which occur, ideally, in that upward order. One or more members of the cycle may be missing but, as Duff and Walton (1962) have shown, the mudstone and seatearth are usually present. The rocks are similar, as regards both lithology and cyclic disposition, to their equivalents elsewhere in the Yorkshire and East Midlands Coalfield. They have been described at some length in other publications (e.g. Eden and others 1957; Smith and others 1967) and it is not therefore proposed to repeat the details here. In general they show that deposition took place in a shallow, fresh or brackish water environment subject to rapid but irregular subsidence. Occasional incursions of the sea were responsible for the distinctive marine bands, which usually consist of dark mudstones and occur at or near the base of more than a dozen cycles. The fossils from these marine bands are the most significant of all those used for Coal Measures correlation, but also of considerable importance are the non-marine bivalves, or mussels, which occur widely in the mudstones, and to a lesser extent in the siltstones and ironstones. The mussels, which occur in characteristic assemblages at a number of horizons, have been used to zone the Coal Measures (pp. 43–6 and Figs. 6, 8 and 10).

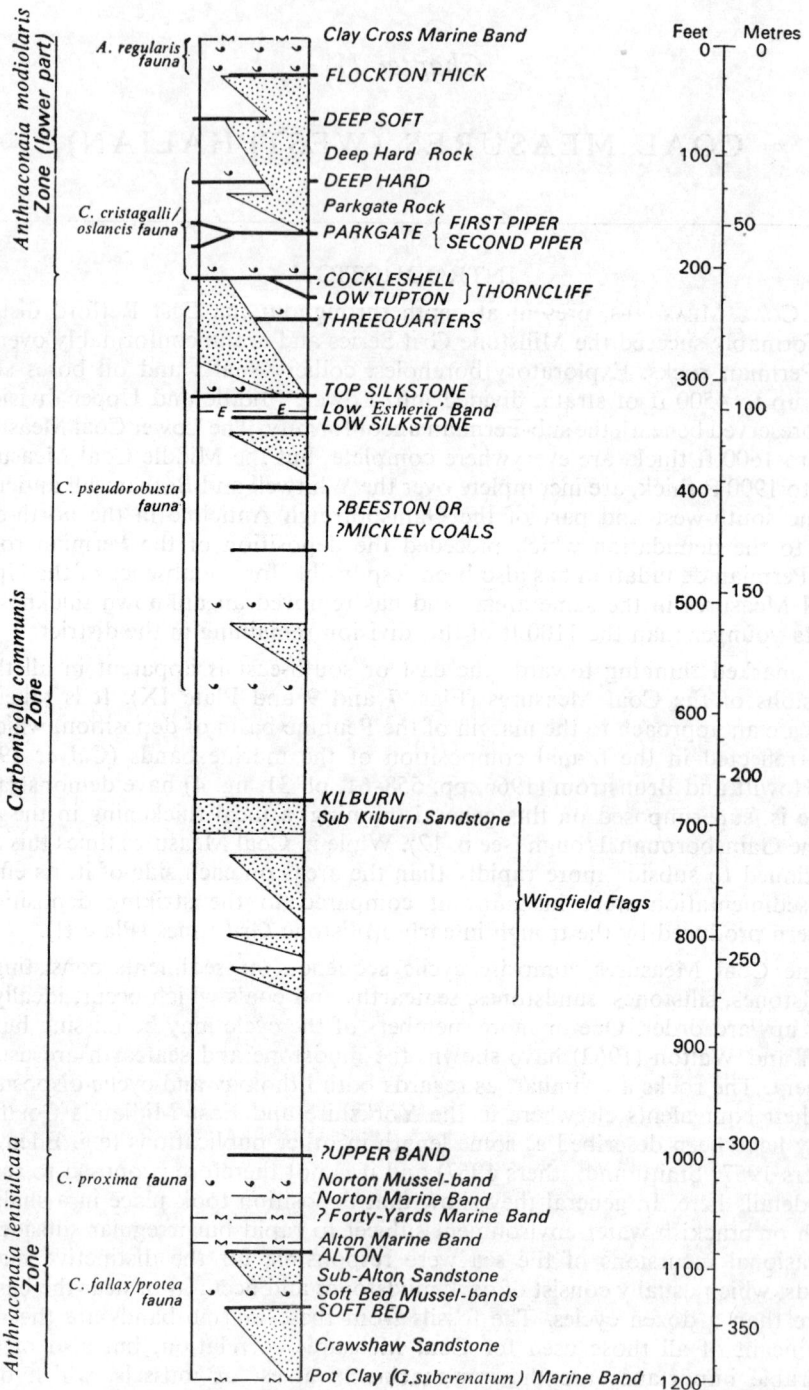

FIG. 6. *Generalized section of the Lower Coal Measures, showing the principal coals, fossil-bands and sandstones*

For key to ornament see Fig. 10

GENERAL STRATIGRAPHY

The following brief account summarizes the main features of the stratigraphy of the Coal Measures. A detailed stratigraphical account is given on pp. 47–99.

LOWER COAL MEASURES

The Lower Coal Measures, corresponding to Westphalian A of the standard European classification, consist of the strata between the base of the Pot Clay (*Gastrioceras subcrenatum*) Marine Band and the base of the Clay Cross (*Anthracoceratites vanderbeckei*) Marine Band (Fig. 6). They are 1600 ft thick in the west and probably less than 800 ft in the east (Fig. 7). As already demonstrated by Howitt and Brunstrom (1966, pl. 32), there is local thickening across the Gainsborough Trough, which, however, was much less active than during Millstone Grit times (p. 12).

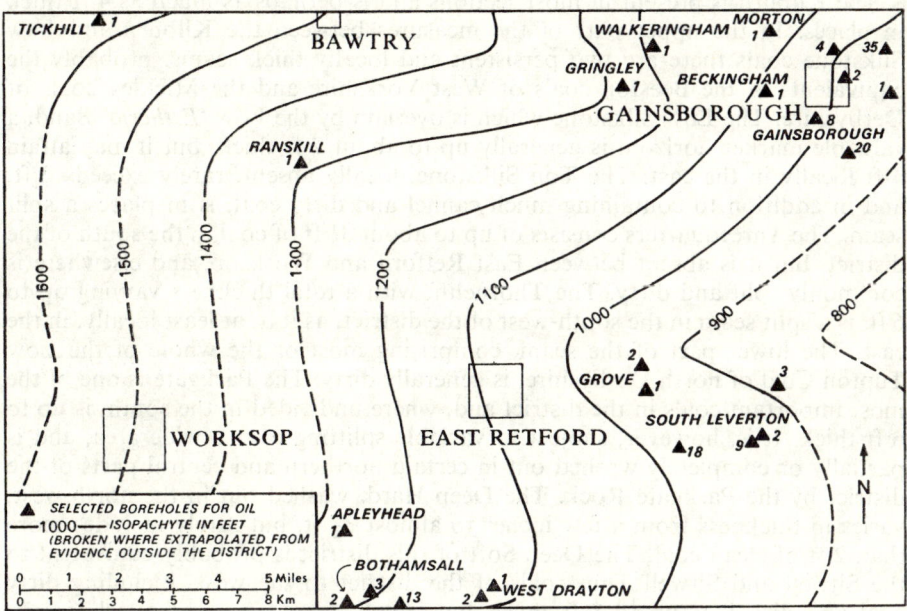

FIG. 7. *Isopachyte map of the Lower Coal Measures*

Marine bands are confined to the lowest part of the sequence. The Pot Clay Marine Band has yielded a goniatite/pectinoid fauna, locally with calcareous brachiopods, in the north-west; elsewhere in the district, except where washed out by the Crawshaw Sandstone in the south, it is of *Lingula* facies. *Lingula* also occurs between the Soft Bed and Alton coals at the horizons of the Holbrook and Second Smalley marine bands, but not at the slightly higher horizon of the First Smalley Marine Band. The Alton (*Gastrioceras listeri*) Marine Band has not yielded goniatites, but the pectinoid facies has been found in one borehole in the north; elsewhere it either contains *Lingula*, locally with sponge spicules, conodonts and foraminifera, or is absent and may then be represented by a fish-band. The Lower and Upper Parkhouse marine bands are probably absent. The Forty-Yards Marine Band is apparently present only locally in the north, where it contains *Lingula*, foraminifera and conodonts, but over most of the district its horizon can be tentatively inferred from the presence of fish debris.

The Norton, Upper Band and Burton Joyce marine bands, which occur elsewhere between the Forty-Yards horizon and the Wingfield Flags, have not been found in the East Retford district, but the Norton is apparently represented by a *Lingula* band in·the Corringham area, not far from Gainsborough.

Coals below the Kilburn horizon are of economically insignificant thickness. The Soft Bed is represented by a trace of coal in the Bothamsall Oilfield, the Holbrook is apparently missing everywhere, and the Smalley coals, an inch or two thick, are seen only in the north-west. The Alton is apparently absent locally and is nowhere more than 1 ft thick; it cannot everywhere be identified with certainty. The Forty-Yards has not been recorded except possibly at West Drayton in the south, where it may be 8 in thick. A 9-in coal in the same borehole is probably the Upper Band, but this seam has not been identified elsewhere in the district.

The Kilburn is present in most sections and is perhaps as much as 4 ft thick in places. In the upper part of the measures between the Kilburn and Low Silkstone coals there are two persistent and locally thick seams, probably the equivalents of the Beeston coals of West Yorkshire and the Mickley coals of Derbyshire. The Low Silkstone which is overlain by the Low '*Estheria*' Band, a valuable marker horizon, is generally up to about 2 ft thick, but it may attain 4 ft locally in the east. The Top Silkstone, locally absent, rarely exceeds 2 ft, and in addition to containing much cannel and dirty coal, is in places a split seam. The Threequarters consists of up to about $3\frac{1}{2}$ ft of coal in the south of the district, but it is absent between East Retford and Worksop, and elsewhere is commonly split and dirty. The Thorncliff, with a total thickness varying up to 6 ft, is a split seam in the south-west of the district, as it is, at least locally, in the east. The lower part of the seam, comprising most or the whole of the Low Tupton Coal of north Derbyshire, is generally dirty. The Parkgate is one of the most important coals in the district and, where undivided in the south, is up to 6 ft thick. It is, however, subject to variable splitting over a wide area, and is partially or completely washed out in certain northern and central parts of the district by the Parkgate Rock. The Deep Hard, washed out in the north-west, varies in thickness from a few inches to almost $3\frac{1}{2}$ ft, but rarely includes more than 2 ft of clean coal. The Deep Soft of this district is probably equivalent to the Sitwell and Sitwell Thin coals of the district to the west. Including dirty coal and dirt, it is up to 8 ft thick, but is split in places, particularly in the extreme west, and is locally washed out. The Flockton Thick is up to 4 ft thick in the north-west, but thin and dirty in the south-west; in the east it has not been identified with certainty.

The principal sandstones are the Crawshaw and Sub-Alton Sandstones, the Wingfield Flags and the sandstones between the Top Silkstone and Thorncliff coals and between the Parkgate and Flockton Thick coals. The Crawshaw Sandstone is present in the south, where, in the Bothamsall Oilfield, it is united with the Rough Rock (p. 25) and is an oil-sand. Only 10 ft thick in the extreme north-west of the district, it is absent elsewhere in the north, and in the east. The Sub-Alton Sandstone is present through most of the district and is up to 70 ft thick at Bothamsall, where it is one of the main oil-sands. The Wingfield Flags, generally thickest in the south, vary widely in thickness up to 175 ft. They include, at the top, the Sub-Kilburn Sandstone, which thickens northwards to 50 ft, and is distinguished by its abundant coloured micas. This sandstone is the most persistent part of the Flags, and in places is the only member present.

At some localities sandstone occupies the whole of the interval between the Top Silkstone and Thorncliff coals, but in others it is present at more than one horizon or may be completely absent. Thick sandstones occur above the Parkgate, Deep Hard and Deep Soft coals. The two former, the Parkgate Rock and the Deep Hard Rock, are commonly united in the north-west, and the last is locally united with the Deep Hard Rock. In a few places virtually all of the measures between the Parkgate Coal and the Clay Cross Marine Band consist of sandstones and siltstones.

MIDDLE COAL MEASURES

The Middle Coal Measures consist of the strata between the base of the Clay Cross (*Anthracoceratites vanderbeckei*) Marine Band and the top of the Top ('*Anthracoceras*' *cambriense*) Marine Band (Fig. 8). In terms of the standard European classification they represent the whole of Westphalian B and the lower part of Westphalian C. They are over 1900 ft thick in the west and probably 800 ft or less in the unproved south-eastern corner of the district. Isopachytes are shown on Fig. 9, which also illustrates the continuing, but subdued, effects of the Gainsborough Trough detected by Howitt and Brunstrom (1966, fig. 4). The succession is incomplete, owing to pre-Permian erosion, over the Kiveton–Bothamsall and Whitwell anticlines in the south-west (Fig. 9), and locally over the Gainsborough Anticline in the north-east.

Marine bands, apart from the Clay Cross Marine Band at the base, occur only in the top half of the Middle Coal Measures. The Clay Cross Marine Band, 2 to $12\frac{1}{2}$ ft thick, has a varied faunal assemblage in some sections, particularly the thicker ones, but contains only *Lingula* in others. The Two-Foot Marine Band, up to $8\frac{1}{2}$ ft thick, contains abundant *Lingula*, but other marine fossils are not numerous. The Clown Marine Band has been found only in one borehole, where it is represented by a thin *Lingula* band close above the Clown Coal. The Haughton Marine Band, 15 to more than 25 ft thick, generally contains only scattered marine fossils. The Sutton Marine Band, known from ten localities in the district and up to about 4 ft thick, normally contains only *Lingula*, *Orbiculoidea* and fish remains. The Mansfield Marine Band, over 30 ft thick in some sections, has a rich fauna, and in many localities includes, in its lower part, the well-known Mansfield 'cank', up to 3 ft thick. The *Edmondia* Band, widely variable in thickness in the ten localities from which it is known, has been washed out over much of the district by the Mexborough Rock (see below). Usually consisting of mudstones paler in colour than in the other marine bands, it contains foraminifera, bivalves, ostracods and fish. The Shafton Marine Band, up to 7 ft thick, has a varying fauna, but the elongate mussel *Anthraconaia spathulata* is characteristic. The Top Marine Band, locally washed out by the Ackworth Rock (p. 42), is more than 20 ft thick in some localities and contains a rich fauna, evidently at two separate horizons within the band.

Lioestheria is common at some horizons in the upper part of the Middle Coal Measures, especially in the Manton '*Estheria*' Band below the Clown Coal, and in the measures between the *Edmondia* Band and the Top Marine Band.

The Middle Coal Measures contain numerous coal seams; many are named, and, in the lower part of the sequence, several are of economic importance. The Second Ell and Lidget coals attain 2 to 3 ft locally, but the former is evidently absent in places. The Haigh Moor and Swallow Wood are closely related complex seams, locally combined. Each comprises three beds of coal with an

D

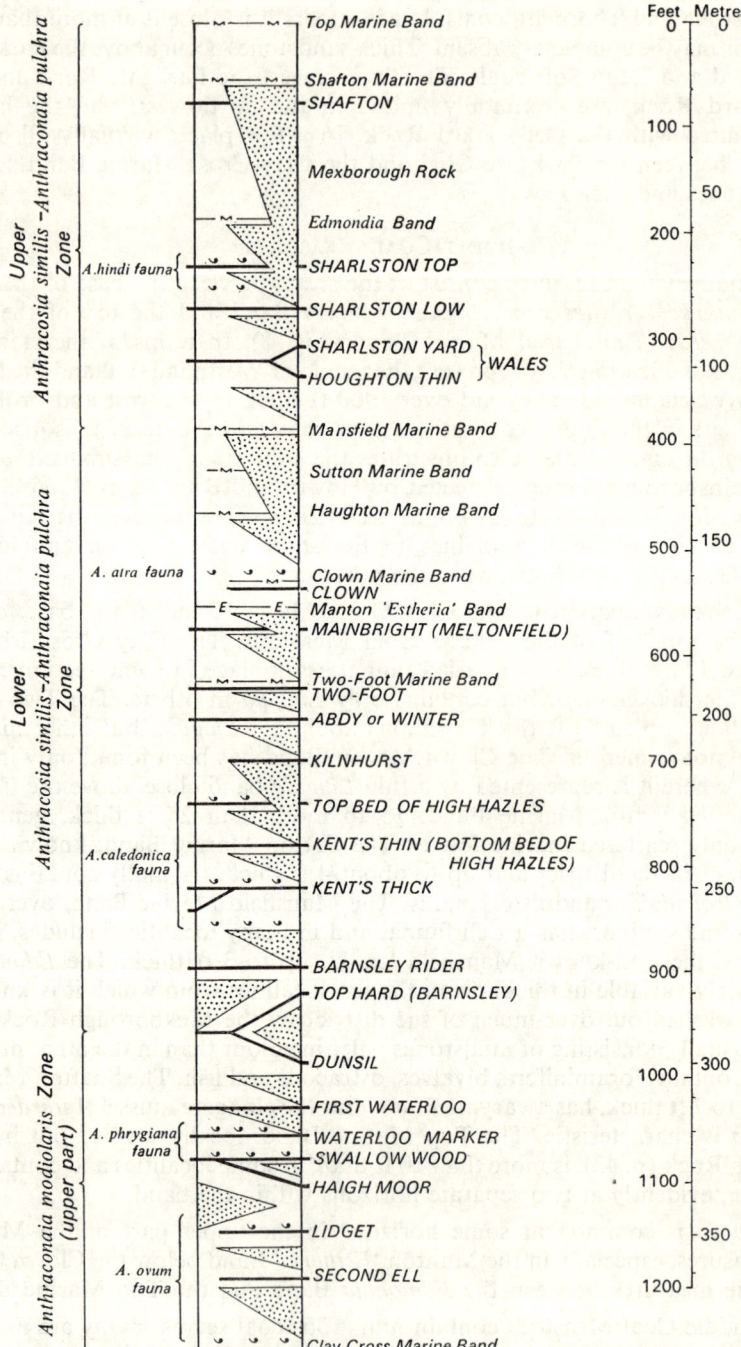

FIG. 8.   *Generalized section of the Middle Coal Measures, showing the principal coals,*
fossil-bands and sandstones

For key to ornament see Fig. 10

overall thickness of up to 4 ft for the Haigh Moor and up to 6 ft for the Swallow Wood. The Waterloo Marker consists in the west of the district of a few inches of coal, frequently in two leaves; it has not been identified in the east. The First Waterloo, also split in some sections, is 2 ft or more thick locally in the south-west and east. The Dunsil is widely variable in thickness up to about 4 ft, but it is joined with the thin overlying Blidworth Coal over much of the district, and in the east this combined seam joins the Top Hard to give a seam 4 to 7 ft thick. The Top Hard, known locally in workings as the Barnsley, is up to 5 ft thick where separate from the Blidworth and Dunsil, and has the typical structure known elsewhere in the coalfield. In the north-west of the district it joins the Blidworth and overlying Barnsley Rider to form the true Barnsley Coal. The Kent's Thick, consisting of up to $4\frac{1}{2}$ ft of coal in the west, is generally split into two beds by measures or a dirt band. The High Hazles,

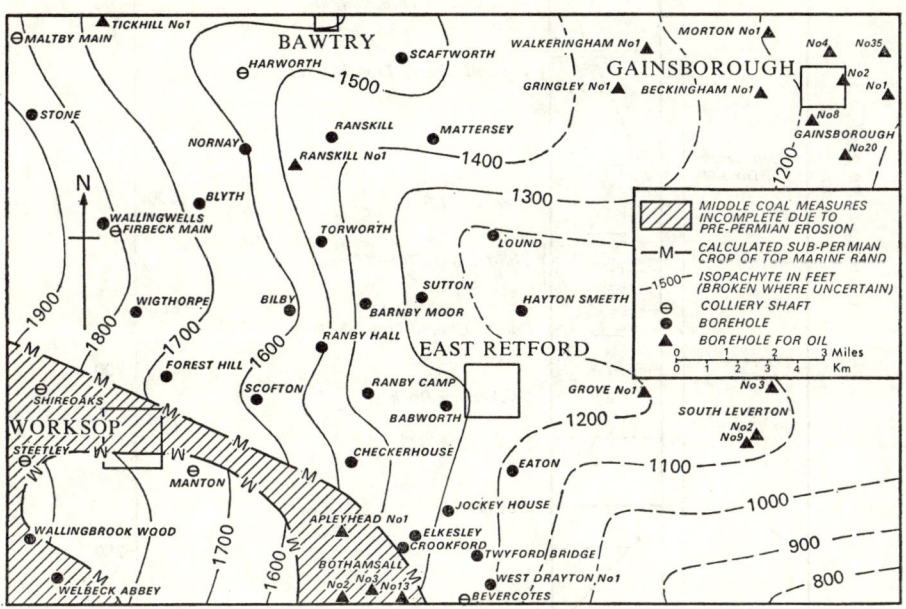

Fig. 9. *Isopachyte map of the Middle Coal Measures*

about 3 ft thick in the south-west of the district, is elsewhere split into two thin seams, the lower of which is the Kent's Thin of Yorkshire. The Kilnhurst is a seam of poor quality and very variable thickness, and the Abdy is a thin coal, especially in the south and east. The Two-Foot is $2\frac{1}{2}$ to 3 ft thick over much of the western part of the district, and may be of comparable thickness in places in the east. Many sections include cannel, which comprises nearly the whole of the seam where it is exceptionally thick in the north-west. The Mainbright, or Meltonfield, is 3 ft thick locally in the south, but elsewhere consists of one or more leaves of coal and dirty coal. The Clown, locally absent in the north of the district, is $2\frac{1}{2}$ to 3 ft thick over much of the west, but attains 4 ft or more in places both in the west and in the east. The Wales is a dirty seam $3\frac{1}{2}$ to nearly 6 ft thick in the south and south-west, but it splits northwards and probably eastwards into two thin coals, the Houghton Thin and the Sharlston Yard.

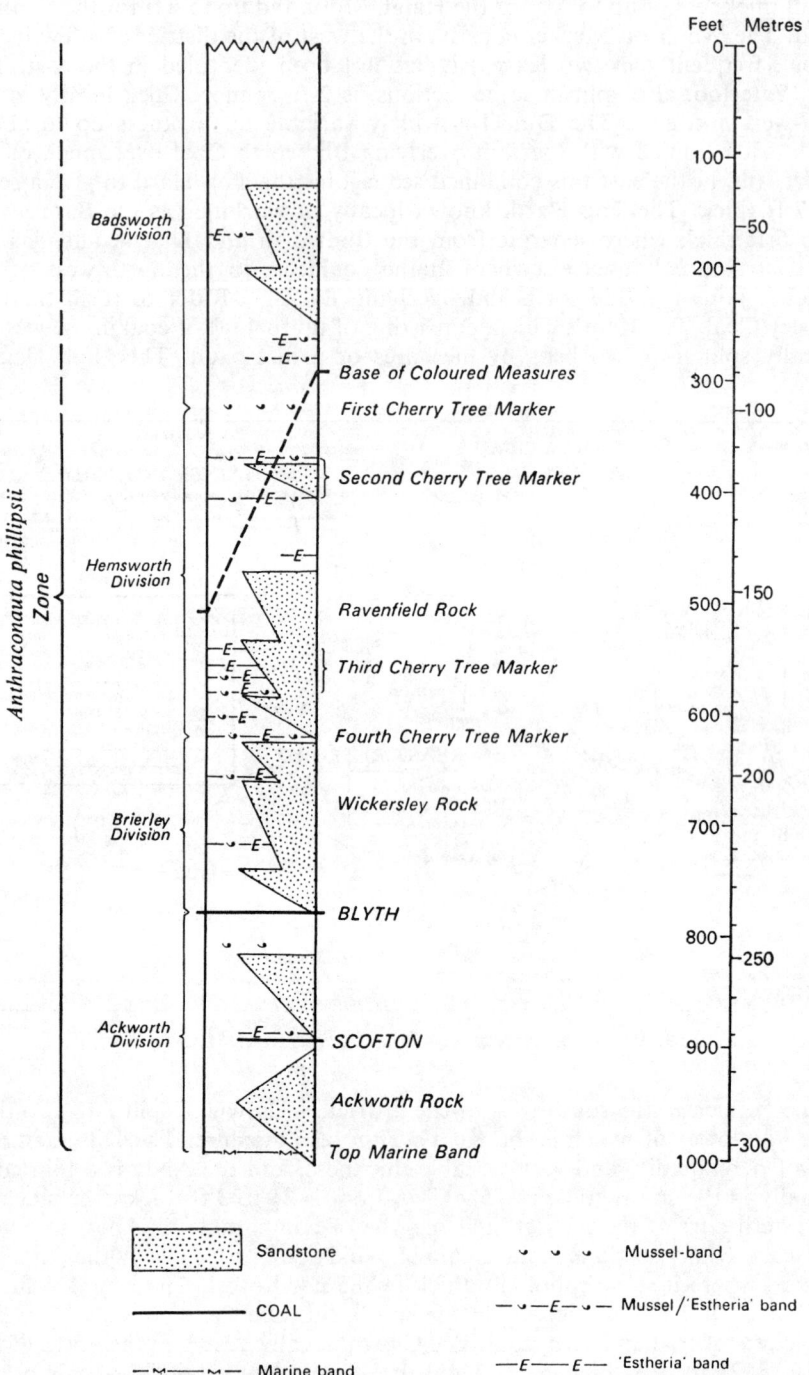

FIG. 10. *Generalized section of the Upper Coal Measures showing the component divisions and the principal marker horizons and sandstones*

The Sharlston Low, a thin and commonly split seam in the west, has not been recognized in the east. The Sharlston Top also is a thin seam of little economic account. The Shafton is present only locally in the west of the district, but is probably more widespread in the east, where it may be up to 3 ft thick.

Sandstones occur in almost every major cycle of the Middle Coal Measures and many attain substantial thicknesses locally. They are associated with washouts which locally cut out a number of seams, including the Haigh Moor, Top Hard and High Hazles, as well as several of the marine bands. Two sandstones between the Clay Cross Marine Band and the Top Hard Coal in the Gainsborough area are productive oil-sands (p. 236).

The most prominent sandstone is the Mexborough Rock, more than 100 ft thick over a wide area and locally, where joined with the sandstone below the horizon of the Sharlston Top Coal, up to nearly 300 ft. It is distinguished by a pink coloration, thought to be of primary origin, in an irregular area in the west of the district. E.G.S.

UPPER COAL MEASURES

The Upper Coal Measures consist of the strata above the top of the Top Marine Band (Fig. 10), and represent the upper part of Westphalian C and possibly part of Westphalian D of the standard European classification. Up to 1100 ft are preserved in the district, but an unknown further thickness has been removed by pre-Upper Permian erosion; and over the Kiveton-Bothamsall and Whitwell anticlines (Fig. 9) and locally over the Gainsborough Anticline the Upper Coal Measures have been completely lost. Correlation of marker horizons (see below) in shafts and cored boreholes in the western part of the district shows that, as in the Lower and Middle Coal Measures, there is considerable sedimentary thinning from west to east. Oil bores in the eastern part of the district have proved up to about 400 ft of Upper Coal Measures (Plate V), but the only identifiable horizon is the Ackworth Rock (see below) at the base.

Detailed sections show that the succession in the East Retford district is closely similar to that described by Goossens and Smith (1973) in the northern outcrops of the exposed Yorkshire Coalfield. Many horizons can be correlated, including the Blyth Coal with the Brierley seam of Yorkshire, and the principal faunal markers—the Fourth, Third, Second and First Cherry Tree markers—have all been identified. These markers consist of individual mussel or '*Estheria*' bands or groups of such bands. The classification of the Upper Coal Measures into the four divisions erected by Goossens and Smith (*ibid.*) in the exposed coalfield is followed here (Fig. 10). The Ackworth Division (base of Upper Coal Measures to base of Blyth Coal) is 150 to about 250 ft thick in provings, the Brierley Division (Blyth Coal to base of Fourth Cherry Tree Marker) is 100 to about 225 ft thick, the Hemsworth Division (Fourth Cherry Tree Marker to base of First Cherry Tree Marker) is 200 to just over 300 ft thick, and the Badsworth Division (measures above the base of the First Cherry Tree Marker) has been proved to 350 ft.

There is only one coal of any economic importance, namely the Blyth seam, which attains a thickness of more than 3 ft in the north-west of the district; elsewhere it is a thin, sometimes split, seam. The only other named coal is the Scofton in the Ackworth Division; it is nowhere very thick, but is significant for correlation purposes.

Sandstones, locally thick, occur through most of the succession. The Ackworth Rock, present nearly everywhere in the west of the district, though locally represented by siltstone, is more than 100 ft thick in some sections in the north-west, and in places washes out the Top Marine Band. In some areas there is also a thick development of sandstone and siltstone in the upper part of the Ackworth Division, between the Scofton and Blyth coals. The thick sandstone that occurs above the Blyth Coal, and locally forms the whole of the Brierley Division, is evidently the Wickersley Rock of the exposed coalfield. The sandstone occurring in the lower part of the Hemsworth Division, and 100 to more than 150 ft thick in several sections, is apparently the Ravenfield Rock. It commonly replaces many of the cycles of the Division, including those composing the Third Cherry Tree Marker, and in places washes out the Fourth Cherry Tree Marker. The local thick development of sandstone in the Badsworth Division is thought to be the Badsworth Rock of the exposed coalfield (*ibid.*, p. 506).

As described on pp. 46–7, the highest Upper Coal Measures are red or mottled, belonging to a primary-coloured facies, the base of which is diachronous.   R.F.G., E.G.S.

PALAEONTOLOGY

The Coal Measures fossils of the East Retford district are closely comparable with those from the Ollerton and Chesterfield districts to the south and south-west, and for a description of their characteristics and a discussion of environments the reader is referred to the account of the latter (Calver *in* Smith and others 1967, pp. 87–97). Such differences or additions as have been noted in the current investigation are stated below or under the relevant stratigraphical details (pp. 47–99).

Earlier references to fossils from the district are given by Wilson (1926) and Edwards (1951, pp. 82–94), particular attention having been paid to the fossils collected by W. H. Dyson from Maltby Main Colliery sinking. The fossils from Sutton Borehole were listed by Wilson (1927, pp. 141–3), and subsequently the non-marine bivalves from this borehole were described by Clift and Trueman (1929, p. 96) and the marine faunas discussed by Edwards and Stubblefield (1948, pp. 217 *et seq.*). Several non-marine bivalves from boreholes are figured or cited in the monographs of Trueman and Weir (1946–56) and Weir (1960–68). Authors of species referred to in the text are listed in the index of fossils.

MARINE FAUNAS

The marine horizons are confined to the Lower and Middle Coal Measures; not all the bands known from the Lower Coal Measures have been recognized, but there is a complete and representative series of the nine bands belonging to the Middle Coal Measures. Different facies are represented by the respective marine horizons, some of which also show some geographical variations; these trends are described in the account of the stratigraphy (pp. 47–99). The facies vary from the thin *Lingula*-bearing bands of the minor incursions to the goniatite/pectinoid or cephalopod/calcareous brachiopod assemblages of such bands as the Pot Clay and Mansfield marine bands, the latter attaining a thickness of over 30 ft. Examples of the intermediate facies, dominated by the *Myalina/Hollinella* community, are known from the Clay Cross, Two-Foot and *Edmondia* horizons. In general terms the marine bands tend to show a regional facies trend consistent with a shore-line lying towards the east (see also Calver 1968).

Non-marine Faunas

The application of the non-marine bivalve classification (Trueman and Weir 1946, p. xxviii) to the East Retford district is shown on Figs. 6, 8 and 10. All the zones are recognized except for the two highest, viz: the *A. tenuis* and *A. prolifera* zones. It is possible that the basal part of the former zone is represented by the highest strata proved, but the evidence is not conclusive (see below). Further subdivision of parts of the sequence depends on the recognition of broadly defined faunal belts within the zones, based on the periods of dominance of distinctive species or groups of species. The approximate vertical range of these successive faunas is also shown on Figs. 6, 8 and 10.

The non-marine bivalve succession is closely comparable to that described from the Chesterfield district (Calver *in* Smith and others 1967, pp. 95–7), but in general the mussels are less abundant, and as a consequence some of the regional mussel-bands are less well represented than in adjacent areas. Other elements of the non-marine fauna such as the ostracods and estheriids show a similar distribution, except that there are numerous localities from which the Low '*Estheria*' Band in the upper part of the *C. communis* Zone has been recorded, in contrast to its absence from the Chesterfield district.

The following account gives the main features of the sequence.

Lower Coal Measures (comprising the *A. lenisulcata* Zone, the *C. communis* Zone and the lower part of the *A. modiolaris* Zone). The widespread mussel-band or series of distinct mussel-bands ranging over a minimum of 20 ft in the basal part of the Lower Coal Measures are correlated with the *Carbonicola fallax*/*C. protea* faunas from the sequence above the Soft Bed Coal of Yorkshire (Eagar 1947, 1952). The characteristic species include forms assigned or related to *Carbonicola artifex, C. declinata, C. fallax, C. haberghamensis, C. limax, C. protea* and *C. rectilinearis;* in addition there are sporadic records of '*Anthraconaia*' *lenisulcata* and *Curvirimula sp. nov.* [fine ornament]. *Geisina arcuata* and fish remains are also present at some levels.

Rare examples of *Carbonicola* aff. *extenuata* and *Curvirimula sp.* occur some 12 ft above the Alton Marine Band and are thought to be the equivalent of the *Carbonicola prisca*/*C. extenuata* band found at the equivalent horizon in the East Midlands and Yorkshire (Eagar 1956, p. 345). The Norton Mussel-band of the area to the south (Eden 1954, p. 97) is probably represented in the western part of the district by rare records of *Carbonicola proxima* from a short distance above the horizon of the Norton Marine Band. This is the highest fauna of the *A. lenisulcata* Zone in this district. In the measures above the presumed horizon of the Upper Band Marine Band, corresponding to the lower half of the *C. communis* Zone, only sparse occurrences of bivalves and other non-marine fossils have been noted; this is a relatively barren part of the sequence in adjacent districts, but the dearth of fossils in the East Retford district is in part due to the lack of cored boreholes through these measures.

The first indication of the *Carbonicola pseudorobusta* fauna is in the mussel-band about 80 ft above the Kilburn Coal; *C. pseudorobusta* and *Curvirimula* aff. *trapeziforma* are recorded. A similar fauna also occurs some 80 ft higher in the sequence.

Between the Beeston Coal horizon and the Low '*Estheria*' Band the mussel-bands contain *Anthracosphaerium sp. nov.* cf. *dawsoni, Carbonicola browni, C.*

cf. *communis*, *Curvirimula candela* and rare examples of *Anthraconaia* recalling *A. fugax*.

A *C. pseudorobusta* fauna is found above the Top Silkstone Coal and above the Threequarters Coal. The higher of these bands is the highest record in the district of typical *C. pseudorobusta*, which is associated with forerunners of the overlying *C. cristagalli* fauna. There is an interesting record of the distinctive *Anthraconaia potoriba* at about the Threequarters horizon in Scofton Borehole. The highest record of *Curvirimula* is some 15 ft above the Threequarters in Nornay Borehole, and as this genus is not known from the *A. modiolaris* Zone, the base of that zone is placed at the Thorncliff Coal.

The beds between the Thorncliff and Parkgate coals contain *Carbonicola cristagalli* and allied forms, but not in the profusion known elsewhere at this horizon. As well as *C. cristagalli*, the assemblage includes *C. rhomboidalis* and *C. oslancis*, together with *Anthracosphaerium* cf. *dawsoni* and abundant *Geisina arcuata*. Above the Deep Hard, *C.* cf. *cristagalli* occurs locally and is the highest record of *Carbonicola* in the district. In the Chesterfield district the measures above the Deep Soft show a marked faunal change from the beds below, notably the incoming of *Anthracosia regularis* and variants. In the present district mussels have not been noted between the Deep Soft and Flockton Thick coals, and the earliest record of the *A. regularis* fauna is above the latter coal. *A. regularis* also occurs between the Joan Coal horizon and the Clay Cross Marine Band, locally in association with *Lioestheria* and *Geisina arcuata*.

Middle Coal Measures (comprising the upper part of the *A. modiolaris* Zone and the Lower and Upper *A. similis–A. pulchra* zones). The faunas between the Clay Cross Marine Band and the First Waterloo Coal are dominated by *Anthracosia spp.* of the *A. ovum* and *A. phrygiana* groups, the latter becoming more important in the beds above the Swallow Wood Coal. Typical species include *A. ovum*, *A. nitida*, *A. beaniana*, *A. phrygiana*, *A. disjuncta* and forms resembling *A. subrecta*. In addition *Anthraconaia salteri* and *Anthracosphaerium* aff. *affine* are known from the Swallow Wood horizon; the characteristic *Naiadites* of these measures is *N. quadratus*.

Only a few mussels have been noted in the beds between the First Waterloo and the Top Hard, and the distinctive Dunsil fauna with typical *A. phrygiana* known elsewhere in the region has not been recognized.

The practice of placing the lower boundary of the Lower *similis-pulchra* Zone at the Top Hard is followed here, although there is no direct evidence from the beds immediately above and below the coal. The lowest distinctive fauna of the Lower *similis-pulchra* Zone occurs just below the Kent's Thick and includes *Anthraconaia pulchella*, *Anthracosia spp.* including *A. caledonica* and *A. simulans*, as well as *Naiadites productus*; this assemblage is considered to represent the widespread *A. pulchella* fauna known from a similar horizon in several Pennine coalfields (Smith and others 1967, p. 173).

The well developed mussel-bands which occur above the Kent's Thick and High Hazles coals can be referred to the *A. caledonica* fauna. In addition to this species, the assemblages include *Anthracosia aquilinoides*, *A.* cf. *elliptica*, *A. sp.* cf. *fulva*, *A. planitumida* and *A. simulans*. Rare examples of *Anthraconaia* aff. *librata* are recorded, and *Anthracosphaerium* cf. *exiguum* and *N. productus* also occur.

Although the Two-Foot Marine Band is known from several localities, the overlying mussel-band appears to be poorly developed compared with adjacent areas (Smith and others 1967, p. 179) and the characteristic *Anthracosia acutella/A. concinna* fauna has not been recognized. Certain elements of this fauna appear slightly higher in the sequence, above the lower leaf of the Main-bright Coal.

Sporadic records of the *Anthracosia atra* group are known from above the Mainbright, but the typical occurrence of this species is in the mussel-band above the Clown Coal, where it is associated with examples of *Carbonita humilis*. The abundance of this ostracod at this horizon is a regional feature (Smith and others 1967, p. 182). Mussels are rare in the higher beds up to the Mansfield Marine Band, but include *Anthracosia atra*, *A. rubida* and *Naiadites* aff. *alatus*.

As in other areas there is a marked faunal change in the beds above the Mansfield Marine Band. *Anthracosia* and *Anthracosphaerium* disappear and *Naiadites* is the dominant genus; *Anthraconaia* still occurs, but is not common except at particular horizons which are thereby distinctive. The characteristic species of *Naiadites* are *N. hindi*, *N. melvillei* and forms resembling *N. productus;* in the highest part of this sequence, i.e. the upper part of the Upper *similis-pulchra* Zone, *Naiadites* cf. *daviesi* is the more characteristic form.

Significant occurrences relate to the presence below and above the Sharlston Top Coal of *Anthraconaia hindi*, which is found with the closely related *A. adamsi* in many other British coalfields at this same general horizon.

The first appearance of *Geisina subarcuata* is in the roof measures of the Sharlston Low Coal, and it occurs intermittently up to the Top Marine Band; other ostracods in these measures are *Carbonita humilis*, *C.* cf. *pungens* and *C. claripunctata*. The lowest occurrence of *Lioestheria vinti* in this part of the sequence is a short distance above the *Edmondia* Band, but it recurs in several bands up to the Top Marine Band. There is also the rare occurrence of *Hemicycloleaia* cf. *minima* in the first cycle above the Shafton Marine Band.

Upper Coal Measures (embracing the *A. phillipsii* Zone and ? basal part of the *A. tenuis* Zone). A further change in the non-marine bivalve fauna takes place near the base of the Upper Coal Measures with the loss of *Naiadites* and the incoming of *Anthraconauta*. The dominant form is *A. phillipsii*, with subsidiary *A. wrighti*. Some examples of *Anthraconauta* from the uppermost beds show an approach to *A. tenuis*, but there are no typical forms and there is no positive faunal evidence that the *A. tenuis* Zone is represented in the district (but see below).

Anthraconaia is represented by *A. pruvosti*, recorded from two horizons some 100 and 350 ft above the Blyth Coal. Not far above this coal *Lioestheria vinti* gives way to the larger *Euestheria simoni*. There are also rare records of the distinctive estheriid *Anomalonema* [*Estheriella*] *defretinae* from some 70 ft above the Blyth (see also Edwards 1951, p. 90). In the highest Coal Measures known from the district, one horizon has yielded *Leaia* at a few localities. In the other Pennine coalfields *Leaia* tends to be more characteristic of the *A. tenuis* Zone than the underlying *A. phillipsii* Zone, so that it is possible that these highest Coal Measures belong to the upper of these zones.

The common ostracod from these measures is *Carbonita humilis*, which occurs in most of the mussel-bands, sometimes accompanied by *C. pungens* or *C.* cf.

agnes. Other fossils which complete this characteristic Upper Coal Measures fauna are *Spirorbis* and fish remains including scales of *Rhabdoderma*, *Rhadinichthys* and *Rhizodopsis*.   M.A.C.

STAINED AND COLOURED MEASURES

There are two types of red Coal Measures, here distinguished as 'stained' and 'coloured' measures. Both owe their colour to the oxidized nature of their iron content, but they are distinctly different in origin. The stained measures are former grey measures which have been subject to secondary oxidation, long after deposition, by weathering at or near the pre-Permian ground surface (see Anderson and Dunham 1953). The coloured measures, on the other hand, evidently experienced oxidizing conditions of deposition, and have thus been red from the time of their formation.

The Stained Measures occur in a narrow zone immediately or close below the Permian, and are independent of stratigraphical horizon. The oldest beds known to be affected in the district are Middle Coal Measures just below the horizon of the *Edmondia* Band in Welbeck Abbey Borehole. Depth of staining varies from perhaps as little as 10 to over 80 ft, though its effects, if any, on primary-coloured measures, where these underlie the Permian, cannot be detected. In many sections the measures immediately underlying the Permian are grey. These grey measures are less than a foot to about 4 ft thick, with an exceptional 8 ft at Ranskill Borehole, and are generally thickest where they consist of sandstone or siltstone. In Ranskill Borehole the top 1 ft of white sandstone is pyritous. Where the grey beds are argillaceous they may be soft and structureless, e.g. at Steetley Colliery, where they are described as 6 in of white pipe clay. The grey beds are believed to have been formerly red, and it is supposed that they owe their present colour to the reducing effects of water percolating downwards, through any Basal Permian Sands that might be present, from the Marl Slate sea (Smith and Francis 1967, p. 18).

The stained measures are seldom wholly red, but are commonly mottled or are selectively reddened in certain beds, and the original cyclic nature of the sediments is still apparent. Nearly everywhere there is a gradual passage downwards into normal grey measures. Ironstone bands and nodules show the most marked staining effects, some having been converted to soft ochre. Generally the argillaceous rocks are more affected than the arenaceous, but dark carbonaceous mudstones show few or no signs of colour change, and the deepest staining effects in some sections are the reddening of micaceous laminae in sandstone.

The Coloured Measures, called 'Red' by Edwards (1967) to distinguish them from the stained measures in the country to the south, are a facies of the Upper Coal Measures (pp. 41–2). Their base is diachronous, occurring slightly higher in the succession towards the north-east, so that it lies in the Hemsworth Division, not far above the Third Cherry Tree Marker, in Torworth and Wigthorpe boreholes, but in the lower part of the Badsworth Division at Stone Borehole and Maltby Main Colliery (Plate IX and Fig. 10). The coloured measures extend upwards to the base of the Permian rocks and their upper parts have therefore been liable to secondary oxidation, and, like the stained measures (see above), the topmost few feet in many sections have been turned grey by still-later reduction. No primary grey beds, like the 'Newcastle Beds' of the Ollerton district (Edwards 1967, p. 62), occur within the coloured Upper Coal

Measures of the East Retford district. Around Ollerton, however, the 'Newcastle Beds' are underlain by 380 to 480 ft of coloured measures (*idem*), and the maximum thickness of coloured measures seen in the present district is just over 400 ft at Harworth Colliery. At Nornay Borehole, where the top 130 ft were not cored, coloured measures occur down to about 470 ft below the sub-Permian unconformity.

The coloured measures are not wholly red except in the Badsworth Division at the top of a few sections in the north-western corner of the district. Generally they are mottled or alternately red and grey like the stained measures, and, as is the case with the stained measures, the reddening can be seen to diminish downwards in borehole cores. Often the first sign of the incoming of coloured measures is an isolated red-mottled seatearth. Lithologically the coloured measures are similar to the grey measures, and generally have as well developed a cyclic pattern. They include substantial thicknesses of red or purple sandstone, contain coals and seatearths, and mussels and '*Estheria*' are found in some of their mudstones. They are regarded as an intermediate facies between normal grey measures and the blocky red mudstones with thin sandstones (espleys) and without coals, fossils or ironstone nodules, which make up the Etruria Marl facies of the Midlands.   R.F.G., E.G.S.

DETAILS OF STRATIGRAPHY

The Upper and Middle Coal Measures and the upper part of the Lower Coal Measures have been explored by shafts and cored boreholes in the western part of the district, but these measures in the east, and the lower part of the Lower Coal Measures everywhere, are known only from oil-bore data. The latter, while invaluable, depend largely on chipping samples and the interpretation of geophysical logs, with only limited coring at selected horizons. Thus few macrofossils are available, and measurement and correlation are much less precise than in a fully cored sequence. The account of the measures below the Low Silkstone Coal is therefore less adequate than that of the measures above, and these higher measures are described much more fully in the west than in the east.

Notes on all fossils collected from the measures below the Low Silkstone Coal and from the Upper Coal Measures are included in the respective accounts, but so many fossils are available from the rest of the Coal Measures that examination and description have been necessarily selective. The complete faunas obtained from four representative boreholes (Eaton, Nornay, Scaftworth and Scofton) are given in Appendix 1, and certain additional palaeontological information is included in the account.

LOWER COAL MEASURES

BASE OF COAL MEASURES TO LOW SILKSTONE COAL

Nearly all the available information on these measures has been obtained from the records of oil bores. Plate III shows comparative sections of selected bores and illustrates the variations in lithology: lines of correlation are, however, in many instances, tentative.

The known range of thickness is from more than 1100 ft in the north-west to less than 700 ft in the east-central part of the district. There are no coals of proved economic value, but certain of the sandstones are oil-bearing locally. The marine bands in the basal part of the group are not well developed and, when present, are typically of *Lingula* facies.   E.G.S.

The Pot Clay (Gastrioceras subcrenatum) Marine Band can be firmly identified only in Tickhill No. 1 and Ranskill No. 1 oil bores, where the goniatite/pectinoid fauna is present. Elsewhere it is dominantly of *Lingula* facies, and is recognized by its position in the stratigraphical sequence, particularly with respect to the underlying *Gastrioceras cancellatum* Marine Band (see p. 23) and the overlying Soft Bed mussel-bands (p. 49).

In Tickhill No. 1 Oil Bore the marine band consists of dark grey mudstone, canky in parts, which, excluding the *Planolites ophthalmoides* phase above, is 4 ft 10 in thick at 4428¼ ft. A typical goniatite/pectinoid fauna has been identified, the full faunal list being: *Planolites ophthalmoides*, *Lingula mytilloides* [5·0 mm], *Caneyella?*, *Dunbarella* cf. *papyracea*, a nuculoid bivalve, *Posidonia sp.*, *Anthracoceratites sp.*, *Gastrioceras subcrenatum*, *Homoceratoides* aff. *divaricatus*, mollusc spat, *Hindeodella sp.*, fish, including palaeoniscid scales.

In Ranskill No. 1 Oil Bore the marine band is at least 2 ft 10 in thick at 4355 ft 2 in. It consists of dark grey to black, largely micaceous mudstone with three thin limestone bands. The mudstone above one limestone band is calcareous and above another is carbonaceous with coalified and pyritized wood. The fauna provides an interesting example of the goniatite/pectinoid facies occurring in association with a benthonic facies containing crinoids, calcareous brachiopods and trilobites, the latter being contained in the calcareous parts of the marine band. The complete faunal list is as follows: sponge spicules, crinoid columnals, *Lingula mytilloides* [6·0 and 8·0 mm], *Orbiculoidea sp.*, *Rhipidomella sp.*, smooth spiriferoid, *Dunbarella* cf. *papyracea*, *?Posidonia gibsoni*, *Huanghoceras falcatum*, *Anthracoceratites?*, *Gastrioceras subcrenatum*, trilobite fragments, platformed conodonts, *Serpuloides sp.*, fish including acanthodian scales and *Rhabdoderma sp.* The retreat stage of the marine band may extend upwards into the overlying dark mudstones with fish debris (see below), which contain *Planolites ophthalmoides* below 4344 ft 1 in.

Elsewhere the Pot Clay Marine Band is evidently of *Lingula* facies, and its horizon has been determined by reference to the distinctive *Gastrioceras cancellatum* horizon below (see p. 23) and the Soft Bed mussel-bands or higher horizons above. The resulting correlation (Plates II and III) does not agree with the gamma-ray log correlations of Howitt and Brunstrom (1966) and Downing and Howitt (1969). A *Lingula* band has been proved at the horizon of the marine band by coring at Walkeringham No. 1 (4606 ft 4 in to 4607 ft 2 in), Beckingham No. 1 (4558¾ to 4559¼ ft) and Gainsborough No. 1 (4614¾ ft to 4615 ft 11 in) oil bores. The only fossils recorded, apart from *L. mytilloides*, are pyritized sponge spicules, *Serpuloides* (including *S. stubblefieldi* at Gainsborough No. 1) and fish fragments. The marine band may be represented by the *Lingula* band at 4169 ft in South Leverton No. 7 Oil Bore and by the mudstones with *Serpuloides sp.* [showing ? attachment discs] and ?fish, between 4649½ and 4653 ft, in Gainsborough No. 7 Oil Bore.

In the Bothamsall and West Drayton oil bores the marine band is evidently washed out below the Crawshaw Sandstone (p. 49) as was suggested by Taylor and Howitt (1965) and Edwards (1967). The marine band at 3347 ft in Bothamsall No. 1 Oil Bore, tentatively recognized as the Pot Clay by Edwards (*ibid.*, p. 68) is now considered to be the *Gastrioceras cancellatum* horizon (p. 23).   M.A.C., E.G.S.

The measures between the Pot Clay and Alton marine bands are just over 160 ft thick in Tickhill No. 1 Oil Bore and perhaps as little as 30 ft in some of the Gainsborough bores; they show a general thinning towards the east (Plate III). The chief named horizons (Fig. 6) are the Crawshaw Sandstone, Soft Bed mussel-bands, Sub-Alton Sandstone and Alton Coal.

At Tickhill there are about 50 ft of measures, largely dark grey silty mudstones, but containing interbanded fine sandstone, between the Pot Clay Marine Band and the Crawshaw Sandstone. Plant debris is common and a single juvenile *Naiadites sp.* was found 37¾ ft above the marine band.

In Ranskill No. 1 Oil Bore, where the Crawshaw Sandstone is absent, less than 15 ft of measures separate mudstones containing Soft Bed mussels (see below) from the Pot Clay Marine Band. These consist of grey to dark mudstones with a few ironstone bands and scattered plant debris. Fish scales occur at several horizons, particularly immediately above the marine band where *Elonichthys sp.* and *Rhadinichthys sp.* are recorded.

The Pot Clay–Soft Bed succession is similar to that at Ranskill in the Gringley, Walkeringham, Morton, Beckingham and Gainsborough oil bores, and probably also at South Leverton and Grove. A *Rhizodopsis* scale is recorded 15 ft above the marine band at Walkeringham No. 1 Bore, and fish scales occur in the 9 ft of cored silty mudstones succeeding the presumed marine band horizon in South Leverton No. 7 Bore.

The Crawshaw Sandstone is present in Tickhill No. 1 Oil Bore and may be about 10 ft thick: most of the core was lost at this horizon, the only recovery being 2 ft 10 in of fine-grained light grey sandstone with dark micaceous laminae. The sandstone is absent in the other oil bores in the northern part of the district from Ranskill to Gainsborough, and probably also at South Leverton and Grove. The sandstone identified as the Crawshaw Sandstone by Taylor and Howitt (1965) and Howitt and Brunstrom (1966) in the oil bores east of Ranskill is here considered to be the Rough Rock (p. 25).

The thick sandstone below the Soft Bed sequence in the Bothamsall Oilfield (Plate III), placed in the Coal Measures by Brunstrom (1963, p. 5) and referred to the Crawshaw Sandstone by Taylor and Howitt (1965, p. 198) and Edwards (1967, p. 68), is now thought to be a combined Crawshaw Sandstone and Rough Rock, as was inferred by Downing and Howitt (1969, p. 247). The upper part of the combined rock is fine-grained to coarse or pebbly, showing coarsening downwards in some sections, and in places is false-bedded or contains silty and carbonaceous partings. The West Drayton sections are similar to those of the Bothamsall bores, but in the Apleyhead bores, where correlation at this horizon is obscure, the Crawshaw Sandstone appears to be present and the Rough Rock absent.

The Soft Bed (or Belperlawn) Coal is present locally as a trace of coal in the Bothamsall Oilfield. At Bothamsall No. 4 Bore, for instance, $\frac{1}{2}$ in of shaly coal at 3295 ft 8 in is separated from the Crawshaw Sandstone by 6 ft 4 in of measures composed of mudstone-seatearth or siltstone with sphaerosiderite. In No. 5 Bore the seatearth is a ganister. The Soft Bed Coal is not identifiable in the oil bores to the north of Bothamsall, but its horizon is indicated by hard grey silty mudstone-seatearth between 4344 ft 4 in and 4345 ft in Tickhill No. 1 Oil Bore.   E.G.S.

In the exposed coalfield to the west of the present district the measures between the Soft Bed and Alton coals, where fully developed, consist of four cycles, namely those above the Soft Bed, Holbrook, Second Smalley and First Smalley (also called the Middle Band or Clay) coals (Eden 1954, p. 85). The three upper cycles contain a *Lingula* band locally. The three lower cycles contain mussels which form the lower, middle and upper divisions of Eagar's (1947, 1952) Soft Bed faunal sequence. This faunal sequence provides one of the key horizons for correlation of the basal Coal Measures in the East Retford district (Plate III). In Tickhill No. 1 Oil Bore all three divisions can be recognized. The only bivalve from the lower division is *Carbonicola protea*, which is associated with fish remains, including *Megalichthys sp.* and palaeoniscid scales, in dark grey shales overlying the Soft Bed seatearth (see above). The middle division is represented by *Carbonicola* aff. *haberghamensis* and *C. rectilinearis*, which occur with ostracods (*Geisina?*) and fish debris including *Rhizodopsis sp.* in 6 ft of dark grey silty shales at $4334\frac{1}{2}$ ft. Immediately below is 1 ft of dark shale with *Lingula mytilloides* [6·0 mm, including broad form], resting on seatearth. There is therefore no coal at the Holbrook horizon. The upper division of the Soft Bed fauna is represented by *C.* cf. *artifex*, *C. pilleolum?*, abundant poorly preserved *C. spp.* and *Naiadites sp.* associated with *Geisina arcuata* and *Rhizodopsis sp.* These fossils occur through $11\frac{3}{4}$ ft of dark grey silty mudstone which rests at $4308\frac{3}{4}$ ft on a $3\frac{1}{2}$-ft *Lingula* band. The latter consists of dark to grey mudstone, becoming black and carbonaceous in the bottom 3 in, and containing *L. mytilloides* [5·0 mm] with, at the base, abundant fish debris including *Elonichthys sp.* and *Rhadinichthys sp.;* pyritized juvenile specimens of *C.* cf. *haberghamensis* were found on a bedding plane at $4310\frac{1}{2}$ ft, not mixed with *Lingula* and evidently indicating a non-marine intercalation in the *Lingula* band. The Second Smalley Coal consists of 3 in of dirty coal at $4312\frac{1}{2}$ ft, immediately below the *Lingula* band. The First Smalley Coal at Tickhill is evidently the seam, at least $5\frac{1}{2}$ in thick and resting on ganister, at 4287 ft.

The only other section in the East Retford district in which the three divisions of the Soft Bed faunal sequence can be recognized is that of Walkeringham No. 2 Oil Bore. Here dark grey mudstones between 4680 ft 1 in and 4693½ ft contain mussels at three horizons separated by fish-bands. The fauna of the lower division is: *Spirorbis sp.*, *C. artifex*, *C. declinata ?*, *C. fallax*, *C. rectilinearis*, *Curvirimula sp. nov.* [fine ornament], *Naiadites sp.*, *G. arcuata*, *Rhizodopsis sp.* and palaeoniscid scales; that of the middle division is: *Carbonicola* cf. *fallax* and *N. sp.* cf. *productus;* and that of the upper division: *C. declinata*, *C. fallax*, *C.* cf. *limax*, *C.* cf. *rectilinearis* and *Naiadites sp.* No coals or seatearths are recorded in the section, but the fish-bands separating the mussel-bands are probably the lateral equivalents of the *Lingula* bands above the Holbrook and Second Smalley coal horizons in Tickhill No. 1 Bore. At Walkeringham the Soft Bed sequence is found close above the horizon of the Pot Clay Marine Band, a situation also obtaining at Ranskill and Gainsborough.

Other oil bores besides Tickhill No. 1 and Walkeringham No. 2 in which mussels assigned to the Soft Bed sequence have been found are: Ranskill No. 1 (between 4330½ and 4338¼ ft), Walkeringham No. 1 (4582½ ft to 4589 ft 11 in), Gainsborough No. 1 (4607 ft to 4607 ft 2 in), South Leverton No. 3 (4206 ft 11 in to 4207 ft 5 in), West Drayton No. 1 (3737 ft), Bothamsall No. 2 (3334 ft 2 in to 3351 ft), Bothamsall No. 4 (3272 ft 10 in to 3288 ft 1 in), Bothamsall No. 8 (3318 ft 8 in to 3323 ft 10 in), Bothamsall No. 11 (3273 to 3276 ft) and Bothamsall No. 13 (3425½ to 3430 ft). The Soft Bed fauna in Bothamsall Nos. 1 and 5 oil bores, just south of the district boundary, has already been described by Edwards (1967, p. 68). In addition to the mussels at Tickhill No. 1 and Walkeringham No. 2 oil bores already listed, *Anthraconaia sp.* (? *modiolaris* group) and '*Anthraconaia*' cf. *lenisulcata* are recorded from this horizon at Ranskill No. 1 and West Drayton No. 1 oil bores respectively. Nowhere except at Tickhill, can the *Lingula* bands associated with the Soft Bed sequence be positively identified, though the *Lingula* band at 4917 ft in Gringley No. 1 Oil Bore (Plate III) may well be at the First Smalley horizon. In some sections, such as Walkeringham No. 2 Oil Bore (see above), fish-bands may represent one or other of the *Lingula* bands. It should be noted that in Corringham No. 1 Oil Bore east of Gainsborough (and in

the area of One-inch Sheet 102) spicules of the sponge *Hyalostelia* were found in dark grey micaceous silty mudstone at 4898½ ft, an horizon lying between *Lingula* bands correlated with the Pot Clay and Alton marine bands, and probably corresponding to one of the marine horizons in the Soft Bed sequence. The occurrence of *Hyalostelia* at a similar horizon is recorded by Strong (*in* Falcon and Kent 1960, p. 53) in the Eakring Oilfield.   M.A.C., E.G.S.

The Sub-Alton Sandstone lies in the cycle above the First Smalley Coal or its horizon, and below the Alton Coal (Smith and others 1967, pp. 105–6). It appears to be present throughout most of the East Retford district (Plate III). At Tickhill No. 1 Oil Bore it is apparently represented by the sandstone about 30 ft thick at 4286½ ft: the 8 ft of cores recovered are of brownish grey, fine- to coarse-grained sandstone with mudstone inclusions at the base. Including ganisteroid seatearth at the top, the Sub-Alton Sandstone is about 18 ft thick in the Walkeringham oil bores. Light grey and greyish brown massive sandstone was cored in No. 1 Bore (4558 ft 10 in to below 4577 ft), and pale grey to white, mainly massive, very fine-grained sandstone in No. 2 Bore (4654¼ to 4671½ ft). The latter cores show a few shaly partings with abundant mica, especially near the base.

The Sub-Alton Sandstone is around 10 ft thick in Ranskill, Beckingham and Morton oil bores and in those Gainsborough bores where it can be recognized. The only cores are from Gainsborough No. 1 Oil Bore where 2¾ ft of sandstone-seatearth with a little sphaerosiderite on 9¾ ft of light grey, fine-grained sandstone with silty mudstone bands, rests on interbanded sandstone and siltstone at 4596 ft. In Corringham No. 1 Oil Bore, east of Gainsborough, the Sub-Alton Sandstone may be represented by 16 in of ganister (at 4882 ft 4 in), and in some of the other Corringham bores even this appears to fail.

In the South Leverton and Grove oil bores the Sub-Alton Sandstone is probably the 15- to 35-ft bed found in most sections some 25 to 80 ft above the presumed Pot Clay horizon, and nowhere cored. In the Bothamsall Oilfield, where it is one of the main oil sands (Taylor and Howitt 1965, p. 198), and at Apleyhead it is up to 70 ft thick and well developed in most sections. Cores show it to be light grey where not oil-stained, fine- to coarse-grained, and massive to flaggy, typically containing dark, micaceous, carbonaceous mudstone partings. In the West Drayton bores the Sub-Alton Sandstone is less than 10 ft thick.

The Alton Coal. A thin coal, presumably the Alton, is recorded, but not cored, above the Sub-Alton Sandstone in Tickhill No. 1 Oil Bore. The seam is apparently absent in the Ranskill, Gringley, Walkeringham and Morton oil bores. In Gainsborough No. 1 Oil Bore the 2 in of coal at 4580½ ft appears to be the Alton, or, because it lies some distance below the Alton Marine Band (see below), a lower leaf of that seam. Elsewhere in the Gainsborough–Beckingham bores it cannot be identified with confidence, and is clearly absent in some sections.

In South Leverton No. 3 Oil Bore the 6-in coal cored at 4186 ft 11½ in is presumed to be the Alton. Several of the sections of the other bores in this oilfield and of the Grove oil bores show a thin coal at this horizon, possibly up to 12 in thick in South Leverton No. 12 Bore.

The Alton Coal is absent in the West Drayton and Apleyhead oil bores and in about half of the Bothamsall bores that lie within the East Retford district. The other half of these Bothamsall bores show a thin coal up to 6 in thick, which is more firmly identifiable as the Alton than in most other sections in the district. In a few sections ganister is recorded at this horizon.   E.G.S.

The Alton (Gastrioceras listeri) Marine Band has not yielded goniatites within the district, and its recognition depends on the position, with respect to other markers, of the *Lingula* band representing it, on its known high gamma-ray content (Ponsford 1955, pp. 34–5; Knowles 1964), and, in some instances, on fossils additional to *Lingula* (Plate III).

In Gringley No. 1 Oil Bore the Alton Marine Band may be represented by black shale between 4886 ft 4 in and 4890½ ft which contains abundant foraminifera including *Ammodiscus sp.*, pyritized sponge spicules, *Lingula mytilloides* [5·0 mm], *Hindeodella sp.* and palaeoniscid scales including *Rhadinichthys sp.* The abundance of foraminifera is evidence that this band is not lower than the Alton horizon (Calver 1968, p. 26), but it may indicate a higher

horizon (see below).

In Walkeringham No. 1 Oil Bore the marine band is indicated by the presence, in the bottom part of 15 ft of dark *Lingula*-bearing mudstone (at 4552 ft), of *Caneyella multirugata*, *Dunbarella papyracea*, *Posidonia gibsoni* and *P. sp. nov.*, which are taken to represent the pectinoid facies of the *G. listeri* horizon. It is the easternmost known extension of this facies.

In Walkeringham No. 2 Oil Bore 8 ft 2 in of dark grey mudstone with *L. mytilloides* [9·0 mm] and fish debris at 4644 ft 4 in are identified as the Alton Marine Band because of the presence, in overlying mudstones, of *Carbonicola aff. extenuata* and *Curvirimula sp.* These bivalves are pyritized and possibly stunted, and they are thought to represent the horizon of the *Carbonicola prisca/extenuata* band which overlies the Alton Marine Band farther west in the Yorkshire and East Midlands coalfields (Eagar 1956, p. 345).

In Gainsborough No. 1 Oil Bore dark grey mudstone and silty mudstone between 4564¾ ft and 4570 ft 7 in contain several large *Ammodiscus sp.* together with *L. mytilloides* [6·0 mm] and fish remains. Large *Ammodiscus* are a feature of the Lower Parkhouse Marine Band of the Chesterfield district (Smith and others 1967, p. 112), but may also be indicative of a higher horizon or of the Alton Marine Band. On stratigraphical grounds (Plate III) and gamma-log evidence (Downing and Howitt 1969, fig. 6) this marine band appears to be the Alton, though it is separated from the thin (? Alton) coal overlying the supposed Sub-Alton Sandstone (see p. 50) by nearly 10 ft of sandstone and mudstone. If this marine band is not the Alton, then the latter is absent in this bore and probably elsewhere in the Gainsborough Oilfield.

There has been no coring at the horizon of the marine band in the Grove oil bores, nor in the South Leverton Oilfield except at No. 3 Bore. Here the marine band is apparently represented by one of the two *Lingula* bands—probably the lower—at 4162 ft and 4175 ft 5 in. Eleven feet below the lower *Lingula* band and forming the roof of the presumed Alton Coal (see above) is 3 in of black canky shale, which may possibly correspond to the "Black Grit" of Strong (*in* Falcon and Kent 1960, p. 32), though no fossils were recorded from it.

In West Drayton No. 1 Oil Bore the dark grey shale between 3707½ and 3709 ft, containing foraminifera including abundant *Ammodiscus sp.*, pyritized sponge spicules, *L. mytilloides* [6·0 mm] and palaeoniscid scales, is thought to be the Alton Marine Band. As in Gringley No. 1 and Gainsborough No. 1 oil bores (see above), however, the possibility of the band representing a higher marine horizon is not excluded.

In Apleyhead No. 1 Oil Bore the marine band at 3391 ft 4 in, giving a marked peak on the gamma-ray log, is correlated with the Alton Marine Band (Downing and Howitt 1969, fig. 5). Foraminifera, including large *Ammodiscus sp.*, together with scales of *Elonichthys sp.*, have been recovered from 3¼ ft of black shale. Fish remains including *Elonichthys sp.* and *Rhadinichthys sp.* occur through several feet of black shale immediately above the marine band. In Apleyhead No. 3 Oil Bore the Alton Marine Band is apparently represented by 3 in of dark grey to black shale with *L. mytilloides* [2·5 mm] and *Rhabdoderma sp.* at 3390¾ ft. Palaeoniscid scales occur in beds of the same lithology from 1 to 3¾ ft above the *Lingula* band, and probable fish remains and a ?crustacean were found in grey micaceous mudstones immediately below the *Lingula* band. Only fish debris was found (3420 ft 7 in to 3423 ft 8 in) at the expected Alton horizon in Apleyhead No. 2 Oil Bore, but coring commenced only a few inches above the fish-bed, and the marine band could be a little higher.

In the Bothamsall Oilfield Edwards (1967, p. 71) has tentatively identified the Alton Marine Band in Nos. 1, 5, 6 and 13 bores, recording a fauna consisting of sponge spicules, *L. mytilloides* and fish debris. The same band has been cored in Nos. 2, 8, 9 and 16 bores (respectively 8 in thick at 3255 ft 2 in, 12 in at 3237½ ft, 1 in+ at 3280 ft 11 in, and thickness not recorded at 3245½ ft). The lithology is grey to dark mudstone, and only *L. mytilloides* [3·0 to 5·0 mm] and fish debris have been found in the band. In No. 9 Bore there is a 1½-in band of hard, silty, pyritous mudstone between the *Lingula*-bearing mudstone and the Alton Coal. No fossils have been found in it, but it is reminiscent of the "Black Grit" of Strong (*in* Falcon and Kent 1960, p. 53), a bed possibly present also in No. 13 Well (Edwards 1967, p. 71). The Alton Marine Band has been identified in

No. 3 Bore, uncored at this horizon, by the peak it gives on the gamma-ray log (Downing and Howitt 1969, fig. 5), a similar peak to that given by the *Lingula* band in e.g. Nos. 1 and 2 bores. The marine band is apparently locally absent in the Bothamsall Oilfield, where cores through its horizon in Nos. 4, 15 and 20 bores show only a fish-band in which *Elonichthys sp.*, *Rhabdoderma sp.* and *Rhadinichthys sp.* have been found.

M.A.C., E.G.S.

The measures between the Alton Marine Band and the Kilburn Coal show an eastward and southward thinning from a maximum of about 500 ft in Tickhill No. 1 Oil Bore to a minimum of about 240 ft in some of the South Leverton oil bores (Plate III). They are about 400 ft thick in Ranskill and Gringley oil bores, 375 to 400 ft at Walkeringham and about 325 ft at Beckingham and Morton. Approximately 260 ft thick in Gainsborough No. 1 Oil Bore, they are apparently less than 300 ft elsewhere in the Gainsborough Oilfield. In the South Leverton Oilfield they are 240 to 260 ft thick, except in the isolated No. 18 Bore to the west, where they may be as much as 350 ft. They are 250 to 275 ft thick in the Grove oil bores, about 300 ft at West Drayton, a little under 400 ft at Apleyhead, and 300 to 375 ft at Bothamsall. Few boreholes have been cored through this part of the succession in the East Retford district.

As in the exposed coalfield to the west and south-west (Eden and others 1957; Smith and others 1967), these measures are conspicuous in that they consist essentially of a thick sequence of argillaceous rocks overlain by a thick sandy sequence (Plate III). The sandy measures are named the Wingfield Flags after their equivalents in north Derbyshire, and are at the same horizon as the Greenmoor (or Brincliffe Edge) Rock and Grenoside Sandstone of the Sheffield (100) district.

In the Chesterfield (112) district of the exposed coalfield the lowest part of the measures between the Alton Marine Band and the Kilburn Coal comprise a number of cycles with named coals and marine bands, viz: Lower and Upper Parkhouse marine bands, Forty-Yards Coal, Forty-Yards Marine Band, Norton Coal, Norton Marine Band, Upper Band Coal, Burton Joyce Marine Band (Smith and others 1967, fig. 9). None of these horizons can be positively identified in the East Retford district, but tentative correlations can be made in several instances.

The Lower and Upper Parkhouse marine bands are probably absent in this district, though, as already noted, abundant foraminifera, characteristic of these bands farther south-west (Smith and others 1967, p. 112), and including large specimens typical of the Lower Parkhouse Marine Band, have been found at several East Retford localities in a bed here correlated with the Alton Marine Band (pp 51–2).

The Forty-Yards Coal and Marine Band. In Gringley No. 1 Oil Bore shale between 4870½ and 4871 ft, and only 15 ft above the presumed Alton Marine Band (p. 51), has yielded *Lingula mytilloides* and a rich conodont assemblage including *Hindeodella sp.*, *Ozarkodina sp.* and *Idiognathodus sp.* Abundant conodonts are indicative of the Forty-Yards and Norton marine bands, and this shale band, on its position in the sequence, is taken to be the former. It is underlain by a seatearth, but no coal is recorded. Seven feet above is a mussel-band containing pyritized *Carbonicola sp.* (*? proxima/crispa* group).

A comparable occurrence is the 6-in bed of grey silty mudstone at 4510 ft 8 inches in Walkeringham No. 1 Oil Bore, which has yielded *L. mytilloides* and a conodont assemblage including *Hindeodella sp.* and platformed types. This bed lies 4 ft above a seatearth and about 40 ft above a marine band correlated with some confidence with the Alton (p. 51). In Walkeringham No. 2 Oil Bore the marine band, upwards of 2½ ft thick, at 4617½ ft is presumed to be the Forty-Yards Marine Band. It consists of dark mudstone with foraminifera including common large specimens of *Ammodiscus sp.*, *L. mytilloides* [5·0 mm] and fish debris including *Rhadinichthys sp.*

No Lower Coal Measures were cored in the Beckingham and Morton oil bores, but the seatearth detected by geologists of the BP Company in chippings at about 4480 ft in Beckingham No. 1 Bore is probably at the Forty-Yards horizon, and this can be tentatively correlated (Plate III) with a gamma-ray peak above a sandstone at about 4515 ft in Morton No. 1 Bore.

The Forty-Yards horizon cannot be identified in the Gainsborough Oilfield. The only bore cored at this level is No. 1, and here there are no marine bands above the presumed Alton horizon at 4570 ft 7 in (see p. 51). Dark mudstone with a 9-in cank band between 4539 ft 10 in and 4543 ft 5 in contains fish remains and pyritized strap-like markings ('fucoids') and is lithologically suggestive of a marine band, but no definite marine fossils were found. This band, however, is thought to be at a higher horizon than the Forty-Yards, which, by comparison with the section of Corringham No. 1 Oil Bore (see below) is probably represented by one of the four seatearths which occur between 4547 and 4560 ft.

In Corringham No. 1 Oil Bore, east of the district boundary near Gainsborough, there are two *Lingula* bands above the supposed Alton Marine Band (with *Serpuloides sp.*, *L. mytilloides* [10·0 mm] and fish) at 4881 ft. The higher of these is thought to be the equivalent of the Norton Marine Band (see p. 54), and the lower that of the Forty-Yards Marine Band. The latter, comprising 6 ft 2 in of dark grey micaceous silty mudstone at 4860 ft 4 in, contains only sparse *L. mytilloides* [3·0 mm] and *Rhadinichthys sp.*

Measures at this horizon were not cored in the South Leverton and Grove oil bores except possibly at South Leverton No. 3, where the 2 ft 1 in *Lingula* band at 4162 ft may represent the Forty-Yards Marine Band.

No marine horizons were recorded in the cores of West Drayton No. 1 Oil Bore above the level of the presumed Alton Marine Band (p. 52), but the 8-in coal at 3690 ft may possibly be the Forty-Yards Coal.

In Apleyhead No. 1 Oil Bore there is no coal, but the marine band may be represented by a fish-band between 3313 and 3317½ ft, which has yielded, in addition to fish, a single conodont (*? Lonchodina*). A few feet above it elongate *Anthraconaia sp.* were collected.

In the Bothamsall Oilfield the Forty-Yards Marine Band is apparently absent, but its horizon may be indicated by a fish-band up to 30 ft above the presumed Alton horizon (p. 52). From it the following have been collected: *Elonichthys sp.*, *Megalichthys sp.*, platysomid scales indet., *Rhabdoderma sp.* and *Rhadinichthys sp.* There is no coal at this horizon, but the seatearth recorded at 3280 ft in No. 15 Bore may be that of the Forty-Yards Coal.

E

The identification of other named horizons between the Forty-Yards Marine Band and Wingfield Flags is extremely tenuous. The only cored sections of these measures are provided by Gainsborough No. 1 and West Drayton No. 1 oil bores, and just east of the district boundary, Corringham No. 1 Oil Bore. In the two former there are no marine horizons, and only at West Drayton is there any coal—a single seam 9 in thick at 3620 ft. This coal may be the Upper Band, while the dark shale with faint impressions of *Carbonicola sp.* associated with fish remains 15 to 17 ft below may be the Norton Mussel-band (Eden 1954, p. 97: Eagar 1956, p. 350; 1962, pp. 325–6). In Corringham No. 1 Bore the grey to dark silty mudstone between $4837\frac{1}{2}$ ft and 4846 ft 7 in, which contains 'fucoids' and fish in the upper part and *L. mytilloides* [4·5 mm] at 4841 ft 2 in, is apparently the Norton Marine Band. *Carbonicola* aff. *proxima* [cf. Eagar 1956, fig. 10d] occurs $8\frac{1}{2}$ to 11 ft higher. This bivalve, while not diagnostic, is characteristic of the Norton Mussel-band (see above). The dark mudstone from 4814 ft 4 in, resting on seatearth at 4818 ft 2 in and containing fish scales and 'fucoids', may well be at the Upper Band horizon.

Fossils from the measures between the Alton Marine Band and the Wingfield Flags, other than those listed above, consist chiefly of fish debris, but *Geisina arcuata* is recorded at 4402 ft, about 160 ft above the Alton Marine Band, in Gainsborough No. 1 Oil Bore, and cores from Grove No. 1 Oil Bore yielded *Planolites* cf. *ophthalmoides* between 4155 ft 11 in and 4158 ft, and large *P. ophthalmoides* [i.e. about 18 mm] between 4170 ft 10 in and 4173 ft 7 in, respectively about 175 ft and about 190 ft below the Kilburn Coal.

Sandstone is not prominent in the measures between the Alton Marine Band and Wingfield Flags except possibly in certain of the South Leverton bores and at Grove. Here, for example in South Leverton Nos. 5, 10 and 18 and Grove No. 1 oil bores, sandstone up to 50 ft thick in one or more bands lies close above the assumed horizon of the Alton Marine Band. It is, however, possible that the marine band horizon is higher and that these sandstones represent a local thick development of the Sub-Alton Sandstone.

In the absence of precise marker horizons between the Alton Marine Band and the Wingfield Flags, it is difficult to correlate and name the sandstones. It may be, however, that the thin sandstone between the presumed Alton and Forty-Yards horizons in Walkeringham No. 1 Oil Bore, which is probably also present at Ranskill, Morton, Gainsborough and West Drayton, is the equivalent of the Loxley Edge Rock of the exposed coalfield (Eden 1954, pp. 94–5; Eden and others 1957, p. 45).

In Gringley No. 1 Oil Bore a thin band of rock at about 4840 ft, some 30 ft above the supposed Forty-Yards Marine Band (p. 53), is described in the original log as a densely pyritic sandstone, which it superficially resembles. Thin sections [E 21505–7] of the rock have been examined by Dr. K. C. (now Sir Kingsley) Dunham, who reports that it is "a siderite mudstone composed of ill-developed spheruliths of impure siderite ($\omega \simeq 1\cdot795$) up to 0·5 mm diameter, in a groundmass composed of clay minerals, clastic mica, chalcedonic silica and tiny angular quartz grains. Parts (E 21505) are very rich in iron sulphide, occurring in irregular ramifying masses. In these the carbonate encloses the pyrite areas, while the clay minerals aggregate occurs in clear pockets between the pyrite–carbonate areas. Large crystals of colourless calcite are also present here."   E.G.S.

The Wingfield Flags comprise a fairly distinctive and persistent sandstone member, the Sub-Kilburn Sandstone[1] at the top and, normally separated from it by mudstones, a very variable and arbitrarily defined sequence consisting largely of sandstones and siltstones below (Plate III). Total thicknesses range up to about 175 ft, the thicker sections, in general, being found in the south. In scattered oil bores throughout the district there is little or no sandstone below the Sub-Kilburn Sandstone.

The name "Wingfield Flags" is used here because of the comparison that can be made with similar beds occurring at this horizon in north Derbyshire (Smith and others 1967, pp. 116–7). The Wingfield Flags of Derbyshire are the equivalent of the combined Greenmoor Rock and Grenoside Sandstone of the Sheffield district (*idem*), and it may

---

[1]The term "Sub-Kilburn Sandstone" is used here for the "Kilburn Sandstone" of the BP Company's logs to avoid possible confusion with the Kilburn Rock (Gibson and others 1908, pp. 70–1) which lies above the Kilburn Coal in parts of Derbyshire.

well be, as suggested by Downing and Howitt (1969, p. 247) that the latter is the correlative of the Sub-Kilburn Sandstone in the present district. Only the lower part of the Wingfield Flags as defined above, i.e. that part approximating to the Greenmoor Rock, is termed "Wingfield Flags" by the BP Company's geologists. There is, however, some evidence (see below) that the Sub-Kilburn Sandstone is impersistent to the south of the East Retford district, so that in more southern districts the usage of the term "Wingfield Flags" may be largely consistent.

The impersistent sandstones forming the lower part of the Wingfield Flags are generally fine-grained, but beds of medium-grained sandstone are recorded from scattered bores. A calcareous cement is not uncommon and chlorite and coloured micas have been noted in several sections.

There are several records of a seatearth horizon and, rarely, of a trace of coal in the mudstone and siltstone between the lower part of the Wingfield Flags and the Sub-Kilburn Sandstone. Fish debris is recorded in Gainsborough Nos. 4 and 6 oil bores, and includes *Rhabdoderma sp.* from the latter (from 4467 ft to 4468 ft 5 in). A distinctive bed of black chamositic siltstone probably occurs in all those Bothamsall oil bores that lie within the district. This would appear to be either the 'Kilburn Marker' of Strong (*in* Falcon and Kent 1960, p. 52), which to the south lies close below the Kilburn Coal, or a similar bed at a lower horizon. If it is the the former, the Sub-Kilburn Sandstone must die out to the south, perhaps within the Bothamsall Oilfield, for in No. 1 Bore a black chamositic bed is found 17 ft above the Wingfield Flags (Edwards 1967, p. 72). This case is supported by the abundance of coloured micas in the Sub-Kilburn Sandstone (see below) and in the uppermost part of the Wingfield Flags (Strong *in idem*) where the latter are below the 'Kilburn Marker'. There is evidence at Bothamsall, however, of more than one chamositic horizon associated with the Wingfield Flags, e.g. in No. 1 Bore. North of Bothamsall there is only one reliable record of chamosite below the Kilburn Coal: this is at Apleyhead No. 2 Oil Bore, where dark shale with chamosite is logged below the Sub-Kilburn Sandstone, some 40 ft beneath the Kilburn Coal.

The Sub-Kilburn Sandstone is 25 to 50 ft over most of the district, with the thicker sections tending to be found in the north. The sandstone is largely fine-grained, but in some sections is medium- or even coarse-grained in parts. The few cores taken of it show that it varies from thinly bedded to massive. Its most distinctive feature is the conspicuous presence of coloured (brown, bronze and green) micas, rare in higher beds, which have made it useful as a lithological marker (Howitt and Brunstrom 1966, p. 553).

The Kilburn Coal rests directly on the Sub-Kilburn Sandstone in some sections, a ganisteroid top to the latter having been noted in a few oil bores at Gainsborough. In other sections up to 20 ft of mudstone and argillaceous seatearth intervene between the coal and the sandstone.

The Kilburn Coal has not been cored in any of the oil bores in the district, and information about it is therefore available only from chipping samples and geophysical logs. Its absence in some sections may be attributed to the difficulties of detecting a thin coal by these methods, but it also seems that the seam is genuinely absent locally. Generally, however, coal is present at this horizon, varying from a trace to perhaps as much as 4 ft.

No coal is recorded in Tickhill No. 1 Oil Bore and there are only 3 inches in Gringley No. 1, but the geophysical logs suggest that there may be 2 ft in Ranskill No. 1. Two feet are also recorded at Walkeringham in No. 1 Bore, although only a trace of coal is reported from No. 2 Bore. Geophysical evidence suggests a coal about 2 ft thick in Morton No. 1 Oil Bore, but does not support the record of 2 ft in Beckingham No. 1. No coal is recorded in No. 4 Bore at Beckingham.

In the Gainsborough Oilfield the Kilburn Coal is recorded in all but four of the bores that reached its horizon. The majority show a thin seam, but in eight scattered bores the recorded thickness exceeds 2 ft, and locally, as in No. 35 Bore, there may well be as much as 4 ft of coal.

At South Leverton the Kilburn is recorded in all but one of the bores drilled through its horizon, but in only three is it likely to be more than 1 ft thick. The records of 1½ and 2 ft of coal in Grove Nos. 1 and 2 oil bores respectively are not reliable. The Kilburn appears to be absent in both of the West Drayton oil bores, and is thin or absent at

Apleyhead and in most of the Bothamsall oil bores that lie within the district. Reputed coal thicknesses of 2 to 3½ ft in Bothamsall Nos. 10, 19 and 20 bores should be treated with circumspection.

The measures between the Kilburn and Low Silkstone coals are perhaps as much as 450 ft thick in Tickhill No. 1 Oil Bore, but they thin eastwards across the northern part of the district to between 220 and 300 ft in the Gainsborough Oilfield. They range from a little under 200 to 235 ft in the South Leverton oil bores, and are about 300 ft thick in the Bothamsall area, apparently with a local swelling to about 350 ft at West Drayton.

Up to ten cycles have been recognized but the number recorded is usually no more than seven. There are several coals and sandstones, but these cannot be named or extensively correlated with any confidence. Coals are most prominent in the upper part of the sequence, and thick seams are recorded locally. No cores have been taken of these, but substantial thicknesses are supported by geophysical evidence in several instances. A coal generally up to 2 ft, but perhaps thicker locally, occurs some 45 to 70 ft below the Low Silkstone in the Gringley, Walkeringham, Beckingham and Gainsborough oil bores, and there appears to be an additional seam of about the same thickness 20 to 30 ft below in some sections. There are two coals in a similar stratigraphical position at South Leverton and Grove, but here the lower is the most prominent seam, and is probably at least 4 ft thick in certain bores. Only one persistent thick seam is recorded in the Apleyhead and Bothamsall oil bores: this lies 100 to 130 ft below the Low Silkstone horizon and may be up to 4 ft thick. West Drayton No. 1 Oil Bore is the only completely cored section of these measures, and here only one coal, 2 in thick at 3116 ft, was proved between the Low Silkstone and Kilburn. In Sutton Borehole the only coal in the 189 ft of measures cored below the Low Silkstone is 1 ft at 3474½ ft, 111 ft below that seam. Despite these discouraging records, it would seem that there are two widespread and locally thick coals 45 to 130 ft below the Low Silkstone in the East Retford district. Their likely equivalents in the exposed coalfield are the Beeston coals of Yorkshire and the Mickley coals of Derbyshire.

Sandstones in these measures are generally impersistent and less than 30 ft thick, but a sandstone up to 45 ft thick, locally oil-bearing, is found within 100 ft of the Kilburn Coal in many of the Gainsborough bores.

Black chamositic siltstone, reminiscent of the 'Kilburn Marker' (p. 55) is recorded 90 ft above the Kilburn Coal in Grove No. 1 Oil Bore. This is unlikely to be the 'False Kilburn' of Strong (*in* Falcon and Kent 1960, p. 52), but may be the middle one of three such horizons referred to as 'false Kilburn markers' by Edwards (1967, p. 77) and occurring 80 to 100 ft above the Kilburn Coal in the Egmanton and Eakring oilfields. The same horizon is evidently represented in the South Leverton Oilfield, where dark grey or brown siltstone or fine sandstone is recorded 80 to 100 ft above the Kilburn in several bores, although chamosite has been identified only in Nos. 2 and 3.   C.G.G., E.G.S.

Several of the cored boreholes for coal were drilled for a short distance below the Low Silkstone Coal. Nearly all show a thin coal or, in its absence, a seatearth 20 to 40 ft below the Low Silkstone. This coal, which is unnamed, is generally 2 to 7 in thick, but is apparently split (dirt and coal 4 in, on dirt 28 in, on coal 5 in) in Eaton Borehole, and is represented by 6 in of cannel in Torworth Borehole. The 11 in of inferior coal and dirt 67 ft below the Low Silkstone in Manton No. 5 Underground Borehole may be the same seam. The roof measures of the coal or seatearth contain mussels, ostracods (often abundant) and fish. The fossil bed is recorded everywhere except at Mattersey, where the coal is also absent, and at Babworth, where the coal has a sandstone roof. In conjunction with the coal or seatearth below, it forms a supplementary marker to the Low '*Estheria*' Band (p. 57). A composite fauna from this band is as follows: *Spirorbis sp.*, *Anthracosphaerium sp. nov.* cf. *dawsoni*, *Carbonicola browni*, *Curvirimula candela*, *C. subovata*, *C.* cf. *trapeziforma*, *Naiadites flexuosus*, *Geisina arcuata;* fish remains including an acanthodian spine, *Rhadinichthys sp.* and *Rhizodopsis sp.*

Sutton Borehole was continued for 189 ft below the Low Silkstone Coal (at 3363 ft 4 in), largely in argillaceous and silty measures. Apart from the above-described mussel/ ostracod/fish band (at 3382 ft) and scattered *Spirorbis sp.*, plants and fish remains, these measures have yielded fossils from two other horizons, viz: *Anthraconaia* cf. *fugax* [juv.] at 3437 ft, and *Anthraconaia sp.*, *Carbonicola sp.* and *Curvirimula* cf. *trapeziforma* in a 2-ft band at 3454 ft. These last two occurrences are 36 and 19 ft respectively above a 1-ft coal (see p. 56), the latter in the cycle immediately above the coal and the former in the succeeding cycle. It may be noted that *Anthraconaia* is rare in the *C. communis* Zone, but is known from the measures between the Low Beeston and Silkstone coals at a few localities outside the district.

M.A.C., E.G.S.

The only fossils collected from these measures in oil bores are from Walkeringham No. 1 and Gainsborough No. 1. At the former, cores from 3986 to 4113 ft, between the ?Beeston coals (see p. 56) and the Kilburn, yielded *Curvirimula sp.*, at 3986 ft 4 in; *C.*

subovata between 3993 ft 10 in and 3996 ft 1 in, *Spirorbis sp.*, *Carbonicola* cf. *pseudorobusta*, *Curvirimula sp.*, *Naiadites ?*, *Geisina arcuata* and palaeoniscid scales including *Elonichthys sp.* between 4013 and 4014½ ft; small *Carbonicola* cf. *bipennis* between 4057½ ft and 4061 ft 11 in; *C.* cf. *pseudorobusta* between 4091 ft 8 in and 4091¾ ft; *Curvirimula trapeziforma* at 4097 ft 10 in; and a *Strepsodus* scale at 4105½ ft. The cores starting at 4250 ft in Gainsborough No. 1 Oil Bore cover the interval between the above succession and the Kilburn Coal. The only fossils found here were a sineoidal impression, possibly made by the telson of a belinurid, at 4284 ft, cf. *Planolites ophthalmoides* at 4297¼ ft, and, in dark, finely micaceous, silty mudstone forming the immediate roof of the Kilburn Coal, megaspores and abundant fish remains including palaeoniscid scales and *Rhabdoderma sp.*

M.A.C.

LOW SILKSTONE COAL TO CLAY CROSS MARINE BAND

These measures are over 500 ft thick in the north-western part of the district, and show a general thinning to less than 350 ft in the south-east. There are no marine bands, but the Low '*Estheria*' Band near the base is an important faunal horizon. Of several named coals, the Parkgate is economically the most important. Correlation of the coals is shown on Plates IV and V, the former also giving details of lithological variation in the measures.

These measures are known in some detail from colliery shafts and cored boreholes in the western part of the district, which is described separately from the eastern part of the district where the only information available is that provided by oil-bore data (see p. 68).

(i)   The Western Part of the District

The Low and Top Silkstone coals. The Silkstone Coal of districts to the west and south is divided, apparently everywhere in this part of the East Retford district, into two seams, the Low Silkstone and the Top Silkstone. These are separated by up to 40 ft of measures which contain the Low '*Estheria*' Band at their base.

The Low Silkstone Coal generally consists of bright coal, dirty coal and dirt. It has a minimum thickness of 3 inches in Manton No. 6 Underground Borehole and a maximum thickness of 28 in (brights 5 in, dirt and dirty coal 6½ in, brights 13 in, on dirty coal 3½ in) in Ranby Hall Borehole. Adjacent to the latter are Ranby Camp, Barnby Moor and Bilby boreholes, with above average thicknesses of 17 to 22 in.

The measures between the Low and Top Silkstone coals comprise one cycle and are thinnest in the south (Eaton 11¾ ft, Manton No. 7 Underground 12 ft 5 in, Manton No. 5 Underground 13 ft 10 in). From Babworth Borehole northwards the thickness is normally about 30 ft, but it is 56 ft 4 inches in Scaftworth Borehole.

*The Low '*Estheria*' Band*, at the base of the measures, can, where complete, be divided into two parts, the lower consisting of dark grey mudstone and the upper of grey mudstones and silty mudstones. The dark grey mudstone varies in thickness from 0 to 6 ft 10 in, being thickest in Nornay Borehole in the north and absent from Babworth, Barnby Moor, Manton No. 5 Underground and Ranby Hall boreholes. It generally contains abundant 'Estheria' sp. nov. preserved as irridescent films, and in some localities fish debris has also been found. The grey mud-stones and silty mudstones contain sporadic ironstone bands, and 'Estheria' is found sparingly throughout for up to 7 ft above the Low Silkstone Coal. In Nornay Borehole rare 'Estheria' have been recorded in siltstones 12 ft above the Low Silkstone Coal and some 8 ft above the dark mudstone of the lower part of the Low 'Estheria' Band. The band is absent from Manton Nos. 6 and 7 and Steetley No. 1 underground boreholes, where silty beds directly overlie the Low Silkstone. In West Drayton No. 1 Borehole no 'Estheria' were recorded in the 2¼ ft of black shale above the supposed Low Silkstone Coal.

                                                          G.H.R.

The specimens of 'Estheria' at this horizon are distinguished from those of the Middle Coal Measures by their larger size and more widely spaced growth lines.

Above the Low 'Estheria' Band, mudstone is generally present and in some localities contains a few ironstone bands. In Scaftworth Borehole, where the interval between the Low Silkstone and Top Silkstone is greater than usual, Carbonicola circinata was found 16 ft above the Low 'Estheria' Band, and Curviri-mula sp. and Geisina arcuata occurred some 17 ft higher. Rare mussels are recorded in Firbeck No. 2 Underground (Curvirimula cf. subovata), Lound (indeterminate fragments), Mattersey (C. cf. subovata) and Ranby Hall (Carbonicola sp.) boreholes. The mudstones pass up into siltstones and sandstones which, in turn, give way to the seatearth below the Top Silkstone Coal.   G.H.R., M.A.C.

The Top Silkstone Coal is generally of poor quality, containing much cannel, dirt and dirty coal. No coal was recorded at this horizon in Babworth, Mattersey, Sutton and Torworth boreholes.

Commonly the seam is split into three leaves (Fig. 11). The lower leaf, separate everywhere except in Firbeck No. 2 Under-ground Borehole, comprises a few inches of coal or dirty coal. In Bilby Borehole it is probably represented by 4 in of canneloid shale. The dirt between the lower and main leaves of the seam varies in thickness up to a maximum of 7 ft 5 in at Ranby Camp Bore-hole, but at Mattersey there are 12 ft of measures including sandstone and siltstone,

between the assumed horizons of these two leaves. The main leaf of the seam, thickest in Firbeck No. 2 Underground, Blyth and Nornay boreholes (3 to 3½ ft), commonly contains a large proportion of cannel, which in Ranby Hall and Lound boreholes passes into canneloid brights. In the south-west, however, no cannel has been recorded, and here the main leaf consists of 1½ to 2 ft of bright and dirty bright coal. The sections at Firbeck No. 2 Underground, Nornay and Ranskill boreholes are notable for the pre-sence of ironstone, up to 2 in thick, in the upper part of the cannel.

The upper leaf of the Top Silkstone consists of a few inches of usually inferior coal within a foot of the main leaf.   R.F.G., G.H.R.

The measures between the Top Silkstone and Threequarters coals vary in thickness between 34¾ ft (Eaton Borehole) and 96 ft 4 in (Scaftworth Borehole), though the range is generally between 50 and 70 ft. The variation has no discernible regional trend.

In Babworth, Ranby Camp, Steetley No. 1 Underground and Sutton boreholes the measures consist largely of sandstones or siltstones commencing a short distance above the Top Silkstone and extending upwards to, and in some cases beyond, the horizon of the Threequarters Coal. Elsewhere, except in Eaton Borehole, the measures generally com-prise two cycles, each of which has argil-laceous beds in the lower part passing up into silty or sandy beds above. Only in Jockey House and West Drayton No. 1 bore-holes is there a thin coal developed at the top of the lower cycle. In Firbeck No. 2 Under-ground and Wigthorpe boreholes the seat-earths within about 17 ft of the base of the Threequarters Coal are thought to be splits off the seam.

The mudstone or shale overlying the Top Silkstone Coal is commonly dark grey, but in places black, and contains numerous iron-stone bands. Fish debris has been noted in many boreholes, and mussels, including Carbonicola declivis, C. pseudorobusta, Curvirimula candela and Naiadites flexuosus, occur sporadically and are in places preserved in shelly ironstone. Geisina arcuata is recorded from Nornay and Ranby Hall boreholes, and ostracods have also been noted at Ranby Camp. The following plants were collected from 3324 to 3329 ft, just

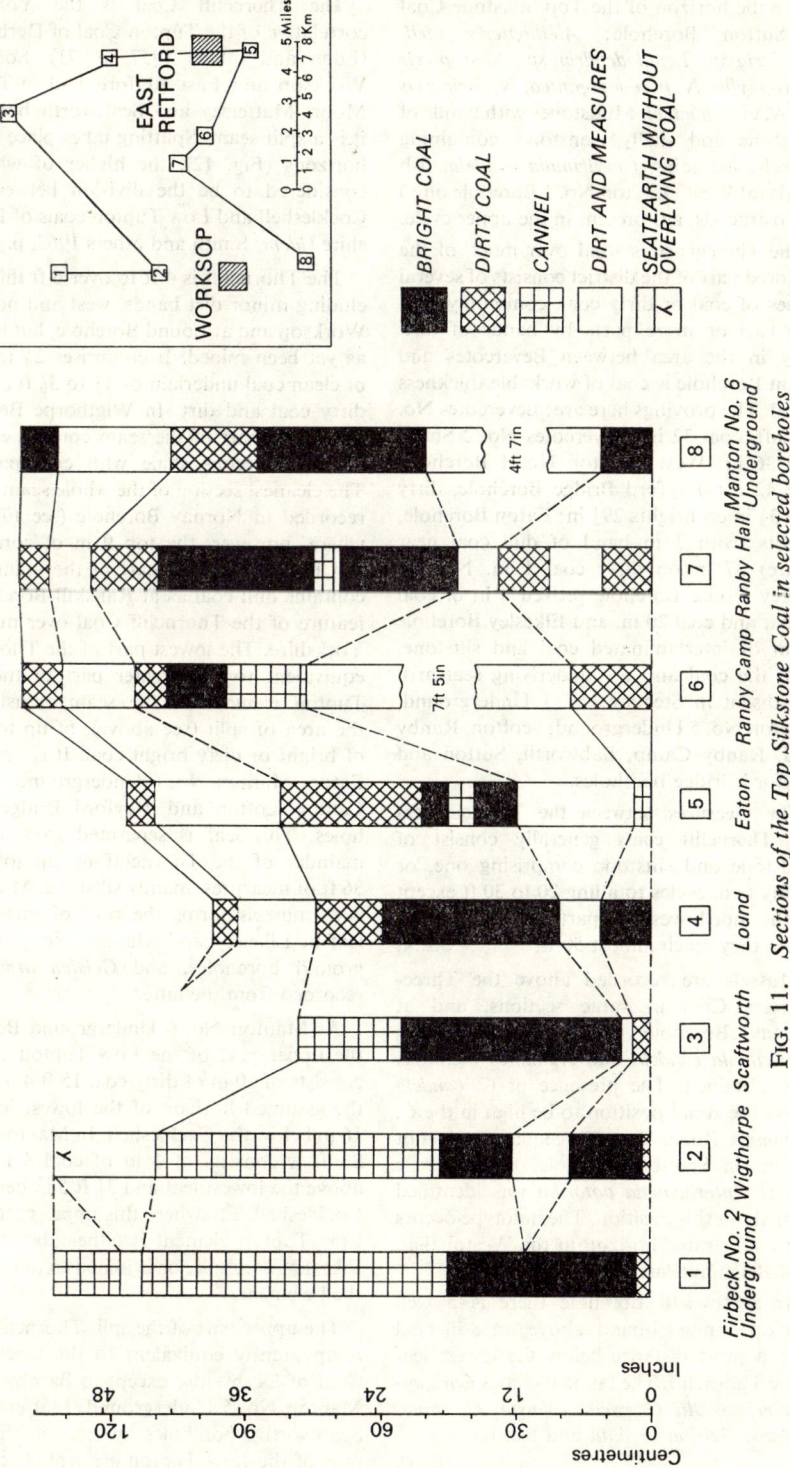

FIG. 11. Sections of the Top Silkstone Coal in selected boreholes

above the horizon of the Top Silkstone Coal in Sutton Borehole: *Alethopteris serli, Annularia sp., Lepidodendron sp., Neuropteris heterophylla, N. pseudogigantea, N. rarinervis* and *N.* cf. *schuetzei*. Mudstones with bands of ironstone and shelly ironstone, containing mussels, including *Curvirimula candela*, fish debris (at West Drayton No. 1 Borehole only) and ostracods, are present in the upper cycle.

The **Threequarters Coal** over much of the explored part of the district consists of several inches of coal or dirty coal, commonly split into two or more parts by bands of dirt. Only in the area between Bevercotes and Eaton Borehole is coal of workable thickness found. The provings here are: Bevercotes No. 1 Shaft, coal 32 in; Bevercotes No. 2 Shaft, coal 36 in; West Drayton No. 1 Borehole, coal 42 in; Twyford Bridge Borehole, dirty coal $3\frac{1}{2}$ in on brights $29\frac{1}{2}$ in; Eaton Borehole, brights (with $\frac{1}{2}$ in band of dull coal near centre) 37 in on dirty coal 1 in. Nearby, Jockey House Borehole proved 8 in of coal on bat and coal 20 in, and Elkesley Borehole 38 in of interlaminated coal and siltstone. Both the coal and the underlying seatearth are absent in Steetley No. 1 Underground, Manton No. 5 Underground, Scofton, Ranby Hall, Ranby Camp, Babworth, Sutton and Twyford Bridge boreholes.

The **measures between the Threequarters and Thorncliff coals** generally consist of mudstone and siltstone comprising one, or locally two, cycles totalling 20 to 30 ft except in the north-western part of the district, where they reach almost 50 ft.   G.H.R.

Mussels are recorded above the Threequarters Coal in some sections, and at Nornay Borehole *Carbonicola communis, Curvirimula candela* and *Naiadites flexuosus* were obtained. The presence of *C. candela* shows the zonal position to be high in the *C. communis* Zone. The Threequarters is not present in Scofton Borehole, but the rare mussel *Anthraconaia potoriba* was identified from about this position. The holotype occurs at a comparable horizon in the Westphalian A of Belgium (Pastiels 1964, p. 70).

At Scaftworth Borehole there is a well developed mussel-band above an 8-in coal lying a short distance below the lowest leaf of the Thorncliff. The fauna includes *Carbonicola cristagalli, C. pseudorobusta, C. rhomboidalis, Geisina arcuata* and fish remains.
                    M.A.C.

The **Thorncliff Coal** is the Yorkshire correlative of the Tupton Coal of Derbyshire (Eden and others 1957, p. 71). South of Worksop and East Retford and in Barnby Moor, Mattersey and Scaftworth boreholes it is a split seam. Splitting takes place at two horizons (Fig. 12), the higher of which is considered to be the division between the Cockleshell and Low Tupton coals of Derbyshire (*idem*; Smith and others 1967, p. 134).

The Thorncliff is 4 ft to over 6 ft thick, including minor dirt bands, west and north of Worksop and at Lound Borehole, but has not as yet been mined. It comprises $2\frac{1}{2}$ to $3\frac{1}{2}$ ft of clean coal underlain by $1\frac{1}{2}$ to $3\frac{1}{2}$ ft of coal, dirty coal and dirt. In Wigthorpe Borehole this lower part of the seam consists entirely of carbonaceous shale with coal partings. The cleanest section of the whole seam is that recorded in Nornay Borehole (see Fig. 12), where, however, the top 9 in of core were lost in drilling. This part of the seam often contains dull coal as at Ranskill Borehole, a feature of the Thorncliff Coal over much of Yorkshire. The lowest part of the Thorncliff, equivalent to the greater part of the Low Tupton element of the seam, consists, in the area of split (see above), of up to 16 in of bright or dirty bright coal. It is absent in Eaton, Manton No. 6 Underground, Ranby Camp, Scofton and Twyford Bridge boreholes. This leaf is separated from the remainder of the Thorncliff by up to about 36 ft of measures, mainly siltstone. Mudstone with mussels forms the roof of this lowest leaf at Elkesley and Manton No. 7 Underground boreholes, and *Geisina arcuata* is recorded from the latter.

At Manton No. 6 Underground Borehole the upper part of the Low Tupton element consists of 10 in of dirty coal 15 ft 4 in above the assumed horizon of the lowest leaf and 16 in below the Cockleshell. In Manton No. 4 Shaft it consists of 1 in of coal 4 ft 11 in above the lowest leaf and 31 ft 5 in below the Cockleshell. Elsewhere this upper part of the Low Tupton element is either absent, as at Elkesley Borehole, or is joined to the Cockleshell element.

The upper part of the split Thorncliff seam is apparently equivalent to the Cockleshell Coal of Derbyshire except in Barnby Moor, Manton No. 5 Underground, Mattersey and Scaftworth boreholes where it includes part of the Low Tupton element. It consists

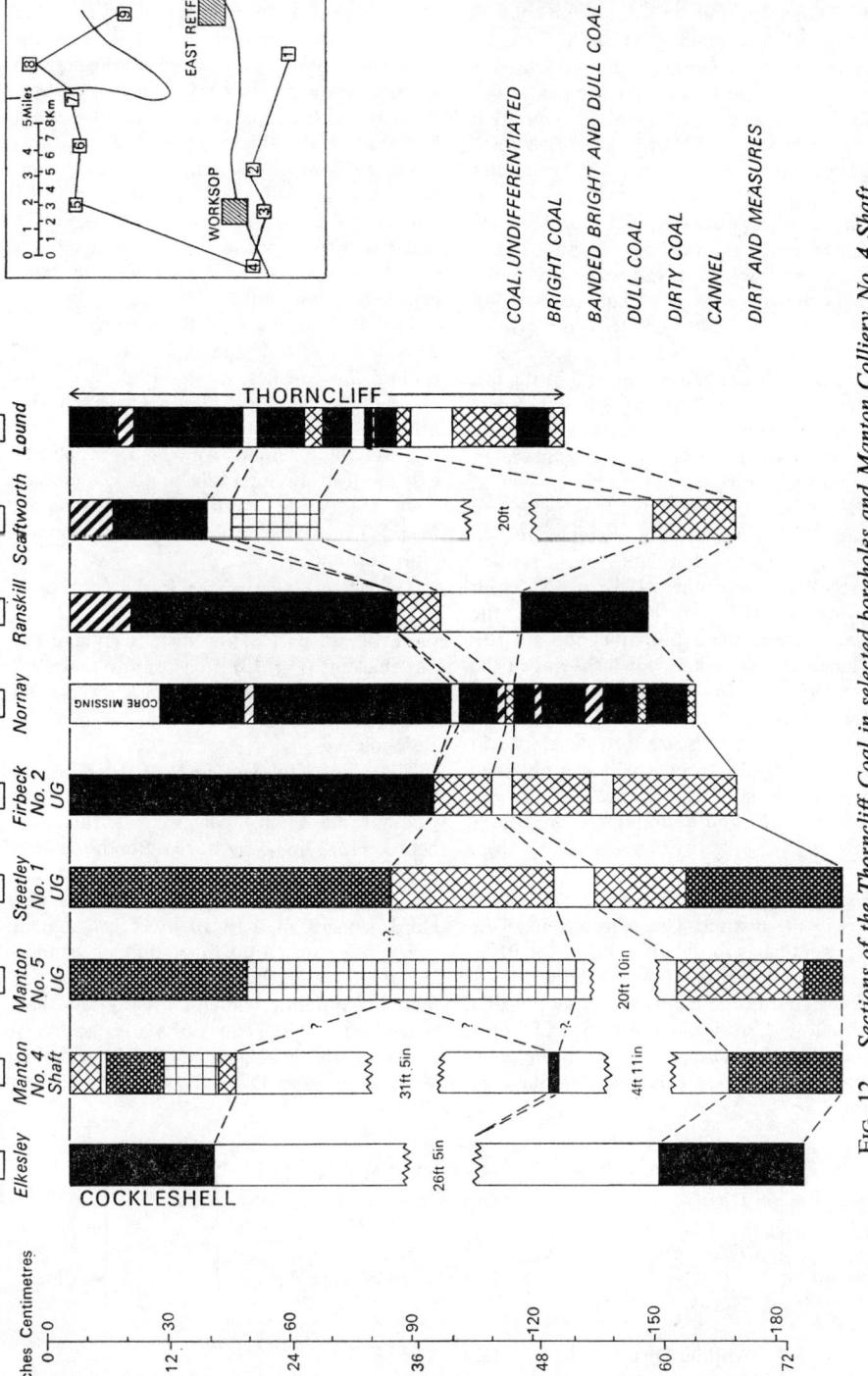

FIG. 12. *Sections of the Thorncliff Coal in selected boreholes and Manton Colliery No. 4 Shaft*

Sections 3, 4 and 5 are from underground (UG) boreholes. Areas in which the Thorncliff is a split seam (in north-east and south) are demarcated on inset map.

of 4 to 49 in of largely bright and dirty bright coal, but at Manton No. 4 Shaft and Manton No. 5 Underground and Scaftworth boreholes it includes cannel. The 32 in of cannel at Manton No. 5 Underground Borehole is probably a local thick development. The Thorncliff is absent, probably owing to a washout filled by siltstones, at Babworth Borehole, and is not recorded in Bevercotes No. 1 Shaft.

The measures between the Thorncliff and Parkgate coals are between 29 and 74½ ft thick, being thickest in the north and west. A majority of the recorded sections fall within the range of 40 to 60 ft. The measures comprise two, or less frequently three, cycles, and commonly contain a thin coal at the top of the lowest cycle. Typically the lower part of each cycle consists of dark mudstones, which usually contain mussels and often have thin shelly ironstones.   R.F.G.

The measures between the Thorncliff and Parkgate coals are usually characterized by bands containing abundant well-preserved mussels (Eden and others 1957, p. 77; Smith and others 1967, pp. 134–5), but in the present district these faunal bands are less prominent. Locally the roof measures of the Thorncliff contain *Carbonicola cristagalli*, *Anthracosphaerium* cf. *dawsoni* and *Geisina arcuata*; higher bands are well developed in Nornay Borehole, from which was obtained the typical fauna of *C. cristagalli*, *C. oslancis*, *C. rhomboidalis* and abundant *G. arcuata*.
                                                      M.A.C.

The Parkgate Coal is one of the major seams in the district. The relationship of its component parts to the divisions of the Piper Coal of Derbyshire is tabulated below.

In the south, dirt bands are generally thin or absent and, as a result, 4 to 6 ft of workable coal are present. To the north the seam is split by dirt at two principal horizons (see above) as shown in Figs. 13 and 14. Thus, in the area north of Worksop, the Middle Dirt of the Parkgate divides the main part of the seam into two beds of coal of little or no economic value, and nearly everywhere in the north where the Roof Coal is present it is separated from the main part of the seam by a considerable thickness of dirt or measures (53 ft at Bilby Borehole). In the north and around East Retford (Fig. 13), where the Parkgate is overlain by sandstone, the seam is partly washed out in most sections, and at Mattersey and Nornay boreholes has been completely removed.

The *Bottom Softs* of the Parkgate Coal consist of up to 29½ in of bright coal. The bottom few inches, corresponding to the Dandies of the Sheffield district (Eden and others 1957, p. 79) are often dirty and have a relatively high sulphur content. In the southeast the Bottom Softs thin to a few inches of dirty coal, as at Twyford Bridge Borehole (Fig. 14), and they are absent in West Drayton No. 1 Borehole.

The *Middle Dirt* of the Parkgate Coal is absent or only a few inches thick in much of the explored part of the district. Where the dirt thickens (Fig. 13) it passes into seatearth and siltstone and has a maximum proved thickness of 17 ft 5 inches in Wigthorpe Borehole.

The *Hards* and *Top Softs* of the Parkgate Coal, which together correspond to the Lower Coal of the First Piper, vary in thickness between 21½ in (Wigthorpe Borehole) and about 48 in (Jockey House Borehole, where the upper limit is not precisely fixed). The Hards consist of 0 to 10 in of bright coal overlain by up to 20 in of dull or banded bright and dull coal. They are notable for their low sulphur content, usually less than 1 per cent. The Top Softs are generally thicker than the Hards, probably reaching 30 inches in West Drayton No. 1 and Jockey

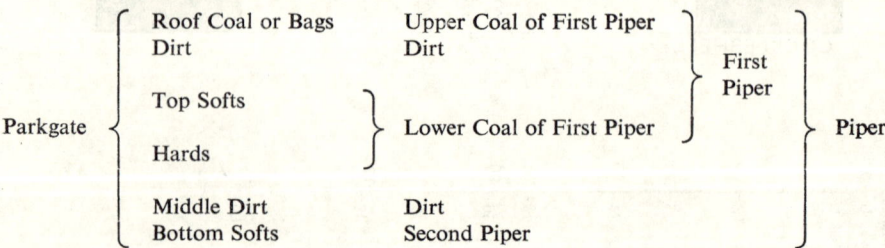

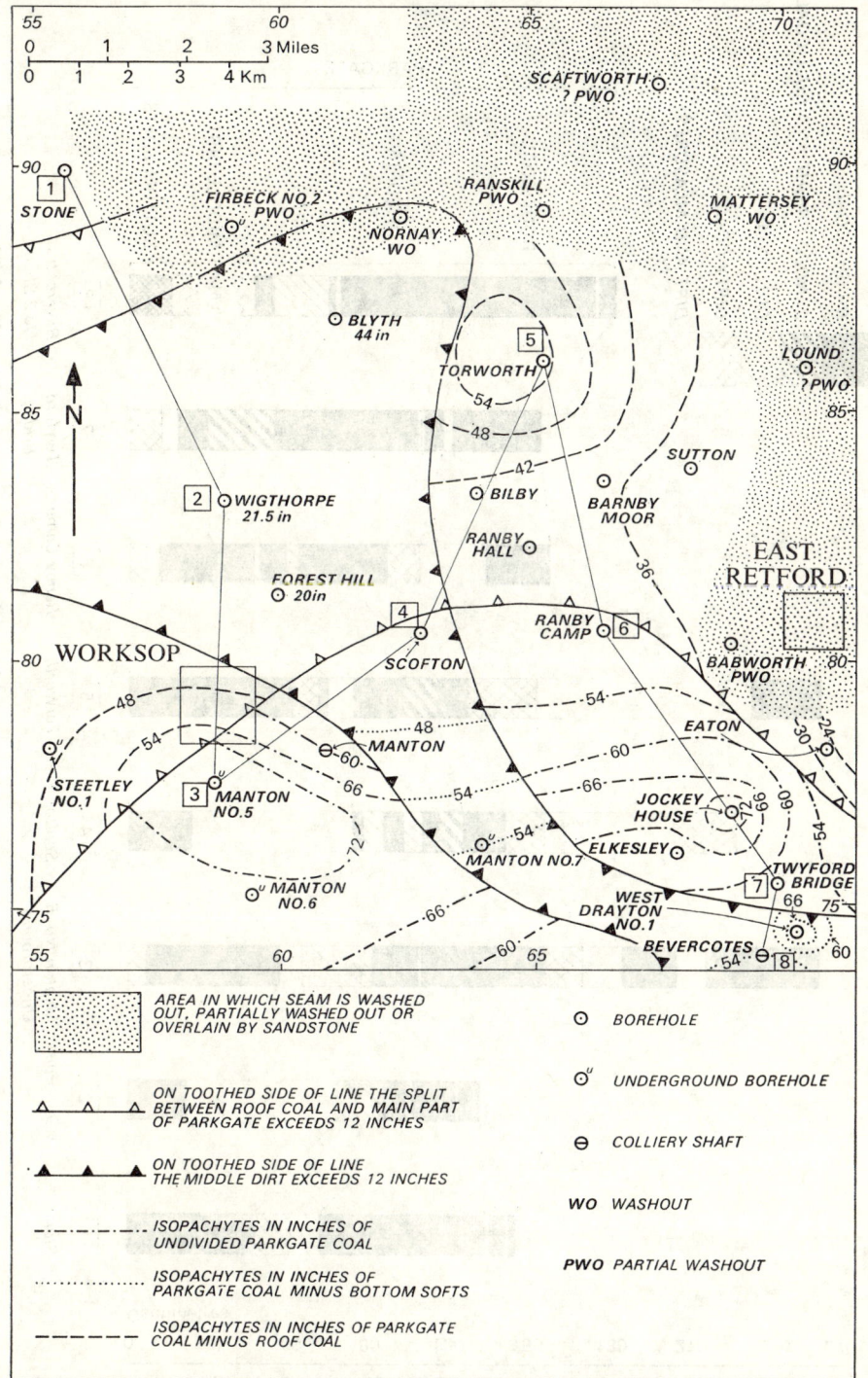

FIG. 13. *Plan of the Parkgate Coal*

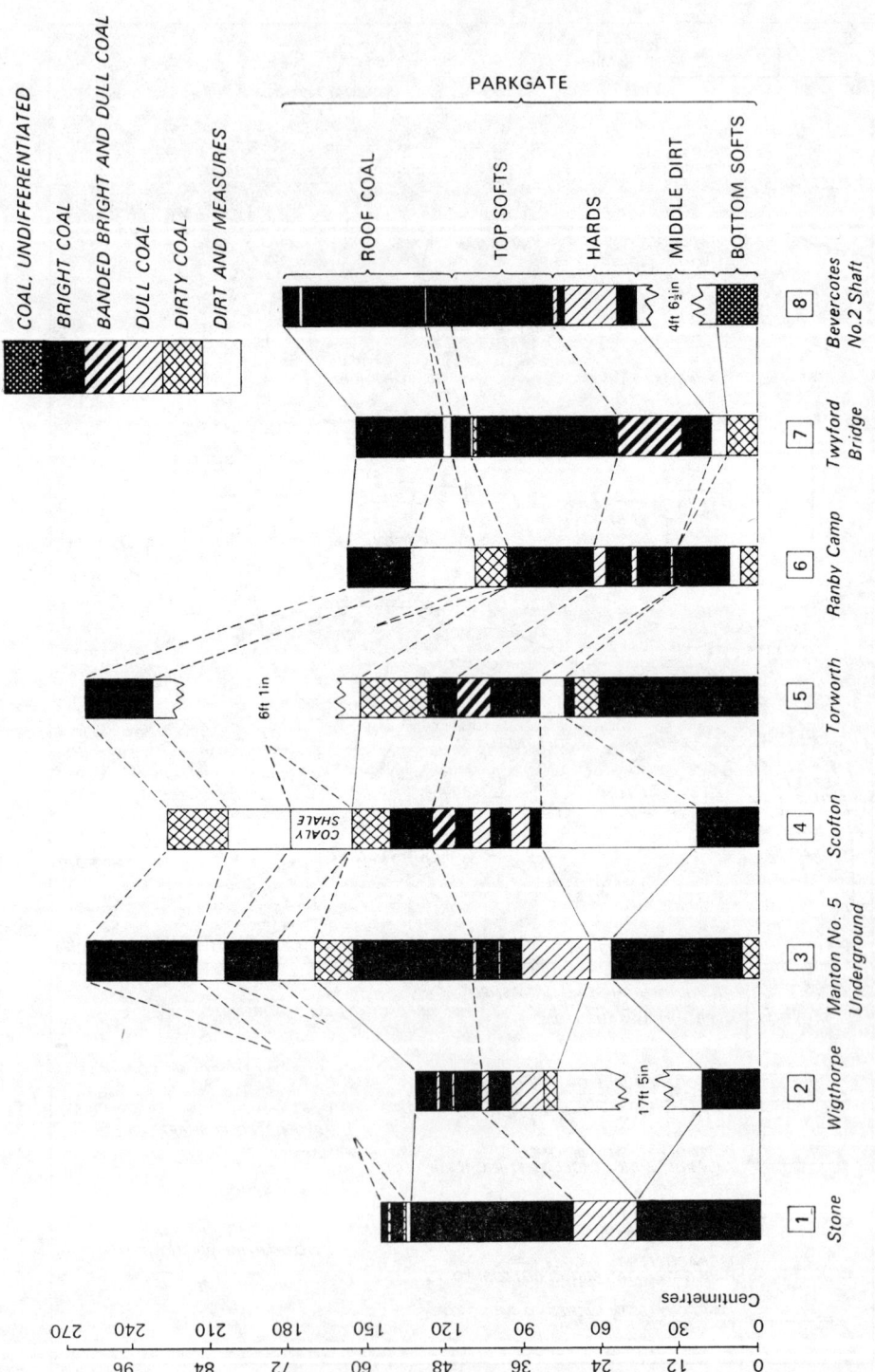

FIG. 14. Sections of the Parkgate Coal in selected boreholes and Bevercotes Colliery No. 2 Shaft
For sites see Fig. 13

House boreholes. The ash content generally increases upwards, and much of the Top Softs is dirty coal in Torworth and Barnby Moor boreholes.

The *Dirt* between the Top Softs and the Roof Coal is very thin or absent in the Manton–Bevercotes area and around Stone Borehole (Fig. 13), but elsewhere it varies in thickness up to 53 ft in Bilby Borehole. The thicker sections comprise seatearth, sandstone, siltstone and mudstone, all of which contain much plant debris.

The *Roof Coal* consists of one or two leaves of brights, totalling up to 24 in. Where it is separate from the main part of the Parkgate it tends to deteriorate in both thickness and quality, and in Barnby Moor, Steetley No. 1 Underground, and Wigthorpe boreholes it is absent. Where both leaves are present, the upper is generally the thicker and cleaner coal, and where only one leaf is recorded that leaf is thought to be the upper.   R.F.G.

The measures between the Parkgate and Deep Hard coals are between 21 and 72½ ft thick in the central and southern areas of the explored part of the district. In the north, except at Stone Borehole, the Deep Hard and Parkgate Rocks are united and much of the measures consist of sandstone and siltstone (Plate IV).

They comprise one or two cycles, and rarely contain a thin coal. Where two cycles are developed there is generally a seatearth at the top of the lower one, though in some instances (e.g. Manton Nos. 5 and 7 underground boreholes) only a few inches of black shale are present.

The Deep Hard Coal is washed out in the north except in Stone Borehole, where it is apparently split into a number of thin seams (Plate IV). Elsewhere it consists of up to 40 in of bright and dirty bright coal with dirt bands, but the main bed of the seam rarely exceeds 24 in. In Manton No. 4 Shaft and Manton No. 6 Underground, Ranby Camp and Scofton boreholes the lowest few inches of the seam form a separate leaf 4 to 14 ft below the main bed. To the south-east of these provings this leaf cannot be traced, but is unlikely to have rejoined the main bed.

Above the Deep Hard and separated from it by up to 35 ft of measures, a thin coal is thought to be the equivalent of the Deep Hard Roof Coal, which splits off from the main part of the Deep Hard in the area east of Chesterfield (Smith and others 1967, pp. 141–6). At Stone Borehole this thin coal appears to have rejoined the Deep Hard.

The measures between the Deep Hard and Deep Soft coals are generally about 60 ft thick, but vary irregularly between 45 ft in Stone Borehole and 115 ft in Manton No. 6 Underground Borehole. No thicknesses are available in the north, where the Deep Hard Coal is washed out (Plate IV). Here sandstone and siltstone of the Deep Hard Rock, united with the Parkgate Rock, attain a maximum thickness of 134¾ ft in Ranskill Borehole. Similarly, thicknesses are not available in the area north of East Retford, where the Deep Soft Coal is washed out (see below). Where thick sandstones or siltstones are absent, the measures contain several thin coals or seatearths. The lowest of these is the Deep Hard Roof Coal (see above), but the rest may be split off the Deep Soft of more southerly districts. Whether or not the lowest coal present in Firbeck No. 2 Underground, Nornay and Scaftworth boreholes is the Deep Hard Roof Coal, is uncertain.

Though fossils are not common in these measures, the dark mudstones overlying the Deep Hard have yielded *Anthraconaia?*, *Carbonicola* cf. *cristagalli* and *Naiadites sp.*, and ostracods have been recorded in Manton No. 7 Underground and Twyford Bridge boreholes. Rare mussels have been recorded above the Deep Hard Roof Coal, and a few mussels, including *C. oslancis* and *Naiadites sp.*, have been found above the thin coal of uncertain correlation mentioned above.

The Deep Soft Coal, so-called, of the East Retford district is probably equivalent to the combined Eckington Deep Soft (or Sitwell) and Sitwell Thin coals of the Sheffield district (Eden and others 1957, pp. 87–9). To the south the seam is a part of the Deep Soft group of coals (Deep Soft, Roof Soft, Top Soft) of the Chesterfield district (Smith and others 1967, pp. 146–53), of which the Top Soft is thought to be the equivalent of the Sitwell.

The seam is very variable in composition and thickness (Fig. 15), and generally a large proportion of the coal has an appreciable ash content. The explored part of the district can be divided into four areas (Fig. 15, inset map). In the west the seam is split into

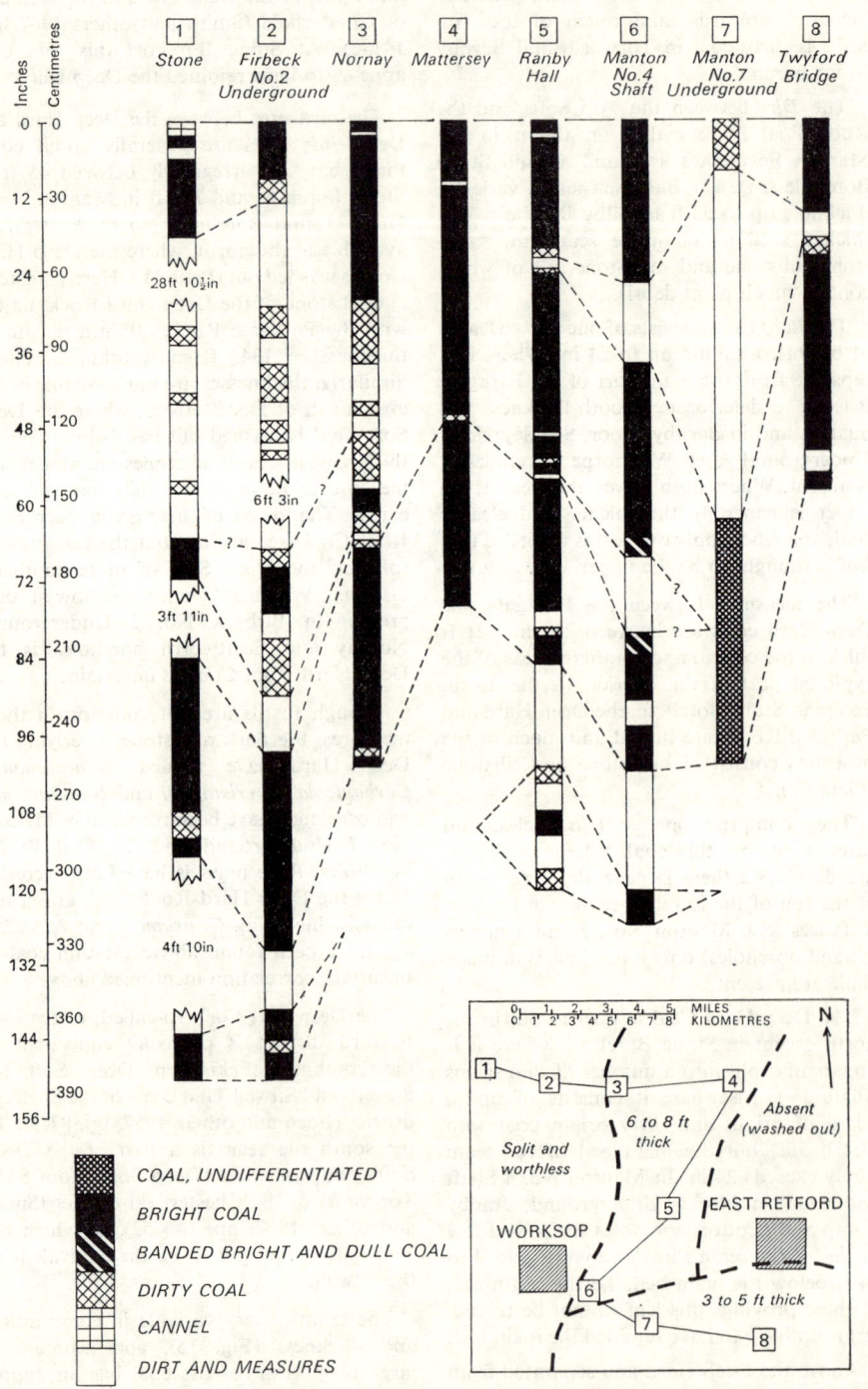

FIG. 15. *Sections of the Deep Soft Coal in selected boreholes and Manton Colliery No. 4 Shaft*

PLATE V

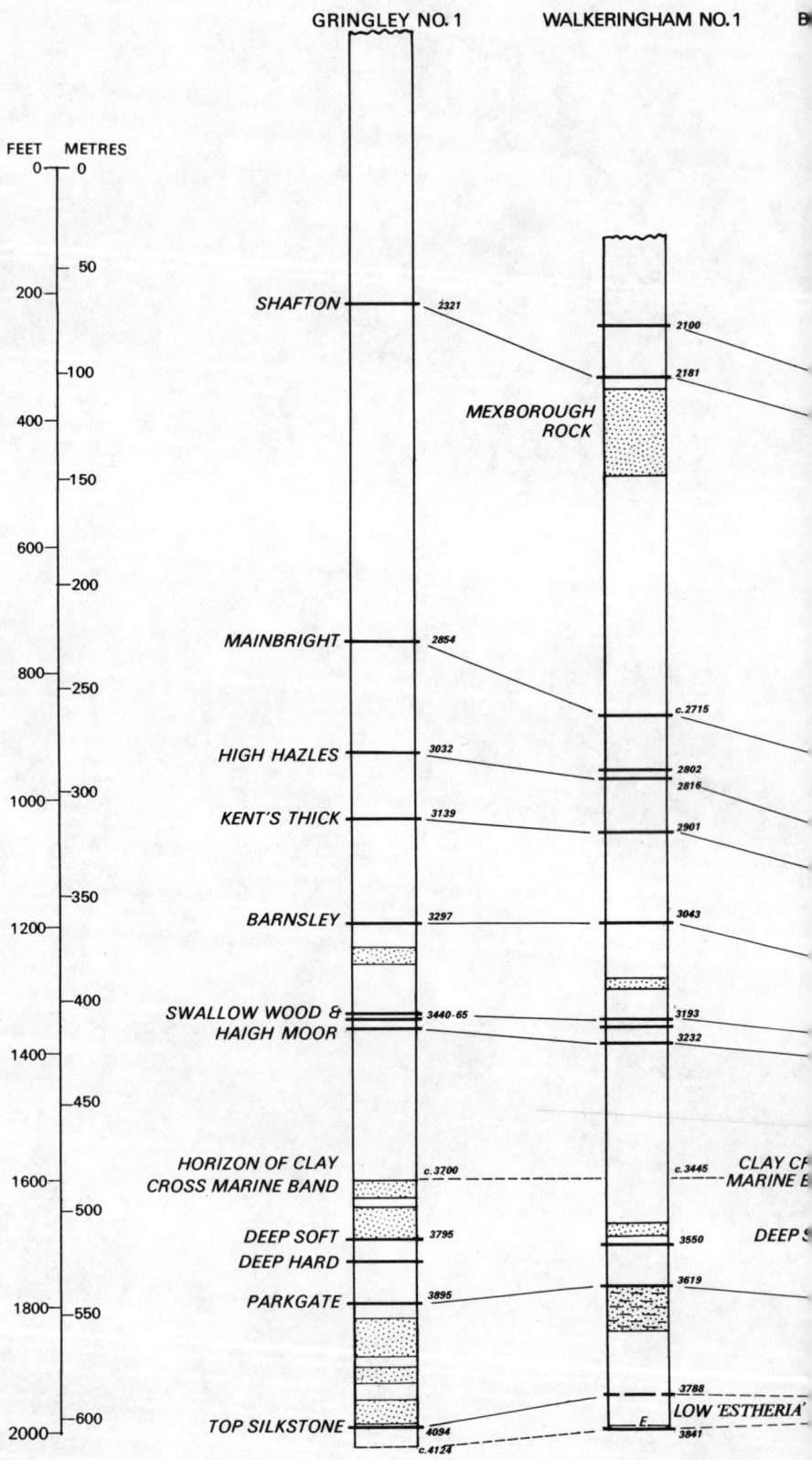

several leaves which may be many feet apart. At Wigthorpe, for example, the 16-in uppermost leaf is separated by 6 ft 11 in of seatearth with coal streaks from a 4-in leaf which lies $48\frac{3}{4}$ ft above the 23 in of coal and dirt comprising the rest of the seam. The $48\frac{3}{4}$ ft of measures here consist of siltstone and sandstone with much plant debris. In Stone Borehole the upper leaf is 18 in thick and lies nearly 30 ft above the remainder of the seam, which is itself widely split. At Steetley No. 1 Underground Borehole, however, the Deep Soft is absent, owing to a washout. In the area extending from Manton Colliery to Scaftworth Borehole the Deep Soft consists of 6 to 8 ft of brights with minor bands of dirty coal and dirt, and here forms one of the most economically important seams in the whole of the Coal Measures. To the east of Manton and south of East Retford the seam is 3 to 5 ft thick, but the correlation of individual leaves with the composite seam to the north is not clear. The Deep Soft is washed out to the north of East Retford in, for example, Babworth, Barnby Moor, Ranby Camp and Sutton boreholes. R.F.G., G.H.R.

The measures between the Deep Soft Coal and the Clay Cross Marine Band are thinnest in Scaftworth Borehole (57 ft 7 in) in the north and thickest in Manton No. 4 Shaft (126 ft 5 in). Thicknesses of over 100 ft persist north-eastwards from the Manton area through Forest Hill, Scofton, Bilby and Torworth boreholes, and eastwards to Babworth Borehole. At these localities and in the intervening Barnby Moor, Sutton, Ranby Hall and Ranby Camp boreholes the measures consist predominantly of siltstones and sandstones resting directly on, or commencing only a short distance above, the Deep Soft Coal, though seatearths and thin coals are developed locally and the Joan Coal or its seatearth is always present near the top. To the west and south of this area of arenaceous deposits the measures comprise up to six cycles, but the only named coals are the Flockton Thick, generally in the first cycle above the Deep Soft, and the Joan.

The Flockton Thick Coal is, at least in part, the equivalent of the Chavery Coal of the Sheffield district (Eden and others 1957, p. 90). It is thickest in the north, in Nornay (dirty coal 17 in on brights 32 in) and Blyth (bright and dull coal 11 in on brights 17 in) boreholes, thinning to 22 in of coal in Firbeck No. 2

Underground Borehole and deteriorating to 43 in of dirt and coal at Wigthorpe. In Steetley No. 1 Underground Borehole 18 in of coal at the base of the seam are separated by $2\frac{3}{4}$ ft of seatearth with coal partings from a further 9 in of coal. Two inches of inferior coal 10 ft 5 in higher in the sequence may represent a split off the top of the seam, the underlying beds being seatearth and shale with plant debris. In the south the seam is thickest in Manton No. 6 Underground Borehole (30 in of inferior coal). It thins eastwards to a few inches, though precise correlation here is difficult.

The Flockton Thick is generally overlain by mudstones, dark in their lower part, which contain ironstone bands and mussels including *Anthracosia regularis*, associated with *Geisina arcuata*. In many localities one or two thin bands of shelly ironstone are present. The lithology and fauna are similar to those found above the Chavery Coal in the Sheffield and Chesterfield districts, though in the present district the ironstone bands are not so well represented. In the south (Eaton, Elkesley, Twyford Bridge and Jockey House boreholes) the probable Flockton Thick is overlain by mudstones and siltstones containing abundant plant debris and extending up to the seatearth beneath the Joan Coal. Elsewhere, and particularly in the west, one or two seatearths develop between the Flockton Thick and Joan coals. The lower seatearth, which may be at the horizon of the Brown Rake Coal of the Chesterfield district (Smith and others 1967, p. 153) is overlain by beds similar to those above the Flockton Thick in that they contain ironstones, mussels and ostracods.

The Joan Coal, absent in some localities, is generally represented by a single bed of coal, with a maximum thickness of $17\frac{1}{2}$ in at Twyford Bridge Borehole, though this may be only one leaf of a split seam. More than one leaf is present in some sections, as at Eaton Borehole, where 16 in of coal are separated by 16 in of seatearth from a lower 5 in of dirty coal.

In the west, as in the Ollerton (113) district to the south (Edwards 1967, p. 91) a variable thickness of shale with a non-marine fauna intervenes between the horizon of the Joan Coal and the base of the Clay Cross Marine Band (Fig. 16). It is $8\frac{3}{4}$ ft thick in Manton No. 6 Underground Borehole, 8 ft 5 inches in Wigthorpe Borehole, 6 ft 7 inches in Scaftworth Borehole, 6¾ ft in Manton No. 5 Underground Borehole and 5 ft in Forest Hill Borehole. Elsewhere the thickness does not exceed 19 in. Where these non-marine shales are thin they are generally dark grey, in places silty, and locally contain '*Estheria*', ostracods, fish debris and mussels. In the thicker sections the dark grey shale is generally confined to the top 18 in or so and is under- lain by grey shales with ironstone bands and scattered mussels. The fauna from this horizon is considered in conjunction with that of the Clay Cross Marine Band (p. 70).

In Barnby Moor, Jockey House, Mattersey, Ranby Camp, Ranby Hall, Scofton, Steetley No. 1 Underground, Stone and Twyford Bridge boreholes the Marine Band apparently rests on the Joan Coal or its seatearth (Fig. 16).   G.H.R.

(ii)   The Eastern Part of the District

A tentative correlation of the measures between the Low Silkstone Coal and the Clay Cross Marine Band is shown on Plate V. Gainsborough No. 1 Oil Bore was cored (see pp. 265–6) from just below the marine band to the Low Silkstone horizon (3642 ft to 4062 ft 4 in). Core recovery was poor, however, and virtually all the coal was lost. Parts of these measures were also cored in a few other oil bores.

The Low Silkstone Coal is recorded in the majority of the oil bores. The thickness is usually 2 ft or less, but exceptionally, as in a few of the bores at Gainsborough, it may be as much as 4 ft.

The measures between the Low and Top Silkstone coals are generally thicker (30 to 45 ft) in the northern oil bores than at Grove and South Leverton (15 to 30 ft). The Low '*Estheria*' Band has been cored at Walkering- ham No. 1 and at Gainsborough Nos. 1, 2, 3 and 4 oil bores. In Gainsborough No. 1 Bore it is only 2 ft 4 in thick; in the other cores it varies from 4 ft to more than 10 ft. The dark lower part of the band (p. 57) has been distinguished only in Gainsborough No. 2 Bore, where it is 4 ft 8 in thick. The fauna of the band is confined to '*Estheria*' *sp. nov.* and fish debris, except possibly in Gainsborough No. 2 Bore where *Anthracosphaerium sp.* occurs, either within or close above the band. Mussels occur above the Low '*Estheria*' Band in at least three of the five cores; in Gains- borough No. 4 Bore, and also in Walkering- ham No. 1 Bore, these included *Curvirimula subovata*. In the northern oil bores there are scattered records of a thin sandstone, 5 to 25 ft thick in cored sections, close below the Top Silkstone Coal. A thin mudstone band within this sandstone yielded ostracods in Gains- borough No. 2 Bore.

The Top Silkstone Coal is almost always present according to the oil-bore records. Thicknesses up to 5 ft are recorded, but few bores show more than 2 ft of coal, and the thicker sections should be treated with reserve. In a number of bores the seam con- sists of two leaves up to about 10 ft apart, and in some others the upper leaf may be represented by a seatearth. At Gainsborough No. 3 Oil Bore, 7 ft of sandstone occur between the two leaves of coal.

The measures between the Top Silkstone and Parkgate coals are thickest (almost 200 ft) at Gringley; in many of the other oil bores they are between 130 and 170 ft, but at South Leverton they are probably in the range 110 to 130 ft. At Gainsborough No. 1 Oil Bore they comprise seven cycles, and the Threequarters Coal (3970½ ft) and the Thorncliff Coal in two leaves (3922 and 3904 ft) can be recog- nized (pp. 265–6). Here the measures between the Top Silkstone and Threequarters coals are mainly argillaceous and constitute two cycles. The lower cycle, 25 ft thick, has a seat- earth at the top and contains numerous iron- stone bands. There is a fish-bed at the base, also recorded in Gainsborough No. 2 Oil Bore, where there are mussels in this cycle. The upper cycle yields mussels at Gains- borough No. 1, including *Curvirimula candela*, characteristic of the highest part of the *C. communis* Zone. More than half of the interval between the Threequarters and Thorncliff coals in this bore is occupied by sandstone, and the total thickness of these measures, here and at Gainsborough No. 5 Oil Bore, compares to the maximum seen in the western part of the district (p. 60). The two leaves of the Thorncliff Coal are very thin, and separated by mudstones. A poorly developed seatearth divides the measures between the Thorncliff and Parkgate coals into two cycles, both of which contain mussels and ostracods, and the upper, in addition, fish (p. 265).

Several uncored oil bores show one or more seams between the Top Silkstone and Parkgate coals, but it is not possible to correlate them with certainty. It is possible that the 36-in coal reported at 3780 ft in Walkeringham No. 2 Oil Bore could be the upper leaf of the Thorncliff, while the 18-in seam reported at 3850 ft in the same bore could be the Threequarters. The 12-in coal reported at 3750 ft in Beckingham No. 1 Oil Bore may be the Thorncliff or a leaf of that seam. There are no persistent sandstones below the Parkgate Coal, but one or more beds occur locally in the 100 ft of measures underlying the coal. e.g. at Gringley, Walkeringham and Grove.

The Parkgate Coal is recorded in all but a few of the oil bores. In the northern bores it is commonly 2 to 4 ft thick and may locally be as much as 5 ft in the Gainsborough Oilfield. Washouts (see below) may account for the seam's absence or below-average thickness in parts of this oilfield. At South Leverton the seam is generally thin, and recorded thicknesses of more than 2 ft at three oil bores appear dubious. At Grove the Parkgate may be absent or, as at Eaton Borehole (p. 261) split into two or more leaves.

The measures between the Parkgate and Deep Soft coals vary in thickness between 95 and 140 ft in those sections where the Deep Soft can be identified. The lowest values appear to be at Gringley, Morton and Beckingham. In Gainsborough No. 1 Oil Bore (p. 265) the bulk of these measures is formed by the Parkgate Rock, 90 ft thick, which rests directly on the Parkgate Coal. Uncored oil bores at Gainsborough suggest that this rock reaches a maximum of about 125 ft in the vicinity of No. 1 Bore, and thins rapidly towards the south and west. It is also present at Morton and Grove, and in some of the South Leverton oil bores. In oil bores to the west of the River Trent at Gainsborough, at Beckingham, Walkeringham and Gringley, and also in many of the South Leverton sections, there is little or no sandstone at this horizon, and here coals appear between the Parkgate and Deep Soft seams. The most prominent is a 12- to 24-in seam about mid-way in these measures, which may be the Deep Hard.

The Deep Soft Coal, like the Parkgate and Top Silkstone, can be correlated in the northern oil bores (Howitt and Brunstrom 1966, p. 557), although it has not been proved at Walkeringham nor in parts of the Gainsborough Oilfield, and the main leaf may be absent at Morton. The records suggest that it is very variable in thickness, up to perhaps as much as 6 ft. At Gainsborough it is absent in twelve bores in the centre of the oilfield but it apparently reaches its maximum thickness in peripheral bores. In a number of Gainsborough bores it evidently deteriorates into a split seam, and it may be washed out in places. At Grove and South Leverton the seam is apparently very thin or absent.

The measures between the Deep Soft Coal and the Clay Cross Marine Band are usually 90 to 115 ft thick, with no discernible regional trend. The maximum thickness of 140 ft at Morton is doubtful because the top leaf of the Deep Soft may be absent here. In many oil bores there is a prominent sandstone directly or close above the Deep Soft Coal; this is poorly developed at Grove, South Leverton and Walkeringham. At Gainsborough No. 1 Oil Bore this sandstone, 82 ft thick, is succeeded by a seatearth which is overlain by dark mudstone with mussels and ostracods. This seatearth/mudstone horizon is interesting because it provides a more prominent gamma-ray peak than does the Clay Cross Marine Band a short distance above. The characteristic gamma-ray peak attributed to the Clay Cross Marine Band by Howitt and Brunstrom (1966, p. 553) may in some oil bores in this area have been confused with the peak at this lower horizon. In some of the bores at South Leverton, the lower of two gamma-ray peaks similarly indicates a seatearth. This seatearth underlies a coal, possibly the Flockton Thick, 40 to 60 ft below the Clay Cross Marine Band. The same seam is perhaps 3 to 4 ft thick at Grove and 1 ft at Beckingham No. 1 Oil Bore, where it lies about 20 ft below the marine band.

At Walkeringham and Gainsborough thin sandstones occur close below the Clay Cross Marine Band. At Beckingham and South Leverton there is a thin seam at the horizon of the Joan, and this coal is apparently more prominent at Grove, although the reported thicknesses of 4 ft or more are unlikely.

C.G.G.

F

MIDDLE COAL MEASURES

The Middle Coal Measures are here divided for convenience at the Top Hard Coal and the Mansfield Marine Band. As in the case of the Lower Coal Measures above the Low Silkstone Coal, and for the reasons stated on p. 57, the western and eastern parts of the district are described separately.

CLAY CROSS MARINE BAND TO TOP HARD COAL

These measures are thickest—over 450 ft—in the north-west, and thinnest—between 200 and 250 ft—in the central and south-eastern parts of the district. The only marine horizon is the Clay Cross Marine Band at the base. There are several named coals, of which the Haigh Moor–Swallow Wood complex of seams and the Dunsil are economically of most importance. The latter unites with the overlying Top Hard Coal in parts of the district. Correlation of the coals is shown on Plates IV and V.

(i)   The Western Part of the District

The Clay Cross Marine Band is recorded in every cored borehole that has penetrated these measures. The marine fossils occur in dark grey, rather silty mudstone with ironstone bands. The band varies in thickness from 1¾ ft in Stone Borehole to 12 ft 7½ inches in Scaftworth Borehole, and is more than 4 ft over a wide area depicted on Fig. 16. This figure also shows the area in which mussels occur within the upper part of the marine band, a feature noted by Edwards and Stubblefield (1948, pp. 214–9). Mussels are commonly present in the marine band to the south of the present district (Edwards 1967, p. 92; Smith and others 1967, pp. 155–7), and it is in this direction that the marine band reaches its maximum thickness of about 15 ft.   G.H.R.

In this district the fauna of the Clay Cross Marine Band varies from a thin *Lingula* phase to a rich assemblage ranging through some 12½ ft of mudstones. An example of the latter development was well displayed in Scaftworth Borehole, where the following sequence through the marine band and associated measures was established (see also p. 296). The thin seatearth at the horizon of the Joan Coal is overlain by a 6 ft 7 in mussel-band containing *Spirorbis sp.*, *Anthracosia regularis*, *Naiadites sp. nov.* and *Geisina arcuata*; fish remains occur throughout but are more common in the upper layers of this unit, which also yield abundant *Lioestheria*. The lowest marine fossil occurs at 3307 ft 4 in, but the incursion appears to have made only a gradual advance, for the beds up to 3304 ft 1 in contain both non-marine and marine fossils in close association. The former are represented by small examples of *Anthracosia* and the ostracods *Carbonita*

and *Geisina*, and the marine element by *Lingula mytilloides*, *Myalina sp.*, *Hollinella* (*Praehollinella*) *claycrossensis* and *Paraparchites sp.* In the overlying beds fully marine conditions were established, as shown by the presence of *Lingula*, the pectinoids *Dunbarella* cf. *papyracea* mut. δ and *Posidonia* cf. *sulcata*, and cephalopods including *Anthracoceratites vanderbeckei*, *Metacoceras* cf. *cornutum* and orthocone nautiloids; this phase ranges through 4 ft 2 in and represents the acme of the incursion. In the succeeding 3 ft, the rich marine fauna disappears leaving only *Lingula*, associated in the uppermost 6 in with small forms of *Anthracosia*, providing evidence of the retreat stage of the incursion and the transition to a non-marine environment. The final replacement of *Lingula* by the mussel fauna occurs at 3294 ft 8 in, and in the succeeding 9 ft of measures a rich assemblage of *Anthracosia* and *Naiadites* indicates the return of typical non-marine conditions.

These and other features of the faunas of the Clay Cross Marine Band and associated measures as developed in the western part of the district are summarized in Fig. 16. This diagram shows that in general the richest fauna is found where the marine band is thickest. Thus the acme of the incursion is represented in the western and southern parts of this area by the goniatite facies, which is a continuation of the predominant facies of this band in the adjacent districts to the south-west (Smith and others 1967, pp. 155–7) and south (Edwards 1967, pp. 91–2). The band thins towards the east and the fauna shows a corresponding transition through the *Dunbarella* or pectinoid facies to a *Lingula* facies. Similarly, in the

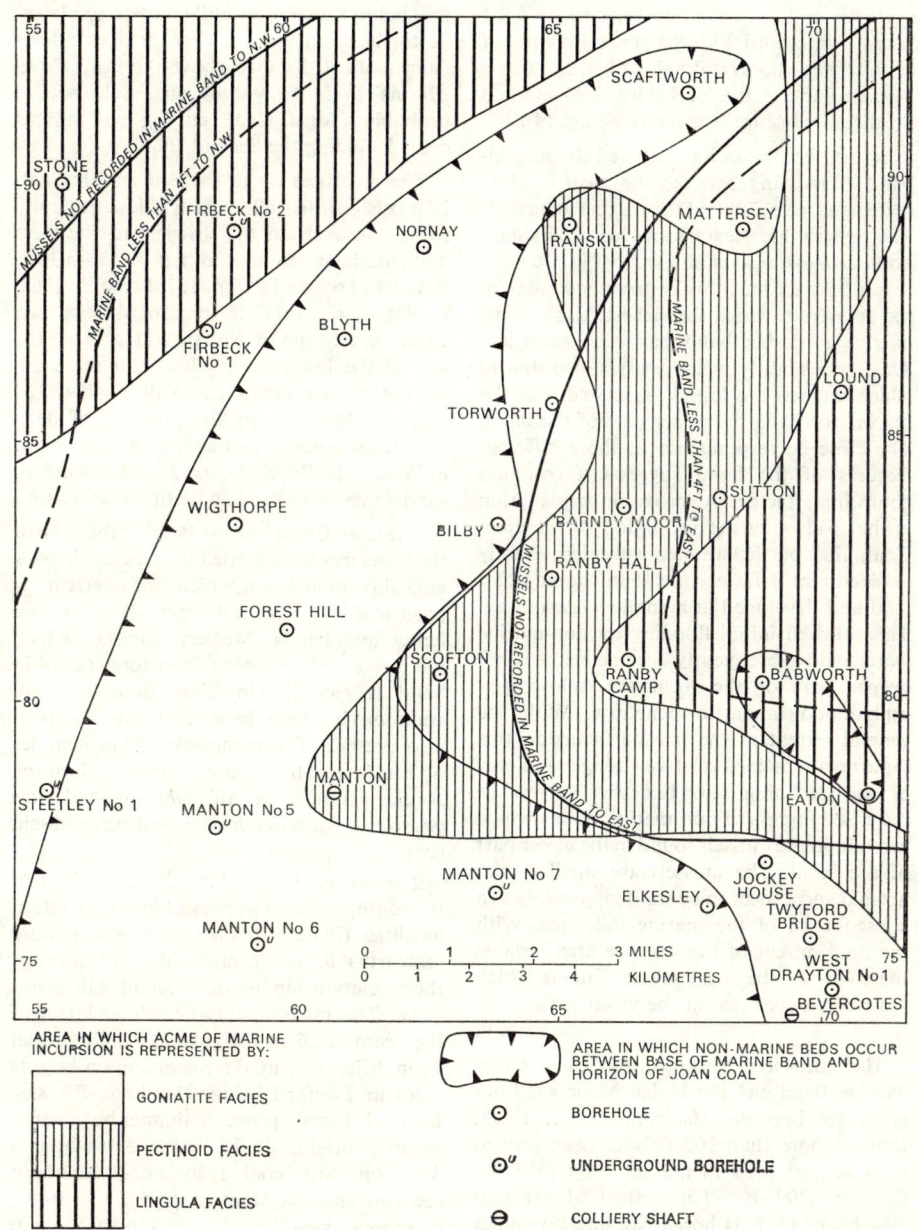

FIG. 16. *Plan of the Clay Cross Marine Band*

extreme north-west of the district the incursion is represented by a *Lingula* facies, although in this area the transition through a pectinoid facies has not been observed. This northern and eastern impoverishment of the fauna is in accord with the general picture of a thin and poorly developed Clay Cross Marine Band in the Yorkshire Coalfield and north Lincolnshire (Calver 1968, fig. 14).

Several boreholes have proved the mussel-band intervening between the Joan Coal or seatearth (see p. 70) and the marine incursion. The geographical extent of this preliminary non-marine phase is shown in Fig. 16, and it can be seen that this is broadly similar to the area from which the richest fauna in the overlying marine band has been recorded. This, or a slightly wider area, is also that in which mussels are found associated with the marine fossils in the upper part of the band. The close correspondence of these different features of the band suggests a common controlling factor. A probable explanation is that following the formation of the Joan Coal, this particular area subsided slightly earlier or at a faster rate than the adjacent land and developed as a shallow crustal sag. This allowed initial flooding by non-marine waters and subsequently became a somewhat deeper part of the local sea, allowing a cephalopod fauna to penetrate. With the general retreat of the marine incursion this same area remained relatively longer as a part of the declining sea, but was subject to increased ingress of non-marine waters which brought in the mussels found in the upper part of the band. The mussels are smaller than normal and appear to be stunted, possibly as a consequence of the marine influence. With the final retreat of the seas the area became the focus of the local mussel faunas which are richly developed in the vicinity.

M.A.C.

The measures between the Clay Cross Marine Band and the Haigh Moor Coal are generally between 160 and 190 ft thick, though more than 200 ft have been proved in a central area including Bilby (215 ft), Scofton (209 ft), Torworth (201 ft) and Wigthorpe (217 ft) boreholes, and boreholes in the east have proved less than 150 ft (147 ft at Eaton, 130 ft at Sutton and 102 ft at Lound).

The measures comprise up to ten cycles in the west, though eastwards only six or seven persist. Several of these cycles contain thin coals, but the only named seams are the Second Ell and Lidget, both of which attain thicknesses of 2 to 3 ft in small isolated areas (see below). In Forest Hill and Scofton boreholes thick siltstones and sandstones extend from a short distance above the Clay Cross Marine Band to within one cycle of the probable Lidget Coal, cutting out several coals including the Second Ell.

The measures up to the Second Ell Coal, in most localities 60 to 70 ft thick, generally form one cycle in the districts to the west and south (Eden and others 1957, p. 95; Smith and others 1967, p. 157; Edwards 1967, p. 92), but within the present district two cycles are common with a seatearth at the top of the lower one. Where this seatearth fails, the two cycles can still be detected in many localities by the presence of dark shales, commonly containing mussels (and, at Wigthorpe Borehole, ostracods), resting on sandstones near the middle of the measures.

The Clay Cross Marine Band at the base of the measures is succeeded by dark mudstones and silty mudstones which are overlain by paler and less silty mudstones with ironstone lenses and bands. Mussels, some of which are preserved in shelly ironstone (a 14-in band is recorded in Bilby Borehole), are common in these beds, and fish debris is also present. The composite fauna includes *Spirorbis sp.*, *Anthracosia beaniana*, *A. ovum*, *A.* cf. *phrygiana*, *A.* aff. *subrecta*, *Naiadites quadratus*, *Megalichthys sp.* and palaeoniscid scales.

The Second Ell Coal can be recognized in the south, and may be present in some western localities (Plate IV). Elsewhere two or more seatearths lie at or near this horizon, and their relationship to the Second Ell is not clear. The maximum recorded thickness of the seam is 26 in (brights $7\frac{1}{4}$ in, dirty coal $7\frac{1}{4}$ in, brights $3\frac{1}{2}$ in, dirty coal 3 in on brights 5 in) in Twyford Bridge Borehole. Elkesley Borehole nearby proved a thinner but cleaner section: brights 16 in, hards 2 in, brights 4 in on dirty coal 1 in. Other southern sections show a seam around 20 in thick, commonly including dirt. In the south-west, Manton Colliery No. 4 Shaft and Manton No. 5 and Steetley No. 1 underground boreholes proved less than a foot of coal.

A maximum of four cycles occur between the Second Ell and Lidget coals in Wigthorpe

Borehole, where there are four seatearths, two with associated thin coals. More generally only two seatearths or coals are present, and these appear to pass into a single seatearth or coal which, to the east (e.g. in Babworth, Eaton and Sutton boreholes), dies out or joins either the underlying Second Ell or the overlying Lidget. Mussels occur at several horizons in these beds, most persistently in the cycle immediately overlying the Second Ell Coal, where *?Anthracosphaerium affine* and *Naiadites* aff. *quadratus* have been recorded.

The Lidget Coal has been identified with a fair degree of certainty in the majority of boreholes, though generally it consists of only a few inches of coal and dirt, and in some localities only a seatearth is present. By far the thickest sections are in the north-west, where 35 in of coal and dirt are recorded in Firbeck No. 2 Underground Borehole (inferior coal 1½ in, dirt 5 in, inferior coal 2 in on mainly bright coal 26½ in), and 31 in of coal in Nornay Borehole, though only 8 in of dirty coal were recovered at the latter. Elsewhere the thickest sections are 23 in of coal and inferior coal in Manton No. 5 Underground Borehole and 17 in of coal in Wigthorpe Borehole.

The 60 to 80 ft of measures between the Lidget and Haigh Moor coals include up to three seatearths each of which may be overlain locally by a few inches of coal. The seatearths are impersistent and it appears likely that the two lower ones combine, and that the top one is a split from the overlying Haigh Moor. The outstanding feature of these beds is the persistent mussel horizon above the Lidget. The dark grey or black mudstone immediately above the coal is often found to be silty and rarely, as in Bilby Borehole, contains fish debris. It passes up into dark grey and grey mudstone with ironstone bands, containing *Anthracosia beaniana*, *A.* cf. *nitida* and *A. phrygiana*. Some of the mussels are preserved in pyrite, and bands of shelly ironstone are not uncommon. *Planolites montanus* has been found in the ironstone bands at several localities. G.H.R.

The Haigh Moor and Swallow Wood coals and the intervening measures. The type area of the Haigh Moor is the hamlet of that name about 4 miles NW of Wakefield. The Swallow Wood is named from a wood in Wentworth Park, ¾ mile SE of Wentworth Woodhouse (One-inch Sheet 87). The seams have been correlated with one another in the past (Mitchell and others 1947, p. 49), but a reassessment of sections in the Wakefield and Barnsley districts has shown that the Haigh Moor is lower in the sequence than the Swallow Wood. Both seams have been recognized in a number of underground boreholes immediately to the west of the East Retford district. The Haigh Moor is at or near the horizon of the Third Waterloo Coal of Derbyshire, and the Swallow Wood Coal is apparently the Second Waterloo Coal (Smith and others 1967, p. 161–4).

In the East Retford district the Haigh Moor and Swallow Wood coals are closely related, and in some sections form a complex of coals separated only by seatearths and bands of dirt (Fig. 17). They are close together in an area including Barnby Moor, Ranby Hall and Scofton boreholes near the centre of the district, and in Bevercotes Colliery No. 1 Shaft, and West Drayton No. 1, Twyford Bridge and Eaton boreholes in the south. Both the Haigh Moor and Swallow Wood are absent in Elkesley and Ranby Camp boreholes, where their horizons are represented by seatearths, and the seams are thin at Babworth and Jockey House. G.H.R., R.F.G.

The Haigh Moor Coal is divisible into three beds, although over most of the western part of the district either the lower two or the upper two are combined. The overall thickness of coal, everywhere brights or dirty brights, in the seam ranges up to almost 4 ft.

The bottom bed of coal is distinct to the north of Worksop, where it is generally 1 to 2 ft thick. A dirt band, up to several inches thick, occurs at various horizons within the bed. The dirt separating the bottom and middle beds of coal has a maximum proved thickness of 54 inches in Bilby Borehole, but decreases in the north to 10½ in at Stone and 2 in at Scaftworth. It disappears to the south of Worksop, where the two lower beds of coal unite.

The middle bed of coal is of similar thickness to the bottom bed. Several inches of dirty coal occur at the base in some sections, but the bed is otherwise dirt-free. To the south of Worksop the two lower beds combine to give a clean seam about 2½ ft thick. North of Worksop, except in Stone and

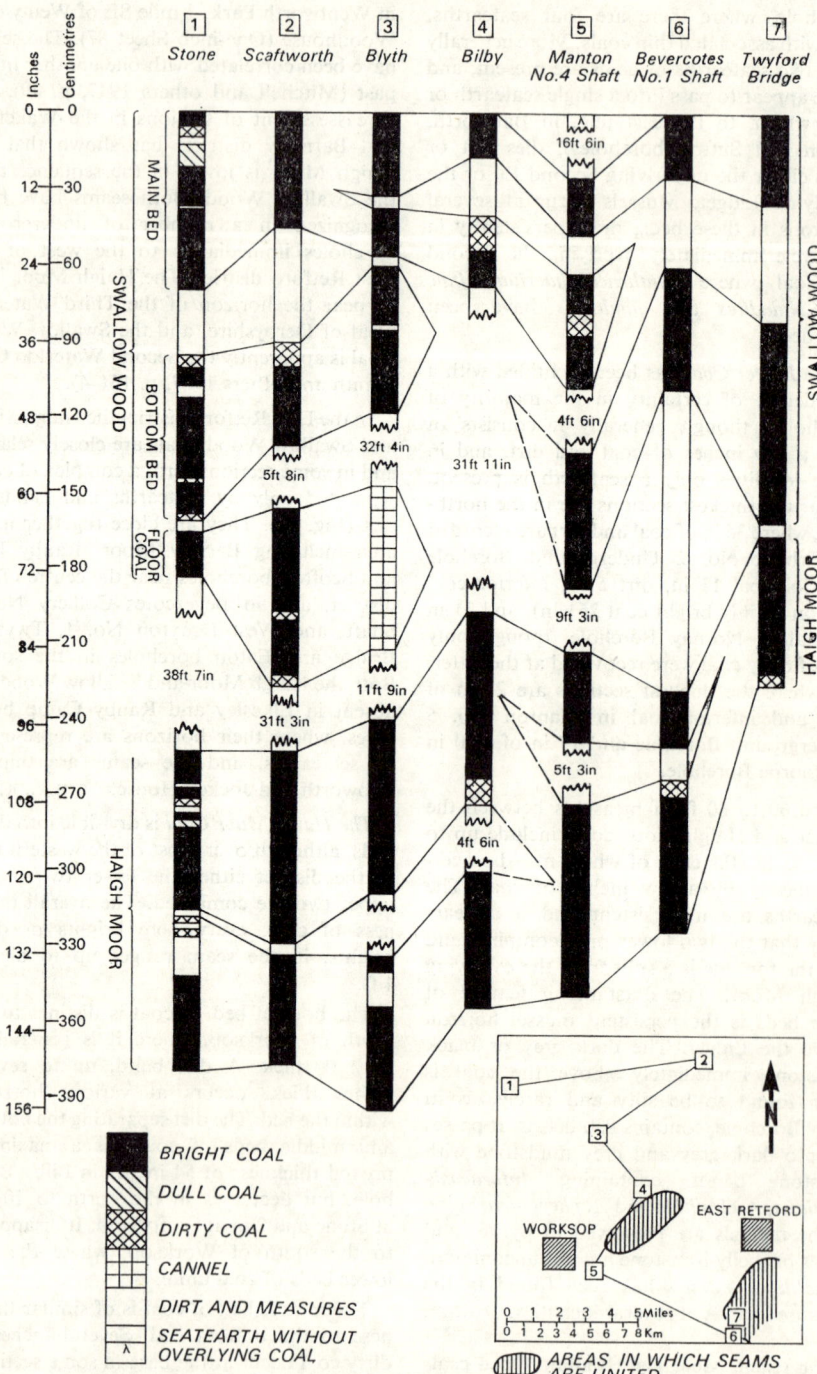

FIG. 17. *Sections of the Haigh Moor and Swallow Wood coals in selected boreholes and colliery shafts*

Wigthorpe boreholes, the middle bed is combined with the top bed to give $1\frac{1}{2}$ to 3 ft of coal. At Stone there is 1 in of intervening dirt and the top bed consists of $8\frac{3}{4}$ in of coal, dirty coal and dirt. At Wigthorpe the dirt is $10\frac{3}{4}$ in thick and the top bed consists of $7\frac{1}{4}$ in of clean coal. South of Worksop the top bed is a separate seam, up to 13 ft 8 in above the combined bottom and middle beds, except in the south-east, where in Eaton, Twyford Bridge and West Drayton No. 1 boreholes it is united with the lower beds to give an unsplit Haigh Moor seam just under 2 ft in thickness and close below the Swallow Wood (see below).

The middle and top beds of the Haigh Moor are washed out in Nornay Borehole; the top bed is washed out in Forest Hill Borehole and absent in Steetley No. 1 Underground Borehole where its horizon, represented by seatearth, is $28\frac{3}{4}$ ft above the lower part of the seam.

The measures between the Haigh Moor and Swallow Wood coals reach a maximum thickness of $47\frac{1}{2}$ ft in Nornay Borehole, where the upper part of the former seam is washed out. The measures comprise one cycle, and in the thicker sections are composed essentially of siltstone and sandstone.

The Swallow Wood Coal is divisible into three parts, the Floor Coal, Bottom Bed and Main Bed. Only in Stone Borehole do the three parts form a single seam, though even here several dirt bands are included in the overall thickness of 73 in. To the south of East Retford, where the Swallow Wood is combined with the Haigh Moor, the Floor Coal is probably absent (see below).

The Floor Coal can only be recognized with certainty in the northern boreholes where it varies in thickness from 7 in (at Stone) to 28 in (at Scaftworth). At Blyth it consists of $25\frac{1}{2}$ in of cannel; elsewhere it is mainly bright coal, with cannel at the top in some sections. A band of dirt or dirty coal is present near the middle of the Floor Coal in most sections.

The Bottom Bed of the Swallow Wood is up to 33 ft above the Floor Coal and at Scaftworth Borehole the intervening measures contain a thin band of inferior cannel which has yielded large examples of both *Anthraconaia salteri* and *Anthracosia* cf. *aquilina*. The Bottom Bed consists of up to about 2 ft

of bright or dirty bright coal, locally with thin dirt bands. It is not clear at Ranby Hall and Scofton whether the basal bed of the Swallow Wood, here immediately overlying the Haigh Moor, is the Bottom Bed or Floor Coal. At other boreholes nearby— Barnby Moor, Bilby, Forest Hill and Wigthorpe—neither of these beds is present. To the south of Worksop and East Retford that part of the Swallow Wood referred to the Bottom Bed (Fig. 17) may include the Floor Coal.

The Main Bed of the Swallow Wood, usually less than 3 ft above the Bottom Bed, is present everywhere, varying in thickness from a few inches to over 2 ft. It consists of bright or dirty bright coal, often with one or more dirt partings. In a number of sections the bed is split into two leaves up to 9 ft apart, and in some others is represented by only the upper of these leaves.

The combined Haigh Moor and Swallow Wood coals occur in two separate small areas, one north-east of Worksop and the other south of East Retford (Fig. 17). In the first area the Swallow Wood component of the combined seam is a mass of dirt and dirty coal 4 to 6 ft thick, while the Haigh Moor consists of two well developed beds, totalling over 4 ft of mainly bright coal, $1\frac{1}{2}$ to 2 ft apart. To the south of East Retford the Haigh Moor component consists of a single bed of bright coal just under 2 ft in thickness, while the Swallow Wood forms the major part of the combined seam. At West Drayton No. 1 Borehole the total thickness of the latter component is 29 in, including a 3-in dirt band. At Twyford Bridge and Eaton boreholes this dirt band thickens to $18\frac{1}{2}$ and 25 in respectively, with a total of 45 in of coal at the former locality.   R.F.G.

The measures between the Swallow Wood and Dunsil coals are commonly about 100 ft thick. The maximum known thickness of 152 ft 4 in is recorded in Stone Borehole, in the north-west, where the measures include a large proportion of sandstone and siltstone. Below average thicknesses are recorded in Ranby Camp Borehole (76 ft $1\frac{1}{2}$ in) and Manton No. 7 Underground Borehole (81 ft). Two named coals are included in these measures, the Waterloo Marker and the First Waterloo, neither of which is of economic importance within the district. The former seam, however, as its name implies, is useful in correlation, its roof containing an assemblage of non-marine fossils associated with bands of ironstone.

The Swallow Wood and Waterloo Marker coals are only 14 ft 2 in apart in Elkesley Borehole, where the intervening measures comprise grey mudstones and siltstones with ironstone bands and mussels. Their maximum recorded thickness is 39 ft 10 inches in Steetley No. 1 Underground Borehole, where dark grey shales with *Spirorbis* and mussels overlying the Swallow Wood give way upwards to grey mudstones and siltstones with ironstone bands and scattered mussels, which are in turn overlain by seatearth. The roof of the Swallow Wood is in general characterized by the presence of *Spirorbis*, which is locally abundant, particularly in 1 ft or so of dark grey to black shale at the base, where fish and *Carbonita humilis* are also recorded. Mussels are generally absent from these bottom few inches, but are numerous in the overlying shales and mudstones up to 20 ft above the coal. They include *Anthraconaia salteri*, *Anthracosia beaniana*, *A. phrygiana*, *Anthracosphaerium* aff. *affine* and *Naiadites quadratus* —a typical upper *A. modiolaris* Zone assemblage.

The Waterloo Marker Coal comprises only a few inches of coal, coal and dirt, or cannel, and in many localities is split into two leaves. The measures between the two leaves of coal are 20½ ft thick in Scaftworth Borehole, where they include mudstones, siltstones and sandstones, with much plant debris, but elsewhere they do not exceed 12¾ ft. In Ranby Hall Borehole they are 5 ft 8 in thick and contain mussels in the upper part. Mussels are also recorded at Scofton Borehole within the top leaf, which consists of cannel 5 in, black shale 5 in on inferior coal 1 in (see p. 305).

The cycle between the Waterloo Marker and First Waterloo coals is, in general, 20 to 30 ft thick, but is over 40 ft in Bilby, Blyth and Elkesley boreholes, where the measures include a large proportion of siltstone and sandstone, and amounts to 50 ft at Jockey House. The black, dark grey and grey mudstones in the lower part of the cycle contain mussels, some of which are preserved in one or more of the bands of shelly ironstone that are so widespread at this horizon. *Spirorbis* is found in some localities, but is not as common as in the roof measures of the Swallow Wood Coal. Fish debris is recorded

in isolated localities. A typical fauna was obtained from Sutton Borehole and included *Spirorbis sp.*, *Anthracosia beaniana*, *A.* cf. *nitida*, *A. phrygiana*, *N. quadratus* and scales of *Rhabdoderma sp.* and *Rhizodopsis sp.* In Eaton Borehole *Carbonita humilis* was also noted at this horizon, and *A. disjuncta* occurs in Nornay Borehole.

The First Waterloo Coal, named from a locality south of Chesterfield, splits into the Top and Bottom First Waterloo coals in the north-east of the Chesterfield (112) district (Smith and others 1967, pp. 165–7). The First Waterloo of the East Retford district may be the equivalent of only the Bottom First Waterloo. The Top First Waterloo is probably largely absent, but may be locally represented by a thin coal beneath the Dunsil, and is possibly the 16-in seam at 80½ ft included in the Dunsil in Steetley No. 1 Underground Borehole (Plate IV).

The First Waterloo is 2 ft or more thick locally in the south-west and here includes cannel in several sections, chiefly at the top. Elsewhere it exceeds 20 in only in Scofton Borehole, and in many localities is absent and represented only by a seatearth. The seam is split in some sections, and in Ranby Hall Borehole there are two thin leaves of coal nearly 40 ft apart (Plate IV). Bilby, a neighbouring borehole, shows only the upper of these two coals, underlain by over 25 ft of beds rich in plant debris.

The measures between the First Waterloo and Dunsil coals exceed 80 ft in Stone Borehole, where they consist almost entirely of siltstones and sandstones. The thickness ranges at random from this figure to a minimum of 13 ft 11 inches in Blyth Borehole. This sequence contains two or three seatearths and thin coals in some localities, and it is not clear which are representatives of the Top First Waterloo and which are splits off the Dunsil Coal.   G.H.R.

The Dunsil Coal takes its name from a hamlet near Chesterfield, in the north of which district the seam splits into two parts (Smith and others 1967, pp. 167–8). It is of little economic consequence and hardly recognizable farther north in the Sheffield district (Eden and others 1957, pp. 98, 102). In the East Retford district the seam is not always readily identifiable because it joins the overlying Blidworth Coal over much of the explored field and, furthermore, these seams join the Top Hard in the east (Plate VII and Fig. 18).

In places the Dunsil is a split seam consisting of an upper leaf of bright coal and a lower leaf composed largely of dirty coal. Sections showing this split are found in Firbeck No. 2 Underground and Blyth boreholes, where the dirty coal is well separated from the bright, and in Torworth and Nornay boreholes, where the two leaves are closer together (Plate VII). A local split within the bright coal is also apparent in some sections. The maximum thickness of the bright coal is in the south-east of the explored field, where 40½ in were found in Eaton Borehole and 27½ inches in Elkesley Borehole. To the west 30 in are recorded at Manton No. 4 Shaft and 26 inches in Scofton Borehole.

The measures between the Dunsil and Top Hard coals are thickest in the vicinity of Firbeck Main Colliery, where 85 ft have been proved in Wallingwells Borehole, and west of Manton Colliery Shaft, where 75 to 80 ft occur in Manton Nos. 5 and 6 underground boreholes. Thicknesses of 50 to 70 ft are found elsewhere throughout the south and west. The measures are thinner where the Dunsil is joined to the Blidworth, but even so, 42 ft occur at Blyth, 36¾ ft at Torworth and 35 ft 4 in at Nornay. East and north of these localities the measures thin out as the combined Blidworth and Dunsil joins with the Top Hard. Sandstones and siltstones are much in evidence in the western sections.

The Blidworth Coal occurs separately from both the Dunsil and Top Hard coals in the west of the district (Fig. 18). Here it generally consists of less than 1 ft of bright coal, and is locally absent, as in Forest Hill Borehole. Where combined with the Dunsil or Top Hard (pp. 76 and 79) it is up to 20 in thick (Plate VII).

Fossils are not common in these measures, but *N.* cf. *triangularis* has been recorded about 10 ft above the combined Blidworth and Dunsil in Bilby Borehole, and *Anthracosia* cf. *nitida* occurs at a similar level in Nornay Borehole.   G.H.R., R.F.G.

(ii)   The Eastern Part of the District

A substantial part of the measures between the Clay Cross Marine Band and the Top Hard Coal (Plate V) was cored in Gainsborough Nos. 4 and 6 oil bores (115 ft of the 320 ft present in No. 4; 151 ft of the 295 ft present in No. 6). Elsewhere few cores have been taken, and these are mainly of sandstone.

There are no cores of the Clay Cross Marine Band, which is identified by its characteristic peak on the gamma-ray logs (Howitt and Brunstrom 1966, p. 553).

The measures between the marine band and the Haigh Moor–Swallow Wood group of coals are more than 200 ft thick at Gringley and Walkeringham, and in some of the South Leverton oil bores. They are about 170 ft thick at Grove. At Beckingham and Gainsborough they range from 130 ft in some bores in the eastern part of the Gainsborough Oilfield, to 180 ft in bores west of the Trent. Most oil bores show no more than six cycles in these lower measures, but in a few bores at South Leverton up to 14 are recorded. This latter number seems excessive, since cored boreholes through these strata in the western part of the district show a maximum of ten cycles (p. 72).

In the south-western and central parts of the Gainsborough Oilfield there is a prominent sandstone close above the Clay Cross Marine Band. The whole (44 ft 10 in) of this rock was cored in Gainsborough No. 6 Bore, where it proved to be fine-grained, with beds 1 to 6 in thick. Its greatest reported thickness is 110 ft at Gainsborough No. 60 Bore, and it is about 40 ft thick at Beckingham, but elsewhere it is thin. Coals are recorded frequently in these lower measures, but most of them are thin. An 18-in coal 50 ft above the horizon of the Clay Cross Marine Band at South Leverton No. 3 Oil Bore may be the Second Ell. Two further seams at South Leverton, either of which may exceed 2 ft, and possibly reach 4 ft, occur about 60 to 100 ft higher, and one of these may be the Lidget. Three or four thin coals are present at about the same level at Walkeringham.

The Haigh Moor–Swallow Wood group of coals usually consists of two or three seams in 10 to 45 ft of strata, but it is impossible to name the individual seams. One or more of these coals may reach 3 ft in thickness. The total thickness of coal in the group appears to be between 2 and 6 ft in most oil bores, but there are exceptional records, probably exaggerated, of 10 ft or more in two of the bores at Gainsborough. Parts of the measures included within this group of coals were cored in Gainsborough Nos. 2, 4 and 12 oil bores, and in each case mussels were obtained. At Gainsborough No. 2 Bore the fauna (3380¼ ft to 3381 ft 8 in) included *Anthracosia aquilina* and *Spirorbis sp.* In No. 4 Bore the roof measures (3365 ft 2 in to 3369 ft 8 in) of the higher of two coals, 11 ft apart and each about 1 ft thick, have yielded *A. nitida, A. ovum, A. phrygiana, Naiadites quadratus* and *Spirorbis sp.* In No. 12 Bore, however, the mussels (3474 to 3481 ft), associated with fish debris, occur close below a seatearth.

The measures between the Haigh Moor–Swallow Wood group of coals and the Top Hard range in thickness in the northern oil bores from about 100 ft to just over 150 ft in some of the wells at Gainsborough, where a prominent oil-bearing sandstone is present. At Grove and South Leverton they appear to be between 80 and 100 ft thick. The largest number of cycles reported is seven (at South Leverton No. 5 Oil Bore), but most bores record one to four. Where there is no thick sandstone in these upper measures, coals are usually recorded, and the thickest seam, reported to be between 1 and 3 ft thick in a

number of bores, and occurring 55 to 100 ft below the Top Hard, may be the First Waterloo. The prominent sandstone in these measures is best developed in the north-eastern part of the Gainsborough Oilfield, where more than 100 ft were logged in three bores. Geophysical records, however, suggest that these thick sections may include some siltstone. The sandstone is also thick in a smaller area in the west of the same oilfield, and at South Leverton, where seven of the southernmost bores record 30 to 45 ft, often in two leaves. The whole of this sandstone was cored in Gainsborough Nos. 4 and 12 oil bores (23 ft 8 in and 73¾ ft respectively), and almost all (71 ft 1 in) in Gainsborough No. 6 Oil Bore. These cores suggest that the sandstone passes laterally and upwards into siltstone, and this may account for the rather abrupt thickness variations. The rock is usually fine-grained, but locally a few feet of it are medium- or even coarse-grained. It may be massive, or may contain micaceous partings which are dark or coaly in the lower part of the sandstone.

Cores of the silty beds between this sandstone and the seatearth of the Top Hard Coal in Gainsborough Nos. 6 and 43 oil bores have yielded only plant fragments. Since there is no prominent seam near the top of these measures, the Dunsil Coal must be either insignificant or, more probably, united with the Top Hard throughout this part of the district.   C.G.G.

TOP HARD COAL TO MANSFIELD MARINE BAND

These measures thin eastwards in the western part of the district from 850 to 550 ft, but in the east the thinning is evidently less marked, and the minimum thickness is probably about 400 ft in the extreme south-east. There are several marine bands, and the Manton 'Estheria' Band is also an important faunal horizon. The Top Hard is the major coal, and other seams of note are the Kent's Thick, High Hazles, Two-Foot and Clown. Plates V and VI give details of the correlation of these measures.

(i) The Western Part of the District

The Top Hard or Barnsley Coal is the most widely worked seam in the district, though it is still untouched over a large part of the explored portion of its field. It has been worked from Maltby Main, Harworth, Firbeck Main, Shireoaks, Steetley and Manton collieries. In the worked area the seam ranges in thickness between 2½ and 7 ft, with the thicker sections in the north. Except at Harworth the Yorkshire Coalfield name

"Barnsley" is used. Although the names "Top Hard" and "Barnsley" have been regarded as synonyms, the Top Hard is in fact only a part of the Barnsley Coal of the Barnsley area.

The true Barnsley is present only in the north-west of the district, where it has been worked in Maltby Main Colliery. A typical section, measured 1500 yd E of the shafts, is given below, together with the names of

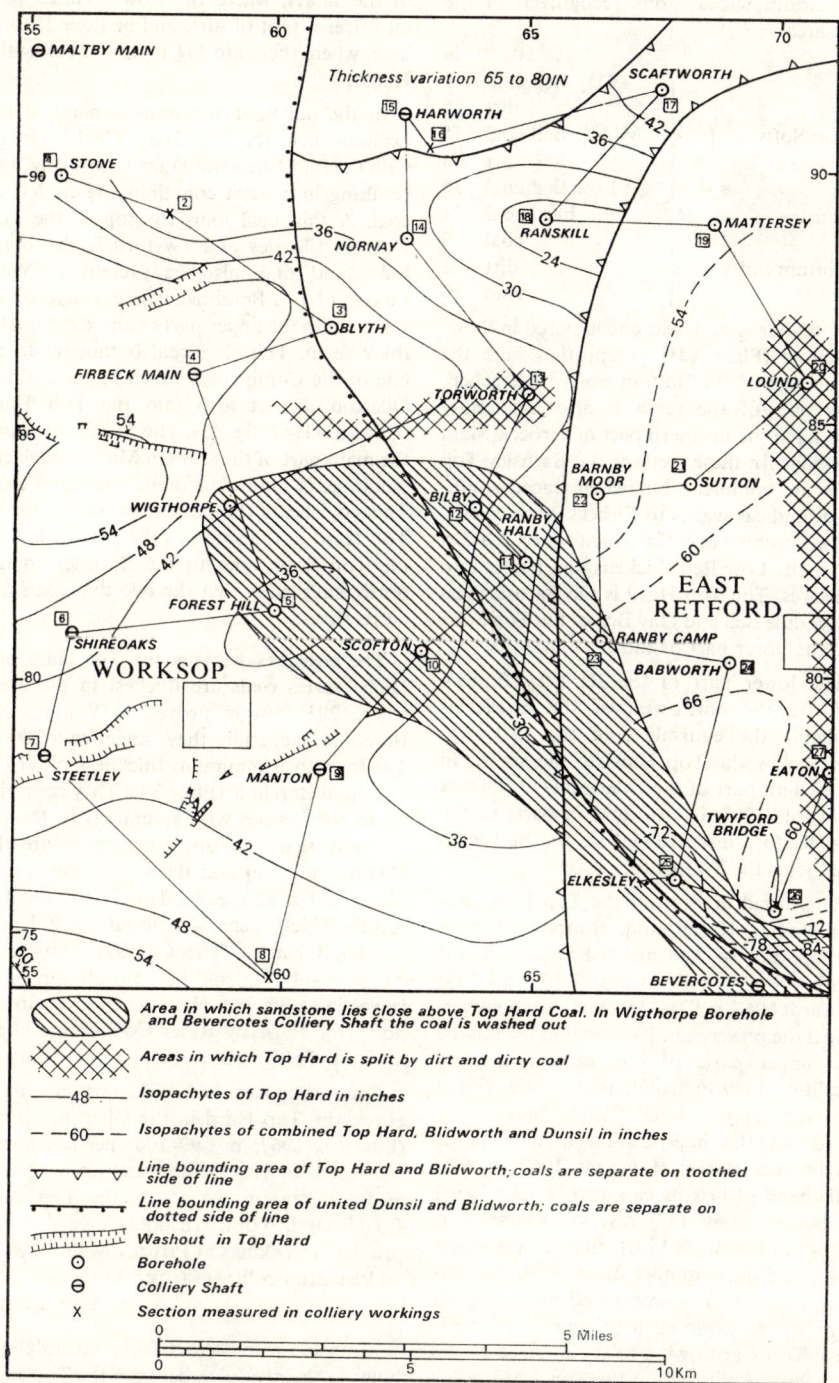

Area in which sandstone lies close above Top Hard Coal. In Wigthorpe Borehole and Bevercotes Colliery Shaft the coal is washed out

Areas in which Top Hard is split by dirt and dirty coal

———48——— Isopachytes of Top Hard in inches

— —60— — Isopachytes of combined Top Hard, Blidworth and Dunsil in inches

Line bounding area of Top Hard and Blidworth; coals are separate on toothed side of line

Line bounding area of united Dunsil and Blidworth; coals are separate on dotted side of line

Washout in Top Hard

⊙ Borehole

⊖ Colliery Shaft

X Section measured in colliery workings

FIG. 18. *Plan of the Top Hard (Barnsley) Coal showing the relationship between this seam and the Blidworth and Dunsil coals*

the various subdivisions recognized in the type area.

		in
Top Softs	Day Bed coal	9
	dirt	6
	Middle Bed coal	15
	dirt	5
	Low Bed coal	15
Hards	hard coal	21
Bottom Softs	coal	7
	dirt	2
	coal	10

A similar section was encountered in Stone Borehole (Plate VII) except that here the lower part of the Bottom Softs is detached. This part of the seam is also apparently missing in the northern part of Firbeck Main workings. In these sections the Barnsley Top Softs are complete, but when traced southwards and eastwards to Firbeck Main shafts, Blyth Borehole and Harworth shafts the dirt above the Low Bed thickens and passes into measures. The Top Hard is the Barnsley less the Middle Bed and Day Bed of the Top Softs and the lower part of the Bottom Softs.

The lower part of the Barnsley Bottom Softs in the north-west of the district is probably the equivalent of the Blidworth Coal below the Top Hard to the south. In the eastern part of the Ollerton (113) district (Edwards 1967, fig. 16) and in parts of the East Retford district (Fig. 18) the Blidworth Coal joins the Top Hard.

The broad division of the Top Hard into Top Softs, Hards and Bottom Softs is perhaps not so well marked as it is to the south (Smith and others 1967, p. 169; Edwards 1967, p. 97). Most sections, however, reveal the presence of hard coal in the middle and upper parts of the seam, and some localities show the subdivision of the Hards into the Upper Hards (Rifler) and Lower Hards by the interposition of a band of bright coal (Gees) (Edwards 1967, p. 97). This band of brights can also be recognized in the Barnsley (87) district, where it is known as the Black List (Mitchell and others 1947, p. 58). A number of sections, particularly those in the central and north-eastern parts of the explored field, show an incomplete Top Hard owing to the absence of the Top Softs. Splits within the Top Hard occur at Torworth Borehole, where 20½ in of dirt are recorded within the Hards, in Firbeck Main workings a mile or so east-south-east

of the shafts, where the Lower Hards pass into over a foot of dirt, and at Lound Borehole, where there are 17½ in of dirt below the Hards.

In the northern and eastern parts of the explored field the Top Hard Coal is united with the Blidworth and Dunsil coals (Fig. 18), resulting in a seam containing 4½ to 7 ft of coal. A thin coal joins the top of the Top Hard in Elkesley and Twyford Bridge boreholes and may also be present in West Drayton No. 1 Borehole, where core recovery was poor in the upper part of an exceptionally thick seam. This thin coal is thought to be one of the Comb coals which in parts of the Ollerton district also join the Top Hard (Edwards 1967, fig. 15). The 15-in coal above the main part of the seam in Manton Colliery workings is almost certainly the same coal. This coal bears the same relationship to the Top Hard in the south as the Middle Bed of the Top Softs does to the Barnsley in the north, suggesting that the two thin coals are correlatives.

The measures between the Top Hard and High Hazles coals are thickest in the west, where 283 ft were proved in Wallingwells Borehole. Generally they range from 200 to 260 ft, with a minimum thickness of 115 ft in Lound Borehole (Plate VI). They comprise up to eight cycles with several coals, though the only seams of any significance are the Comb (Coombe) and Barnsley Rider, up to about 120 ft above the Top Hard, and the Kent's Thick, generally about 50 ft below the High Hazles. The Comb is subject to complex splitting, and in the south one of its leaves joins the Top Hard (see above). In the north the Barnsley Rider Coal occurs at an horizon slightly above that of the Comb.

The prominent bed of sandstone noted above the Top Hard in the Ollerton district (Edwards 1967, pp. 99–100) persists north-westwards to Wigthorpe Borehole, where, as at Bevercotes in the south, the Top Hard Coal is washed out. Washouts have also been proved in workings at Firbeck Main, Steetley and Manton collieries (Fig. 18).

R.F.G., G.H.R.

Mussels are recorded from several boreholes in the measures between the Top Hard and Kent's Thick coals. The following details are taken from Nornay Borehole in the north-western part of the district. The roof measures

PLATE VII

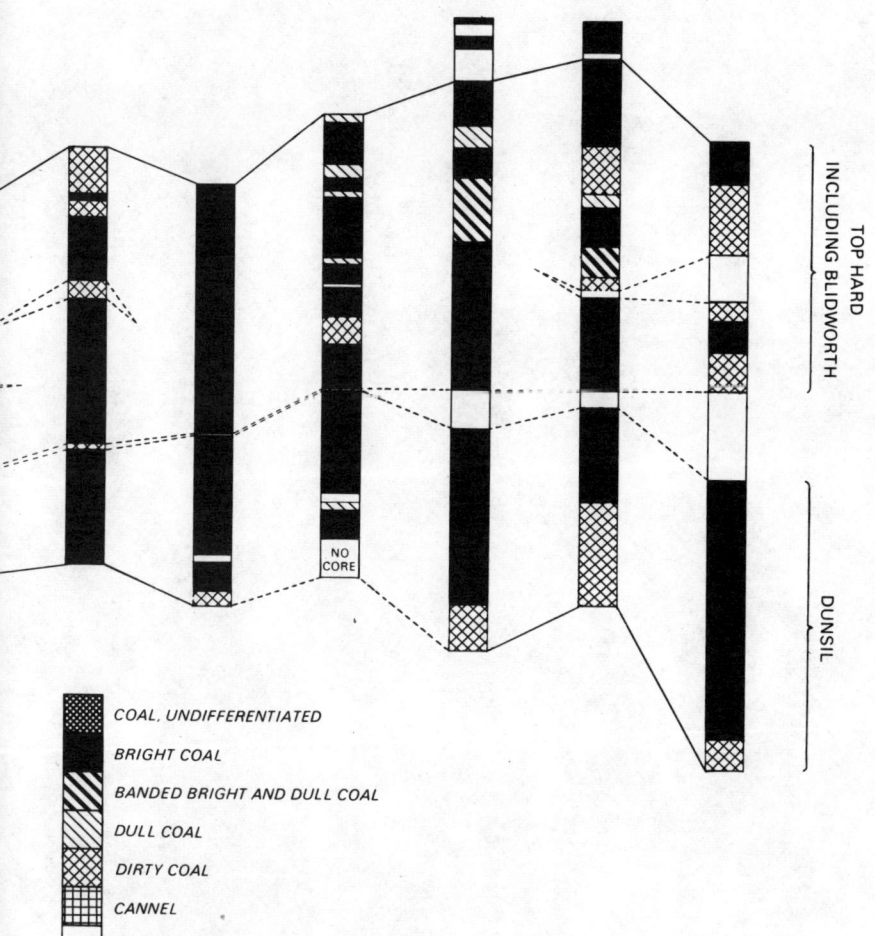

22 Barnby Moor
23 Ranby Camp
24 Babworth
25 Elkesley
26 Twyford Bridge
27 Eaton

TOP HARD INCLUDING BLIDWORTH

DUNSIL

NO CORE

COAL, UNDIFFERENTIATED
BRIGHT COAL
BANDED BRIGHT AND DULL COAL
DULL COAL
DIRTY COAL
CANNEL
DIRT AND MEASURES

of the Top Hard Coal contain *Anthraco-sphaerium?* and *Naiadites productus*; some 80 ft higher is a shelly ironstone with abundant *Anthracosia spp.* including forms intermediate between *ovum* and *lateralis*. Mussels are common above a thin coal lying 20 ft below the Kent's Thick, but the preservation is poor; the fauna includes *Spirorbis sp.*, cf. *Anthraconaia pulchella, Anthracosia simulans*, cf. *Anthracosphaerium propinquum* and *Naiadites* cf. *productus*. Elsewhere in the district this bed yields *Anthraconaia pulchella* and *Anthracosia caledonica*. This horizon is considered to be the equivalent of the widespread *A. pulchella* fauna occurring near the base of the Lower *similis-pulchra* Zone in several Pennine coalfields (Smith and others 1967, p. 173). M.A.C.

The Kent's Thick Coal generally comprises two beds of coal separated by a band of dirt up to about 2 ft thick. In the north the middle dirt is only 5 to 9 in thick in an area including Mattersey, Ranskill and Scaftworth boreholes, where the seam contains 3 to $4\frac{1}{2}$ ft of coal. A southern area in which the middle dirt is thin, or indeed absent, has also been proved; no dirt was recorded in the 4-ft coal in Jockey House Borehole or in the 3-ft coal in Bevercotes Colliery shafts, and in the nearby Elkesley Borehole only $3\frac{1}{2}$ in of dirt were found between an upper 21 in of coal and a lower $21\frac{1}{2}$ in of coal and dirty coal. In the western and central parts of the explored field the dirt band passes into measures up to 25 ft thick separating coals that are generally dirty, though at Maltby Main Colliery the upper coal is 35 in thick and includes only $\frac{1}{2}$ in of dirt. G.H.R.

In Sutton Borehole the following flora (determined by Dr. R. Crookall) was recorded from above the lower leaf of the Kent's Thick Coal: *Annularia* cf. *microphylla, Asterophyllites longifolium* var. *striata*, cf. *Bothrodendron minutifolium, Eupecopteris* cf. *volkmanni, Lepidodendron lycopodioides, L. ophiurus, Neuropteris pseudogigantea, N.* cf. *schuetzei, N. tenuifolia, Sigillaria ovata* and *S. ovata* var. *essenia*.

Several mussel bands are recorded in the measures between the Kent's Thick Coal and the bottom bed of the High Hazles Coal, the most prominent being within about 30 ft of the Kent's Thick. They occur in dark grey, commonly silty mudstone with fairly numerous ironstone bands, some of which are shelly. The composite fauna obtained from the district is: *Anthraconaia* aff. *librata, Anthracosia* cf. *aquilinoides, A.* cf. *faba, A. simulans, A.* cf. *variabilis, A. sp.* cf. *phrygiana, Anthracosphaerium* cf. *exiguum, Carbonita humilis* and fish remains. The shells are not uncommonly preserved in pyrite, and an association with vague 'worm' tracks is recorded at a few localities.

A slightly higher mussel-band contains *Anthracosia sp.* cf. *fulva* and *A.* cf. *planitumida*.

An example of *Palaeoxyris* cf. *helicteroides* was found at $2434\frac{1}{2}$ ft, 20 ft below the bottom bed of the High Hazles Coal, in Eaton Borehole. M.A.C.

The High Hazles Coal is of workable thickness only in the south-west of the district, where it is mined from Steetley and Whitwell (One-inch Sheet 100) collieries. In Steetley Colliery Shaft 36 in of coal are recorded, and in Shireoaks Colliery Shaft there is coal 6 in, warrant 7 in on coal 31 in. Elsewhere in the west of the district the seam deteriorates and splits, as in the neighbouring districts of Sheffield (100) to the west (Eden and others 1957, pp. 106–7) and Ollerton (113) to the south (Edwards 1967, pp. 100–1). The splitting results in the formation of two thin, and in most cases inferior, coals separated by up to 55 ft of measures. In Manton Colliery No. 4 Shaft, where the beds are only 15 ft apart, the top bed consists of 36 in of inferior cannel and the bottom bed of 24 in of coal. Elsewhere the bottom bed is generally thicker than the top bed, though it exceeds 2 ft in thickness only where dirt is included.

This bottom bed of the High Hazles is referred to, over the greater part of the Yorkshire Coalfield, as the Kent's Thin. In the north-west of the East Retford district the name "Kent's Thin" has been given to the 17-in coal on 4 in of coal and dirt at 2259 ft 11 inches in Maltby Main Colliery Shaft (Edwards 1951, pp. 57, 205), and to the 19-in coal on 6 in of dirt and inferior coal at 2254 ft 5 inches in Stone Borehole.

The measures between the two beds of the High Hazles consist largely of siltstones and sandstones, except in the east of the explored part of the field, where mudstones are prominent in many sections. The mudstones contain mussels, including *Anthracosia caledonica* and *Naiadites productus*, and fish debris; and in the few inches immediately above the bottom bed of coal vague 'worm' tracks are common, and are sometimes associated with banded light and dark sediment.

The measures between the High Hazles and Two-Foot coals are thickest in the south-west, where 178 ft were encountered in Steetley Colliery Shaft. Eastwards they thin to around 100 ft. Up to six cycles are present, but coals, apart from the Abdy, are generally thin or impersistent. The *Kilnhurst Coal*, approximately in the middle of the measures and generally the first seam below the Abdy, is the Kilnhurst New of Kilnhurst Colliery, and is thought to be equivalent to the upper part of the Stanley Main of the Wakefield (78) district (Edwards and others 1940, pp. 71–4). The seam is recognizable in the west and north, but is of poor quality and very variable thickness. The thickest section is coal 28 in, dirt and coal 6 in on coal 8 inches in Firbeck Main Shaft.

The Abdy (*Winter or Furnace*) *Coal* is a persistent seam lying approximately 30 ft below the Two-Foot Coal. The best section is the 29 in of brights recorded at Maltby Main Colliery. Southwards and eastwards the seam deteriorates markedly and consists of a variable thickness of inferior coal or coal and dirt.

The only prominent sandstone is that occurring locally between the Kilnhurst and Abdy coals, over 40 ft thick in Eaton Borehole.   G.H.R.

Mussels are found in some abundance above the top bed of the High Hazles in the east of the explored part of the field, e.g. in Scaftworth Borehole, from which *Anthracosia* cf. *aquilina*, *A. sp.* cf. *fulva* and *Naiadites sp.* were obtained. Worm-like tracks and burrows are also common in the roof of this coal. Mussels occur above some of the thin coals or seatearths in these measures, particularly in a band 10 to 20 ft above the High Hazles top bed, from which the following were collected in Eaton Borehole: *Anthraconaia sp.*, *Anthracosia caledonica?*, *A.* cf. *elliptica*, *A. sp.* cf. *fulva*, *A. lateralis* and *Anthracosphaerium sp.* In Nornay Borehole cf. *Anthracosia simulans* was recorded at this horizon. Some of the shells in the measures are preserved in pyrite and some in ironstone bands, which are locally numerous.

G.H.R., M.A.C.

The Two-Foot Coal is 2½ to 3 ft thick over the greater part of the western half of the district. Only at Maltby Colliery in the extreme north-west is it much thicker, and here the seam consists largely of cannel: bastard cannel 23 in, cannel 43¼ in on coal 4¾ in. Nearby, in Stone Borehole, only 21 in of coal were found. Less than 2½ ft are recorded also at Lound (26½ in) and Sutton (22½ in) and in the boreholes between Bevercotes Colliery and Eaton in the south, where the thickness ranges from 8 in at Jockey House to 21 in at Eaton. At Elkesley the seam appears to split into an upper 18 in of coal separated by 5 ft 5 in of measures from a lower 8-in coal. The seam generally consists of bright coal which commonly contains pyrite in its top few inches. Cannel occurs in the top few inches of many of the western sections.

The measures between the Two-Foot and Clown coals range in thickness from 50 ft at Jockey House Borehole in the south-east to 139 ft at Wallingwells Borehole in the west. They generally comprise three cycles, though locally there may be five or more. The lowest cycle contains the Two-Foot Marine Band at its base, and is overlain by the Mainbright Coal. The uppermost cycle contains the Manton 'Estheria' Band at its base. At and north of Bevercotes, where the measures are thinnest, the upper cycle has wedged out locally, leaving only two cycles.

The Two-Foot Marine Band occurs in the lower part of the dark grey to black, generally silty, mudstone or shale that overlies the Two-Foot Coal. It rests directly on the coal, except possibly in Bilby and Scofton boreholes, where no marine fossils are recorded in the bottom 4 in of mudstone and 2 in of pyritous shale respectively, and at Maltby Main Colliery (see below). It is up to 8 ft 7 in thick, the maximum being in Stone Borehole, in the north-west. The same thickness has also been recorded in Maltby Main Colliery (Dyson 1911, p. 610), but re-examination of the specimens from this locality reveals marine fossils in the top 4 ft only, the beds down to the coal containing only fish scales. Thicknesses of 4 to 7 ft, however, are more general for the district, and marine fossils are not numerous except at the base of the band, where *Lingula* is generally abundant. Pyritized plant debris is common and 'fucoids' are recorded in some sections. The marine band is only a few inches thick in Barnby Moor, Eaton, Elkesley, Jockey House, Ranby Camp, Torworth and Twyford Bridge boreholes and in Manton Colliery No. 4 Shaft. The dark mudstones above these thinner sections contain much pyritized plant debris, suggesting that they may be equivalent to shales with marine fossils at other localities.

G.H.R.

The Two-Foot Marine Band shows a variable faunal development in the district, the richest assemblage having been obtained from Maltby Main Shaft, where the following are recorded: *Lingula mytilloides*, *Euphemites anthracinus*, *Strobeus sp.*, *Dunbarella carbonarius*, cf. *Edmondia goldfussi*, goniatite or coiled nautiloid, *Hollinella sp.* and fish debris (Edwards 1951, p. 89). Elsewhere the fauna is largely represented by *Lingula*, except in the neighbourhood of Scofton and Ranby Hall, where evidence of the *Myalina* facies is provided by the occurrence of foraminifera, *Myalina compressa*, *Hollinella sp.* and *Paraparchites sp.* in addition to *Lingula*. The band is very thin or absent in the east of the explored part of the district and no records are available for the extreme south-west.

M.A.C.

The dark grey mudstones above the marine band in most sections gradually pass upwards into paler mudstones, which eventually give way to siltstones and sandstones beneath either the Mainbright Coal or one of the lower subsidiary seatearths. The mudstones, particularly the darker ones, contain sporadic *Anthracosia sp.* and *Naiadites sp.*, and fish are recorded in some sections. Pyrite is fairly common.

The Mainbright or Meltonfield Coal lies 15½ to 70 ft above the Two-Foot Coal. The thickest beds of coal occur in the south and extreme north-west. In the south, 37 in of coal were recorded in Bevercotes No. 1 Shaft and 30 in of coal in Eaton Borehole. The thick sections in the north-west are in Maltby Main Shaft and Stone Borehole, where there are respectively 53 and 52 in of coal, dirty coal and dirt. Elsewhere the seam is split into two or more beds of coal with, in places, further division indicated by seatearths. In the east of the explored part of the field the division of the seam is such that at Mattersey and Lound it is uncertain which seatearths and thin coals represent the Mainbright.

In Scaftworth Borehole mussels are common in the roof of the seam believed to be the lower leaf of the Mainbright at 2473 ft 7 in. They include *Anthraconaia librata*, *Anthracosia* cf. *aquilina*, *A. aquilinoides*, *A. concinna* and *Naiadites sp.*, a fauna similar to that normally found in the beds overlying the Two-Foot Marine Band (Smith and others 1967, p. 179). The Mainbright Coal is overlain by dark to black shale or mudstone commonly containing fish debris and rarely, as in Eaton Borehole, *Lioestheria*. The dark beds pass up into paler mudstones which generally contain numerous ironstone bands. Mussels occur only rarely, but *Anthracosphaerium* aff. *propinquum* is recorded in Scofton Borehole, and *Anthracosia* cf. *atra* occurs in Scaftworth Borehole. The upper part of this cycle generally contains siltstone and sandstone.

The Manton 'Estheria' Band was named by Mr. R. F. Goossens, who examined Manton No. 4 Shaft during its sinking in 1946–52. It occurs at the base of the cycle immediately below the Clown Coal, resting on a seatearth or thin coal. It is up to 33 in thick, but has a very patchy distribution, having been found in only nine of the 33 shafts and boreholes penetrating these measures. Most of the localities are in the south-west—Manton Colliery No. 4 Shaft and Forest Hill, Scofton, Wallingbrook Wood, Welbeck Abbey and Wigthorpe boreholes. Three other localities, Babworth, Ranby Camp and Eaton boreholes, are in the south, and only Blyth Borehole has proved the horizon in the north. *Lioestheria*, and the fish debris commonly found with it, occur in black and silty shale, either resting on the seatearth or coal, or separated from it, as in Wigthorpe Borehole, by up to 15 in of unfossiliferous black silty shale and dark grey siltstone.

The cycle containing the Manton 'Estheria' Band at its base has a maximum thickness of 32 ft 10 inches in Wigthorpe Borehole, but is absent in the Bevercotes area, where the underlying seatearth appears to have joined that of the Clown Coal. The Manton 'Estheria' Band is overlain by mudstone, much of which is dark and which generally contains ironstone bands and nodules. Sparse mussels have been found here, including *Naiadites angustus* in Scofton Borehole. The mudstone becomes paler upwards, and in places passes into siltstone.

In Welbeck Abbey Borehole a 3-in band of pale cream dolomite is recorded near the top of the mudstones, and beneath it is a 1-in grey tonstein. The tonstein is 797 ft 11 in from the surface and $14\frac{1}{4}$ ft above the base of the Manton '*Estheria*' Band. A tonstein in a similar position is recorded in Bilby Borehole at a depth of 2001 ft 8 in. In this case the tonstein is 4 in thick and lies close below the seatearth of the Clown Coal (Eden and others 1963, p. 54).

The **Clown Coal** is $2\frac{1}{2}$ to 3 ft thick over much of the western part of the district. It is, in general, thickest in the south, but Welbeck Abbey Borehole proved only 25 in, and in the Bevercotes area there are only 12 in of coal in West Drayton No. 1 Borehole, 8 inches in Jockey House Borehole and 18 inches in Bevercotes No. 1 Shaft (though No. 2 Shaft encountered 34 in). At Shireoaks Colliery an exceptional thickness of 52 in is recorded. The seam is worked here, and though the thickness proved in the shaft does not persist, sections of up to 45 in of coal are recorded more than 2000 yd to the south-east. North-eastwards from Wallingwells and Torworth the coal thins markedly, and in Scaftworth, Mattersey and Lound boreholes it is absent. The seam consists largely of brights, with bands of hard coal in the lower part. In Eaton and Ranby Camp boreholes the main part of the seam is overlain by 12 in and 14 in of seatearth on which are a further 3 in and 4 in respectively of coal.

The measures between the Clown Coal and the Haughton Marine Band are 102 ft thick in Wallingbrook Wood Borehole and Manton No. 4 Shaft, and thin irregularly eastwards to about 50 ft. Their thickness is not determinable in many sections in the west, where the Haughton Marine Band has been washed out (see below). There are up to six thin coals or seatearths in these measures, but three or four are more usual.

The Clown Coal is overlain by dark grey and black shale and mudstone up to 10 ft thick containing a mussel-band with typical *Anthracosia atra* associated with *Carbonita humilis* and scales of *Rhadinichthys sp.* and *Rhizodopsis sp.*

The Clown Marine Band, developed locally in the roof of the Clown Coal in neighbouring districts (Edwards 1967, pp. 103–4; Smith and others 1967, p. 182; Eden and others 1957, p. 114), has been found only in Bilby Borehole, where *Lingula mytilloides*, associated with fish debris including *Rhizodopsis sp.*, is recorded 4 in above the coal.

At most localities a seatearth overlies the dark mudstones above the Clown Coal, and is in places, as at Eaton and Wallingwells boreholes, topped by a few inches of coal. In the south-west this thin coal passes into cannelly shale, which in Wallingbrook Wood Borehole has no underlying seatearth. The horizon has not been traced in the north of the district, and the evidence afforded by Nornay and Ranskill boreholes suggests that it dies out rather than joins the underlying Clown seam. The roof of the thin coal generally consists of dark mudstone locally containing *A. atra*. The mudstone becomes paler upwards, and the cycle is terminated by a distinctive seatearth consisting of variegated brown, fawn or green mudstone which, almost everywhere, is rich in sphaerosiderite in its lower part. This seatearth is commonly found about one-third of the distance up from the Clown to the Swinton Pottery, and between it and the latter seam are a further one or two cycles which, in places, contain coals and sporadic *A. atra* and *N. angustus*.

The Swinton Pottery Coal, the first seam beneath the Haughton Marine Band, is almost invariably thin, and contains much dirty coal. It is best developed in the north-east at, for example, Scaftworth Borehole: coal 10 in, dirt 4 in, dirty coal $4\frac{1}{2}$ in on coal $13\frac{1}{2}$ in. Many sections show two or three seatearths, with thin coals above them, at this horizon, and it is probable that they result from the splitting of the Swinton Pottery.

In many places the Swinton Pottery and the Haughton Marine Band are separated by measures consisting mainly of dark silty shale, which locally contain fish debris and mussels and rare pyritic worm tracks. These measures are generally a foot or so thick, but they thicken in the south-west to 15 ft at Manton No. 4 Shaft. In the north, at Maltby Main Colliery, the measures between the Swinton Pottery and the Haughton Marine Band are nearly 26 ft thick and consist largely of sandstone and siltstone. In Bilby, Blyth, Forest Hill, Nornay, Torworth, Wallingwells and Wigthorpe boreholes and Firbeck Main and Harworth shafts, sandstone at the horizon of the Oaks Rock cuts out the Haughton

Marine Band, and in some cases extends downwards almost to the Clown Coal.

The Haughton Marine Band was first recorded in the district by Dyson (1911, p. 611) at 1888 ft 1 inch in Maltby Main Shaft. Since that time it has been recorded in Manton No. 4 Shaft and in many boreholes. It consists of dark mudstone, silty in its upper part, with thin ironstone bands and lenses that are locally abundant. The maximum thickness of 26 ft 8½ in is recorded in Ranby Camp Borehole, and the band is probably more than 15 ft thick where present in the western part of the district. It is cut out by faulting in Ranby Hall and Stone boreholes and has been washed out in several sections (see above). Marine fossils, though fairly numerous at some levels in the lower part, are, in general, thinly scattered throughout the marine band, which extends in many localities up to the base of the overlying seatearth, and comprises the greater part of one cycle.   G.H.R.

The typical fauna from the district comprises: sponge spicules, *Lingula* cf. *elongata*, *L. mytilloides, Orbiculoidea* cf. *nitida*, bellerophontoid indet., pleurotomarioid indet., *Polidevcia sp., Dithyrocaris sp., Serpuloides stubblefieldi* and cf. *Tomaculum sp.* In addition there are collections from Maltby Main Shaft referred to this horizon which are more typical of the Mansfield Marine Band; they include *Myalina compressa, Pernopecten carboniferus, Euphemites anthracinus, Naticopsis sp., Straparollus sp. nov., Huanghoceras costatum* and *Solenochilus* cf. *cyclostomus*.
.M.A.C.

The measures between the Haughton and Mansfield marine bands are thickest in Wallingbrook Wood Borehole in the south-west, where 128 ft are recorded. They thin markedly eastwards and have a minimum proved thickness of 43 ft in Sutton Borehole. Five cycles are generally present, though six are found in some places and only four in others. A few inches of coal occur in some of the cycles, and many of the seatearths show shades of fawn and green and are rich in sphaerosiderite. Mudstones are the main constituents of these measures, except locally,

where there is a thick development of sandstone and siltstone. At Harworth and Bilby the succession from the Clown to the Mansfield Marine Band consists largely of sandstone, and individual cycles are unrecognizable. Mussels occur only sparingly in these measures, and are not generally present in more than two cycles at any one locality. In Scaftworth Borehole the thin mussel-band in the cycle above that containing the Haughton Marine Band has yielded *A. atra?* and *Naiadites sp., Carbonita humilis, C. scalpellus* and fish remains.

The Sutton Marine Band overlies the second seatearth above the Haughton Marine Band and has been identified at nine localities, including the type locality, Sutton Borehole (Edwards and Stubblefield 1948, p. 225). It is 1½ in to about 4 ft thick and consists of dark to black mudstone or shale. In many localities it is directly overlain by a seatearth, which in some sections where the marine band is absent (e.g. Babworth and Jockey House boreholes) appears to have joined the seatearth below, so that the complete cycle is missing. The marine band has not been recorded in the north-west of the district.   G.H.R.

In the Sutton Marine Band *Orbiculoidea* cf. *nitida* is characteristically associated with *Lingula mytilloides* and fish remains. An unusual development was found in Scaftworth Borehole, where sparse foraminifera occur with elongate *Anthraconaia*, the latter resembling the *A. spathulata* from the Shafton Marine Band (p. 91). The occurrence of similar forms of *Anthraconaia* has been noted in the retreat stage of several marine incursions of the Middle Coal Measures, and is thought to indicate a facies transitional from marine to non-marine (Calver 1968, pp. 54–5).

A mussel-band occurs in several boreholes up to 20 ft above the Sutton Marine Band. In Eaton Borehole this band contains *Anthracosia* cf. *atra, A. rubida, Naiadites sp.* and fish remains; in Scaftworth and Sutton boreholes respectively *N.* aff. *alatus* and *Anthracosphaerium sp.* were collected.

M.A.C.

(ii) The Eastern Part of the District

Very few cores have been taken of the measures between the Top Hard Coal and the Mansfield Marine Band (Plate V), but the lowest 109 ft, covering almost the whole interval between the Top Hard and Kent's Thick coals, were cored in Gainsborough

G

No. 43 Oil Bore. There are no cores of the marine bands but the positions of two of them, the Two-Foot and Haughton, can be inferred from peaks on the gamma-ray logs.

The Top Hard Coal, probably united with the Dunsil (p. 78), is recorded in almost all of the oil bores. In many bores it is reported to be between 2 and 4 ft thick, and geophysical evidence suggests that it rarely exceeds 4 ft. A few logs give much greater thicknesses, e.g. 10 ft in Gainsborough No. 29 Oil Bore, but these should be treated with reserve. The seam is generally thin at South Leverton, where only two bores record more than 2 ft of coal.

The measures between the Top Hard and Kent's Thick coals range in thickness between about 100 and 160 ft. Except in certain bores at Gainsborough, where thin sandstones occur a short distance above the Top Hard, the measures are argillaceous and usually comprise three to six cycles. A few oil bores, at Morton and Gainsborough, show three or four coals 25 to 100 ft above the Top Hard, but most sections have only two. The lower of these is said to range up to 5 ft in thickness, but it makes little show on some electric logs, possibly because it is a split seam. In Gainsborough No. 43 Oil Bore it consists of two leaves separated by 14 in of mudstone. It is insignificant at Grove and South Leverton. These two coals are 25 to 50 ft apart, and the higher, 45 to 100 ft above the Top Hard, and possibly the equivalent of the Barnsley Rider (p. 79), is normally 2 to 3 ft thick, locally reaching 4 ft. In Gainsborough No. 43 Bore a seatearth between this coal and the Kent's Thick is overlain by mudstone with fish remains. The remainder of the core has yielded only plant debris.

The Kent's Thick Coal is recorded in most oil bores, usually as a seam 1 to 2 ft thick. At Walkeringham, Gainsborough and South Leverton there is evidence in some bores of a double seam, the two leaves being 10 to 25 ft apart.

The precise interval between the Kent's Thick Coal and the Two-Foot Marine Band is not known in this part of the district. It is about 240 ft at Gringley, thinning to less than 150 ft at Gainsborough, and is apparently 200 to 230 ft thick at Grove and South Leverton. Most oil bores record three to seven cycles in these measures.

At Grove a fine- to medium-grained sandstone, 47 ft thick at No. 1 Oil Bore occurs close above the Kent's Thick. In most oil bores one or two coals, each 1 to 3 ft thick, are recorded in the 120 ft of measures overlying the Kent's Thick. Where both seams are present, they are usually less than 40 ft apart. Reports of coals up to 6 ft thick at this level in a few bores at Gainsborough are dubious. These coals are thought to be the top and bottom beds of the High Hazles (p. 81). At Grove and South Leverton an additional higher seam, up to 3 ft thick, occurs within 30 ft of the top bed, and may represent the Kilnhurst. A prominent sandstone, locally more than 100 ft thick, occurs above the High Hazles horizon in the eastern part of the Gainsborough Oilfield; in twelve bores it cuts out the top or both beds of the High Hazles. At Grove and South Leverton the Winter and Two-Foot coals can be recognized, but the former is thin, and reported thicknesses of 5 to 6 ft for the Two-Foot in some bores at South Leverton are probably greatly exaggerated.

The Two-Foot Marine Band has not been proved, and the peak attributed to it on gamma-ray logs is often poor.

The measures between Two-Foot and Mansfield marine bands appear to vary in thickness from about 160 to about 240 ft. The greatest thicknesses are probably in the north at Walkeringham and Beckingham, but there is no discernible regional trend. Up to ten cycles are reported in these measures.

The Mainbright Coal is recorded in many oil bores within 50 ft of the Two-Foot Marine Band. Except locally (e.g. at Grove), where the seam may reach 3 to 4 ft, it is unlikely to exceed 2 ft in thickness. In a few oil bores it is reported to be split into two or more leaves. The Clown Coal can be recognized at Grove, where it may be as much as 4 ft thick. The Clown Marine Band may perhaps be identified in the gamma-ray logs of some bores at Grove, South Leverton and Walkeringham by a peak 30 to 70 ft above that of the Two-Foot Marine Band. The Haughton Marine Band, the base of which lies 45 to 95 ft below the Mansfield Marine

Band, makes a well-marked peak on many gamma-ray logs, but there is no evidence that the Sutton Marine Band is present. A coal up to about 3 ft thick occurs close below the Mansfield Marine Band in many oil bores. C.G.G.

MANSFIELD MARINE BAND TO TOP MARINE BAND

These measures are more than 700 ft thick in the north-west, but in the eastern part of the district are everywhere less than 450 ft, with a minimum of about 300 ft in the extreme south. They are incomplete in the south-west (Fig. 9) and locally near Gainsborough, where the sub-Permian unconformity cuts down below the Top Marine Band. In addition to the Mansfield and Top marine bands, the measures contain the *Edmondia* Band and the Shafton Marine Band: '*Estheria*' occurs at several levels above the former. There are several named coals, but none is of economic importance. Plates V and VIII give details of correlation.

(i) The Western Part of the District

The Mansfield Marine Band, where complete, ranges in thickness from 12 ft 10 in at Lound to over 30 ft in Babworth, Bilby and Torworth boreholes. At Ranby Camp Borehole the upper part of the marine band has probably been washed out, leaving only 8¼ ft of marine sediments, and at Barnby Moor Borehole all but 2 ft 11 in of the marine shales have been faulted out.

The marine fossils are found mainly in dark shale or mudstone, which, in general, becomes paler in colour upwards, in which direction also there is commonly an increase in the number of ironstone nodules and bands. The Mansfield 'cank' (see below) is found in the lower part of the mudstones. In the north this rests on, or is close above, the seatearth or coal at the top of the underlying cycle; in the south up to about 5 ft of dark, in places silty and sandy, mudstone, is interposed between the top of the previous cycle and the 'cank'. In some sections marine fossils have been recorded throughout the mudstones below the 'cank', but more often the lowest few inches contain only fish debris, which is commonly associated with pyrite and carbonaceous matter. In Eaton Borehole no fossils were found in the 2 ft 7 in of mudstone lying beneath the 'cank'.

The Mansfield 'cank' (Dunham *in* Edwards and Stubblefield 1948, pp. 251–3) varies irregularly in thickness and is not as persistent as in adjoining districts (Eden and others 1957, p. 117; Edwards 1967, pp. 106–7; Smith and others 1967, pp. 185, 187). The maximum recorded thickness is 36 inches in Eaton Borehole, with 29 inches at Steetley Colliery and in West Drayton No. 1 Borehole, and 28 inches in Checkerhouse Borehole—all southern localities. In the north it

is absent from Blyth, Sutton and Ranby Camp boreholes and represented by shaly 'cank' or 'canky' mudstone in Lound, Mattersey, Nornay and Scaftworth boreholes. G.H.R.

The varied fauna of the marine band is well displayed in the district, the assemblage comparing closely with that known in adjacent areas. The richest fauna occurs in the lower part of the band, particularly in the 'cank', where cephalopods and calcareous brachiopods are more common. The highest part of the band is notable for the abundance of foraminifera; in these upper layers the marine fauna is less varied and more characteristic of the *Lingula* or *Myalina* facies of less well developed marine bands.

The following list is representative of the fauna found in the district:

(a) Lower part of band (including 'cank'):

Sponge spicules, *Crurithyris sp.*, *Lingula mytilloides*, *Orbiculoidea* cf. *nitida*, *Rugosochonetes skipseyi*, *Tornquistia diminuta*, *Coleolus sp.*, *Euphemites anthracinus*, *Platyconcha hindi*, *Retispira?*, *Soleniscus sp.*, *Aviculopecten sp.*, *Dunbarella* cf. *macgregori*, *Nuculopsis aequalis*, *Palaeoneilo taffiana*, *Polidevcia* cf. *acuta*, *P. attenuata*, *Posidonia sulcata*, *Prothyris carbonaria*, *Schizodus antiquus*, *Streblochondria fibrillosa*, orthocone nautiloid, *Ephippioceras* cf. *clitellarium*, *Metacoceras* cf. *cornutum*, *M.* cf. *perelegans*, *Huanghoceras?*, *Solenochilus* aff. *cyclostomus*, '*Anthracoceras*' *hindi*, *Politoceras politum*, *Cypridina sp.*, *Hollinella sp.*, crinoid columnals, *Conularia sp.*, *Serpuloides* cf. *stubblefieldi*, *Listracanthus sp.*, *Megalichthys sp.*, *Petrodus sp.*, *Rhabdoderma sp.*, *Rhizodopsis sp.*, cf. *Tomaculum sp.*

(b) Upper part of band:

Foraminifera including *Ammodiscus sp.*, *Glomospira sp.*, *Tolypammina sp.*, sponge spicules, *Lingula mytilloides*, *Orbiculoidea* cf. *nitida*, *Donaldina?*, *Myalina* cf. *compressa*, *Polidevcia acuta*, *Hollinella sp.*, *Megalichthys sp.*, *Rhabdoderma sp.*, pyritized strap-like markings ('fucoids'). In addition a specimen of *Belinurus* cf. *bellulus*, usually considered to be more typical of non-marine sediments, was found in the upper part of the band in Eaton Borehole.   M.A.C.

The complete cycle containing the Mansfield Marine Band ranges in thickness between 46 ft at Harworth Colliery and 109 ft at Steetley Colliery. The variations in thickness are independent of those of the marine strata, which generally comprise less than half the cycle.

Above the marine band, dark mudstones persist, gradually giving way to paler mudstones; ironstone bands are common in the dark mudstones, which also contain *Naiadites*, including *N.* cf. *productus,* cf. *Geisina sp.* and, in some sections, fish debris. The non-marine mudstones are generally overlain by siltstone and sandstone, and many sections show an additional mudstone phase occurring between the top of the sandstone and the overlying seatearth, or within the upper part of the sandstone. This upper mudstone contains *Naiadites*, locally associated with *Carbonita humilis* and a bivalve referred to *Gen. et sp. nov.*

The Wales, Houghton Thin and Sharlston Yard coals. The First and Second Wales coals, generally present in the type area near Sheffield (Eden and others 1957, p. 120) persist into the East Retford district where, in the west, they have been equated with the Sharlston Yard and Houghton Thin respectively of the Wakefield (78) district (Edwards 1951, p. 70). The correlation of these seams is complicated by splitting, and many sections show three or four seatearths or thin coals at this general horizon. An attempt at carrying the correlation through the whole of the western part of the district has been made in Plate VIII, which also shows a relatively undivided Wales Coal in a few localities, mainly in the vicinity of Bevercotes Colliery.

None of the coals in this group is of workable quality and thickness. The thickest

sections are: Wales Coal; Jockey House Borehole, dirt and coal 10 in, coal 22 in, shale 3 in on coal 22 in: Houghton Thin Coal; Manton Colliery No. 4 Shaft, dirty coal 42 in: Sharlston Yard; Maltby Main Colliery, coal 24 in.

The Houghton Thin and Sharlston Yard are generally between 10 and 20 ft apart, with a maximum of over 40 ft at Wallingbrook Wood Borehole. The intervening measures are predominantly mudstones, from which mussels and fish scales are recorded in some of the western sections.

The measures between the Wales Coal and the Edmondia Band include the Sharlston Low and Sharlston Top coals of Yorkshire, the latter being the broad equivalent of the Highmain of Nottinghamshire. Eleven seatearths, some with thin coals above, are recorded in Forest Hill Borehole: some of these seatearths result from splitting of seatearths present in the more typical sequences, which comprise seven or eight cycles. Only five cycles are present in the thin succession in West Drayton No. 1 Borehole. The measures are 260 ft thick at Steetley Colliery and 240 ft in Stone Borehole, but only 133 and 139 ft in Eaton and West Drayton No. 1 boreholes respectively. Over much of the western part of the district, however, the thickness is between 180 and 220 ft.

Locally the roof measures of the Sharlston Yard Coal contain *Naiadites* and fish remains.

The Sharlston Low Coal lies 30 to 85 ft above the Wales. It is commonly a split seam with leaves up to about 10 ft apart. The thickest section of coal is in Steetley Colliery Shaft: coal 18 in, dirt 4 in on coal 14 in. In Eaton Borehole 76 in of dirt and coal probably represent the seam.

The cycles between the Sharlston Low and Sharlston Top coals commonly contain a high proportion of silty and sandy measures. In Barnby Moor, Mattersey and Torworth boreholes and Steetley Colliery Shaft the cycles have been replaced entirely by sandstones and subordinate siltstones which form a prominent rock. In Barnby Moor Borehole the rock is joined to the overlying Mexborough Rock, forming an almost uninterrupted sequence of arenaceous beds between the Sharlston Low and the Shafton Marine Band. In localities where the rock is thin or absent, mussels, commonly associated with ostracods and fish, are present in the lower parts of many of the cycles. The following fossils have been collected from these measures: *Spirorbis sp.*, *Anthraconaia hindi*, *Naiadites hindi*, *N. melvillei*, *N. sp.* cf. *productus*, *N.* cf. *triangularis*, *Carbonita humilis*, *C.* cf. *pungens*, *Geisina subarcuata*, *Elonichthys sp.*, *Rhadinichthys sp.*, *Rhizodopsis sauroides*, platysomid scales.

A 2-in tonstein is recorded in Ranskill and Stone boreholes at 32 ft 10 in and 38¼ ft respectively below the Sharlston Top Coal. These occurrences appear to be at the same horizon, though at Stone the tonstein rests on seatearth, which is absent at Ranskill. It is probable that the tonstein is the same as that recorded below the Highmain Coal elsewhere (Eden and others 1963, pp. 54–6).

The Sharlston Top Coal appears to be equivalent to the upper part of the Highmain Coal of Nottinghamshire. It is of little account in the East Retford district, except in West Drayton No. 1 Borehole and Firbeck Main Shaft, where it is a 29-in and 28-in coal respectively. Correlation of the horizon has been attempted in Plate VIII, from which it will be seen that in some sections only a seatearth is developed, while in others four or more thin coals are present. In Barnby Moor and Hayton Smeeth boreholes the seam is washed out by the Mexborough Rock.

The measures between the Sharlston Top Coal and *Edmondia* Band generally comprise two cycles, though in some sections (e.g. Forest Hill Borehole) there are, as a result of splitting, up to four seatearths. Mussels are recorded from mudstones in these measures in the majority of the boreholes, the most significant band being that close above the Sharlston Top. This band contains *Anthraconaia hindi*, *A.* aff. *stobbsi*, *Naiadites* aff. *melvillei*, *Geisina subarcuata* and fish remains including palaeoniscid scales, *Rhabdoderma sp.* and *Rhizodopsis sp.* Higher bands contain *Spirorbis sp.*, *Naiadites* aff. *daviesi*, *N. melvillei*, *N. sp.* cf. *productus*, *Carbonita humilis* and fish remains. At the top of these measures is a coal which is 27 in thick in Harworth Colliery Shaft and consists of 25 in of mainly inferior coal in Stone Borehole. It is, however, only a few inches thick in many sections, and is represented by a seatearth at Ranskill and some southern

localities (Plate VIII). Only in two boreholes is there any appreciable thickness of mudstone or shale between the coal or seatearth and the *Edmondia* Band: West Drayton No. 1, 6 ft 8 in, of which the bottom 3 ft 8 in is dark grey slickensided mudstone, and Jockey House, 5 ft. At the latter locality 4¼ ft of pale grey splintery shale rests on 6 in of siltstone on 3 in of black shale with plant debris; 3 ft above the base of the black shale *Naiadites* cf. *triangularis* is recorded (Edwards and Stubblefield 1948, p. 230). In Eaton Borehole *Naiadites melvillei* occurs in black cannel-like shale 2 in below the *Edmondia* Band, but here there is no underlying coal or seatearth.

The greater part of the measures between the Sharlston Top and the *Edmondia* Band is cut out by the Mexborough Rock in Firbeck Main Colliery Shaft and in Nornay, Scaftworth and Sutton boreholes; in Barnby Moor and Hayton Smeeth boreholes the washout is complete, the Mexborough Rock being apparently continuous with the underlying sandstone.

The Edmondia Band has been seen in ten sections in the East Retford district: Crookford, Eaton, Jockey House, Mattersey, Ranskill, Stone, Wallingbrook Wood and West Drayton No. 1 boreholes and in Maltby Main and Manton No. 4 shafts. Checkerhouse Borehole and the shafts of Bevercotes and Steetley collieries probably passed through the horizon, but it was not recorded. In the remaining nineteen shafts and cored boreholes that might have been expected to encounter it the *Edmondia* Band has been cut out by the Mexborough Rock.

Though dark shales are recorded in Eaton and Stone boreholes and in Manton No. 4 Shaft, the usual lithology of the *Edmondia* Band is a grey mudstone in which thin bands and nodules of ironstone are fairly common, particularly in the upper part. Pyrite is common in the lower part. The band is only 7 in thick in Mattersey Borehole, but it is 12 ft in Stone, Wallingbrook Wood and West Drayton No. 1 boreholes, and 14 ft 8 inches in Crookford Borehole. The fauna is well developed in Eaton Borehole, where it consists of foraminifera, *Edmondia*, *Myalina*, ostracods and fish (see p. 256).

The measures between the Edmondia Band and the Shafton Marine Band are dominated by the Mexborough Rock, which, though absent in the extreme south, is prominent throughout the greater part of the district (see below). Where the rock is thin or absent —Bevercotes Colliery and Crookford, Eaton, Jockey House and West Drayton No. 1 boreholes—six or seven cycles are developed, many of them with thin coals. Above the second coal, or seatearth, the Main '*Estheria*' Band of Edwards and Stubblefield (1948, pp. 231–2) is found. *Lioestheria* generally occurs in some abundance in a foot or two of dark grey shale, and in Crookford Borehole is found in 4 ft 2 in of sediment including part of an overlying seatearth. In Wallingbrook Wood Borehole, in the south-west, *Lioestheria* is found above each of two closely spaced seatearths, the upper of which is nearly 100 ft above the base of the *Edmondia* Band. These last occurrences are likely to be at a higher horizon than the Main '*Estheria*' Band and may have their equivalents in Yorkshire (Goossens 1952, p. 192).

In Eaton Borehole *Naiadites sp.* is associated with *Lioestheria vinti* and fish scales at the Main '*Estheria*' Band horizon. Some 14 ft higher in the same borehole *N.* cf. *hindi*, *Carbonita* cf. *pungens* and fish remains were collected.

The Mexborough Rock of the East Retford district is locally the exact equivalent of the sandstone of that name in the Barnsley (87) and Sheffield (100) districts (Mitchell and others 1947, pp. 79–81; Eden and others 1957, p. 122), but over considerable areas sandstone occurring between the horizons of the Shafton Coal and Shafton Marine Band is included. Fig. 19 shows that thicknesses in excess of 200 ft are encountered at Firbeck Main Colliery and in Barnby Moor Borehole. Indeed, in the latter locality 285 ft of rock are recorded, but some of this exceptional thickness is due to the coalescence of the rock with sandstone beneath the Sharlston Top Coal. In much of the area where it is over 100 ft thick, the Mexborough Rock replaces all the cycles between the Shafton Coal and the *Edmondia* Band, and in some sections even cuts out the latter (Plate VIII). It can be seen (Fig. 19) that the area of maximum thickness extends from west to east across the middle of the district. To the south the Rock is inconspicuous; to the north it thins to about 70 ft.

The Rock is commonly fine- to medium-grained with breccia and conglomerate bands, which occur mainly in the lower part and contain fragments of mudstone, siltstone and ironstone. Medium- to coarse-grained sandstone is recorded in Blyth, Forest Hill, Scofton, Stone and Wigthorpe boreholes in the west, and also at Jockey House. The pinkish or reddish brown colour so characteristic of the rock near its outcrop in the west (Eden and others 1957, p. 122) persists into part of the East Retford district (Fig. 19). This coloration is considered to be of primary origin, since it can affect sandstone hundreds of feet below the zone of sub-Permian secondary reddening (p. 46).

In the north, the top of the Mexborough Rock is separated from the Shafton Marine Band by up to 80 ft of measures containing a maximum of three cycles. The coal locally developed at the top of the second cycle is the Shafton. In Scaftworth Borehole this seam, lying some 25 ft below the Shafton Marine Band, comprises 32 in of coal and dirt; in Ranskill and Torworth boreholes it is 26 in and 18 in respectively. *Lioestheria* is present above the coal in all three localities, and is also found immediately above the Mexborough Rock in the latter two and at Stone Borehole. In Nornay Borehole this fossil is found 41 ft below the Shafton Marine Band. Fish debris and ostracods, including *Carbonita humilis*, are associated with *Lioestheria* in some of these localities, and rare mussels including *N.* cf. *hindi* are also present in the measures. In Scaftworth Borehole the roof measures of the Shafton Coal contain, in addition to *L. vinti*, megaspores, *N. spp.*, *C. humilis*, *C. pungens*, *C.* cf. *salteriana*, *Geisina subarcuata* and *Rhabdoderma sp.*

The Shafton Marine Band has been recorded in all the cored boreholes passing through its horizon except Checkerhouse and Hayton Smeeth, where it is also likely to be present. Observed thicknesses vary up to 7 ft 2 inches in Lound Borehole. At most localities the marine band is found in the roof of a coal, which is dirty and rarely more than 2 ft thick. The marine fossils extend down to within an inch or two of this coal or the seatearth at its horizon, except at Lound and Mattersey, where 11 in and 26 in respectively of unfossiliferous dark mudstone were recorded. The marine band generally consists of grey

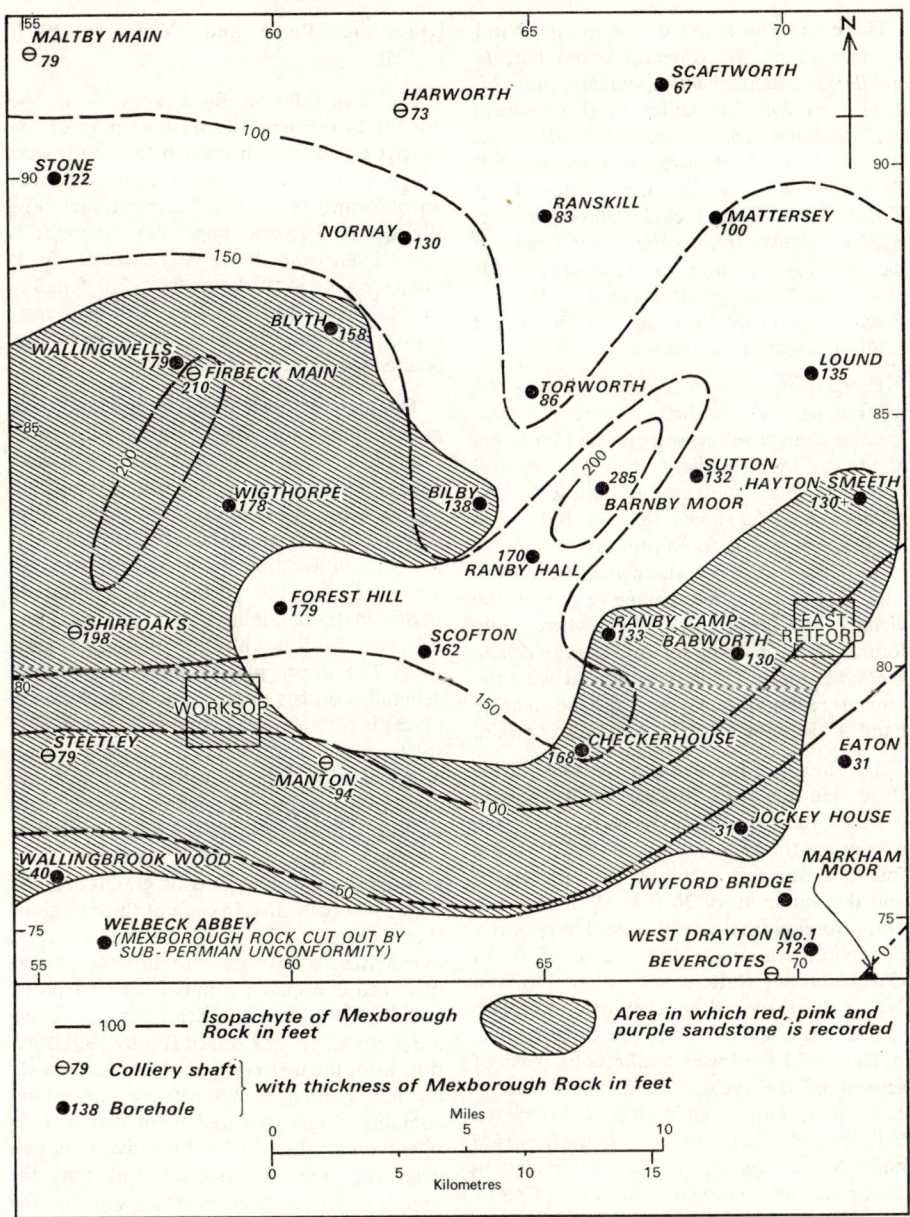

FIG. 19. *Isopachyte map of the Mexborough Rock*

or light grey mudstone resting on dark mudstone or shale, the whole containing ironstone bands. *Anthraconaia spathulata* is commonly found in the upper, light coloured, part of the marine band, and only rarely in the lower part. The dark mudstone at the base of the band normally contains only *Lingula* and fish remains. *Lioestheria* is commonly associated with the marine fossils, and 'fucoid' markings are recorded in most sections.

G.H.R.

The composite fauna of the marine band is: *Ammodiscus sp.*, *Glomospira sp.*, *Lingula mytilloides*, *Anthraconaia spathulata* [*olim* cf. *A. pruvosti* Weir and Leitch *pars*], *Edmondia sp.*, *Myalina compressa*, *Hollinella sp.*, *Lioestheria vinti*, *Rhabdoderma sp.* and cf. *Tomaculum sp.* In addition, collections from Maltby Main Shaft include '*Anthracoceras*' *hindi*, cf. *Domatoceras sculptile*, *Listracanthus sp.* and *Megalichthys sp.* (Edwards 1951, p. 89). An unusual occurrence is that of *Belinurus sp.* in the lower part of the marine band in Wigthorpe Borehole.

From north-west to south-east across the western part of the district there is a progressive change in facies from the rich fauna at Maltby Main Colliery, through a central region in which the acme is represented by *Myalina* and *Edmondia* (e.g. at Bilby and Scofton boreholes) to an area where *Lingula* is dominant (e.g. at Babworth Borehole). This change is carried a stage further in the vicinity of Eaton Borehole, where only foraminifera, *Lioestheria* and *Anthraconaia spathulata* occur. This is in accord with the known regional behaviour of the marine band (Calver 1968, pp. 54–6, fig. 20). M.A.C.

The measures between the Shafton Marine Band and the Top Marine Band show a marked thinning from west to east, the maximum thickness of about 142 ft being found in Stone Borehole in the north-west, and the minimum of 30 ft in West Drayton No. 1 Borehole in the south-east. The majority of sections, however, show 50 to 95 ft. In Checkerhouse, Sutton and West Drayton No. 1 boreholes they comprise only two cycles, but elsewhere there are three to five. A feature of the lower argillaceous parts of several of the cycles is the abundance of *Lioestheria vinti*, which in places is associated with mussels, including *Naiadites* cf. *daviesi* and *N.* cf. *hindi*, ostracods (including *Carbonita* cf. *claripunctata* and *Geisina subarcuata*) and fish debris (including acanthodian spines, palaeoniscid scales, platysomid scales, *Rhabdoderma sp.* and *Rhizodopsis sp.*). Noteworthy is the presence of *Hemicycloleaia* cf. *minima* in the first cycle above the Shafton Marine Band, recorded by Mr. R. F. Goossens at 1173 ft 10 inches in Forest Hill Borehole and at 1157½ ft in Scofton Borehole. It is significant that this fossil has been recorded at an identical horizon in the Prestwich area of Lancashire (Poole and Whiteman 1955, p. 303).

The Top Marine Band appears to be present throughout the western part of the district except for an area in the south-west where it is cut out by the sub-Permian unconformity (Fig. 9), and an area, including Blyth and Torworth boreholes, where it is washed out beneath the Ackworth Rock. It ranges widely in thickness from 2 ft 5 inches in Mattersey Borehole and 3 ft in Ranby Camp Borehole to over 20 ft in Bilby, Scaftworth and Lound boreholes.

The marine fossils occur in grey and dark grey mudstones, both lithologies being present in the majority of sections. The dark mudstone, which is commonly pyritic and contains ironstone bands, occurs at the base, resting directly on a coal or seatearth in most localities, though in places there are a few inches of intervening dark, commonly pyritic shale containing only plant debris and fish and, in Ranskill Borehole, *Lioestheria vinti*. The upper part of the marine band generally consists of grey mudstone, which in places is silty and may contain thin bands of siltstone and sandstone (e.g. at Lound Borehole), and in which ironstone bands are generally common. The thick sections recorded in Bilby, Scaftworth and Lound boreholes contain comparatively little dark mudstone and are largely composed of grey mudstone with siltstone bands. In each of these sections there is a band several feet thick in which no marine fossils have been found; the marine band is thus apparently in two separate parts. Details of the section in Scaftworth Borehole are given on pp. 289–90. In Ranby Hall Borehole also, the marine band, 11 to 12 ft thick, includes a median 2 ft of grey mudstone containing only pyritized plant debris. It is possible that the division into two parts was originally more widespread and that the upper part has been washed out by the Ackworth Rock in many localities. It is noteworthy that an upper marine phase has been recorded in the Chesterfield (112) district (Smith and others 1967, p. 191), where the only recorded section through the Top Marine Band proved marine ostracods 22 ft 1 in above the main part of the band.

G.H.R.

The composite fauna of the marine band is: foraminifera, including *Glomospira sp.*,

Spirorbis sp. [attached to *Dunbarella* fragment], *Crurithyris sp.*, *Lingula mytilloides*, *Orbiculoidea* cf. *nitida*, *Coleolus sp.*, *Platyconcha* aff. *hindi*, *Retispira?*, *Dunbarella* cf. *macgregori*, *Myalina sp.*, *Palaeoneilo sp.*, *Polidevcia* cf. *stilla* (Hind *non* McCoy), *Schizodus?*, *Coelogasteroceras dubium*, '*Anthracoceras*'?, *Politoceras kitchini*, *Cypridina 'phillipsii'*, *Serpuloides sp.*, *Planolites ophthalmoides*, cf. *Tomaculum sp.*, faecal pellets, 'fucoids' and palaeoniscid scales. This rich assemblage is characteristic of the *Anthracoceras*/pectinoid + productoid facies (Calver 1968, fig. 21). It is widespread in the district and shows no obvious geographical differences. M.A.C.

(ii) The Eastern Part of the District

No cores have been taken of the strata between the Mansfield and Top Marine Bands in this part of the district (Plate V), except in South Leverton Borehole, of which, however, there is no reliable record of the Coal Measures. Information is available from cutting samples and gamma-ray and electric logs of uncored oil bores, though there are sometimes discrepancies between sampling and geophysical evidence.

The Mansfield Marine Band generally gives a good peak on the gamma-ray logs, but this is less prominent in some oil bores at Grove and South Leverton.

The strata between the Mansfield Marine Band and the *Edmondia* Band appear to vary from about 150 ft (at Gainsborough) to about 250 ft (at South Leverton). Grove No. 1 Oil Bore records twelve coals in these strata, and there may be ten at South Leverton No. 18 Oil Bore; but elsewhere not more than six are reported, and most bores show only one to four. In some oil bores one or two seatearths without coal are recorded in addition. The seams are usually thin, but locally one or more may be as much as 3 ft thick. Scattered reports of thicknesses between 4 and 8 ft should be treated with reserve. It is impossible to name or correlate these coals with any confidence, but it is assumed that two of them occurring within 50 ft of the Mansfield are at the Wales horizon, and may be the Houghton Thin and the Sharlston Yard (p. 88). A persistent gamma-ray peak in the seatearth of a seam 35 to 70 ft below the *Edmondia* Band could indicate the tonstein below the Sharlston Top (p. 89). A sandstone in the cycle above the Sharlston Yard is prominent in the southern part of the South Leverton Oilfield, where it is 70 ft thick in No. 2 Bore. Twenty-five oil bores at South Leverton and Gainsborough record a sandstone between the ?Sharlston Top and the *Edmondia* Band. This rock is stained pink and green at South Leverton No. 3 Oil Bore, where it reaches its greatest recorded thickness of 74 ft. Coloured micas are reported in this sandstone in some bores at South Leverton, apparently their first significant occurrence above the Sub-Kilburn Sandstone (p. 55).

Except at Grove, the gamma-ray peak attributed to the *Edmondia* Band is poor, and the exact horizon of this marine band is often uncertain.

The strata between the *Edmondia* Band and the Top Marine Band are variable in thickness, probably ranging between about 200 and about 260 ft. They are thickest in the northern oil bores, where they largely consist of the Mexborough Rock. This sandstone is more than 100 ft thick in Walkeringham No. 1 and Beckingham Nos. 1 and 2 oil bores, and in some bores in the northern and western parts of the Gainsborough Oilfield, reaching a maximum of 150 ft at Gainsborough No. 44 Bore. It is usually white or grey and fine to medium grained, with angular or subangular grains. It is absent at Gringley, and in the eastern part of the Gainsborough Oilfield, and thin at South Leverton. In these areas up to seven thin coals or seatearths are recorded in the equivalent strata, locally including seams reputed to be as much as 4 ft thick.

Above the Mexborough Rock, the probable Shafton Coal is recorded in almost all of the oil bores, lying 60 to 100 ft below the Top Marine Band. In the northern bores it is usually 2 to 3 ft thick, but at South Leverton it probably does not exceed 2 ft. In some bores there is also a thin coal directly beneath the Shafton Marine Band.

The gamma-ray peaks attributed to the Shafton and Top marine bands, 40 to 60 ft apart, are poorly defined, but that of the Top Marine Band is somewhat more distinctive.

In some oil bores one or more thin seams are reported between the marine bands. The sub-Permian unconformity approaches very close to the Top Marine Band in the Gainsborough Oilfield, and may possibly cut it out in No. 1 and a few adjacent bores.   C.G.G.

UPPER COAL MEASURES

These measures have been divided (following Goossens and Smith, 1973) into the Ackworth, Brierley, Hemsworth and Badsworth Divisions (see p. 41, Fig. 10 and Plate IX).

The Ackworth Division thins eastwards from over 260 ft at Maltby Main Colliery to a little over 150 ft in Hayton Smeeth Borehole. It comprises up to twelve cycles, but sections other than those provided by Maltby Main No. 2 Shaft and Blyth, Ranby Hall and Stone boreholes, show between three and eight.

The division contains six coals at Blyth Borehole and up to five elsewhere. The 24-in seam 58 ft below the Blyth Coal in Jockey House Borehole is correlated with the 30 in of inferior coal 68 ft below the Blyth at West Drayton No. 1 Borehole; it is probably represented elsewhere, e.g. by the 19-in coal 63 ft below the Blyth at Harworth Colliery, but is clearly absent in many sections. This seam may well be the Elmsall Coal of the exposed coalfield (Goossens and Smith 1973, p. 497).

The only other seams of significance are the two thin coals above the Ackworth Rock, which correlate with the thin coal and a higher cannel or seatearth in the exposed coalfield (*ibid.*, pp. 496–7). The higher of these coals is here named the Scofton. This seam, or its seatearth in the absence of coal, can be traced throughout the western part of the East Retford district (Plate IX). There are 14 in of coal in Wigthorpe Borehole, but elsewhere the seam does not exceed 12 in. Mussels occur in the mudstones overlying this horizon at several localities, and at Nornay, Scofton and West Drayton No. 1 boreholes *Lioestheria vinti* has been found. This is the only '*Estheria*' band recorded in the Ackworth Division. At Wigthorpe Borehole there is an unusual occurrence of *Leaia?*, associated with mussels, in this position. Another distinctive feature of the Scofton horizon is the abnormal thickness of its seatearth in many sections, e.g. 19 ft in Eaton Borehole. At West Drayton No. 1 and Wigthorpe boreholes this seatearth shows colour mottling.

At Stone Borehole the Scofton and underlying coal are separated by less than 2 ft of seatearth, and at Maltby Main Colliery the two seams evidently unite to form the 16½-in coal at 976 ft 3½ in (Plate IX). Elsewhere the lower coal horizon, where it can be recognized, is normally about 25 ft below the Scofton. It consists of 16 in and 14 in of coal in Eaton and Ranby Hall boreholes respectively, and at Firbeck Main Colliery there are 12 in of coal and dirt on 21 in of coal. Dirt is a feature of the seam in many sections, and in a few the seam is split—most widely at Manton Colliery No. 4 Shaft, where an upper 3-in leaf of coal is separated by 13¼ ft of seatearth from 9 in of dirty coal on 36 in of inferior cannel with ironstone bands. East of Blyth and Barnby Moor and north of Eaton this lower coal horizon can be identified only in Mattersey Borehole, where it is represented by a seatearth (Plate IX).

The most important sandstone in the division is the Ackworth Rock, close above the Top Marine Band. It has cut out the marine band in Blyth and Torworth boreholes and probably also at Maltby Main Colliery, and at these localities the base of the Rock is taken as the base of the Upper Coal Measures. The Ackworth Rock is thickest in the north, where 105 ft of sandstone are recorded at Maltby Main Colliery, and 121 ft of sandstone with siltstone in Torworth Borehole. Southwards much of the sandstone passes into siltstone, of which there are substantial thicknesses, up to 80 ft, in some of the boreholes between Worksop and East Retford. Only at Barnby Moor and Wallingwells boreholes is no Ackworth Rock recorded.

At several localities, e.g. Ranby Camp and Scofton boreholes, virtually all the measures between the Scofton Coal horizon and the Blyth Coal consist of siltstone and sandstone. In most sections, however, discrete bands of sandstone or siltstone are developed, and prominent among these is the sandstone in the cycle immediately above the Scofton horizon (Plate IX), 83 ft thick at Maltby Main Colliery. Higher in the division there is a 40- to 50-ft sandstone in Lound and Wigthorpe boreholes and at Firbeck Main Colliery, though this is not necessarily at precisely the same horizon in each case.

Ironstone with white ooliths has been recorded in two sections: in Ranby Camp Borehole at 1327 ft 7 in, about 20 ft above the Scofton Coal; and in Lound Borehole at 1838 ft 10 in, immediately above the Scofton Coal. In both sections the ironstone is 2 in thick, and at the former locality contains much galena and pyrite.

With the exception of *Anthraconauta wrighti* from 1525 ft, $68\frac{1}{2}$ ft above the Scofton horizon, in Sutton Borehole, all the mussels recorded from the Ackworth Division are referred to *A. phillipsii* and variants. They occur in several of the cycles, and the most notable horizons are perhaps the roof measures of the Scofton Coal (see above) and a band some 25 to 50 ft below the Blyth, at e.g. Blyth, Mattersey, Nornay, Ranskill and Scaftworth boreholes, where ostracods are also common. The following ostracods have been recorded from this and other horizons in the Ackworth Division: *Carbonita humilis*, *C. pungens*, *C. rankiniana*, *C. salteriana* and *Geisina subarcuata*, the last species having been found only in the lower part of the Division at Barnby Moor and Jockey House boreholes. Apart from plants, the only other fossils are *Lioestheria vinti* and *Leaia?* (see above), *Planolites*, *Spirorbis*, fish, including *Acanthodes wardi*, *Petalodus hastingsi*, *Rhabdoderma sp.* and *Rhizodopsis sp.*, and a tuberculate arthropod fragment (from 1441 ft 10 inches in Torworth Borehole). *Planolites* is represented by *P. ophthalmoides* and *P. montanus*, the former occurring only in the grey mudstones succeeding the Top Marine Band and extending over $4\frac{3}{4}$ ft at Wigthorpe Borehole, and the latter having been found at one horizon, above the thin representative of the Ackworth Rock, in Barnby Moor Borehole. Recorded plants include: *Alethopteris valida*, *Cordaites principalis*, *Lepidodendron acutum*, *Neuropteris heterophylla* and *Sigillaria tessalata*.

The Brierley Division shows an eastward thinning from 226 ft at Maltby Main Colliery to less than 100 ft in Hayton Smeeth and Lound boreholes. Sixteen cycles can be recognized in Scofton Borehole, but elsewhere there are between three and nine, with no apparent relationship between the thickness of the Division and the number of cycles developed. Coal occurs locally in all the major cycles, but the Blyth seam at the base is the only one of economic interest.

The Blyth Coal has been identified only in the western part of the district. Here it is thickest in the north, where it provides considerable workable reserves. In the northwest there is cannel at the top of the seam and a thin dirt band towards the bottom (Maltby Main Colliery, cannel 2 in, brights 24 in, dirt 2 in, brights 8 in; Stone Borehole, cannel $7\frac{3}{4}$ in, coal $13\frac{3}{4}$ in, dirt 1 in, dirty coal $0\frac{1}{2}$ in, coal $10\frac{1}{2}$ in), but other sections in the north show dirt-free bright coal with streaks and thin bands of hard coal. In the area of clean coal, sections analysed by the Coal Survey Laboratory give an average ash content of 2·0 to 3·7 per cent and a sulphur content of 1·5 to 2·3 per cent. The seam is 39 in thick in Blyth Borehole, 38 in at Nornay, 36 in at Torworth and Mattersey and in Harworth Colliery No. 1 Shaft, and 33 in at Ranskill. Mr. E. Skipsey of the N.C.B. reports that the seam was experimentally worked in 1958 at Harworth Colliery: it consisted of 3 ft or slightly less of good-quality bright coal with 4 per cent ash. In Scaftworth Borehole, where the seam could have been expected to be more than 36 in thick, it is washed out.

Southwards from the above area the Blyth thins to 30 in at Ranby Hall, 29 in at Bilby and Barnby Moor, 26 in at Firbeck Main Colliery, 25 in at Ranby Camp and 21 in at Wigthorpe. Accompanying the thinning is a deterioration in quality, with the ash content rising to 5·6 per cent at Wigthorpe and 7·6 per cent at Ranby Camp. Farther south the seam rapidly becomes worthless, consisting of $16\frac{1}{2}$ in and 12 in of dirty coal at Babworth and Scofton respectively, and only 10 in and 8 in of coal at Eaton and Jockey House.

East of Torworth and Barnby Moor, the Blyth Coal was not recorded at Sutton and is apparently absent at Hayton Smeeth, though there may have been drilling losses at this horizon in both of these boreholes. An eastward deterioration of the seam, however, is shown by Lound Borehole, where the Blyth is evidently split into two leaves almost

8 ft apart. The upper leaf consists of 16 in of bright coal with an ash content of 5 per cent, and the lower leaf of 18 in of bright coal with dirt partings near the top and an ash content of 7·2 per cent. The Coal Survey Laboratory records partings of ankerite in both leaves, a mineral also reported from the seam in Bilby and Torworth boreholes.

A coal occurring about 60 to 125 ft above the Blyth can be correlated in many of the sections in the east and north-east of the explored part of the district. It is commonly a split seam and attains its maximum thickness of 32½ in (coal 10¼ in, dirt 9¾ in on coal 12½ in at 1465 ft 1 in) in Scaftworth Borehole.

Only in a few sections in the north-west is there a coal at the top of the Division, immediately below the Fourth Cherry Tree Marker or its horizon. This is represented by 13 in of bat with coal in Firbeck Main No. 1 Shaft and by an inch or two of coal elsewhere.

Sandstone or siltstone occurs from place to place in almost all cycles, but is most common in the lower part of the Division. A thick, locally coarse-grained, sandstone, replacing the lower cycles, extends in a belt south-eastwards from Maltby Main to East Retford. Including some siltstone, this rock is 177 ft thick in Maltby Main Colliery No. 2 Shaft, and 170 ft at Stone Borehole; carbonaceous streaks are common in the sandstone at the latter locality, and conglomerate occurs at several levels. At Babworth Borehole the whole of the Brierley Division, excepting a seatearth at the top and the Blyth Coal at the base, consists of sandstone. This thick sandstone is evidently the Wickersley Rock of the district to the west (Eden and others 1957, pp. 126–7 and fig. 26), and is the broad equivalent of the Brierley Rock of the exposed coalfield to the north-west (Goossens and Smith 1973, p. 499).

The seatearth at the top of the Division, below the Fourth Cherry Tree Marker, is thick in several sections, but is not so prominent an horizon as the seatearth below the Second Cherry Tree Marker (see below).

Mussels, completely absent from some sections, occur at a number of horizons in others; identifiable specimens are referred to *Anthraconauta phillipsii* or *A. wrighti* or

their variants. The recorded ostracods all belong to the genus *Carbonita*—*C.* cf. *agnes*, *C. humilis* and *C. pungens*. There are two locally developed '*Estheria*' bands, the more prominent some 30 to 60 ft from the top of the Division and the other in the lower part. With two exceptions the identified estheriids belong to *Euestheria simoni: Anomalonema defretinae* occurs at the higher horizon (1409 ft) in Eaton Borehole, and has been found at the lower horizon (992 ft) in Blyth Borehole. Other fossils recorded from the Brierley Division include *Cochlichnus kochi*, *Spirorbis*, fish debris and the plants *Calamites suckowi* and *Neuropteris heterophylla*.

The Hemsworth Division thins eastwards from 313 ft in Stone Borehole to probably less than 200 ft in Hayton Smeeth Borehole. The number of recognizable cycles varies from six to a maximum, in Stone Borehole, of eighteen, and is largely dependent on the thickness of Ravenfield Rock present (see below).

There are eight thin coals in Maltby Main No. 2 Shaft and Stone Borehole, but elsewhere no more than four are recorded. The thickest and most persistent seam is that underlying the Second Cherry Tree Marker, 18 inches in Blyth and Nornay boreholes and 17 in at Maltby Main Colliery. It contains cannel at Harworth Colliery, and consists of 8 in of inferior cannel in Sutton Borehole. A distinctive feature of this coal horizon is its seatearth, which is commonly of abnormal thickness—over 20 ft at several localities. There is a coal at the base of the Third Cherry Tree Marker in the north-west of the district; this is usually dirty, and is thickest (dirty coal 11 in, dirt 9 in on dirty coal 8 in) at Firbeck Main Colliery.

Sections in the north and north-east of the explored part of the district show a thick sandstone in the lower part of the Division. This sandstone, evidently at the horizon of the Ravenfield Rock of the Barnsley (87) district (Mitchell and others 1947, pp. 90–1), has a maximum proved thickness of over 150 ft in Nornay Borehole and is more than 100 ft in several other sections. It replaces many of the cycles of the Division, including those containing the Third Cherry Tree Marker (Plate IX), over a wide area. Conglomerate, especially at the base of the Ravenfield Rock, provides evidence at several localities of washing out: the Fourth

Cherry Tree Marker has been removed by this means in a number of sections.

The Fourth Cherry Tree Marker is a mussel/ '*Estheria*' band consisting of black or dark grey shale or mudstone, normally with overlying grey mudstone. At Torworth Borehole a 26-in band of siltstone is included in the marker. A 1- to 2-in ironstone band occurs in Ranskill, Stone and Wigthorpe boreholes, in several other sections there are ironstone lenses, and at Eaton Borehole some of the mussels are preserved in ironstone. The total recorded thickness of the marker varies widely up to 7 ft 7 in at Wigthorpe Borehole. The horizon of the marker has been determined in all sections (Plate IX), but the band is completely washed out beneath the Ravenfield Rock in Babworth, Bilby, Mattersey, Ranby Camp, Ranby Hall and Scaftworth boreholes and at Harworth Colliery, and partially washed out at Barnby Moor and Nornay.

No fossils have been recorded from the marker at Lound Borehole, and only fish fragments were found in the incomplete section at Barnby Moor. *Anthraconaia pruvosti* occurs in Eaton Borehole, but elsewhere mussels are represented by *Anthraconauta phillipsii* with, in addition, *A.* aff. *tenuis* in Wigthorpe Borehole. *Euestheria simoni* has been found at Blyth, Eaton, Forest Hill, Scofton and Stone boreholes. The only other fossils recorded from the marker in the district are *Spirorbis sp.*, *Carbonita humilis* (Torworth and Wigthorpe boreholes) and *Rhabdoderma?*.

The Third Cherry Tree Marker is well seen in Wigthorpe Borehole, where its five component cycles have been recognized. Details of this section at $384\frac{1}{4}$ ft are as follows:

	ft	in
Fifth Cycle		
Mudstone, grey; thin ironstone bands; scattered *E. simoni* ..	1	5
Mudstone, dark grey; scattered *A.* aff. *phillipsii*..	0	1
Fourth Cycle		
Siltstone-seatearth, grey	1	0
Siltstone; sandstone streaks ..	5	6
Mudstone, grey; scattered rootlets	1	0
Shale, black; plants	0	2
Coal, dirty	0	3

	ft	in
Third Cycle		
Mudstone-seatearth, dark at top	4	7
Siltstone with 9-in sandstone band	4	0
Sandstone, grey and red, coarse ..	2	6
Mudstone, grey and purple; scattered '*Estheria*'	2	9
Second Cycle		
Mudstone-seatearth, grey and red	0	9
Sandstone, grey	2	3
Siltstone; sandstone streaks ..	1	9
Mudstone, grey; ironstone bands; scattered *E. simoni*	4	0
Mudstone, dark; ironstone bands; *Spirorbis sp.*, *A.* aff. *phillipsii*, *E. simoni*	2	9
Shale, black; *A.* aff. *phillipsii* ..	0	9
Siltstone, black	0	3
First Cycle		
Sandstone-seatearth	2	3
Siltstone, grey	5	6
Mudstone, grey; ironstone bands; *Spirorbis sp.*, *Anthraconaia* aff. *pruvosti*, *Anthraconauta phillipsii*, *E. simoni*..	3	9
Shale, black; plants	1	0
	48	3

This is by far the thickest section of the Third Cherry Tree Marker proved in the district. In Stone Borehole it is $25\frac{1}{4}$ ft thick and comprises three cycles, which are probably the lowest three of the marker. Incomplete sections with one or two cycles have been proved in Bilby (14 ft), Forest Hill (1 ft 2 in) and Torworth (14 ft 10 in) boreholes. *A. phillipsii* and *E. simoni* occur in all these boreholes, and *Spirorbis* is also present except at Bilby. In addition *Carbonita pungens* was found at Bilby, and in the lower of the two cycles present at Torworth *Anthraconaia* cf. *pruvosti*, *C. humilis* and fish remains were collected. In other boreholes in the district the marker has evidently been washed out, and replaced by the thick sandstone occurring at the Ravenfield Rock horizon (see above).

The Second Cherry Tree Marker, comprising parts of two cycles, has been recorded in seven boreholes, in five of which, Blyth ($43\frac{1}{2}$ ft), Forest Hill ($44\frac{1}{2}$ ft), Nornay (47 ft 2 in), Ranskill (43 ft 7 in) and Stone ($63\frac{3}{4}$ ft), both cycles are fossiliferous. The best section is that provided by Blyth Borehole:

	ft	in
Upper Cycle		
Mudstone, grey; thin ironstone bands; scattered *Anthraconauta sp.* and *E. simoni*	2	3
Mudstone, dark; thin ironstone bands; *E. simoni*	0	9
Lower Cycle		
Coal, dirty	0	2
Mudstone-seatearth, black ..	0	7
Mudstone-seatearth, grey, with yellow mottling in lower part; ironstone nodules	11	6
Mudstone-seatearth, dark at top	2	0
Siltstone and sandstone	10	9
Mudstone, grey; silty in upper part	10	3
Mudstone, dark; *E. simoni* ..	0	9
Mudstone, grey; ironstone bands; *A.* aff. *phillipsii*..	3	3
Mudstone, dark	1	2
Ironstone	0	1
	43	6

Mussels are not recorded in the upper cycle elsewhere in the district. In the lower cycle *Anthraconaia* aff. *pruvosti* occurs at Nornay, *Planolites montanus* at Ranskill, and ostracods at Forest Hill (*C. humilis*) and Stone.

Outside the Cherry Tree Markers, few fossils have been found in the Hemsworth Division. A mussel-band with *Anthraconauta sp.* occurs between the Fourth and Third Cherry Tree Markers at Blyth, Stone and Wigthorpe, and there are associated estheriids in the two former boreholes. *E. simoni* is recorded some 50 ft below the Second Cherry Tree Marker in Forest Hill Borehole, and *P. montanus* occurs in a similar position at Scaftworth.

The following plants have been recorded from the Division in Maltby Main Colliery No. 2 Shaft (W. H. Dyson Collection): *Calamites cisti, C. suckowi, Cordaites principalis, Lepidodendron simile, Neuropteris heterophylla, N.* cf. *osmundae, Stigmaria ficoides.*

The Badsworth Division is found in all sections north of a line from Carlton in Lindrick to East Retford, except in Scaftworth Borehole. The maximum thickness present is about 354 ft at Nornay Borehole, but the topmost 127 ft were not cored here. In Harworth Colliery No. 1 Shaft nearly 303 ft were met, and in Stone Borehole, which provides the thickest cored section in the district, 290 ft were proved. The limited information provided by Blyth, Nornay and Stone boreholes indicates an easterly thinning comparable to that exhibited by the lower divisions. The greatest number of cycles recorded is fifteen at Stone Borehole.

A 23-in coal is recorded 132 ft above the base of the Division at Harworth Colliery, and there are 24 in of coal and dirt at a lower horizon in Maltby Main No. 2 Shaft. Elsewhere seams are impersistent and only a few inches thick. The 23-in seam at Harworth Colliery rests on a thick seatearth, and this is also recognizable in Blyth and Stone boreholes, where the coal is absent.

There is a 76-ft sandstone in Hayton Smeeth Borehole. The same bed in Ranskill Borehole is 109 ft thick and is partly coarse-grained, as are the 33- and 50-ft sandstones, nearly 40 ft apart, in Nornay Borehole, and the 42-ft sandstone at Blyth. The Blyth sandstone and the upper sandstone at Nornay may well be the equivalent of the Badsworth Rock of the exposed coalfield (Goossens and Smith 1973, p. 506). The lower sandstone at Nornay is correlated with the 38-ft sandstone at Maltby Main Colliery. It is thought that the two sandstones at Nornay unite to form the thick bed at Hayton Smeeth and Ranskill.

Except in Maltby Main No. 2 Shaft and Stone Borehole, where the lowest beds are entirely grey, the whole of the Badsworth Division has been classed as coloured measures (Plate IX). Only the highest beds, however, are predominantly red, the lower part of the so-called coloured measures being distinguished by sporadic red, brown and purple mottling, which is particularly characteristic of the seatearths (p. 46).

The First Cherry Tree Marker, recognized in eight sections (Plate IX), is a prominent mussel-band consisting of up to about 4 ft of mainly black mudstone with bands of dark grey ironstone. The mussels are referred to *Anthraconauta phillipsii* and variants, pyritized specimens of which are a distinctive feature of the marker (*ibid.*, pp. 493, 505–6), and have been recorded from Blyth, Nornay, Ranskill and Stone boreholes. Other fossils found in the marker are *Carbonita humilis*, *Spirorbis sp.*, fish teeth and scales, including *Ctenoptychius apicalis* in Maltby Main No. 2 Shaft, and, at Ranskill Borehole, an incomplete insect wing.

Fossils recorded from the Badsworth Division above the First Cherry Tree Marker include *Euestheria simoni*, occurring in three cycles at Stone Borehole. The lowest of these horizons has also been found in Barnby Moor, Blyth, Nornay and Ranskill boreholes, and possibly also in Sutton Borehole. *Leaia sp.* is associated with *Euestheria* in Barnby Moor and Blyth boreholes. Mussels, recorded from a number of levels in the Division, include *A. phillipsii*, *A.* cf. *tenuis* and *A. wrighti*. Other fossils comprise *Carbonita humilis*, *Cochlichnus kochi*, *Spirorbis sp.* and the plants *Calamites suckowi*, cf. *Lepidostrobus* and *Neuropteris gigantea*.   R.F.G., E.G.S.

REFERENCES

ANDERSON, W. and DUNHAM, K. C. 1953. Reddened beds in the Coal Measures beneath the Permian of Durham and south Northumberland. *Proc. Yorks. geol. Soc.*, **29**, 21–32.

BRUNSTROM, R. G. W. 1963. Recently discovered oilfields in Britain. *Wld Petrol. Congr.* Frankfurt. Section 1, 1–10.

CALVER, M. A. *in* SMITH, E. G., RHYS, G. H. and EDEN, R. A. 1967. *q.v.*

—— 1968. Distribution of Westphalian marine faunas in northern England and adjoining areas. *Proc. Yorks. geol. Soc.*, **37**, 1–72.

CLIFT, S. G. and TRUEMAN, A. E. 1929. The sequence of non-marine lamellibranchs in the Coal Measures of Nottinghamshire and Derbyshire. *Q. Jnl geol. Soc. Lond.*, **85**, 77–108.

DOWNING, R. A. and HOWITT, F. 1969. Saline ground-waters in the Carboniferous rocks of the English East Midlands in relation to the geology. *Q. Jnl engng Geol.*, **1**, 241–69.

DUFF, P. McL. D. and WALTON, E. K. 1962. Statistical basis for cyclothems: a quantitative study of the sedimentary succession in the East Pennine Coalfield. *Sedimentology*, **1**, 235–55.

DUNHAM, K. C. *in* EDWARDS, W. and STUBBLEFIELD, C. J. 1948. *q.v.*

DYSON, W. H. 1911. The occurrence of marine bands at Maltby. *Rep. Br. Ass. Advmt Sci.*, 1910 (Sheffield), 610–11.

EAGAR, R. M. C. 1947. A study of a non-marine lamellibranch succession in the *Anthraconaia lenisulcata* Zone of the Yorkshire Coal Measures. *Phil. Trans. R. Soc. B.*, **233**, 1–54.

—— 1952. The succession above the Soft Bed and Bassy Mine in the Pennine region. *Lpool Manchr geol. Jnl*, **1**, 23–56.

—— 1956. Additions to the non-marine fauna of the Lower Coal Measures of the North Midlands coalfields. *Lpool Manchr geol. Jnl*, **2**, 328–69.

EDEN, R. A. 1954. The Coal Measures of the *Anthraconaia lenisulcata* Zone in the East Midlands Coalfield. *Bull. geol. Surv. Gt Br.*, No. 5, 81–106.

—— STEVENSON, I. P. and EDWARDS, W. 1957. Geology of the country around Sheffield. *Mem. geol. Surv. Gt Br.*

—— ELLIOT, R. W., ELLIOTT, R. E. and YOUNG, B. R. 1963. Tonstein bands in the coalfields of the East Midlands. *Geol. Mag.*, **100**, 47–58.

EDWARDS, W. 1951. The concealed coalfield of Yorkshire and Nottinghamshire. 3rd edit. *Mem. geol. Surv. Gt Br.*

—— 1967. Geology of the country around Ollerton. 2nd edit. *Mem. geol. Surv. Gt Br.*

—— WRAY, D. A. and MITCHELL, G. H. 1940. Geology of the country around Wakefield. *Mem. geol. Surv. Gt Br.*

—— and STUBBLEFIELD, C. J. 1948. Marine bands and other faunal marker-horizons in relation to the sedimentary cycles of the Middle Coal Measures of Nottinghamshire and Derbyshire. *Q. Jnl geol. Soc. Lond.*, **103**, 209–60.

FALCON, N. L. and KENT, P. E. 1960. Geological results of petroleum exploration in Britain 1945–1957. *Mem. geol. Soc. Lond.*, No. 2.

GIBSON, W., POCOCK, T. I., WEDD, C. B. and SHERLOCK, R. L. 1908. The geology of the southern part of the Derbyshire and Nottinghamshire Coalfield. *Mem. geol. Surv. Gt Br.*

GOOSSENS, R. F. 1952. Marine bands proved in the new borings at Wentbridge and Darrington, Yorkshire. *Proc. Yorks. geol. Soc.*, **28**, 188–220.

—— and SMITH, E. G. 1973. The stratigraphy and structure of the Upper Coal Measures in the exposed Yorkshire Coalfield between Pontefract and South Kirkby. *Proc. Yorks. geol. Soc.*, **39**, 487–514.

HOWITT, F. and BRUNSTROM, R. G. W. 1966. The continuation of the East Midlands Coal Measures into Lincolnshire. *Proc. Yorks. geol. Soc.*, **35**, 549–64.

KNOWLES, B. 1964. The radioactive content of the Coal Measures sediments in the Yorkshire–Derbyshire Coalfield. *Proc. Yorks. geol. Soc.*, **34**, 413–50.

MITCHELL, G. H., STEPHENS, J. V., BROMEHEAD, C. E. N. and WRAY, D. A. 1947. Geology of the country around Barnsley. *Mem. geol. Surv. Gt Br.*

PASTIELS, A. 1964. Les lamellibranches non marins de la zone à *Communis* (Westphalian A) de la Belgique. No. 9. *Centr. nat. Géol. Houill.*

PONSFORD, D. R. A. 1955. Radioactivity studies of some British sedimentary rocks. *Bull. geol. Surv. Gt Br.*, No. 10, 24–44.

POOLE, E. G. and WHITEMAN, A. J. 1955. Exploratory boreholes in the Prestwich area of the south-east Lancashire Coalfield. *Trans. Instn min. Engrs*, **114**, 291–318.

SMITH, D. B. and FRANCIS, E. A. 1967. Geology of the country between Durham and West Hartlepool. *Mem. geol. Surv. Gt Br.*

SMITH, E. G., RHYS, G. H. and EDEN, R. A. 1967. Geology of the country around Chesterfield, Matlock and Mansfield. *Mem. geol. Surv. Gt Br.*

STRONG, T. M. W. *in* FALCON, N. L. and KENT, P. E. 1960. *q.v.*

TAYLOR, F. M. and HOWITT, F. 1965. Field meeting in the U.K. East Midlands oilfields and associated outcrop areas. *Proc. Geol. Ass.*, **76**, 195–209.

TRUEMAN, A. E. and WEIR, J. 1946–56. A monograph of British Carboniferous non-marine lamellibranchia, Pts 1–9. *Palaeontogr. Soc.* [*Monogr.*].

WEIR, J. 1960–68. A monograph of British Carboniferous non-marine lamellibranchia, Pts 10–13. *Palaeontogr. Soc.* [*Monogr.*].

WILSON, G. V. 1926. The concealed coalfield of Yorkshire and Nottinghamshire. 2nd edit. *Mem. geol. Surv. Gt Br.*

—— 1927. The eastern boundary of the concealed coalfield of Yorkshire and Nottinghamshire. *Summ. Prog. geol. Surv. Gt Br. for 1926*, 138–46.

Chapter V

PERMO-TRIASSIC

INTRODUCTION

APART FROM a small area of Lias in the north-east, the solid rocks at surface in the East Retford district belong entirely to the Permo-Triassic. They rest unconformably on the Coal Measures and are conformably succeeded by Jurassic rocks to the east. Fig. 20 shows the generalized stratigraphical succession of the Permo-Triassic rocks. The Rhaetic is included as the topmost division of the Triassic though it has faunal affinities with the Jurassic and is included in that system by some authors (e.g. Sherlock 1947; Swinnerton and Kent 1949). The Permo-Triassic has been traditionally divided at the junction of the Lower Mottled Sandstone and Upper Permian Marl into Triassic above and Permian below, but detailed work, primarily by Sherlock (1911, 1926, 1947) has shown that there is lateral passage between rocks of the two systems as thus defined. Indeed, the Lower Mottled Sandstone of the East Retford district is considered to be essentially a sandy facies of the Upper Permian Marl. The base of the Trias in this district is now thought to approximate to the base of the Bunter Pebble Beds, though it may be somewhat higher or a little lower locally. There are many other examples of diachronism in the Permo-Triassic (see below), a not unexpected feature of rocks which for the most part were laid down at or near the fluctuating margin of a depositional basin.

The Basal Permian Sands and Breccia, the Lower Permian Marl and the lowest parts of the Lower Magnesian Limestone crop out to the west of the district, but the easterly dip (Fig. 34) is responsible for the successive north–south outcrops of all the other Permo-Triassic formations within the district's boundaries. The hard, generally well-bedded Lower and Upper Magnesian Limestones produce wide dip-slopes, but the other divisions, consisting largely of mudstones and sand, give, in an area of regular dip and low relief, outcrops roughly proportionate to their thicknesses. Thus the Keuper Marl and Bunter Pebble Beds outcrops are approximately equal in area, together occupying five-sixths of the total area of the district (Fig. 2). Details of the concealed Permian rocks are known, however, at well over a hundred localities in borings for coal and oil.

CLASSIFICATION, CORRELATION AND
CONDITIONS OF DEPOSITION

Permian Rocks. The Permian rocks of the East Retford district, despite the use of the terms 'Lower', 'Middle' and 'Upper' in the nomenclature of the various divisions, are, with the possible exception of the Basal Permian Sands and Breccia, all of Upper Permian (or Zechstein) age. Many authors regard the Basal Permian Sands and Breccia as basal Zechstein deposits, but Wills (1956, pp. 22, 103) thought they were older, and Smith and Francis (1967, p. 92),

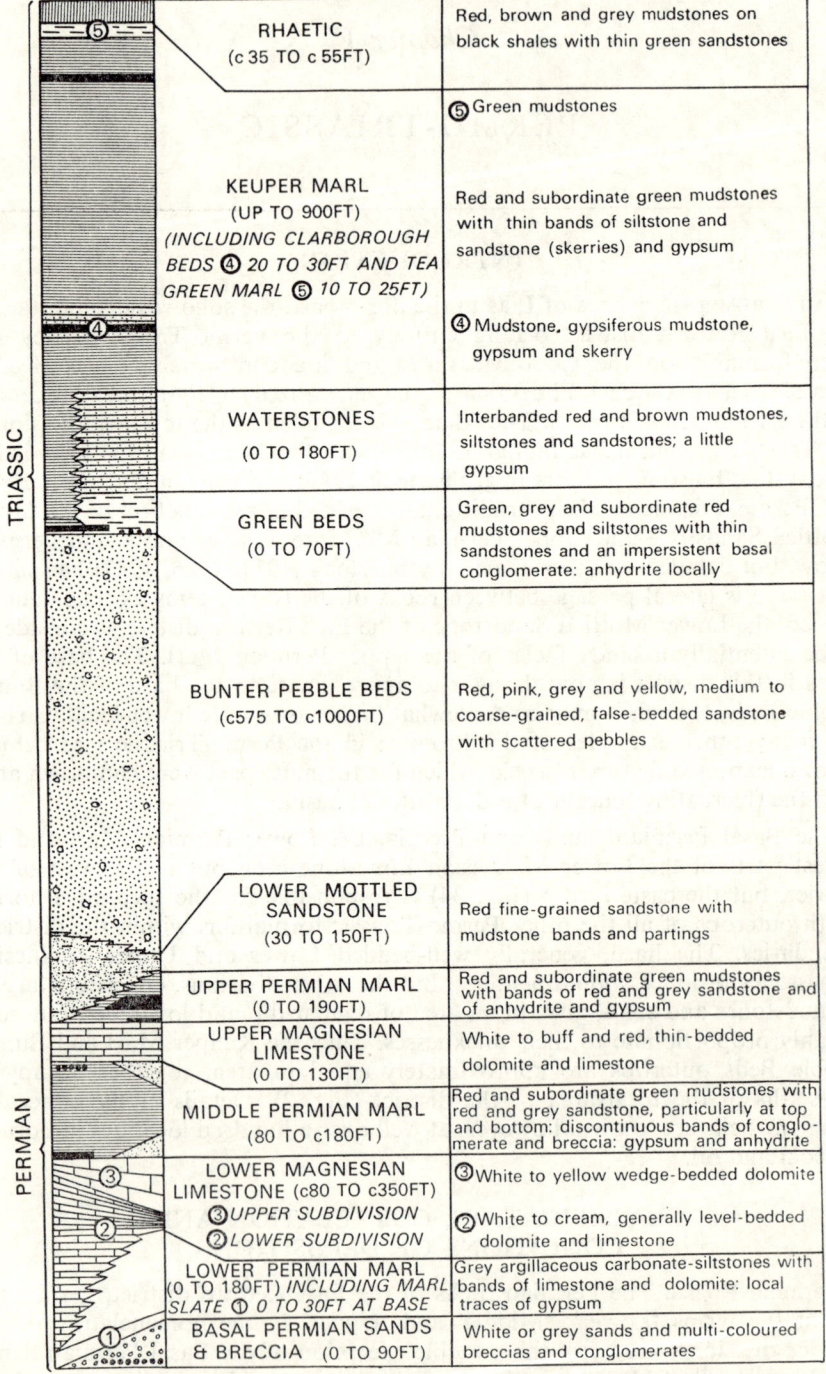

	RHAETIC (c 35 TO c 55FT)	Red, brown and grey mudstones on black shales with thin green sandstones
		⑤ Green mudstones
	KEUPER MARL (UP TO 900FT) (INCLUDING CLARBOROUGH BEDS ④ 20 TO 30FT AND TEA GREEN MARL ⑤ 10 TO 25FT)	Red and subordinate green mudstones with thin bands of siltstone and sandstone (skerries) and gypsum
		④ Mudstone, gypsiferous mudstone, gypsum and skerry
	WATERSTONES (0 TO 180FT)	Interbanded red and brown mudstones, siltstones and sandstones; a little gypsum
	GREEN BEDS (0 TO 70FT)	Green, grey and subordinate red mudstones and siltstones with thin sandstones and an impersistent basal conglomerate: anhydrite locally
	BUNTER PEBBLE BEDS (c575 TO c1000FT)	Red, pink, grey and yellow, medium to coarse-grained, false-bedded sandstone with scattered pebbles
	LOWER MOTTLED SANDSTONE (30 TO 150FT)	Red fine-grained sandstone with mudstone bands and partings
	UPPER PERMIAN MARL (0 TO 190FT)	Red and subordinate green mudstones with bands of red and grey sandstone and of anhydrite and gypsum
	UPPER MAGNESIAN LIMESTONE (0 TO 130FT)	White to buff and red, thin-bedded dolomite and limestone
	MIDDLE PERMIAN MARL (80 TO c180FT)	Red and subordinate green mudstones with red and grey sandstone, particularly at top and bottom: discontinuous bands of conglomerate and breccia: gypsum and anhydrite
	LOWER MAGNESIAN LIMESTONE (c80 TO c350FT) ③ UPPER SUBDIVISION ② LOWER SUBDIVISION	③ White to yellow wedge-bedded dolomite ② White to cream, generally level-bedded dolomite and limestone
	LOWER PERMIAN MARL (0 TO 180FT) INCLUDING MARL SLATE ① 0 TO 30FT AT BASE	Grey argillaceous carbonate-siltstones with bands of limestone and dolomite: local traces of gypsum
	BASAL PERMIAN SANDS & BRECCIA (0 TO 90FT)	White or grey sands and multi-coloured breccias and conglomerates

TRIASSIC

PERMIAN

FIG. 20. *Generalized section of the Permo-Triassic rocks*
The principal gypsum and anhydrite horizons are shown in black

describing equivalent rocks in Durham, suggested that they are of Lower Permian age by analogy with the Russian type area, where the Upper Permian commences at the base of the transgressing marine Kazanian deposits.

The Basal Permian Sands and Breccia were laid down under continental conditions by aeolian and intermittent fluviatile activity on an arid low-lying peneplain composed of folded and faulted Coal Measures rocks. There followed the rapid transgression of the Zechstein Sea across the area and the deposition of the marine Marl Slate. Most of the district continued to experience marine conditions almost until the end of the Permian period, but there was a shore-line in or near the south-western part after the end of Lower Magnesian Limestone times, and the Upper Magnesian Limestone and Upper Permian Marl are absent from this area. The presence of plant fragments and miospores in the marine Zechstein deposits shows that during their formation there was vegetation, probably largely coniferous, on the land surrounding the sea. During much of Upper Permian times, therefore, there was an amelioration of climatic conditions compared with those prevailing during the deposition of the Basal Permian Sands and Breccia.

The Permian rocks show a thickening across the district from less than 400 ft in the south-west to over 800 ft in the north-east (Fig. 21), i.e. towards the centre of the basin of deposition. That the thickness of the deposits is a direct reflection of the amount of subsidence of the basin floor is indicated by the lack of evidence, away from the immediate littoral area, of any appreciable deepening of the sea—indeed the marine deposits point to a wide, relatively shallow shelf sea. For much of the time this sea was highly saline, as is shown by the restricted faunas and the presence of evaporites.

Smith (1970) has recognized five evaporite cycles in the Permian succession of Durham and Yorkshire. In the East Retford district the Lower Magnesian Limestone together with the anhydrite or gypsum at the base of the Middle Permian Marl apparently correspond to the first cycle (EZ1). There is, however, the possibility that only the Lower Subdivision of the Lower Magnesian Limestone represents EZ1, in which case the Upper Subdivision of the Limestone with overlying sulphates would correspond to the second cycle (EZ2). Smith, however, infers that the carbonates of EZ2 are absent in the present district unless they are represented by thin bands of dolomite in the Middle Permian Marl. The Upper Magnesian Limestone and the overlying locally developed sulphates, apparently equivalent to the Billingham Main Anhydrite of County Durham, are the deposits of the third cycle (EZ3), and the Upper Anhydrite of the Upper Permian Marl (see p. 157) was formed during the fourth cycle (EZ4). The fifth cycle (EZ5) does not appear to be represented in the East Retford district.

Variations in the sequence of the Permian rocks are illustrated by Fig. 21. Here it can be seen that the lower part of the Lower Magnesian Limestone passes laterally into Lower Permian Marl, and there is also a general upward passage of Marl into Limestone. In the south-west of the district there is a thick development of sandy dolomite at the top of the Lower Magnesian Limestone. This may represent an area of carbonate deposition which was receiving quantities of sand from the nearby land, or it may originally have been sand of Middle Permian Marl age which has been extensively dolomitized. It is notable in the latter context that, in the same general area, a considerable part of the Middle Permian Marl succession is locally represented by sand. Breccias or conglomerates at several horizons in the Middle Permian Marl are evidence of

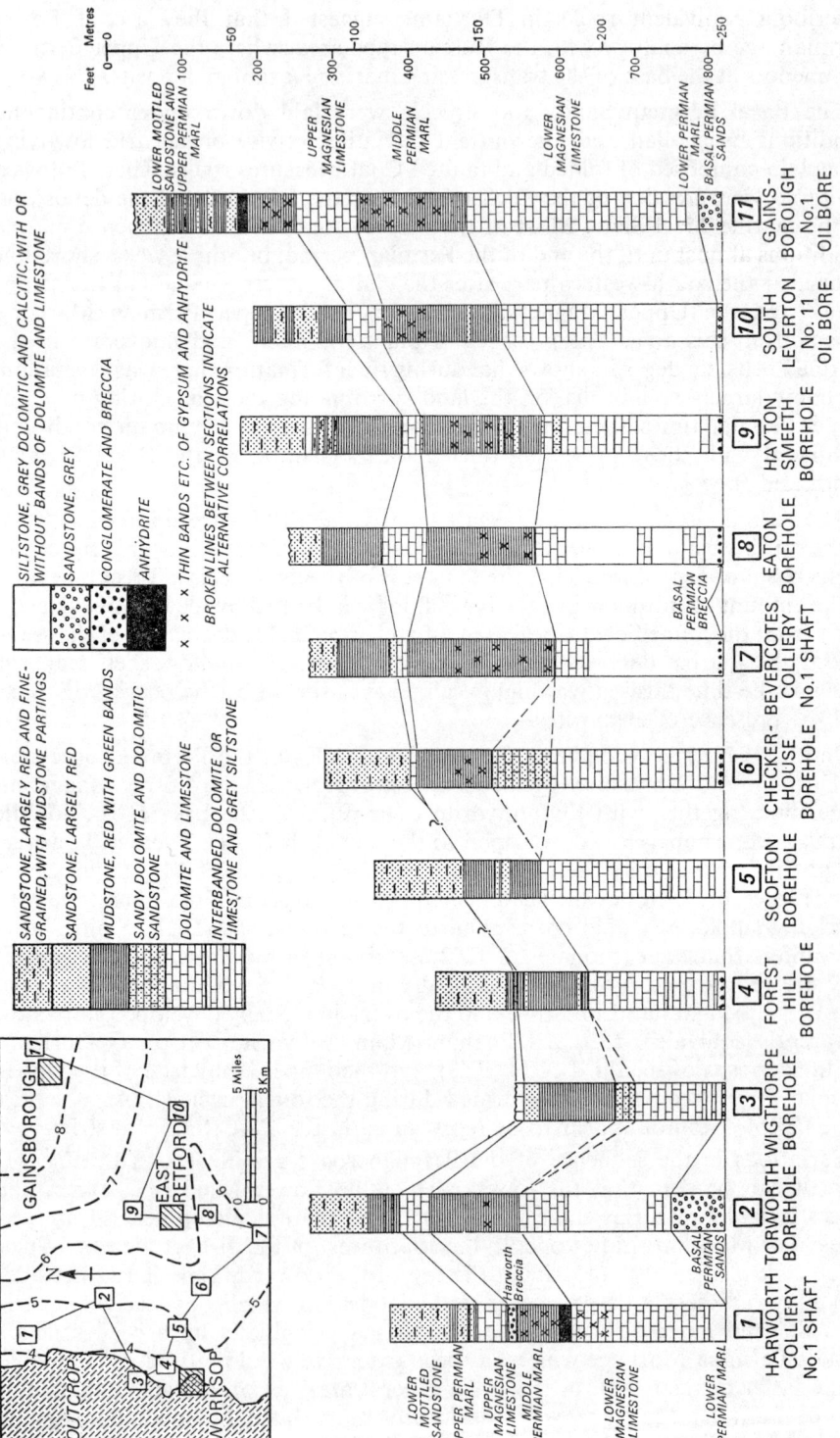

FIG. 21. *Comparative sections of the Permian rocks*

Broken lines on the inset map are isopachytes in hundreds of feet of the Permian rocks (including up to 150 ft of Lower Mottled Sandstone)

non-sequences in this formation, but whether individual horizons are widespread or of only local significance has not been determined. It is not clear whether there is a non-sequence in the south-western area where the Upper Magnesian Limestone is absent, or whether there are sands and mudstones, included with the Lower Mottled Sandstone or Middle Permian Marl, which are the age-equivalents of the Limestone.

The lateral passage which takes place between the Upper Permian Marl and the Lower Mottled Sandstone has already been referred to above. These deposits represent the final silting-up of the Zechstein basin and indicate the establishment of arid continental conditions which were to persist into Triassic times.

Triassic Rocks. The boundary between the Permian and Trias is here taken at the base of the Bunter Pebble Beds, though, as explained above, this is an arbitrary lithological line.

The Triassic period has been divided into a number of stages—in ascending order Scythian, Anisian, Ladinian, Carnian, Norian and Rhaetian (see Fig. 22)—which were established in the marine Alpine facies and have been correlated with the north European sequence with its classical tripartite division into Bunter, Muschelkalk and Keuper. Thus the Bunter rocks of Germany are considered to be of Scythian age, the Muschelkalk (following e.g. von Bubnoff 1935, table vi; Ricour 1963, table i; Klaus 1965) of Anisian and early Ladinian age, and the Keuper rocks of late Ladinian to Rhaetian age.

In Britain there are no rocks of Muschelkalk facies, and the rocks of Bunter and Keuper facies have been generally equated with the Bunter and Keuper rocks of Germany. Wills (1910, pp. 268–9), however, suggested a correlation of

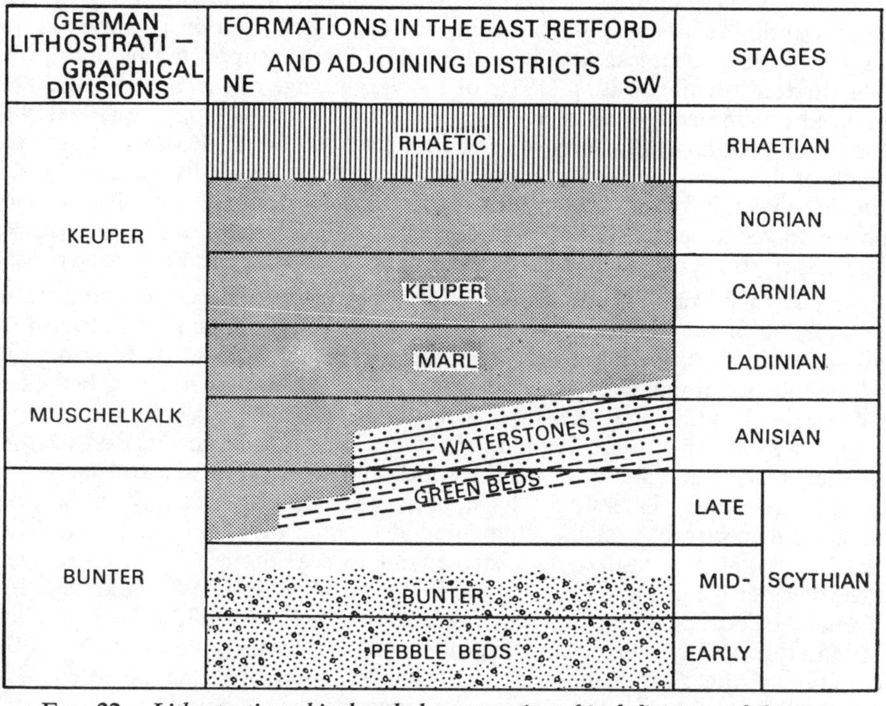

GERMAN LITHOSTRATI – GRAPHICAL DIVISIONS	FORMATIONS IN THE EAST RETFORD AND ADJOINING DISTRICTS		STAGES	
	NE	SW		
KEUPER	RHAETIC		RHAETIAN	
			NORIAN	
	KEUPER		CARNIAN	
	MARL		LADINIAN	
MUSCHELKALK	WATERSTONES		ANISIAN	
	GREEN BEDS		LATE	SCYTHIAN
BUNTER	BUNTER		MID-	
	PEBBLE BEDS		EARLY	

FIG. 22. *Lithostratigraphical and chronostratigraphical divisions of the Trias*

the Waterstones of Bromsgrove, Worcestershire, with the Muschelkalk, and Warrington (1967, 1970) showed that part of the Waterstones and the Keuper Sandstone of the Bromsgrove area were probably of Anisian age. The discovery of *Lingula* in the Waterstones at Stonish Hill, Eakring, (Rose and Kent 1955) suggested that rocks of equivalent age to the Muschelkalk might also be present in Nottinghamshire.

Smith and Warrington (1971) have shown that the Green Beds, Waterstones and the lower part of the Keuper Marl of the East Midlands are facies formations ranging in age from late Scythian to Carnian. The Green Beds are of late Scythian age in the East Retford district, but farther south they are entirely Anisian. The Waterstones are probably partly late Scythian and partly Anisian in age in this district, but are apparently no older than the Anisian in the south and may range up into the Ladinian. In other words, the base of the Keuper Marl is of late Scythian age in the northern part of the East Retford district, but of late Anisian age in the south, where the Waterstones facies is present; still farther south it is evidently Ladinian. There is evidence (*ibid.*) that the evaporite horizon in the Green Beds or basal Keuper Marl and that of the Clarborough Beds (Figs. 20 and 30) are not diachronous, the former being apparently of late Scythian age everywhere and the latter occurring constantly near the Ladinian–Carnian boundary. The boundary between the Norian and Rhaetian stages has not been fixed in Britain: it is assumed to approximate in this district to the base of the Rhaetic rocks (Fig. 22), but may be lower (see p. 198).

As has been pointed out by Lamplugh and others (1908, p. 36) farther south, there is evidence that the Bunter Pebble Beds are unconformably overlain by the younger Triassic rocks. The local conglomerate at the base of the Green Beds and differences in strike between the Pebble Beds and overlying rocks are suggestive of a stratigraphical break, which is further supported by Smith and Warrington's demonstration (1971) of the varying age of the Green Beds. This inferred unconformity has been correlated by Geiger and Hopping (1968) with the Hardegsen Disconformity of the continent (Trusheim 1963), which was the result of mid-Scythian earth movements. If, as seems highly likely, this correlation is correct, it follows that in the East Retford district the Bunter Pebble Beds are restricted to an early Scythian and early mid-Scythian age, and rocks of late mid-Scythian age, equivalent to the Solling Gruppe of Germany, are absent.

The Bunter Pebble Beds are believed to have been deposited under arid continental or semi-continental conditions, probably as delta fans in transitory lakes fed by fast-flowing rivers from the south or south-west. Considerable subsidence accompanied their deposition up to the mid-Scythian, when there was evidently uplift and the upper part of the Pebble Beds was probably removed by erosion. In late Scythian times the sea returned to the East Retford district, though not far away to the south and west there was still land. The Green Beds and Waterstones have been interpreted as fluvial and littoral deposits respectively, which accumulated at the margin of this land area, while the Keuper Marl was deposited farther out to sea, though not necessarily in deeper water (Smith and Warrington 1971). At first in the East Midlands this Triassic sea occupied much the same area as the final phase of the Zechstein Sea, but later, probably by Anisian times, it had spread much wider, far beyond the confines of the East Retford district. The sea was, like the Zechstein Sea, hypersaline, and again the presence of miospores, particularly abundant and accompanied by much plant debris in the Green Beds facies, shows that the surrounding land supported vegetation. At times the sea may have partially dried up, leaving local lacustrine conditions. The anhydrite and some of the gypsum occurring at the three principal sulphate horizons in the Keuper Marl (Figs. 20 and 30) may represent primary evaporite deposits, but the gypsum at the two higher horizons is fibrous in outcrop sections and clearly of secondary origin. The source of the fibrous gypsum is uncertain, but it was probably pre-existing sulphate.

The Lower Rhaetic deposits indicate near-normal marine conditions, but those of the Upper Rhaetic suggest that the Triassic period closed in this district with a local return to Keuper Marl-type conditions.   E.G.S.

BASAL PERMIAN SANDS AND BRECCIA

Basal Permian deposits are entirely concealed below younger rocks, but they have been encountered in many boreholes and shafts. They are very variable in composition, including round- and angular-grained sands, pebbly sands, conglomerates and breccias, with admixture and interbanding of several types at many places. Generally, however, the thicker developments consist, wholly or largely, of sands with well-rounded grains. The sands occur chiefly in the west and south-west of the district, where they show a similar development to that noted in the Sheffield district (Eden and others 1957, pp. 138–40) and in the northern part of the Chesterfield district (Smith and others 1967, pp. 195–7), but they are also found in a belt extending east-north-eastwards from the Wigthorpe area to Gainsborough. They vary irregularly in thickness up to 72 ft at Torworth and a reputed 90 ft at Bilby. Breccia and conglomerate are more widespread, occurring alone and also commonly underlying the sands. In the western part of the district they are typically only a few inches thick, but in the central and eastern parts there are generally a few feet. For example, 2 ft 10 in were recorded at West Drayton No. 1 Borehole, 3 ft 4 in at Hayton Smeeth Borehole, and 3 ft 10 in, interbanded with sandstone, at Sutton Borehole. Records of oil bores, based on chippings, are less reliable than those of cored holes, but the breccia is apparently up to 8 ft thick in the Walkeringham bores, at least 5 ft in several bores at Gainsborough, and probably more than 10 ft in some of the Bothamsall bores. The Basal Permian Sands and Breccia appear to be absent in the north-west of the district—they are not recorded in the shafts of Maltby Main, Harworth and Firbeck Main collieries—and in scattered localities elsewhere.

The sands are generally white or pale grey in colour, in some instances with a greenish tinge, and are largely quartzose, though locally feldspathic. They vary from fine- to coarse-grained, and from poorly sorted to equigranular. The grains vary in shape between angular and well rounded, but are usually subangular. The well rounded sands are usually the coarsest and the individual grains show signs of frosting. The sands may be loose and friable, or cemented by calcite or dolomite. Pyrite is generally abundant, especially in the deeper sections provided by oil bores in the east of the district. False bedding has been observed in cores from several boreholes. Upward passage of the sands into Lower Permian Marl by interbedding has been noted in the records of oil bores at Beckingham, Gainsborough and South Leverton. This phenomenon has not been seen in cored boreholes within the district, but it occurs in Barlborough

No. 1 Borehole, some 4 miles to the west (Eden and others 1957, pp. 140 and 181).

The breccia and conglomerate consist of angular or rounded fragments, up to an observed 1 inch in diameter, of sedimentary rocks (probably largely Carboniferous and including sandstone, mudstone, limestone and ironstone), acid and basic igneous rocks and metamorphic rocks (perhaps partly from the Charnwood area), quartz and jasper, set in a matrix of sandstone, siltstone, mudstone, limestone or dolomite. Pyrite is common in aggregates and as infillings of vughs; there are limonite grains, and well rounded quartz grains occur whatever the matrix.

Mr. R. K. Harrison notes that the sand-grade components of both the sands and the breccia are probably of polycyclic origin in part, and that both major and minor mineral suites indicate an original granitic provenance. Accessory minerals, in addition to widespread pyrite and two examples of marcasite, include anatase, apatite, garnet, ilmenite, leucoxene, monazite, rutile, tourmaline and zircon. Of these, garnet is relatively common and is taken to indicate derivation from a local metamorphic source. Apatite is less significant than in succeeding Permo-Triassic sandstones.

The breccia apparently represents the rock waste accumulated on the pre-Permian desert plain, while the conglomerate and associated sands are evidence of occasional stream activity. However, the bulk of the sand, like the 'Yellow Sands' of Durham (Smith and Francis 1967, pp. 92, 97–8), is believed to be of aeolian origin, though there was probably some subaqueous redistribution during Marl Slate times.   E.G.S.

DETAILS

Basal Permian deposits were not recorded in Harworth, Firbeck Main and Maltby Main Colliery shafts, nor in Nornay and Wallingwells boreholes, though their non-recognition in the boreholes may be due to the fact that no cores were taken at this horizon. Records from other boreholes for coal and water and from colliery shafts are as follows (depth to base of deposit in brackets following name):

Babworth Borehole (at 1197 ft): 2 in of conglomerate composed of pebbles, up to $\frac{1}{4}$ in diameter and consisting mostly of quartzites and Coal Measures rocks, in a matrix of pale grey, coarse-grained sandstone.

Barnby Moor Borehole (at 944 ft 10 in): 2 ft 4 in of soft, white, fine- to medium-grained sandstone, false-bedded in top 13 in with dips up to 17°.

Bevercotes Colliery: conglomerate with rounded and subangular pebbles up to 1 in diameter: 10 in thick at 1182 ft 5 inches in No. 1 Shaft; $7\frac{1}{2}$ in at 1185 ft 8 inches in No. 2 Shaft.

Bilby Borehole (at 726 ft): coring commenced in the Basal Permian Sands which are estimated to be 90 ft thick by the driller. Twenty-six feet of soft, light grey sandstone, fine-grained with medium- and coarse-grained sandstone lenses, were seen. The sand grains are mostly of quartz, but some are of feldspar; a few tiny pebbles were seen at the base. Strong false bedding was seen in places. Pipes infilled with Basal Permian Sands extend down into the underlying Coal Measures.

Blyth Borehole (at 416 ft): 7 in of grey sandstone with coarse, rounded grains, on 2 in of breccia.

Checkerhouse Borehole (at $965\frac{3}{4}$ ft): 7 in of breccia.

Crookford Borehole (at $1107\frac{1}{4}$ ft): 3 in of breccia composed of scattered small pebbles up to $\frac{1}{2}$ in diameter in a matrix of grey limestone with a few rounded sand grains.

Eaton Borehole (at 1299 ft 10 in): 10 in of conglomerate composed of unsorted, rounded pebbles up to $\frac{1}{2}$ in diameter and scattered pellets of green mudstone in a matrix of medium-grained sandstone; 1-in calcite vein.   E.G.S.

A thin section (E 33539) of a specimen from the base of this deposit is of fine gravel in a matrix of medium to coarse, rounded sand (igneous quartz, cherty silica, feldspars) partly replaced and cemented by fine rhombic aggregates of dolomite, with scattered iron sulphide granules, and veined by calcite. The pebbles, up to 1 cm in greatest diameter, show mainly low sphericity and lie in roughly parallel orientation. They include finely cryptocrystalline silica (?chert), calcitic sandstone, vein-quartz, siltstone, quartzitic conglomerate, chloritized and leucoxenized metasediment, sheared quartzite, and turbid quartz aggregates.   R.K.H.

Elkesley Borehole (? at 1137 ft): not cored, but Schlumberger log suggests that there is 1 ft of sandstone at this horizon.

Forest Hill Borehole (at 568¾ ft): 1 ft 1 in of breccia.

Harness Grove Borehole: 31 ft of grey sandstone not bottomed at 235 ft.

Hayton Smeeth Borehole (at 1379 ft 4 in): 3 ft 4 in of breccia.

Jockey House Borehole (at 1113 ft): 4 in of conglomerate. A specimen examined in thin section (E 22565) by Mrs. J. E. Morey showed rounded and angular fragments of sandstone, quartzite, granophyre, vein-quartz, rhyolite, ash and calcareous mudstone in a matrix of grey micaceous siltstone, calcareous in parts, with quartz grains. Aggregates of pyrite and grains of limonite also occur.

Lound Borehole (at 1308 ft): 3 ft 1 in of light grey sandstone with rounded grains; small pebbles which are concentrated in the top 9½ in and bottom 9 in.   E.G.S.

A specimen from the base proved, on examination in thin section (E 33537) to be a dolomitic conglomerate. It consists of scattered, well rounded gravel up to 1 cm mean diameter in a pale greyish green sandy matrix. The gravel, which is partly aligned parallel to the bedding, includes cherty silica, leucoxenized quartzite, dusty quartzite aggregates charged with micro-inclusions, quartz-porphyry, siltstone and sandy limestone, the last containing conspicuous heavy minerals. The sandy matrix consists mainly of composite grains of quartz and potash feldspar, clear igneous quartz, quartzite, chert and sandstone, cemented by coarse, interlocking anhedral dolomite, which partly replaces both sand and gravel. Heavy minerals (Fig.

28), iron sulphide apart, include conspicuous angular and subhedral zircon (colourless and purple), step-etched pale pink garnet, minor reddish brown rutile, leucoxene, yellowish anatase and sparse tourmaline, apatite and ilmenite.   R.K.H.

Manton Borehole (at 629 ft 5 in): 1 ft of 'soft blue gritstone'.

Manton Colliery: 1 ft 2 in of hard grey sandstone and fine pebble conglomerate with small angular fragments, at 654 ft 2 inches in No. 4 Shaft. In No. 2 Shaft 1 ft of grey sandy shale and stone-bind were recorded at the base of the Permo-Trias (619 ft 10 in).

Mattersey Borehole (at 1125 ft 10 in): 5 in of breccia, generally composed of angular fragments of shale and ironstone up to ¾ in diameter in a matrix of grey mudstone. However, a specimen (E 33542) from the top of the deposit is a gravelly sandstone consisting of poorly sorted, well rounded gravel, up to 1 cm diameter, scattered amongst coarse sand which includes frosted grains. Quartz, rock particles and feldspars are the main components, and the rock is cemented in parts by calcite and extensively replaced by iron sulphide.

Ranby Camp Borehole (at 1031 ft 7 in): ½ in of conglomerate composed of quartz pebbles in a matrix of grey mudstone.

Ranby Hall Borehole (at 889½ ft): 4 ft 1 in of calcareous and dolomitic sandstone on 5 in of breccia. The sandstone is mostly medium-grained, the grains being subangular, but thin beds of ill-sorted sand which include coarse, rounded grains occur in the top 1 ft. The breccia is composed of angular fragments, largely of brown and green mudstone, in a matrix of argillaceous, calcareous sandstone.
                E.G.S.

Specimens of the sandstone (from 888 ft) and conglomero-breccia (from 889 ft 5 in) were examined in thin section (E 33538A and E 33538 respectively). The former is a well consolidated sandstone with moderately sorted clastic grains averaging about 0·2 mm, and ranging from rounded to subangular. Apart from mainly igneous quartz, common rock particles include quartzite, cherty silica, jasper, mudstone, siltstone and chloritic metasediments. Feldspars include microcline and orthoclase. A little illite occurs in the matrix, but dolomite ($\omega = 1\cdot684$) and ferroan dolomite ($\omega = 1\cdot701$) with calcite are conspicuous and extensively replace the clastic grains. Secondary sulphide was identified by an X-ray powder photograph (NEX 269)[1] as marcasite. Heavy detrital minerals (Fig. 28) include well rounded to euhedral, colourless and purple zircon, rounded, dark reddish brown rutile, well rounded apatite, bluish green tourmaline, much pinkish step-etched garnet and staurolite. Leucoxene is probably largely authigenic. The conglomero-breccia consists of poorly sorted angular to subrounded grey and buff phenoclasts, up to $2\cdot5$ cm along their major axes, set in a fine ($0\cdot1$ mm) sandy matrix cemented with calcite. The grey phenoclasts include siltstones and mudstone composed (NEX 282) of illite, kaolinite, quartz and chlorite. The buff phenoclasts (NEX 283) also include mudstones composed of illite, kaolinite and quartz. Less common are fragments of porphyry, cherty silica, limestone, quartzite, calcitic siltstone and argillaceous metasediments. Calcite extensively replaces clastic grains.   R.K.H.

Ranskill Borehole (at $845\frac{1}{4}$ ft): 3 in of conglomerate composed of well rounded to subangular small pebbles, chiefly of quartz, in a matrix of pale grey, fine-grained sandstone. Pyritous vughs occur.

Scaftworth Borehole (at 1201 ft 2 in): 2 in of grey sand with rounded grains, containing sporadic pebbles and subangular fragments of quartzite. The top 2 in of the underlying Coal Measures have cracks infilled with this sand.

Scofton Borehole: Lower Permian Marl rests directly on Coal Measures, but seatearth at the top of the latter contains cracks infilled with fine conglomerate of Basal Permian type.

Shireoaks Colliery (at 214 ft 10 inches in one record; 223 ft 11 inches in another): 1 ft 8 in of 'grey hard rock'.

South Leverton Borehole (at $1959\frac{1}{2}$ ft): 9 in of breccia.

Steetley boreholes: Steetley Wood proved $30\frac{1}{2}$ ft of soft grey sandstone with rounded grains, not bottomed at 160 ft; Steetley Colliery No. 2 proved 41 ft of 'light sand rock' at 191 ft.

Steetley Colliery (at $223\frac{3}{4}$ ft): 47 ft 5 in of 'lights and rock'.

Stone Borehole (at $126\frac{1}{4}$ ft): $3\frac{1}{4}$ ft of soft grey sandstone with wind-rounded grains.
                                                                E.G.S.

A specimen from 126 ft (E 33372) is rather friable and moderately graded ($0\cdot06$ to $1\cdot0$ mm), with a fairly high proportion of well rounded clastics, scattered in an abundant carbonate matrix. Igneous quartz predominates amongst the clastic constituents, with subordinate metamorphic quartz, microcline, orthoclase and particles of quartzite, cherty silica, siltstone and carbonate-cemented sandstone. Illite and kaolinite are conspicuous in the matrix. Carbonates—fine rhombic aggregates of dolomite and anhedral calcite—are the main intergranular cement and exhibit replacement of the clastics. A little granular iron sulphide is present. A modal analysis gave in volume per cent: quartz (67); feldspars (5); illite, muscovite, kaolinite (6); carbonates (20); chlorite ($0\frac{1}{2}$); iron sulphide, composite grains (1). Heavy minerals, ignoring the sulphide, include conspicuous pinkish step-etched garnet and well rounded to subhedral, colourless, and less commonly purplish, zircon: subordinate dark reddish brown rutile, minor well rounded apatite, tourmaline, leucoxene and a trace of monazite also occur.   R.K.H.

Sutton Borehole (at 1096 ft 10 in): $5\frac{1}{4}$ ft of conglomerate and sandstone divisible, according to Mr. G. V. Wilson, into: black conglomerate (small pebbles) 2 in, grey conglomerate (small fragments in matrix of sand and limestone) 2 ft 10 in, sandstone (silica grains cemented by limestone) 1 ft 5 in, on grey conglomerate as above 10 in.   E.G.S.

Two specimens were examined in thin section—from $1091\frac{3}{4}$ ft (E 33541A) and $1096\frac{1}{2}$ ft (E 33541). The former is a gravelly sandstone consisting of well rounded, fine-grained (averaging 2 mm diameter) gravel and coarse sand (1 mm) scattered through a sandy and micaceous matrix. The coarser clastics comprise quartzitic aggregates, pyritized cherty silica, sandstone, porphyry, tuff and dolerite. The matrix consists of subrounded to subangular fine sand ($0\cdot1$ mm mainly) with abundant quartz, conspicuous

---

[1]Numbers prefixed NEX refer to X-ray powder photographs in the Northern England collection of the Institute of Geological Sciences at Leeds.

rock particles (cherty silica and quartzite) and potash feldspars. Finer constituents include brownish illite and chlorite, carbonate, goethite and leucoxene. Garnet is sparingly present.

The other specimen is a dolomitic gravelly sandstone consisting predominantly of grey, cross-bedded fine sandstone-siltstone, with intercalations of coarse sand and fine gravel and partings of dark grey mudstone. The coarser constituents are mainly well rounded. The predominant clastic constituents range between 0·01 and 0·4 mm, and comprise mainly rounded to subangular igneous quartz, orthoclase, microcline, brownish cherty silica, fused and sheared quartzite, argillaceous siltstone and mudstone. The main cement consists of coarse anhedral plates of ferroan dolomite ($\omega = 1\cdot701$). Heavy minerals include conspicuous euhedral to rounded grains of zircon; step-etched, pinkish garnet; dark reddish brown rounded to subangular rutile; minor well rounded apatite; brown tourmaline, and ilmenite.   R.K.H.

Torworth Borehole (at 864 ft): 72 ft 2 in of soft, white, medium-grained, feldspathic sandstone, marly at top.

Twyford Bridge Borehole (at 1172 ft): not cored, but 2 ft of sandstone were recorded by the driller.

Wallingbrook Wood Borehole (at about 200 ft): most of the core was lost and the thickness is not recorded. A few pebbles and samples of round-grained sand were recovered.

Welbeck Abbey Borehole (at 291¾ ft): 4 in of breccia composed of dark pellets up to ¼ in diameter in a grey calcareous matrix.
                E.G.S.

A specimen from 291½ ft (E 33540) is a calcitic gravelly sandstone; it is charged with rounded fine gravel, coarse sand and rounded to irregular iron sulphide nodules averaging 2 to 3 mm in diameter. The coarse components include calcarenite, fine carbonate-cemented sandstone, sandstone and limestone impregnated with sulphide prior to deposition,

polygranular quartz—chert, quartzite, vein-quartz and feldspar-porphyry. The matrix consists of fine sand, brownish illite, carbonate and sulphide.   R.K.H.

West Drayton No. 1 Borehole (at 1223¼ ft): 2 ft 10 in of well cemented conglomerate. Thin sections (E 22563–4) of two specimens, 1 ft apart, were examined by Mrs. J. E. Morey; the pebbles and fragments (which vary from subrounded to angular) range in size up to 2½ cm diameter and consist of quartzite, sandstone, vein-quartz, felsite, ?slate, limestone, dolomitic limestone, calcite, ash, rhyolite, decomposed basalt and granodiorite. One pellet of iron-stained baryte was found. The matrix is composed of granular and well crystallized calcite with a varying proportion of dolomite, small quartz grains and aggregates of pyrite.

Wigthorpe Borehole (at 287 ft): 2½ ft of soft grey sandstone with coarse, rounded grains.

Details of the Basal Permian Sands and Breccia in oil bores[1], none of which was cored through the Permo-Triassic, are based on the observations of British Petroleum Company's geologists.

No Basal Permian deposits were recorded in the Tickhill, Gringley and Grove No. 1 bores.

Ranskill No. 1 proved about 3 ft of white medium-grained sandstone with rounded grains, mostly consisting of quartz with rare particles of igneous rocks.

Walkeringham. 3 to about 9 ft of breccia, consisting of pebbles of quartz, quartzite and igneous rocks in a matrix of siltstone or fine- to coarse-grained, pyritic sandstone, were recorded.

Morton No. 1. 7 ft of white medium-grained sandstone were recorded at the base of the Permian.

Beckingham. The sands vary from a thin development in No. 1 Oil Bore to between 40 and 60 ft in No. 2, where the upper part is apparently interbedded with Lower Permian

---

[1]In each section of the Details of this chapter the oil bores are normally dealt with in the order: Tickhill, Ranskill, Gringley, Walkeringham, Morton, Beckingham, Gainsborough, South Leverton, Grove, West Drayton, Apleyhead, Bothamsall; i.e. as if connected by a line running roughly eastwards from Tickhill to Gainsborough and thence generally south-westwards to Bothamsall.

Marl. They are white or grey, fine-grained or silty, and, in some instances, calcareous and well cemented. Except in No. 1 Bore, conglomerate with pebbles of quartz and igneous rocks is recorded at the base.

Gainsborough. Basal Permian deposits are recorded in all the oil bores, varying from a trace to probably 30 ft. Breccia or conglomerate is up to at least 5 ft thick and may occur above or below a development of sands. Fragments vary from angular to well rounded and are commonly composed of quartz, quartzite or green igneous rocks, though metamorphic rocks, sandstone, limestone, chert, jasper and chalcedony are also recorded. The matrix may be sandy, calcareous or clayey. The sands are largely quartzose, and white or pale grey in colour, but are otherwise extremely variable: they may be very fine- to coarse-grained, with grains ranging from angular to well rounded, and are poorly sorted to well sorted and friable to strongly cemented. In several bores the upper part of the sands is apparently interbedded with siltstone and mudstone. Pyrite is common, and in some cases abundant in both the sands and the rudaceous deposits.

South Leverton. Conglomerate is recorded in ten oil bores; this is overlain by a few feet of sand in No. 9 Bore, and sand, probably about 20 ft thick, occurs alone in two other bores, at one of which there is interbedding with marl in the upper part of the deposit. The sand is white, fine- to medium-grained, and varies from loose to well cemented. In the other bores sand is either absent or cannot be distinguished with certainty from Coal Measures sandstone. Recorded pebbles from the conglomerate are composed of quartz and green igneous rock.

Grove. About 10 ft of white coarse-grained, poorly cemented sand with pyrite were recorded in No. 2 Oil Bore.

West Drayton. In No. 2 Oil Bore 2 or 3 ft of breccia were recorded.

Apleyhead. No Basal Permian deposits were recorded in No. 1 Oil Bore, but there is a thin pyritic breccia in Nos. 2 and 3.

Bothamsall. Basal Permian deposits, not recorded in two oil bores, attain a maximum thickness of about 17 ft (not confirmed by geophysical logging). Most of the bores show breccia or conglomerate, in some cases overlain by poorly sorted sand, and in only one bore (No. 13) is there sand alone. Recorded fragments and pebbles are composed of quartz, sandstone, and igneous and metamorphic rocks in a sandy or calcareous matrix. Pyrite is common.   E.G.S.

LOWER PERMIAN MARL

The Lower Permian Marl, entirely concealed by younger rocks, is inferred from underground information to be present throughout the district except locally in the north—where it is absent at Maltby Main Colliery and probably in a few uncored boreholes. The wide thickness variations illustrated in Figs. 21 and 23 are principally accounted for by local passage of the Lower Magnesian Limestone into Marl.

The Lower Permian Marl of this district consists partly of grey, finely laminated, argillaceous and carbonaceous dolomite-siltstone with a characteristic fossil assemblage (see below) consisting of plants, *Lingula* and fish, and partly of more massive argillaceous silt-grade equivalents of the Lower Magnesian Limestone. The former type is comparable to the well known Marl Slate of Durham which is equated (Sedgwick 1835, p. 75) with the German Kupferschiefer. In oblique section the rock displays the typical flecked appearance observed in the Durham Marl Slate (Magraw and others 1963, p. 169), but it is generally less carbonaceous and bituminous, and no bituminous odour (Smith and Francis 1967, p. 102) has been noticed. In addition to dolomite, the fine aggregate of carbonate composing the bulk of the rock includes, in some specimens, calcite, ankerite and siderite. It may be that calcite is predominant over dolomite in some sections, but ankerite and siderite are generally sparse, though nodules of sphaerosiderite occur. The non-carbonate fraction consists, in addition to carbonaceous matter and clay minerals (mainly chlorite and illite), of silt-grade quartz and microcrystalline pyrite. Much of this pyrite is present as spheres or framboids (Deans 1950, p. 347; Love 1962; Love and Amstutz 1966), 1 to 7 microns in diameter, and believed to be of bacterial (Schneiderhöhn 1923) or other microbiological origin (Love 1962). Pyrite also occurs megascopically, as does sphalerite, galena and chalcopyrite. Sulphides are, however, generally less abundant than in the Durham Marl Slate. Most authorities (Schouten 1946; Deans 1950; Dunham 1960, 1961; Love 1962; Hirst and Dunham 1963) agree that most or all of the pyrite in the Marl Slate and Kupferschiefer is syngenetic, but there is a difference of opinion as to whether the other metallic sulphides are of syngenetic or later origin.   E.G.S.

The Marl Slate of Durham has an enrichment of some metallic elements compared to average shales, and in this respect is geochemically similar to the Kupferschiefer (Deans 1950; Hirst and Dunham 1963). A semi-quantitative analysis by X-ray fluorescence spectrography was therefore carried out on selected specimens of Marl Slate from the East Retford district, and the results are expressed in Table 2 as ranges (cols. 1 and 2), and compared with ranges of values obtained by Hirst and Dunham (1963, table v) on samples of Marl Slates from Durham (col. A), and with averages and ranges for shale (Rankama and Sahama 1949, table 5.52; Krauskopf 1955, p. 416) (col. B).

Except for Zn, Pb and Mo, the metals reported in the East Retford specimens are not in abnormal quantities compared with the ranges and means for shales (col. B). Compared with published data for the Durham Marl Slate (col. A), the values lie towards the lower limits of the ranges given. The Pb and Zn anomalies are clearly accounted for by galena and sphalerite respectively. The latter appears to be post-depositional compared with the pyritospheres, thus supporting earlier paragenetic ideas (Hirst and Dunham *ibid.*, p. 921). The Mo may perhaps be concentrated by sorption on or in the bituminous matter, as suggested by Hirst and Dunham (*ibid.*, p. 930).   R.K.H.

In the East Retford district interbanding of the basal beds of the Marl Slate with Basal Permian Sands has been noted in several oil bores in the east, and there is a passage by admixture of sand and marl between the two divisions in Torworth Borehole. Interbanding of Marl Slate with basal Lower Magnesian Limestone occurs at Scaftworth Borehole (see p. 132). Bands of dolomite and limestone and of unlaminated marl occur within the Marl Slate at some localities, and these have a fauna like that of the Lower Magnesian Limestone.

In some sections in the western and central parts of the district the whole of the Lower Permian Marl is of Marl Slate lithology. The thickest section proved in a cored borehole is just under 13 ft at Ranby Camp, and it is probable that in this western and central region the Marl Slate is 8 to 20 ft thick over wide areas. In the east, oil bores apparently show that it reaches 30 ft in places. In some sections where the Marl Slate is overlain by blocky, unlaminated marl, there is evidence of a transition between these two basic types of lithology, making it difficult to draw a precise top to the Marl Slate. Generally, however, the apparently transitional rocks have more in common with the massive marl, including the occurrence of elements of the Lower Magnesian Limestone fauna.

The massive type of Lower Permian Marl is generally of a grey colour, similar to or of a lighter shade than the Marl Slate, but brown and greenish

TABLE 2. *Trace elements (in parts per million) in the Marl Slate, determined by X-ray fluorescence spectrography.*

	1	2	A	B
Cu	<20	<20	13–754	30–150
Rb	20–100	20–100	15–213	250–700
Zn	1000	n.d.–1000	0–31000	50–300
Sr	20–100	20–100	152–360	170
Y	n.d.	n.d.–20	—	28
Ti	100–500	100–500	2397; 3297*	4300
Zr	20–100	20–100	20–170	120–1000
Pb	20–100	n.d.–1000	23–2590	20
Cr	<20	<20	—	410–680
Mo	20–100	n.d.–100	0–825	1
Mn	20–100	20–500	170–4400	620
Co	n.d.	n.d.	0–229	10–50
Ni	<20	<20	33–445	20–100

n.d. not detected
—   no data

1. Marl Slate; E 35658: Barnby Moor Borehole at 935 ft.
   Analyst: Mrs. M. M. White.
2. Marl Slate; limits of determination for three specimens:
   E 35659: Ranby Hall Borehole at 884¾ ft.
   E 35797: Babworth Borehole at 1190 ft.
   E 35798: Ranskill Borehole at 840¾ ft.
   Analyst: Mrs. M. M. White.
A. Marl Slate, SE Durham; limits of 87 analyses in Hirst and Dunham 1963, p. 926.

*two oxide analyses recalculated to elemental proportions.

B. Average values and ranges for shale (Rankama and Sahama 1949, table 5·52; Krauskopf 1955, p. 416).

varieties are sometimes found and red bands have been recorded in the South Leverton area. It is composed of silt-grade dolomite or calcite with subordinate clay and quartz fractions. Lamination, not visible in hand specimen, can be seen to be weakly developed in some thin sections, where it is caused by streaks of probable carbonaceous matter and sparse framboidal pyrite. Traces of gypsum are recorded at a few localities. Bands of hard, dense, pale grey dolomite and limestone are common in many sections, particularly towards the top, where the boundary between the Lower Permian Marl and the Lower Magnesian Limestone has to be decided quite arbitrarily by the relative proportions of marl and dolomite or limestone present. Thus in adjacent sections equivalent beds may be allocated in the one case to the Lower Permian Marl and in the other to the Lower Magnesian Limestone. The marl beds that occur within the Lower Magnesian Limestone (pp. 122–3 and Fig. 21) are lithologically identical with the upper part of the Lower Permian Marl, and may even in parts approach a Marl Slate lithology. When the limestone between these beds and the Lower Permian Marl fails, the whole of the resulting marl is classed as Lower Permian Marl, which accounts for the apparently excessive thicknesses of that formation in some sections (e.g. the 180 ft at Bevercotes Colliery). At Markham Moor Borehole on the southern boundary of the district and in several nearby sections to the south the whole of the Lower Magnesian Limestone has become a variable sequence of interbanded marl and limestone (Edwards 1967, p. 123) which could well be designated Lower Permian Marl. It follows that thickness variations of the Lower Permian Marl above the Marl Slate are of little regional significance, and in Fig. 23 the Lower Permian Marl and Lower Magnesian Limestone have been treated as one unit for isopachyte purposes. The Marl Slate has been included only because it is not satisfactorily delimited or cannot even be separately identified in many of the borehole and shaft records.

The transgression of the Zechstein Sea at the beginning of Lower Permian Marl times was evidently sufficiently rapid to avoid much redistribution of the Basal Permian Sands (Smith and Francis 1967, p. 104). The Kupferschiefer and the Durham Marl Slate are believed to have been deposited in a barred basin of this sea, which had stagnant and foul bottom conditions (see e.g. Love 1962; Hirst and Dunham 1963). That the Marl Slate of the East Midlands accumulated under less euxinic conditions is indicated by the reduced concentrations of sulphide and bituminous material and by the intercalation of more normal Zechstein deposits. Schuchert (1915, pp. 266–9) estimated the depth of the Kupferschiefer sea at 300 to 600 ft, but these figures are perhaps excessive for the present area in view of the intimate association of the Marl Slate and the massive marl, which passes rapidly into Lower Magnesian Limestone of presumed shallow-water origin (see p. 123), and in view of the sporadic intercalation of the Marl Slate with the Lower Magnesian Limestone itself. It would seem, however, that the water must have been deep enough to submerge the Basal Permian Sands, which may have stood up as much as 90 ft above the general ground level at Bilby (see p. 108).

The upper part of the Lower Permian Marl was deposited in quiet water which may have been deeper than that in which the bulk of the Lower Magnesian Limestone carbonates formed, and shallower than in the Marl Slate sea, though the lithological differences between the massive marl and the Marl Slate could be accounted for by the removal of the basin bar or by more rapid carbonate precipitation.   E.G.S.

The only fossils yielded by the rocks of Marl Slate lithology in this district are fragmentary plant remains and a fauna consisting of *Lingula credneri* and palaeoniscoid fish. Sufficient palaeoniscoid remains have been collected, mainly from the basal 5 ft, to suggest that the fish fauna was similar to that represented by the well known collections from the Marl Slate outcrop in County Durham and from the German Kupferschiefer. Schuchert (1915) concluded that the Kupferschiefer fish were freshwater forms which were either washed into the sea or which lived in an upper layer of fresh water above saline bottom waters. Westoll (1941), on the other hand, considered that these fossils are the autochthonous remains of fish which inhabited "fairly shallow lagoons, perhaps of somewhat brackish nature, near the shores of the invading Zechstein sea".

The presence of *Lingula* is indicative of a marine environment. The absence of other invertebrates in the Marl Slate, however, suggests an analogy with the *Lingula* phase in the early stages of Upper Carboniferous marine transgressions in which the fauna was limited by bottom conditions inimical to most benthonic forms, or by the brackish nature of the water.

Marl Slate plants collected from this district have been described by Walton (1928) and Stoneley (1958). Those identifiable are nearly all conifer species which have been recorded elsewhere in the Marl Slate and Kupferschiefer. These coniferous remains were washed into the sea from their original environments, which Schweitzer (1962) suggests were dry, sunny slopes near the sea.

All the fossils collected from the Lower Permian Marl are listed in Plate X. Those from beds not of Marl Slate facies are discussed with the Lower Magnesian Limestone fauna on pp. 123–5.   J.P.

DETAILS

The Lower Permian Marl is apparently absent at Maltby Main Colliery, but it has been proved in all the other colliery shafts, though in only one has Marl Slate been identified. At Bevercotes 179 ft 7 in were recorded in No. 1 Shaft and 175 ft 10½ inches in No. 2, excessive thicknesses which are accounted for by the complete failure of the lower part of the Lower Magnesian Limestone, so that the thick development of marl usually found within the Limestone in this area (Fig. 23) and the Lower Permian Marl proper come together and are indistinguishable (Fig. 21 and p. 130). The whole of the section in No. 2 Shaft is described as grey marl, whereas in No. 1 Shaft the bottom 50 ft are said to be dominantly grey marl and the rest largely green and brown marl with bands of sandstone, dolomite and siltstone. In the latter shaft a ¼-in thick lens of cannel is recorded near the base of the formation. Records of the Lower Permian Marl in the other shafts are as follows: Firbeck Main, 36 ft 2 in of 'stone-bind'; Harworth No. 1, 10 ft of 'stone-bind' on 12 ft 4 in of 'grey bind and marl'; Manton No. 2[1], 22½ ft of 'blue shale'; Shireoaks, 33¾ ft of 'blue bind'; Steetley, 17 ft 5 in of 'dark blue bind with ironstone [sic]' on 43 ft 2 in of 'soft blue bind'.

The majority of the exploratory boreholes for coal have been cored through the Lower Permian Marl (for thicknesses see Fig. 23). Marl Slate has been recognized in almost half of these, and may be present in some or all of the rest. Details of the Lower Permian Marl in those boreholes where Marl Slate has been positively identified are as follows (thickness and depth to base of Lower Permian Marl given in brackets):

Babworth Borehole (13 ft 10 in at 1196 ft 10 in). The Marl Slate is at least 6 ft 10 in thick and contains chalcopyrite and galena near the base. Abundant plant fragments are recorded in the Lower Permian Marl as a whole, and there are carbonate bands which become more frequent upwards until the formation passes into the basal beds of the Lower Magnesian Limestone.   E.G.S.

A thin section (E 35797/P) of Marl Slate from 1190 ft is a laminated, pale grey dolomite- and calcite-siltstone containing films of opaque to translucent, bituminous material. Framboidal pyritospheres, 1 to 3 microns in diameter, are concentrated along the bituminous films. The last-named show minor crenulations due, in places, to intervening coarse grains of quartz and carbonate. In reflected light, pyrite is the predominant reflecting mineral with perhaps a trace of galena. There is also a trace of leucoxene. The results of a semi-quantitative X-ray fluorescence analysis are given in Table 2.   R.K.H.

Barnby Moor Borehole (11 ft 10 in at 942½ ft). The whole of the Lower Permian Marl is of Marl Slate lithology; no carbonate bands were recorded.   E.G.S.

A thin section (E 35658) of a specimen from 935 ft consists essentially of a fine aggregate (8 microns) of carbonate (predominantly dolomite with sparse siderite and ankerite) with interspersed flaky clay minerals oriented parallel to the bedding and responsible for the fissility of the rock. Yellowish brown carbonaceous flakes are also arranged parallel to the bedding. An X-ray powder photograph (NEX 793) of the whole rock, taken by Mr. K. S. Siddiqui, showed dolomite,

---

[1]At Manton Colliery No. 4 Shaft 92½ ft of 'hard grey shale with limestone bands' are recorded, but a comparison with nearby sections, including that of No. 2 Shaft, suggests that the larger part of these measures should be placed in the Lower Magnesian Limestone. A specimen from about 23 ft above the base is of Marl Slate lithology with *Lingula* cf. *credneri*.

illite, chlorite and quartz. A polished thin section confirmed that many of the discrete opaque constituents consist of framboidal pyrite ranging from 1 to 7 microns in diameter. Individual grains compare with the ordered framboids (types I, II) of Love and Amstutz (1966, p. 276). Sphalerite forms relatively coarse brown crystals up to 2 by 0·5 mm, transgressing bedding planes. In addition there is very fine, non-reflecting opaque material which is presumed to be carbonaceous. The results of a semi-quantitative X-ray fluorescence analysis are given in Table 2. The specimen is comparable with published descriptions of Marl Slate (Hirst and Dunham 1963) and with specimens (E 33058 and E 34401) from Bishop Auckland, County Durham, and North Pier, Cullercoats, Northumberland.   R.K.H.

Crookford Borehole (4 ft+ at 1107 ft). The basal 4 ft of the Lower Permian Marl, consisting of Marl Slate with thin carbonate bands, were recovered, but the succeeding 14 ft of core, some of which may have been of Marl, were lost.

Eaton Borehole (24 ft at 1299 ft). The basal beds of the Lower Permian Marl, at least 2 ft 1 in, and possibly as much as 10 ft 4 in, thick consist of Marl Slate. Carbonate bands make up about 30 per cent of the higher beds, the proportion of carbonate to marl increasing upwards so that there is a passage into the Lower Magnesian Limestone. A band of pink fibrous gypsum, $\frac{1}{8}$ to $\frac{1}{4}$ in thick, is recorded at 1275$\frac{1}{4}$ ft, and there are calcite veinlets in parts. Plant debris is abundant throughout the Lower Permian Marl, much of it pyritized.

Forest Hill Borehole (8 ft 8 in at 567 ft 8 in). The whole of the Lower Permian Marl has been described as grey shale, although a specimen collected from 559$\frac{1}{2}$ ft is a hard carbonate rock. A specimen from 563 ft has a Marl Slate lithology, but specimens from 6 in and 1 ft lower are of unlaminated marl with foraminifera. The two latter may be representative of the lowest Lower Permian Marl or be from a band or bands of unlaminated material within the Marl Slate.

Ranby Camp Borehole (12 ft 8$\frac{1}{2}$ in at 1031 ft 6$\frac{1}{2}$ in). The whole of the Lower Permian Marl, excepting the carbonate bands, is of Marl Slate lithology. A nodule from 1030 ft 10 in (E 35682) consisted of aggregated brown sphaerosiderite, the individual spheroids averaging 0·4 mm diameter. Secondary goethite was seen coating joints. An X-ray powder photograph (NEX 798) showed siderite, illite, chlorite, quartz and dolomite.

Ranby Hall Borehole (13$\frac{1}{4}$ ft at 885 ft). The lower part at least, and possibly the whole of the Lower Permian Marl, is of Marl Slate lithology. A thin layer of pyrite was noted at 876 ft 4 in.   E.G.S.

An X-ray powder photograph (NEX 789) of a specimen from 884$\frac{3}{4}$ ft taken by Mr. Siddiqui, showed dolomite, illite, chlorite and quartz. A thin section (E 35659) is very similar to that described from Barnby Moor Borehole (see above), except that here the carbonaceous and sulphide constituents are less concentrated and the sphalerite is lacking. The rock has a cellular texture, with pools of clear carbonate, which may represent a degree of recrystallization, separated by turbid carbonate-clay mineral aggregate and opaque dust. Clastic silty resistates form a weakly graded layer. The results of a semi-quantitative X-ray fluorescence analysis are given in Table 2.   R.K.H.

Ranskill Borehole (21$\frac{1}{4}$ ft at 845 ft). The Lower Permian Marl is of typical Marl Slate lithology in parts (e.g. at 840$\frac{3}{4}$ ft), less well laminated in others (e.g. at 843$\frac{3}{4}$ ft) and un-laminated with bivalves in still others (e.g. at 844$\frac{1}{2}$ ft). Carbonate bands occur in the top 11$\frac{3}{4}$ ft, where there are also veins and vughs containing calcite and chalcopyrite, and there is again an upward passage into the Lower Magnesian Limestone.   E.G.S.

A thin section (E 35798/P) from 840$\frac{3}{4}$ ft is a grey siltstone with marked conchoidal fracture. In section normal to the bedding there is a finely laminated structure, dolomite-silt being intercalated with dark brown to opaque bituminous films, which anastomose in places. The silty laminae are composed of fine (2 to 6 microns) dolomite, clay minerals and mica flakes, with coarser clastic particles —quartz and feldspar—which tend to disturb the evenness of the laminae. Opaque specks of framboidal pyrite are abundantly scattered throughout, and are especially concentrated along the bituminous flakes. Opaque minerals comprise predominant pyrite (mainly pyritospheres), with traces of probable galena. There are also minute non-reflecting opaque, metallic grains which could not be identified. In addition there are granules of probable leucoxene. The results of a semi-quantitative

H

X-ray fluorescence analysis are given in Table 2.   R.K.H.

Scaftworth Borehole (about 10½ ft at 1201 ft). The whole of the Lower Permian Marl may be Marl Slate, and argillaceous bands of this lithology occur in the overlying basal 2 ft of the Lower Magnesian Limestone. Thin carbonate bands, forming up to 50 per cent of the beds at the top, diminish in frequency downwards and are rare in the bottom 8 ft. Veinlets of galena were recorded in a carbonate band at 1192 ft 1 in.

Scofton Borehole (13¼ ft at 651½ ft). A specimen 4 ft from the base of the Lower Permian Marl is of Marl Slate lithology. No carbonate bands are recorded.

Sutton Borehole (7 ft 8 in at 1091 ft 7 in). The Lower Permian Marl is recorded simply as 'shale'. Specimens from 1090 ft, however, are of Marl Slate lithology, and it is probable that these are representative of the whole formation.

The presence of Marl Slate can be inferred in a few other boreholes from which no specimens are extant; e.g. the 15 ft 5 in of Lower Permian Marl in West Drayton No. 1 Borehole is recorded as finely banded grey limy shale with sporadic limestone partings, and the lowest 7 ft 2 in of the 10 ft of Lower Permian Marl (at 791 ft 10 in) at Torworth Borehole is described as finely banded grey mudstone. At Torworth the basal beds pass down, by admixture with sand, into the Basal Permian Sands, and galena is recorded at 785 ft 2 in.   E.G.S.

A specimen (E 35655) of Lower Permian Marl from 784 ft, just above the supposed Marl Slate, in Torworth Borehole is a poorly bedded, grey dolomite-siltstone with conchoidal fracture. Micaceous clay minerals (illite mainly) are aligned along the bedding planes, with carbonaceous flakes, ferric oxide streaks and spicules of organically derived brown collophane up to 1·5 mm length. An X-ray powder photograph (NEX 790) shows dolomite, illite, gypsum, chlorite and quartz.

This and several specimens from other boreholes show that there is a gradation of rock-types between Marl Slate and the massive, relatively carbon- and sulphide-free varieties of Lower Permian Marl. The other specimens examined in thin section are as follows:

Blyth Borehole (depth 409½ ft) (E 35679). This is a grey laminated, slightly cross-bedded carbonaceous dolomite-siltstone with abundant fossils—foraminifera, ? bryozoa and thin-shelled bivalves. It is composed of a very fine aggregate (4 to 8 microns) of carbonate (predominantly dolomite) with interspersed clay minerals, subparallel carbonaceous films, mica flakes, strings of opaque granules (including framboidal pyrite) and scattered silty resistates. There are also sporadic elongated collophane particles. Coarser patches of clearer carbonate are probably secondary. Galena occupies a trident of tapering veinlets, each ranging up to 1 mm thick and 1·5 cm long, transecting bedding planes. An X-ray powder photograph (NEX 797) shows galena, dolomite, quartz and illite.

Blyth Borehole (410 ft) (E 35656). Fragments of *Lingula?* occur in this grey, wavy-bedded, dolomitic siltstone, which consists of fine (4 to 24 microns) carbonates (dolomite predominant, minor ankerite), and interspersed clay minerals oriented along the bedding, with flakes of carbonaceous matter, streaks of ferric oxide, silty resistates and particles of collophane. An X-ray powder photograph (NEX 792) shows dolomite, illite, chlorite and quartz.

Welbeck Abbey Borehole (291 ft 1 in) (E 35680). This is a pale grey, fine-grained (4 to 8 microns) dolomitic siltstone, consisting of an aggregate of carbonate, micas and clay minerals (oriented along bedding planes), with streaks of brown carbonaceous material, silty resistates, and sporadic detrital clay particles (up to 0·3 mm length). Granular opaque constituents include framboidal pyrite spheres (up to 0·2 mm diameter). An X-ray powder photograph (NEX 796) shows illite, dolomite, chlorite and quartz.

Wigthorpe Borehole (279¼ ft) (E 35657). This is a dense, medium grey bedded dolomite-siltstone, shaly in part, bearing fragments of *Lingula?*. The fine-grained dolomite mosaic (4 to 24 microns) is charged with clay minerals aligned along the bedding with carbonaceous filaments, much opaque dust, and a little silty quartz. An X-ray powder photograph (NEX 791) shows dolomite, illite, chlorite and quartz.

Wallingbrook Wood Borehole (193 ft 11 in) (E 35681). This specimen is fairly typical of the massive type of Lower Permian Marl. It consists of massive, pale grey carbonate-siltstone composed of fine-grained calcite and dolomite (averaging 4 microns), bearing shell fragments and microfossils. Coarser patches of probable secondary carbonate occur sporadically. Clay minerals appear to be subordinate. Laminations are less well defined than in the specimens described above, though there are scattered subparallel ?carbonaceous films and streaks of sulphide, including framboids. Under high magnification ($\times$ 500) the carbonate matrix exhibits a chain-like microstructure. An X-ray powder photograph (NEX 795) shows calcite, illite, kaolinite, quartz and traces of ?chlorite and dolomite.   R.K.H.

No cores of the Lower Permian Marl have been taken in any of the oil bores of the district, but chipping samples have been examined by the geologists of the BP Company and the following account is based on their records.

In *Tickhill* No. 1 and *Ranskill* No. 1 oil bores respectively, 3 to 4 ft of dark grey and black siltstone and about 13 ft of grey calcareous siltstone and mudstone are assigned to the Lower Permian Marl.

There may be no Lower Permian Marl at *Gringley* No. 1 Oil Bore, in which case the red and grey shales underlying the Lower Magnesian Limestone belong to the Coal Measures. If Lower Permian Marl is present it is thin.

In the *Walkeringham* oil bores there are about 6 ft of Lower Permian Marl, described in No. 2 Bore as calcareous, irregularly laminated silty mudstone and mudstone, suggesting that it may all be Marl Slate.

In *Morton* No. 1 Oil Bore the Lower Permian Marl is represented by 8 ft of light grey calcareous silty mudstone.

At *Beckingham* 20 ft of grey calcareous siltstone with carbonaceous material are recorded in No. 4 Bore, and a similar thickness of argillaceous or silty measures occurs in No. 1 Bore. The Lower Permian Marl is apparently absent in Nos. 2 and 5 bores, though at the former interbedded red, brown and grey mudstone and siltstone occur in the upper part of the Basal Permian Sands and the lower part of the Lower Magnesian Limestone.

Lower Permian Marl is not recorded in a few of the *Gainsborough* oil bores, but in the rest it is up to about 95 ft thick. In the majority of these Marl Slate, described as grey to black, or brown, laminated shale or siltstone, commonly micaceous and carbonaceous, is readily identifiable. In many bores the whole of the Lower Permian Marl is represented by Marl Slate, generally 10 to 20 ft thick, but varying from under 5 ft up to, exceptionally, 30 ft. The Marl Slate may contain thin limestone bands, and may be interbedded in the lower part with Basal Permian Sands. In a number of bores the Marl Slate is overlain by up to 60 ft of grey, or in some instances brown or red, mudstone or siltstone, which is commonly calcareous and contains limestone bands. This general description applies to the whole of the Lower Permian Marl in those bores where Marl Slate is not recorded, but whether the Marl Slate is absent in these cases or, more probably, has been insufficiently described, is not clear.

In only one (No. 9) of the 15 oil bores at *South Leverton* is Lower Permian Marl not recorded. In nine bores Marl Slate—described as laminated or finely banded, grey or brown, calcareous siltstone or mudstone—can be recognized, varying in thickness from less than 10 ft up to about 30 ft, and forming the whole or only the lower part of the Lower Permian Marl. The rest of the Lower Permian Marl consists of grey, generally calcareous, mudstone and siltstone with limestone bands. Red bands are noted in several bores, pyrite in a few, and there were traces of gypsum in No. 5. These measures are very variable in thickness, up to a maximum of about 155 ft above 10 ft of Marl Slate at No. 11 Bore, and it is obvious at South Leverton that beds of the same age are arbitrarily referred to the Lower Permian Marl or Lower Magnesian Limestone according to whether mudstones and siltstones or limestones predominate in them.

In the *Grove* oil bores the Lower Permian Marl is represented by about 15 ft of measures described as grey and brown, fine-grained, laminated siltstone—evidently all Marl Slate.

West Drayton No. 2 Oil Bore shows 20 ft of 'grey striped marls' below the Lower Magnesian Limestone; presumably these are Marl Slate.

In the *Apleyhead* and *Bothamsall* oil bores the Lower Permian Marl is represented by

about 7 to 22 ft, but generally 10 to 15 ft, of
Marl Slate—grey, finely laminated, calcareous
siltstone and shale with carbonaceous

material. Siderite and pyrite, and in one case
limestone bands, are recorded at Bothamsall.
                                                              E.G.S.

LOWER MAGNESIAN LIMESTONE

The Lower Magnesian Limestone crops out along the western edge of the
district, where it forms a gentle dip-slope dissected by easterly-flowing streams.
In addition there are, in the Gildingwells area west of Firbeck Main Colliery,
several small inliers (Fig. 23) formed by depositional domes of dolomite protruding through the Middle Permian Marl (see below). As has already been
explained (pp. 112–5) the bulk of the Lower Permian Marl consists of argillaceous carbonaceous dolomite-siltstones that are the age-equivalents of Lower
Magnesian Limestone elsewhere, and, because of rapid facies changes, thickness
variations are only meaningful when the two formations are considered
together. In Fig. 23 isopachytes have been drawn for the combined formations,
and show a thickening from less than 150 ft in the west of the district to over
350 ft in the north-east. The apparently anomalous thick development of over
300 ft around Checkerhouse Borehole in the south-west is referred to below.

In the west the Lower Magnesian Limestone is readily divisible into two parts,
the upper of which corresponds to the Upper Subdivision of south Yorkshire
(Edwards and others 1940, pp. 124–6; Mitchell and others 1947, pp. 114–5;
Edwards and others 1950, p. 39; Eden and others 1957, p. 142), and the lower
of which, together with much of the Lower Permian Marl as is the case in the
Chesterfield district (Smith and others 1967, pp. 200–1), corresponds to the Lower
Subdivision. Recently a distinctive thin group of dolomites and mudstones
associated with a disconformity have been recognized at the junction of the two
subdivisions throughout the outcrop from Ripon to Nottingham (Smith 1968).
These beds—the Hampole Beds—can be seen at several outcrop localities in the
East Retford district and can be tentatively identified in a few western boreholes.
In the central and eastern parts of the district, however, the Hampole Beds
cannot be recognized, but whether this is due to facies change or to inadequacy
of the borehole records is not clear. It may be that where the two subdivisions
can be distinguished in the east by the lithological distinctions outlined below,
the junction is not at the Hampole Beds horizon.

The Lower Subdivision is only seen at outcrop where valleys and quarries cut
through the overlying Upper Subdivision. Typically it consists of white, grey or
cream to brown, thinly bedded dolomites, dolomitic limestones and limestones,
which vary from compact to saccharoidal and from fine- to coarse-grained, and
which contain ooliths and, less commonly, pisoliths and oncoliths. Cross-
lamination, ripple-marks and stylolites are common. The bedding, viewed
broadly, is level and regular, but in detail it is often uneven, 'ropy' or nodular,
especially in the lower parts of the Subdivision, a feature that may be due to
pressure-solution, as suggested by Woolacott (1919, p. 164). No reefs have been
seen in the district, but they are known to the north (Mitchell 1934, pp. 136–8)
and to the west (Eden and others 1957, p. 142), and debris of reef-forming
algae is found in the Limestone. Partings and bands of grey marl (argillaceous
carbonate-siltstone), identical with the bulk of the Lower Permian Marl, are
numerous and tend to become more frequent towards the base, where there is
usually passage by alternation into the Lower Permian Marl. At a few localities
where the Limestone rests on Marl Slate there is interbanding of the two.

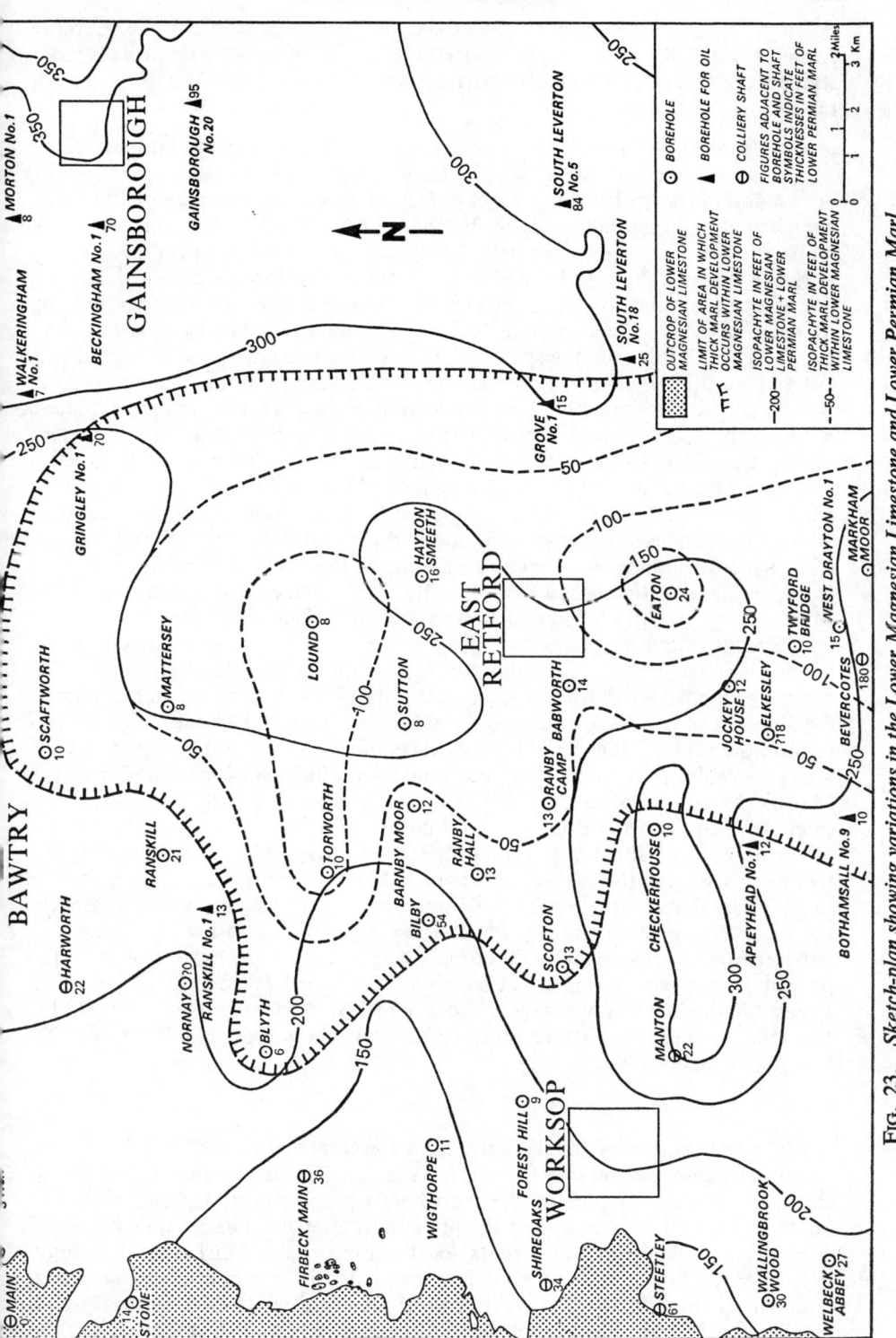

FIG. 23. Sketch-plan showing variations in the Lower Magnesian Limestone and Lower Permian Marl

Partings of green or greenish grey clay also occur in the Limestone, more commonly in the Lower than in the Upper Subdivision, and usually coat stylolitic surfaces. Carbonaceous partings are also recorded in the lower part of the Lower Subdivision.

The Upper Subdivision forms most of the Lower Magnesian Limestone outcrop within the district. It consists of white to brown or yellow and occasionally pink or red dolomites, which range from porcellanous to coarse-grained and exhibit a variety of bedding, but are typically coarsely saccharoidal and wedge-bedded. Apparently it is mostly composed of recrystallized oolites and pisolites. Locally in the west where the Upper Subdivision is thinly developed—25 ft or less—it is lithologically similar to the Lower Subdivision and can be distinguished only by its position above the Hampole Beds. Generally the Subdivision is less than 100 ft thick, though the lower boundary is obscure in many sections, and it is everywhere thinner than the combined Lower Subdivision and Lower Permian Marl. Sandy dolomite with a high proportion of quartz occurs in several places, particularly at the top of the Subdivision, and is a feature of the south-western part of the district—the same area in which the Middle Permian Marl contains much sand (p. 141 and Fig. 25). The thick development of the Lower Magnesian Limestone in the Worksop–Checkerhouse area (Fig. 23) is apparently due to an additional thickness of sandy dolomite and dolomite-sandstone, the incidence of which corresponds to an area of thinning in the Middle Permian Marl (Fig. 25). This has led Edwards (1951, p. 101) to place the sandy dolomite in the Middle Permian Marl. Thin sections show that there has been considerable replacement of quartz grains by dolomite (see p. 133), and it could be that the deposit was originally a sandy development of the Middle Permian Marl which has been extensively dolomitized. On the other hand it may be that dolomite-replacement of quartz is a less significant factor, and that the deposit accumulated in the Lower Magnesian Limestone sea into which quantities of sand were washed, resulting in sandy dolomites and dolomitic sandstones similar to the Mansfield Sandstone (see e.g. Smith and others 1967, p. 201). In places, for example in some of the Bothamsall oil bores, there is interbanding of the dolomite with red mudstone at the top of the Lower Magnesian Limestone, suggestive of a passage into the Middle Permian Marl. In most places, however, the junction is sharp, and in the Gildingwells area dome-shaped structures in the Limestone project into the Marl. These are believed to have originated as banks of ooliths, probably submarine dunes similar to those forming at the present time on parts of the Great Bahama Bank (Smith 1968, p. 477; Newell and Rigby 1964). They show a general north-north-westerly alignment comparable to similar occurrences farther south (Eden and others 1957, p. 146 and fig. 27; Smith and others 1967, p. 201).

In the central part of the district a thick development of grey 'marl' is found within the Lower Magnesian Limestone (Fig. 23). This is generally composed of carbonate-siltstone similar to the marl bands already described in the Lower Subdivision, and to the Lower Permian Marl, but it contains thin bands of dolomite and limestone, and parts locally approach a Marl Slate lithology. It is pyritous in places, and abundant plant debris occurs at some horizons. Over 100 ft thick in the Torworth–Lound area and south of East Retford, it reaches nearly 140 ft in Lound Borehole and nearly 160 ft at Eaton. On the southern boundary of the district, between Bevercotes and Markham Moor, the 'limestone'[1] separating the marl from the Lower Permian Marl fails and the two marls are indistinguishable (see p. 114). Generally the marl is a facies of the Lower Subdivision though dolomites of the Upper Subdivision apparently pass into marl locally, and at Markham Moor Borehole (Edwards 1967, pp. 123, 232) virtually the whole of the Lower Magnesian Limestone is marl with dolomite and limestone bands. In such cases, however, it may be that the Upper Sub-division is thin or absent.

In addition to dolomite and calcite and, locally, quartz, the Lower Magnesian Limestone contains a little ankerite in places, and there are accessory clay minerals, feldspars, micas, rock particles and oxides and hydroxides of iron. The last-named are responsible for the local pink or red coloration of the dolo-mites. Heavy minerals recorded include, in addition to iron ores, garnet, tour-maline and zircon. Pyrite occurs as specks and aggregates, and is also reported in pellet form: it is found most commonly in the lower part of the Limestone, and particularly at the base. Epigenetic pyrite, calcite and sphalerite occur in thin veins and as crystals infilling vughs, and small amounts of galena are recorded as specks and joint-coatings.

Anhydrite and gypsum have been noted in several underground sections, usually occurring as streaks, sometimes associated with red or grey mudstone bands, but gypsum is found in the cavities of honeycombed dolomite at Maltby Main Colliery, and what may be nodular anhydrite is recorded at South Lever-ton Borehole (p. 131). These evaporites occur most commonly high in the Lower Magnesian Limestone, apparently in, or at the base of, the Upper Subdivision.

The Lower Magnesian Limestone was clearly laid down in shallow saline water, and there is even evidence (p. 130) that occasional deposits accumulated under supratidal conditions. The large-scale false bedding common in the Upper Subdivision suggests that, on the whole, the water was somewhat deeper in Upper than in Lower Subdivision times.   E.G.S.

Palaeontology. In this district it has been found impossible to recognize any difference between the faunal content of the Lower Magnesian Limestone and that of the Lower Permian Marl of non-Marl Slate lithology, and they are discussed here together. All the identified macrofossils from these formations are listed in Plate X. The only previous records of macrofossils from the Lower Permian Marl and Lower Magnesian Limestone of the East Retford district were made by Walton (1928), Edwards (1951, pp. 98–101) and Stoneley (1958, p. 300) and these are incorporated in Plate X.

The difficulty of distinguishing in boreholes between the Upper Subdivision of the Lower Magnesian Limestone on the one hand and the Lower Subdivision and Lower Permian Marl on the other (see p. 120) makes it impossible to be precise about the fossils each has yielded. It is clear, however, that a very large proportion of the collections listed on Plate X are from the Lower Subdivision and the Lower Permian Marl. These beds contain a characteristic Zechstein fauna, but are generally lacking in important reef-building elements such as calcareous algae and the bryozoan *Synocladia* which are well known from the lower parts of the Middle Magnesian Limestone reef of County Durham

---

[1]In many records the term *limestone* has been used for dolomite and dolomitic limestone as well as for rocks composed of calcite. Where the term appears in quotes it should therefore be interpreted in this broad sense, i.e. as connoting carbonate rock.

EXPLANATION OF PLATE X

MACROFOSSILS RECORDED FROM BOREHOLES THROUGH THE LOWER

MAGNESIAN LIMESTONE AND LOWER PERMIAN MARL

Depths are given in feet from surface. Underlined figures refer to occurrences in rocks of Marl Slate lithology. Figures followed by (?) indicate a doubtful record.

The only macrofossils collected from the Lower Magnesian Limestone and Lower Permian Marl of this district not included here are (a) *Bakevellia sp.* from Steetley Quarries, (b) cf. *Strophalosia morrisiana* at 600 ft and *Lingula* cf. *credneri* at about 630 ft in Manton Colliery No. 4 Shaft, (c) *Schizodus obscurus* at 2242 ft in Gainsborough No. 1 Oil Bore, (d) Fish plate indet. at 647 ft in Scofton Borehole, (e) *Pterospirifer alatus* at about 675 ft and *Lingula credneri* at 720 ft in Harworth Borehole (on the site of one of the shafts at Harworth Colliery).

Markham Moor Borehole is almost exactly on the boundary between the East Retford and Ollerton districts; details of the rocks in the borehole are given by Edwards (1967, pp. 232–6).

The microfossils found in the Lower Magnesian Limestone and Lower Permian Marl of the East Retford district include miospores, microplankton, foraminifera and ostracods. Recorded miospores and microplankton are listed in Fig. 24 and discussed on pp. 136–9; the ostracods and foraminifera are discussed on pp. 135–6.

(Trechmann 1925; Smith and Francis 1967). However, together with strata at similar horizons in adjoining areas of Derbyshire and Nottinghamshire (see e.g. Anderson 1964; Edwards 1967), they have yielded a larger and more varied foraminiferal fauna as well as some brachiopod and bivalve forms, such as *Craspedalosia lamellosa, Dasyalosia goldfussi* and *Aviculopinna? pinnaeformis*, commonly recorded from the German Zechstein but hitherto little known from the Upper Permian marine strata of this country.

The most characteristic fossil assemblage found in the Lower Subdivision of the Lower Magnesian Limestone and the Lower Permian Marl is an association of brachiopods such as *Horridonia, Lingula* and *Strophalosia*, with foraminifera and rarer bryozoa. Only locally, especially in the western part of the district, e.g. between 220 and 230 ft in Welbeck Abbey Borehole and 283 and 284 ft in Wigthorpe Borehole, are bivalves the commonest element in the fauna, with *Bakevellia binneyi* the most abundant form. The general fauna contrasts with that of the lower part of the Lower Magnesian Limestone along its outcrop from central Yorkshire to Nottingham, including the area immediately west of the present district. In these outcrop areas bivalves, especially *Bakevellia* and *Schizodus* and, to a lesser extent, gastropods are the commonest fossils (Kirkby 1861; Eden and others 1957, p. 144). It is comparable, however, with the fauna recorded by Malzahn (1957) from equivalent horizons in Zechstein 1 argillaceous and carbonate rocks of north-western Germany.

Probably the only identifiable animal fossils collected from the Upper Subdivision of the Lower Magnesian Limestone in this district are *Schizodus obscurus* from 2242 ft in Gainsborough No. 1 Oil Bore, bivalves and foraminifera from 850 ft in Sutton Borehole, and *Bakevellia sp.* from Steetley Quarries. The last is the only fossil so far recorded from the outcrop of the Lower Magnesian Limestone in the East Retford district. The plants, *Lingula*, gastropods and *Bakevellia* from above 1100 ft in Markham Moor Borehole just beyond the southern boundary of the district may also be from the Upper Subdivision.

The invertebrate remains in the Lower Permian Marl and Lower Magnesian Limestone of this district are generally well preserved, and the presence of bivalves retaining both valves in position, productoids with fragile spines intact, and fenestrate bryozoa in large fragments suggests that they remain at or near the site where they lived, which in most cases was probably a soft sediment-covered sea-bottom. Most of the identifiable plant remains other than miospores (see below) are fragments of transported terrestrial conifers similar to those in the rocks of Marl Slate lithology (see p. 116). Carbonaceous, mostly strap-like, algal remains are common however and are probably of marine origin. Palynological studies have been carried out only on samples of the Lower Magnesian Limestone from Sutton Borehole. The assemblages, consisting of abundant miospores and sporadic microplankton, are described on pp. 136–9.   J.P.

DETAILS

Exposures in the area around Maltby Main Colliery have been examined by Dr. G. H. Mitchell. In the Limekiln Lane quarry [5629 9287][1], now flooded, he measured 8 ft of white to buff, thinly bedded saccharoidal dolomite, harder and close-grained in the top 2 ft and friable in places towards the bottom of the section. In the large overgrown quarry

---

[1]National Grid References are given in this form throughout the memoir. All fall within the 100-km grid square SK.

EXPLANATION OF PLATE XI

PERMIAN MACROFOSSILS

The specimens figured are from the East Retford district unless otherwise stated.

1. *Schizodus obscurus* (J. Sowerby): right valve, Upper Magnesian Limestone, Oldcoates, Nottinghamshire. (B1021) × 2½.

2. *Tubulites permianus* (King); Upper Magnesian Limestone, old quarry on south side of St Mark's Church, Oldcoates [SK 5869 8860]. (PJ 3108) × 4.

3. *Liebea squamosa* (J. de C. Sowerby); left valve, Upper Magnesian Limestone, old quarry, 1700 yd at 312° from St John's Church, Balby, Yorkshire [SE 5523 0260]. (Doncaster Sheet 88). (PJ 3149) × 2½.

4. *Permophorus costatus* (Brown); Lower Magnesian Limestone. Dalton Borehole, depth 1474 ft, Nottinghamshire [SK 7575 7322] (Ollerton Sheet 113). (Bm 2886) × 2½.

5. *Streptorhynchus pelargonatus* (Schlotheim); Lower Magnesian Limestone, Barnby Moor Borehole, depth 903 ft, Nottinghamshire. (YPF 4168) × 2½ See p. 135.

6. *Strophalosia morrisiana* King; Lower Magnesian Limestone, Sutton Borehole, depth 1072 ft, Nottinghamshire. (B 1419) × 2½.

7. *Nucula ? tateiana* King; Lower Magnesian Limestone. Welbeck Abbey Borehole, depth 223 ft 7 in, Nottinghamshire. (Bq 5274) × 4.

8. *Acanthocladia anceps* (Schlotheim); Lower Magnesian Limestone, Alverley Hall Quarry, near Doncaster, Yorkshire [SK 5538 9918] (Doncaster Sheet 88). (PJ 3192) × 2½.

9. *Horridonia horrida* (J. Sowerby) [morphological variety *hoppeianus* Eisel]; Lower Magnesian Limestone, Scaftworth Borehole, depth 1140 ft 7 in, Nottinghamshire. (Bq 1096) × 1½.

10. *Lingula credneri* Geinitz; Lower Permian Marl, Babworth Borehole, depth 1194 ft, Nottinghamshire. (Bl 5466) × 2½.

11. *Aviculopinna ? pinnaeformis* (Geinitz); posterior part of left valve; Lower Permian Marl, Sherwood Colliery, Nottinghamshire [SK 5370 6247] (Chesterfield Sheet 112). (WG 669) × 1½. See pp. 135–6.

All specimens figured are in the collections of the Institute of Geological Sciences, Leeds: registered numbers are given in brackets.

PLATE XI

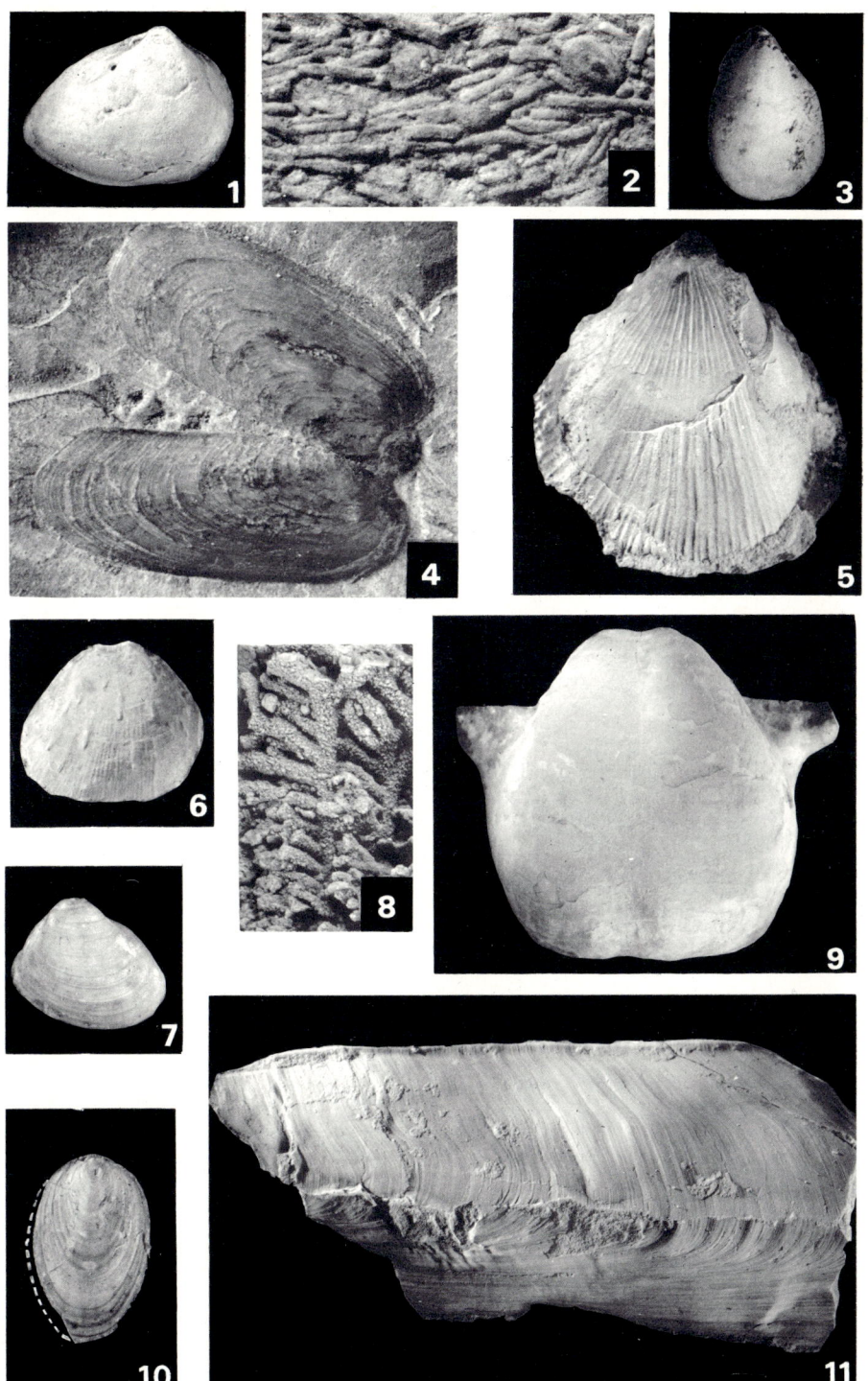

PERMIAN MACROFOSSILS

[563 923] on the south side of Tickhill Road, north of Hanging Holt, he saw fine saccharoidal dolomite, friable in parts and generally showing thin, regular bedding, but with slight wedge bedding in the western face. The railway cutting [551 919] south-west of Maltby Station and the scattered small quarries on Maltby Far Common showed white to cream-coloured, fine saccharoidal dolomite. Wedge bedding, with dips mostly to the north-east and south-west, but also to the north-west, is extensively developed in the cutting, and on the south side of the road bridge a lens of red marl, up to 6 in thick and more than 50 yd in lateral extent, was observed by Dr. Mitchell in the dolomite [5517 9200].

Numerous small exposures can be seen in the steep-sided valley which is occupied by Firbeck Dike and which runs eastwards from the grounds of Roche Abbey to Stone [555 898]. They show that the dolomite, white throughout, is well bedded, and in places flaggy in the upper part, and thickly bedded or massive in the lower.

Old quarries [5506 8835, 5560 8825, 5572 8760] south-west of Firbeck formerly showed up to 14 ft of white, bedded or massive 'limestone', but some of the sections are now obscured. The old quarry [5520 8734] in Lamb Lane, 500 yd SSE of Thwaite House, at least 10 ft deep, was partly filled by 1965, but a few feet of pale buff to white, thinly, and in places irregularly, bedded saccharoidal dolomite of sand grade were still visible. A specimen taken from the quarry before filling, shows in thin section (E 22573) finely crystalline dolomite (ω = 1·680 to 1·690) with a little calcite, rare angular grains of quartz, and specks and stains of limonite.

In the Gildingwells area the old quarry [550 855] in Red Quarry Plantation was filled at the time of the survey in 1946, but many fragments of fine-grained red or pink dolomite were in evidence in the vicinity. A thin section (E 22572) of one such fragment showed a finely crystalline dolomite (ω = 1·680 to 1·700) with some calcite and a little ankerite (ω = 1·715). A few grains of quartz, limonite and magnetite occur and there are traces of former organic structures and ooliths. Limonite staining accounts for the red colour of the dolomite at this locality.

Around Gildingwells the Lower Magnesian Limestone has an uneven surface with hummocks or 'knolls' which are generally elliptical in plan and which, east of the main outcrop, produce small inliers by protruding through the Middle Permian Marl (Fig. 23). The long axes of the majority of the 'knolls' trend north-north-westerly to northerly, though southwards the dominant trend becomes north-westerly to westerly. The 'knolls' are thought to represent depositional dome-structures in the dolomite (see p. 122), though there is little supporting evidence from the only two that provide exposures. The small quarry [5632 8588] in the 'knoll' nearly ½ mile NE of Gildingwells shows 4 ft of buff and pink, almost horizontally bedded, dolomite, and the old quarries [570 844] in the 'knoll' about 450 yd NW of Wallingwells show up to 10 ft of hard, pale grey, bedded dolomite. The steep-sided nature of the latter 'knoll' is demonstrated by a well, 40 yd to the south-east and with a surface level 20 ft below that of the quarries, in which the top of the Lower Magnesian Limestone was encountered at 5 ft. That other 'knolls' are similarly steep-sided is shown by augering in the Marl around them.

Old quarries in the Lindrick Common–Cotterhill Woods area have recently been examined by Mr. D. B. Smith. He gives the following general section for the western end of Lindrick Common Quarry [553 827], where 15 ft of dolomite were visible when Mr. W. N. Edwards surveyed the area in 1940:

	ft
Dolomite, buff, granular, evenly bedded	0¾+
Dolomite, buff, in 3 or 4 beds; cross-laminae foreset to south-east ..	1½
Dolomite, buff-grey, in slightly irregular beds 1½ to 3 in thick; ?algal lamination .. .. .. 3½ to	4
Dolomite, buff-grey, hard finely crystalline; traces of cross-lamination and ripple marks; discontinuous green clay film at top 6 in to	1
Clay, grey .. .. 1½ in to	0¼
Dolomite, buff, generally thinly bedded, with stylolites; ?algal lamination	3
Dolomite, grey, thickly bedded ..	2

At the eastern end of the quarry the succession below the thin bed of grey clay is:

	ft	in
Dolomite, weakly laminated, finely crystalline, in regular 2- to 4-in beds; green clay films on stylolitic bedding planes	2	0
Clay, greenish grey .. ..¼ in to		0½
Dolomite, greyish white, in a single bed		9
Clay, grey .. .. .. 0 to		0¼
Dolomite, buff-grey, massive saccharoidal porous; extensive traces of low-angle cross-lamination (base not seen)	3	6

Mr. Smith considers that the whole of the section probably belongs to the Upper Subdivision of the Lower Magnesian Limestone, but admits the possibility that the 1½- to 3-in bed of grey clay and the overlying 6-in to 1-ft bed of dolomite may represent the Hampole Beds (see p. 120) at the junction of the Lower and Upper subdivisions.

The old quarry [556 826], east of Lindrick Common, shows dolomite dipping north-north-westwards at 20 to 30° in the eastern part and overlain by Middle Permian Marl in the north (Plate XII). The upturned beds are responsible for a marked topographical ridge which extends south-westwards through the Club House of Lindrick Common Golf Club and thence westwards through Stubbings Lathe (see p. 211). A fault, probably of small throw and evidently parallel to the ridge, was observed at the south-eastern edge of the quarry by Mr. Edwards in 1940, but it is no longer visible. Mr. D. B. Smith gives the following section of the quarry, of particular interest because it includes the whole of the Upper Subdivision of the Lower Magnesian Limestone, which, down to the bottom of the probable Hampole Beds, is 25 to 26 ft thick at this locality. Wedge bedding, a typical feature of the Upper Subdivision, is almost entirely lacking here.

	ft	in
Mudstone, red, with sharp base (*Middle Permian Marl*)	6	0
Dolomite, yellowish buff, fairly hard, medium- and thick-bedded (mainly even), sand-grade saccharoidal, porous.. 3 ft 6 in to	4	0

	ft	in
Dolomite, pale greyish buff, irregularly laminated, cellular, with traces of small-scale cross bedding; weathering of the laminae gives a striped appearance .. .. .. 1 ft 9 in to	2	6
Clay, grey		film
Dolomite, as 4-ft bed above but partly cellular at top; many small vesicular cavities in top few inches	1	9
Mudstone, greyish green .. 0 to		1
Dolomite, pale buff-yellow..3 in to		4
Mudstone, greyish green ..1 in to		2
Dolomite, pale buff-grey, hard, finely crystalline, medium- and thick-bedded (more or less evenly), with thin bands of greyish green clay	17	0

Probable HAMPOLE BEDS

	ft	in
Dolomite, greyish buff finely saccharoidal, in 2 beds 3 to 4 in thick and locally lenticular; films of green clay at top and bottom and between beds .. ..7 in to		8
Dolomite, buff-grey hard finely crystalline, in irregular beds about	2	0
Dolomite, yellowish buff and buff-brown, in thick, slightly irregular beds; thin beds of grey clay towards base (base not seen) ..	25	0

The old quarry [559 829] on the south side of Cotterhill Woods Farm was being worked for road metal, building stone and agricultural lime in 1940. About 28 ft of Lower Magnesian Limestone were then exposed, but the quarry was extended before abandonment and over 40 ft can now be seen. The lowest 20 ft consist of thick-bedded to massive dolomite, but they have not been examined in detail. Mr. D. B. Smith gives the following section of the upper beds, measured in the north-west corner of the quarry:

	ft	in
Dolomite, white-cream, sand-grade saccharoidal, porous, with many shallow ripple marks; low-angle small-scale cross-lamination and widespread large-scale wedge bedding; traces of ooliths in least-altered parts .. about	10	0

PLATE XII

LOWER MAGNESIAN LIMESTONE SOUTH-SOUTH-EAST OF WOODSETTS (L 561)

Bedded dolomite forming the uppermost beds of the Lower Magnesian Limestone, and including the whole of the Upper Subdivision, is locally upturned on the right of the picture. The top 6 ft of the section on the left of the picture consists of Middle Permian Marl, which is largely overgrown.

HAMPOLE BEDS

	ft	in
Clay, greyish green, plastic, weakly laminated .. 0 to		0¼
Dolomite, cream-buff, fine sand-grade saccharoidal, in 2- to 3-in beds; slight discontinuity at base		8
Dolomite, as immediately above but in 3- to 12-in beds; ripple marks, low-angle cross-lamination and stylolitic bedding planes .. .. about	12	0

The Hampole Beds here lack the usual development of 1-in clay beds (Smith 1968), but are distinguished by their characteristic thin-bedding and their occurrence at the junction of the wedge-bedded and evenly bedded parts of the Lower Magnesian Limestone sequence.

There are extensive abandoned quarries at Shireoaks and Lady Lee, the former falling partly in the Sheffield (100) district to the west. A section in the main Shireoaks Quarry [549 810] measured by Mr. Edwards in 1939 showed 10 to 12 ft of pale buff fine-grained dolomite in well defined beds generally less than 6 in thick and with several thin layers of grey clay. Thin sections (E 18312–7) of Lower Magnesian Limestone from the Shireoaks quarries examined by Dr. J. Phemister are of granular and crystalline, fine- to medium-grained dolomite with scattered quartz grains and a variable iron content. Open pores are common. In Lady Lee Quarries [5615 7976] Mr. Edwards saw the top 30 ft of the Lower Magnesian Limestone, consisting of wedge-bedded dolomite and surmounted by an outlier of Middle Permian Marl.

The most extensive quarries in the district are at Steetley, where the Steetley Company formerly worked the Lower Magnesian Limestone for processing into basic refractories (see p. 237). The original quarry—Steetley Quarry [549 791]—extends northwards and westwards from the old Brickworks buildings; south-east and east of it is Wood Quarry [554 790], situated in and to the north-east of Steetley Wood; and to the south-west, extending almost to the main Worksop–Chesterfield road and lying partly in the adjacent Sheffield (100) district, is the vast Armstrong Quarry [547 787].

The north-eastern face [5505 7924] of Steetley Quarry shows 6 to 8 ft of flaggy white to buff dolomite on up to about 16 ft of thickly bedded, cream to buff dolomite. Mr. D. B. Smith notes that the upper, flaggy, part of the dolomite is of fine sand grade with a saccharoidal texture, while the lower part is soft and porous with scattered small acicular cavities. Traces of ripple marks and cross bedding have been seen in the top 4½ ft of the flaggy beds, with stylolitic bedding planes below. He describes the thickly bedded dolomite as being of sand grade and saccharoidal, with traces of cross bedding and, in the top 6 ft, of ripple marks. Traces of ooliths, pisoliths, oncoliths and intraclasts were found in the lower part, where there are also films and thin beds of green clay. The Hampole Beds have not been identified, but Mr. Smith considers that their horizon is at 6 ft from the top of the thickly bedded dolomite.

A section [5553 7918] measured in Wood Quarry by Mr. Edwards showed 20 ft of pale buff dolomite in beds about 18 in thick in the lower part, and somewhat thinner in the upper part. At the western end of this quarry the upper beds are represented by a crumbling mass of dolomite grains. In the south-western part of Steetley Wood a water borehole [5510 7898], drilled in an abandoned part of the quarry in 1944, showed:

	Thickness ft	Depth ft
Made ground	8	8
Dolomite..	42	50
Dolomite with mudstone partings (on Lower Permian Marl) ..	44	94

Fossils collected from dolomite and mudstone at 81 ft are listed in Plate X.

Armstrong Quarry, generally about 35 ft deep, is now being rapidly filled with colliery waste. A general section of the quarry, employing the divisions used by the quarrymen, is as follows:

	ft
Top Rag—weathered flaggy dolomite .. .. 6 to	8
Dolomite, pale brown massive; ripple-marked at top .. .. 3 to	7
Blue Stone, which passes laterally into *Greystone* and thence into *Crossly* (see below) .. .. .. 0 to	10

ft

White Bed—pale brown dolomite, with clay-coated stylolitic bedding planes; beds up to 3 ft thick (the best quality stone) .. .. 8 to 10

Bottom Bed—massive, pale brown dolomite 7

The Blue Stone, best seen in the south-east of the quarry, consists of pale grey, hard, compact, well bedded, impure dolomite with thin beds of grey clay between stylolitic surfaces. A specimen (E 20589) examined by Dr. Phemister is composed of dolomite grains (0·5 mm average diameter) with a few grains of quartz and calcite, the former interlocking with the dolomite. Accessory minerals are feldspar and muscovite. Up to 4 ft thick in the south, the Blue Stone thins and disappears to the north-east. Westwards it passes, apparently together with some of the overlying dolomite, into Greystone, a grey, less impure dolomite, up to 10 ft thick. This in turn passes, in the Steetley Farm [5443 7876] area, into Crossly, a grey or pale brown dolomite with abundant stylolites and a brecciated appearance. Like the Blue Stone, the Crossly dies out north-eastwards. In the abandoned western face of the quarry, southwards from Steetley Farm, Mr. D. B. Smith has identified the Hampole Beds, apparently within the Crossly described above; they consist of:

in

Clay, green .. .. .. 0 to 2

Dolomite, grey to buff, porous (small horizontally elongated cavities), cross-laminated, granular .. 3 to 8

Clay, green .. .. .. 0 to 3

Dolomite, grey to buff, hard saccharoidal, fine sand grade, with many small cavities and traces of ?shell casts; thin hard dense zone at top .. .. 2 to 8

Immediately below the Hampole Beds is the Hampole Discontinuity (see p. 120 and Smith 1968). The dolomite immediately below this has, in places, a hardened top like that of the lowest division of the Hampole Beds. These hardened zones are believed to be due to subaerial hygroscopic carbonate precipitation. R.F.G., E.G.S.

Sections of the Lower Magnesian Limestone provided by colliery shafts are given below.

At Bevercotes Colliery the bulk of the formation has apparently passed laterally into Lower Permian Marl, leaving only the topmost part as dolomite (45 ft 7 in at 1002 ft in No. 1 Shaft, and 38¼ ft at 1009 ft 5 inches in No. 2 Shaft). Firbeck Main Colliery shows 71 ft 11 in of dolomite, described as strong and grey for the most part but honeycombed and broken in the bottom 12½ ft, at 195½ ft. In Harworth Colliery No. 1 Shaft there are 185½ ft at 698 ft 4 in or 195½ ft at 708 ft 4 in depending upon whether 10 ft of 'stone-bind' at the base are included or placed in the Lower Permian Marl (see also p. 116). Edwards (1951, p. 178), who gives the known details of the section, adopts the latter alternative. Streaks of anhydrite are recorded in the top 51 ft 8 in. Maltby Main Colliery No. 2 Shaft starts in Lower Magnesian Limestone and proves the base at 157 ft 5 in. Blue clay partings are recorded in the upper part; there are 19 ft of honeycombed dolomite containing gypsum at 117 ft 4 in and this is underlain by 29 ft 2 in of 'laminated limestone'. At Manton Colliery the combined Lower Magnesian Limestone and Lower Permian Marl are 297 ft 4 in thick at 620 ft 10 inches in No. 2 Shaft and 300½ ft at 653 ft in No. 4 Shaft. In No. 2 Shaft the bottom 22½ ft, at least, belong to the Lower Permian Marl; in No. 4 Shaft the division is obscure. Included at the top of the Lower Magnesian Limestone here are 130 ft 7 in of red and blue 'limy sandstone' at No. 2 Shaft and 131 ft 4 in of red and grey 'limestone and limy sandstone' at No. 4 Shaft, which may be of Middle Permian Marl age (see p. 150). At Shireoaks Colliery the Lower Magnesian Limestone is about 123 ft thick at 188½ ft, and at Steetley Colliery there are 79 ft 10 in at 115 ft 11 in (for details see Lamplugh and Smith 1914, p. 117).

Boreholes completely cored through the Lower Magnesian Limestone are as follows (thickness and depth to base in brackets): Checkerhouse (308¾ ft at 954¾ ft), Hayton Smeeth (217 ft at 1360 ft), Manton (280½ ft at 615¼ ft), Scofton (232¾ ft at 638¼ ft), South Leverton (250 ft 1 in at 1914¼ ft), Wallingbrook Wood (127 ft 8 in at 170 ft), Welbeck Abbey (165 ft at 264¾ ft), Wigthorpe (130 ft at 273 ft). In addition all but the top 8 ft were cored at Jockey House (238½ ft at 1100½ ft), and all but the top 15 ft at West Drayton No. 1 (240¼ ft at 1205 ft).

At *Checkerhouse Borehole* the 308¾ ft of rocks consist of alternate beds of 'limestone' and shale 229 ft, succeeded by 79¾ ft of red and grey sandy 'limestone'. The latter has alternatively been considered to be of Middle Permian Marl age (pp. 122 and 149). At *Hayton Smeeth Borehole* the Lower Magnesian Limestone consists largely of grey 'limestone' with shale bands (for details see Wilson 1926a, pp. 199–200), but there are 84½ ft of shale with 'limestone' bands and 9½ ft of shale and 'clunch', respectively 11 and 125 ft from the base: possibly the former could be regarded as Lower Permian Marl. At *Manton Borehole* 178 ft 4 in of grey 'limestone' are succeeded by 99 ft 4 in of 'limy sandstone', 'limestone' and sandstone (see p. 149). The 'limestone' at *Scofton Borehole* is split into two nearly equal parts by a 9½-ft band of hard grey marl with thin beds of 'limestone' 126¼ ft from the base. At *South Leverton Borehole* the Lower Magnesian Limestone is subdivided as follows: 'limestone' 85 ft 1 in, on 'conglomerate with large anhydrite pebbles' 2 ft, 'limestone' with bands of shale 50 ft 5 in, purple and grey marl with 'limestone' ribs 83 ft 4 in, on grey sandy 'limestone' 29¼ ft. The 127 ft 8 in of Lower Magnesian Limestone at *Wallingbrook Wood Borehole* is divisible into 64 ft 8 in of cream-coloured dolomite, largely close-grained, but granular in parts (? the Upper Subdivision), on 63 ft of grey dolomite with bands and partings of grey mudstone which form about 50 per cent of the bottom 23½ ft. The upper part contains stylolites, which have coatings of green clay, and there are regular bands and partings of green or greenish grey mudstone in the top 5 ft 8 in; sporadic vughs also occur, and specks of galena were seen in the bottom 2 in (i.e. at 107 ft). At *Welbeck Abbey Borehole* the top 58 ft 2 in of Lower Magnesian Limestone consist of dolomite which is generally buff to yellow in colour but pink in parts, is compact and fine- to medium-grained, and contains stylolites coated with greenish grey clay in its upper part. These beds, probably representing the Upper Subdivision, rest on 24½ ft of light grey dolomitic limestone which becomes 'earthy' at the base and rests on 82 ft 4 in of light grey, largely fine-grained, but in places medium-grained, calcitic limestone containing bands of grey mudstone which form between 10 and 50 per cent of the rock, the proportion increasing downwards. The 130 ft of Lower

Magnesian Limestone at *Wigthorpe Borehole*, consist of 1½ ft of pale buff 'limestone', on 23½ ft of reddish grey 'limy sandstone' (see p. 150), on 11 ft of fawn 'limestone' (all possibly belonging to the Upper Subdivision), on 94 ft of grey 'limestone' with bands and partings of marl.

In *Jockey House Borehole* there are 74 ft of marl with bands of marly and crystalline 'limestone' (including one 10½ ft thick) at 1033 ft (i.e. 67½ ft from the base of the Lower Magnesian Limestone). A specimen of marl from 970 ft (E 22566) has been examined by Mrs. J. E. Morey. It is composed of angular and subangular grains of quartz together with particles of limonite, pyrite and carbonaceous matter, in a fine carbonate matrix composed largely of calcite with some dolomite and a little ankerite. The 'limestone' below 1033 ft is described as grey, compact and marly with partings of grey marl. What is described by Mr. W. N. Edwards as 'ropy' bedding resembling enterolithic structure (Grabau 1932, p. 758) is common. The 'limestone' above the marl (i.e. above 959 ft) varies from a dense crystalline to a soft sandy or marly rock. Grey or brown marl occurs as partings throughout, and as bands up to 3 ft thick in the bottom 49 ft. Specimens (E 22568–71) taken at intervals were examined by Mrs. Morey who reports that they are mostly crystalline and arenaceous dolomites. A specimen at 940 ft contains a considerable quantity of quartzose silt, and there is a band of silt composed of angular quartz grains (average 0·1 mm diameter) with micas, and cemented in places by dolomite, at 910 ft. Siliceous cement occurs in small patches and as an infilling of cracks in dolomite at 870 ft. Small plates and aggregates of gypsum occur on bedding planes in dolomite at 876 ft. Other minerals noted were calcite, pyrite, limonite, plagioclase, zircon, tourmaline, ilmenite and leucoxene. A specimen of crystalline dolomite from 920 ft (E 22569) showed relict ooliths outlined by finely crystalline pyrite. A marl band at 950 ft (E 22567) consists of minute clastic quartz grains with flakes of muscovite and chlorite and specks of pyrite and carbonaceous matter, in a carbonate—mostly dolomite—matrix. Distinct bedding is accentuated by variations in limonite staining and by grading.

In *West Drayton No. 1 Borehole* the beds between 1024 and 1137 ft consist of grey, greenish grey and brownish grey, locally micaceous, marl with subordinate bands of grey, crystalline and marly 'limestone'. A hydrocarbon pebble was recorded at 1053 ft, and a thin section (E 18874) from this depth, examined by Dr. Phemister, consists of fine-grained granular calcite with, in parts, detrital grains of dolomite and a few grains of quartz. Carbonaceous streaks, sporadic pyrite nodules and probable algal flakes are also present. The 'limestone' below the marl (i.e. from 1137 to 1205 ft) is described as grey, compact and marly, with abundant fossils (see Plate X), numerous thin, irregular marl bands and distorted bedding. The 'limestone' cored above the marl is grey, crystalline in part, and with partings and thin bands of marl over much of the section. A specimen (E 18873) from 1006 ft examined by Dr. Phemister proved to be a fine-grained granular dolomite containing many sub-rounded pieces of dolomitized limestone and dolomitic mudstone. Other small rounded bodies composed of mud and carbonate may be faecal pellets; there is one piece of algal debris, and detrital quartz grains are common. Thin partings of gypsum are recorded between $979\frac{3}{4}$ and 990 ft. Two feet of anhydrite were recorded at 992 ft, but this is probably a misidentification of barytic dolomite. A specimen (E 18871) from 992 ft proves to be granular dolomite with abundant large crystals of baryte. The latter enclose dolomite grains and apparently replace that mineral. There is an 18-in band of grey sandstone with disturbed bedding at 996 ft, a thin section (E 18872) of which was examined by Dr. Phemister. This is composed of well assorted angular grains of quartz and alkali feldspar with subordinate worn rhomboid grains of dolomite. Accessory minerals are mica (as a finely divided aggregate forming a cement and matrix, and as detrital flakes of musco-vite), chlorite, garnet, tourmaline and zircon.

The cored Lower Magnesian Limestone at *Scaftworth Borehole* is grey to pale grey with a few brown bands, largely fine-grained or marly between 1087 and about $1190\frac{1}{2}$ ft and mostly granular above 1087 ft. Bands and partings of grey marl occur at frequent intervals except in the top few feet of core, and include argillaceous bands of Marl Slate lithology in the bottom 2 ft (see p. 118). There are 24 ft 2 in of marl with 'limestone' bands at 1140 ft 2 in. A few small vughs with calcite crystals occur in the bottom part, and the

veins of calcite at around 1181 ft also contain sphalerite and pyrite. Veinlets of galena occur at 1887 ft 8 in.

Cores of the uppermost part of the Lower Magnesian Limestone were obtained from *Sutton Borehole* (where *Schizodus obscurus* was collected $12\frac{1}{2}$ ft from the top) and from a number of wells and boreholes for water on and near the outcrop. In a well at Lindrick Common Golf Club, 42 ft of con-glomerate, marl and sandstone are recorded below 23 ft of typical Middle Permian Marl and above 13 ft of dark and grey 'limestone' with clay bands. According to Dr. A. H. G. Mackintosh (*in litt.*), who saw the cores, the so-called conglomerate was 'limestone' with a brecciated appearance, and these beds are therefore assigned to the Lower Magnesian Limestone.

Boreholes at Worksop Brewery end in Lower Magnesian Limestone, the section of No. 2 Borehole being as follows:

	Thickness ft	Depth ft
MIDDLE PERMIAN MARL	—	178
LOWER MAGNESIAN LIMESTONE		
Dolomitic sandstone, pink, hard and close-grained, with spor-adic micaceous part-ings at top; passes down to pink sandy dolomite which in turn passes down to under-lying dolomite (*14 ft of core missing above 204 ft*)	$39\frac{1}{2}$	$217\frac{1}{2}$
Dolomite, grey, with stylolites and, between 267 and 276 ft, nu-merous vughs ..	$62\frac{1}{2}$	280
Core lost (said to be close-grained 'limestone' with large vughs near base) to bottom of borehole	20	300

E.G.S.

A specimen (E 33365) of fine-grained hematitic and dolomitic sandstone from 210 ft is reddish brown and evenly grained (averaging 0·1 mm), with flakes of muscovite on bedding planes. The clastics are sub-angular, of low sphericity, and consist of mainly igneous quartz, reddened orthoclase, microcline and conspicuous rock particles, cherty silica, illite aggregates and quartzites. Muscovite, chlorite and hematite occur interstitially. Zircon and tourmaline (0 = dark green) are sparse accessories. The abundant cement consists of dolomite (ω = 1·684) which appears to have extensively replaced the clastics, especially feldspar. Hematitization preceded dolomitization. Palimpsests of quartz and feldspar within dolomite grains are common. The ratio of dolomite to unreplaced clastics was estimated as 45:55. A specimen (E 33366) of sandy dolomite at 212 ft is more patchily stained by ferric iron and contains vughy structures infilled with white calcite. It differs also by having a considerably higher proportion of carbonate, which forms a mosaic of interlocking pale brown rhombs (dolomite ω = 1·685), averaging 0·2 mm. Resistates (quartz, feldspars) occur mainly as subordinate, irregular grains without clastic margins and apparently form replacement residuals. There is evidence to suggest that some replacement of clastics by dolomite has occurred, with residua of quartz in optical continuity. Indistinct zones of fine hematite in dolomite rhombs appear to have been inherited from completely replaced hematitized quartz. R.K.H.

The 39½ ft of pink dolomitic sandstone and sandy dolomite at the top of the Lower Magnesian Limestone are apparently a similar development to that recorded at Manton Colliery and Manton, Checkerhouse and Wigthorpe boreholes (see above and p. 122).

Where cores of the Lower Magnesian Limestone have been taken in exploratory boreholes for coal other than those mentioned above, they are from the lower part and in most, if not all, cases from the Lower Sub-division only. Generally the records show pale grey, fine-grained, thinly, and often irregularly, bedded dolomite, dolomitic limestone and limestone, with numerous bands and partings of grey marl, which may be micaceous and occasionally pyritic. Calcite veining is common, and crystals of calcite and dolomite occur fairly commonly in vughs. Galena, in addition to the occurrence in Wallingbrook Wood Borehole (p. 131), was found as traces on joints at 780 ft in Torworth Borehole; sphalerite was noted in Barnby Moor (between 910 and 913½ ft) and Eaton (at 1255 ft 8 in) boreholes; and there were patches of pyrite in the bottom 35 ft of the formation in Mattersey Borehole. The thick development of marl within the Lower Magnesian Limestone (see p. 122 and Figs. 21 and 23) was cored at Mattersey (58 ft 11 in) and partially cored at Babworth, Eaton and Ranby Camp (the bottom 12 ft 5 in, 45 ft and 11 ft 7 in respectively). This marl was described as grey or greenish grey with numerous bands of grey, brownish grey and, occasionally at Babworth, pinkish grey 'limestone', and with locally abundant plant debris.

E.G.S.

A specimen of laminated silty marl from 1246 ft in Eaton Borehole is seen in thin section (E 35660) to consist essentially of calcite. Coarse (3 mm) calcite disrupts bedding planes and appears to be post-depositional. The finer calcite is intermixed with clay minerals and opaque dust. Patchy reorganization of calcite is indicated by cellular disruption of laminae. An X-ray powder photograph (NEX 794) shows chlorite, illite, dolomite, calcite and quartz.

R.K.H.

Fossils are fairly common in most of the borehole cores from the lower part of the Lower Magnesian Limestone; identifiable specimens collected are listed in Plate X.

The only cores of Lower Magnesian Limestone from oil bores in the district were obtained near the top of the formation in Gainsborough No. 1 Bore. Mr. G. D. Gaunt's log is as follows:

	Thickness ft	Depth ft
MIDDLE PERMIAN		
MARL (*not cored*) ..	—	to 2210
LOWER MAGNESIAN LIMESTONE		
Not cored	30	2240
Limestone, light grey, fine- to medium-grained, dolomitic; a few marl partings; faint cross bedding; vughs and shells ..	10¾	2250¾
Marl, grey, sandy in places, with thin limestone bands ..	4¼	2255

J

Shells from 2242 ft 1 in have been identified by Mr. J. Pattison as *Schizodus obscurus*.

Details of the Lower Magnesian Limestone in oil bores are otherwise dependent on the careful examination of chipping samples by the geologists of the BP Company, supplemented by geophysical logs.

Tickhill No. 1 Oil Bore shows 175 ft of dolomitic limestone, apparently interbedded with red and grey fine-grained sandstone in the bottom 80 ft. In *Ranskill* No. 1 Oil Bore there are almost 200 ft of Lower Magnesian Limestone, described as dolomite at the top, passing down to dolomitic limestone which rests on crystalline and silty limestone. The dolomite and dolomitic limestone, generally granular or arenaceous and together about 95 ft thick, probably represent the Upper Subdivision of the Lower Magnesian Limestone. Immediately below are 15 ft or so of limestone described as grey and silty and with a higher gamma-ray count than any other part of the Lower Magnesian Limestone. These may well be the Hampole Beds. Below this horizon there is a marked increase in resistivity of the Limestone as shown by electrical logging, reflecting the fine-grained, close-textured nature of the lower beds.

In *Gringley* No. 1 Oil Bore there are about 250 ft of Lower Magnesian Limestone, recorded as dolomitic limestone in the top 35 ft and as limestone or impure limestone below. Bands of brown and grey shale are noted, some of which contain gypsum.

In the *Walkeringham* oil bores there are 300 ft (No. 1) to 324 ft (No. 2) of dolomite and limestone with several recognizably oolitic horizons. Pyrite is recorded at several places in No. 2 Bore, and traces of anhydrite were found at the base and at about 90 ft below the top. The basal part in this section is estimated to contain up to 30 per cent of grey silty mudstone, probably in thin bands; and in No. 1 Bore streaks of red and grey mudstone are recorded just over 100 ft below the top of the Limestone.

In *Morton* No. 1 Oil Bore 345 ft of Lower Magnesian Limestone were logged. Only the top 15 ft are recorded as being dolomitic and oolitic, much of the rest being described as crystalline limestone, with a high proportion of argillaceous limestone in the bottom 100 ft.

The *Beckingham* oil bores show 314 to 360 ft of Lower Magnesian Limestone. Only the topmost beds are described as dolomitic, the bottom 100 to 150 ft being silty and the rest mainly crystalline limestone. A band of red and grey mudstone, siltstone or silty limestone, 10 to 20 ft thick, occurs near the top in three of the sections.

In the *Gainsborough* oil bores the Lower Magnesian Limestone varies from 235 to 365 ft thick, but much of this variation is accounted for by the association of the thinner sections with thick developments of Lower Permian Marl into which the lower beds of the Limestone have passed laterally. Thus the combined thickness of Limestone and Marl varies only between just over 300 ft and about 385 ft. In general the top 30 to 100 ft, but exceptionally as much as 160 ft, consist of dolomite and dolomitic limestone, and the lower part chiefly of limestone. The dolomite may be coarsely crystalline, but both dolomite and limestone are commonly fine-grained. Granular and saccharoidal rocks are recorded, also oolites and, rarely, pisolites. Sandy bands occur, especially in the dolomite, and bands and partings of grey marl are found at all horizons, but especially towards the bottom of the limestone, which itself tends to become marly and silty downwards. Interbanding of the carbonate rock with reddish brown and subordinate green mudstone is noteworthy in a few bores, where it is characteristic of the upper part and shows a passage into the Middle Permian Marl, but also occurs at other horizons— almost throughout the succession at No. 3 Bore. Pyrite is common, especially at the top and towards the base of the formation; it occurs as specks and aggregates and in the form of pellets. Traces of anhydrite and gypsum are noted in several bores. These, with the exception of gypsum associated with bands of red and grey mudstone in the lower part of the succession at No. 38 Bore, are confined to the upper part, where they occur in the dolomite or at the junction of the dolomite and limestone.

In the *South Leverton* oil bores there are about 100 to almost 300 ft of Lower Magnesian Limestone, but the boundary with the Lower Permian Marl is in many cases arbitrary (see also p. 119), and the combined thickness of the two formations varies only between about 270 and 350 ft. The upper 30 to 100 ft generally consist of porcellanous to coarse-grained and granular dolomite and dolomitic limestone in which ooliths and pisoliths are occasionally recognizable, and which contain sandy bands in some sections. A 20-ft band of calcareous sandstone is recorded 50 ft from the top of the formation in No. 18 Bore. The lower part of the formation consists for the most part of fine-grained grey or greyish brown limestone which becomes marly and develops marl bands downwards, and in many cases passes gradually into the Lower Permian Marl. Carbonaceous partings are noted in this lower part in places.

The *Grove* oil bores proved about 275 ft of Lower Magnesian Limestone, largely fine-grained, crystalline carbonate rocks, dolomitic at the top and calcitic in the lower part. Micaceous siltstone with carbonaceous debris, about 30 ft thick and containing a band of limestone less than 10 ft thick, was recorded in the middle of the formation in No. 1 Bore.

In *West Drayton* No. 2 Oil Bore there are 230 ft of Lower Magnesian Limestone comprising 100 ft of interbanded dolomite, sandy dolomite and dolomitic sandstone, on 50 ft of grey marl and siltstone with some dolomite and limestone, on 80 ft of predominantly grey, but also reddish, limestone.

There are about 200 ft of Lower Magnesian Limestone in the *Apleyhead* oil bores. The top 75 ft in No. 3 Bore are described as sandstone and sandy limestone, while the top 40 to 60 ft at Nos. 1 and 2 bores are described as dolomite, partly sandy, and saccharoidal dolomite respectively. The rest of the formation consists of fine-grained to saccharoidal limestone with, in No. 2 Bore, a 40-ft band of marl and siltstone, with limestone, at the top. Pyritic limestone was also noted in this section, about 20 ft above the Lower Permian Marl.

In the *Bothamsall* oil bores the Lower Magnesian Limestone varies from a little over 200 to a little over 250 ft. Generally the top 30 to about 140 ft consist of dolomite and dolomitic limestone, sandy in parts in many sections, and containing, in some bores, substantial thicknesses of sandstone—e.g. almost 60 ft at No. 12 Bore. The uppermost beds in No. 16 Bore consist of interbanded dolomite, sandy dolomite, sandstone and red mudstone, suggesting a passage into the

Middle Permian Marl, and the occurrence of red dolomite associated with sandy dolomite and sandstone at the top of the Lower Magnesian Limestone in some other sections may be indicative of a Middle Permian Marl age for those deposits (see p. 122). The lower part of the Lower Magnesian Limestone consists largely of grey fine-grained limestone and argillaceous limestone with, at the top, between it and the dolomitic upper beds, up to 40 ft of grey marl or siltstone. Pyrite has been recorded in several bores, chiefly in the lower part of the formation.   E.G.S.

Notes on Palaeontology (see Plate X). Schweitzer (1962, p. 351) considers the species *Hiltonia rivuli* Stoneley inseparable from *Ullmannia bronni* Göppert and his interpretation has been followed here.

Foraminifera are common in the Lower Permian Marl of non-Marl Slate lithology and in the Lower Subdivision of the Lower Magnesian Limestone. They include species of the genera *Agathammina*, *Ammobaculites*, *Ammodiscus*, *Cyclogyra*, *Dentalina*, *Hyperammina* and *Nodosaria*.

Eisel (1909) described six morphological varieties of '*Productus*' *horridus* J. Sowerby (*Horridonia horrida*) occurring in the Zechstein of Gera in central Germany. Only two of these, *initialis* and *hoppeianus*, of which *hoppeianus* is apparently the more common form (Plate XI, fig. 9), have been recognized from this district.

The two specimens of *Streptorhynchus pelargonatus* from Barnby Moor Borehole are relatively large shells, with ventral valves 17·0 mm wide, one 18·5 mm high and the other at least 16·0 mm high. These dimensions are comparable with those of the few other examples of *S. pelargonatus* collected from the Lower Magnesian Limestone of north-eastern England, but contrast with the average shell size of this species in the lower part of the Middle Magnesian Limestone reef of County Durham, in which the ventral valve is normally only 5·0 to 9·0 mm high. It is doubtful, however, whether close examination of this size variation would justify the recognition of two separate species.

The fragmentary specimen from the Lower Permian Marl of Torworth Borehole doubtfully assigned to *Aviculopinna? pinnaeformis* Geinitz 1848 appears to be the external mould of part, including some of the ventral margin, of a large right valve. The strong growth lines are similar to those on the larger fragment referred to this species from the Lower Permian Marl of Sherwood Colliery, near Mansfield (Smith and others 1967, p. 198; Plate XI, fig. 11 of the present account). Similar forms have been recorded from several localities in the German Zechstein but have been assigned by some authors to the species *Aviculopinna prisca* Münster 1837. The latter name is based on bivalves from the German Trias and, as it has not been established that this species and *A. pinnaeformis* are conspecific, it seems preferable at present to assign the Permian forms to the Geinitz species.

Ostracods are fairly common in the Lower Permian Marl of non-Marl Slate lithology and in the Lower Magnesian Limestone of this district. Most are preserved as internal moulds. The commonest forms are similar to '*Cythere' jonesiana* Kirkby or are at least closely related to it. This species has been assigned by several authors to the genus *Macrocypris* but this conclusion was considered by Sohn (1960) to be erroneous.   J.P.

Palynology of the Lower Magnesian Limestone of Sutton Borehole. Miospore assemblages (Fig. 24) were obtained from samples of grey marl within the Lower Magnesian Limestone in Sutton Borehole at depths of 902, 928, 1065 and 1080 ft. The assemblages from 902 and 1065 ft also contained small numbers of microplankton (Plate XIII), while fragments of the internal walls of foraminifera were observed in the lowest assemblage. The samples fall into two groups; the two lowest are from the basal 30 ft of the Lower Magnesian Limestone while the higher two are from 64 and 90 ft below the top of the formation. The following miospores have only been observed in the lower part of the formation (Fig. 24):

Nuskoisporites klausi, Crustaesporites globosus (Plate XIII), *Protohaploxypinus* cf. *samoilovichii,* cf. *Striatopodocarpites fusus,* cf. *Illinites gamsi,* cf. *I. klausi, I. parvus,* cf. *I. unicus, Jugasporites lueckoides* (Plate XIII), *Vestigisporites minutus* and *Platysaccus papilionis.*

The following miospores were only found in samples from the higher part of the formation (Fig. 24):

Taeniaesporites angulistriatus, Strotersporites wilsoni, Jugasporites schaubergeroides (Plate XIII), *Gardenasporites oberrauchi* (Plate XIII), cf. *Microcachryidites fastidiosus,* cf. *Monosulcites minimus,* cf. *Cycadopites sp. R., Nuskoisporites* cf. *rotatus, Limitisporites moersensis* and *Alisporites nuthallensis.* All these species were observed in very small numbers and usually formed less than 2 per cent of an assemblage (based on counts of 250 grains).

Material from Sutton has also been examined by R. F. A. Clarke (1963 and *in litt.*) who recorded similar miospore assemblages to those listed in Fig. 24. However, Clarke also recorded the presence of *Striatopodocarpites cancellatus, Taeniaesporites bilobus, Vittatina hiltonensis, Illinites tectus* and *Platysaccus radialis.* In Clarke's study *Taeniaesporites angulistriatus, Protohaploxypinus* cf. *samoilovichii, Striatopodocarpites fusus* and *Vestigisporites minutus,* species observed only in the lower parts of the Lower Magnesian Limestone in the present study, were found in both the basal and higher parts of the formation.

Of the species found throughout the Lower Magnesian Limestone, *Lueckisporites virkkiae* (Plate XIII) was the most abundant, forming between 14·6 per cent (from 1080 ft) and 22 per cent (from 928 ft) of the assemblages. *Taeniaesporites spp., Protohaploxypinus spp.* and *Striatoabietites richteri* (Plate XIII) occur in both the lower and higher parts of the formation, but in relatively small numbers. *Illinites delasaucei, Labiisporites granulatus* (Plate XIII), *Falcisporites zapfei* and *Klausipollenites schaubergeri* (Plate XIII) occur throughout the Lower Magnesian Limestone in fairly high numbers, *F. zapfei* and *K. schaubergeri* being the commonest.

The miospore assemblages obtained from the Lower Magnesian Limestone of Sutton Borehole are closely comparable with Upper Permian or Zechstein assemblages from Europe (e.g. Potonié and Klaus 1954; Klaus 1955, 1963; Leschik 1956; Grebe 1957; Grebe and Schweitzer 1962; Schaarschmidt 1963). The Zechstein microfloras display considerable constancy and uniformity in composition both geographically and stratigraphically, and are dominated by striate and non-striate bisaccate miospores. Monosaccate and monosulcate forms are rare and trilete asaccate forms are almost entirely

1080	1065	928	902	← DEPTH IN BOREHOLE (FEET)	
			●	*Calamospora sp.*	MIOSPORES
●	●	●	●	*Perisaccus granulosus* (Leschik 1955) Clarke 1965	
cf	●	cf	cf	*Nuskoisporites dulhuntyi* Potonié & Klaus 1954	
●	●	●	●	*Lueckisporites virkkiae* (Potonié & Klaus 1954) Clarke 1965	
cf	cf	cf	●	*Taeniaesporites alatus* Klaus 1963	
	●			*T. albertae* Jansonius 1962	
			?	*T. angulistriatus* (Klaus 1963) Clarke 1965	
cf	●			*T. labdacus* Klaus 1963	
●	●	●		*T. noviaulensis* Leschik 1956	
●				*T.samoilovichii* var. *pantii* (Jansonius 1962) Klaus 1963	
●	●	●	●	*Protohaploxypinus chaloneri* Clarke 1965	
cf	●	●	●	*P. jacobii* (Jansonius 1962) Hart 1964	
	●	●	●	*P. microcorpus* (Schaarschmidt 1963) Clarke 1965	
	●	●	●	*Striatoabietites richteri* (Klaus 1955) Hart 1964	
			●	*Strotersporites wilsoni* Klaus 1963	
●	●	●	●	*Illinites delasaucei* (Potonié & Klaus 1954) Grebe & Schweitzer 1962	
	●		●	*Jugasporites paradelasaucei* Klaus 1963	
			●	*J. schaubergeroides* Klaus 1963	
			●	*Vittatina sp.*	
	●		?	*Paravesicaspora splendens* (Leschik 1956) Klaus 1963	
		●	●	*Gardenasporites oberrauchi* Klaus 1963	
●	●	●	●	*Labiisporites granulatus* Leschik 1956	
			cf	*Microcachryidites fastidiosus* (Jansonius 1962) Klaus 1964	
●	●	●	●	*Falcisporites zapfei* (Potonié & Klaus 1954) Leschik 1956	
●	●	●	●	*Klausipollenites schaubergeri* (Potonié & Klaus 1954) Jansonius 1962	
			cf	*Monosulcites minimus* Cookson 1947	
			cf	*Cycadopites sp. R* Jansonius 1962	
		cf		*Nuskoisporites rotatus* Balme & Hennelly 1956	
	●	●		*Illinites bentzi* (Klaus 1955) Klaus 1963	
		●		*Limitisporites moersensis* (Grebe 1957) Klaus 1963	
		●		*Alisporites nuthallensis* Clarke 1965	
	●			*Nuskoisporites klausi* Grebe 1957	
	●			*Crustaesporites globosus* Leschik 1956	
●	●			*Protohaploxypinus* cf. *samoilovichii* (Jansonius 1962) Hart 1964	
	cf			*Striatopodocarpites fusus* (Balme & Hennelly 1955) Potonié 1958	
	cf			*Illinites gamsi* Klaus 1963	
	●			*I. klausi* Clarke 1965	
	●			*I. parvus* Klaus 1963	
	cf			*I. unicus* Kosanke 1950	
●	●			*Jugasporites lueckoides* Klaus 1963	
●				*Vestigisporites minutus* Clarke 1965	
●				*Platysaccus papilionis* Potonié & Klaus 1954	
			●	*Baltisphaeridium debilispinum* Wall & Downie 1963	MICRO-PLANKTON
	●		●	*Veryhachium reductum* (Deunff 1959) Jekhowsky 1961	
?			●	Tasmanitid	

FIG. 24. *Distribution of miospores and microplankton in the Lower Magnesian Limestone of Sutton Borehole*

EXPLANATION OF PLATE XIII

PERMIAN MIOSPORES AND MICROPLANKTON

1. *Striatoabietites richteri* (Klaus) Hart 1964. (MPK. 457).
2. *Protohaploxypinus jacobii* (Jansonius) Hart 1964. (MPK. 458).
3. *Taeniaesporites samoilovichii pantii* (Jansonius) Klaus 1963. (MPK. 459).
4. *Labiisporites granulatus* Leschik 1956. (MPK. 460).
5. *Illinites bentzii* (Klaus) Klaus 1963. (MPK. 461).
6. *Gardenasporites oberrauchi* Klaus 1963. (MPK. 462).
7. *Lueckisporites virkkiae* Var. A (Potonié & Klaus) Clarke 1965. (MPK. 463).
8. *L. virkkiae* Var. B (Potonié & Klaus) Clarke 1965. (MPK. 464).
9. *Jugasporites lueckoides* Klaus 1963. (MPK. 465).
10. *Protohaploxypinus microcorpus* (Schaarschmidt) Clarke 1965. (MPK. 466).
11. *Jugasporites schaubergeroides* Klaus 1963. (MPK. 467).
12. *Klausipollenites schaubergeri* (Potonié & Klaus) Jansonius 1962. (MPK. 468).
13. *Taeniaesporites labdacus* Klaus 1963. (MPK. 469).
14. *Perisaccus granulosus* (Leschik) Clarke 1965. (MPK. 470).
15. *Nuskoisporites dulhuntyi* Potonié & Klaus 1954. (MPK. 471).
16. *Crustaesporites globosus* Leschik 1956. (MPK. 472).
17. *Veryhachium reductum* (Deunff) Jekhowsky 1961. (MPK. 473).
18. *Baltisphaeridium debilispinum* Wall & Downie 1963. (MPK. 474).
19. Tasmanitid (MPK. 475).

All illustrations magnified × 500. All specimens figured are in the Institute of Geological Sciences Palynological Collection at Leeds; registered numbers are given in brackets.

PLATE XIII

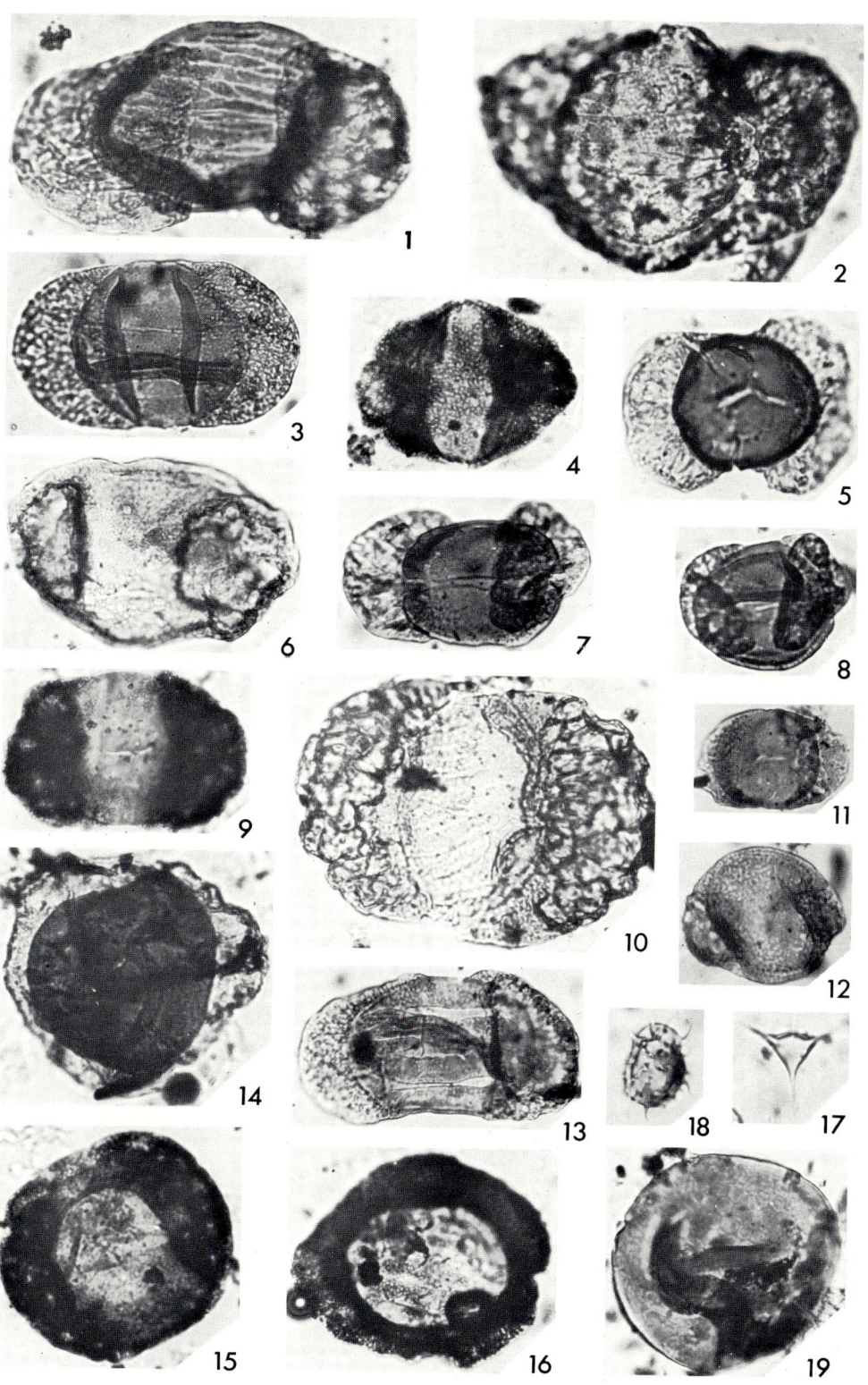

PERMIAN MIOSPORES AND MICROPLANKTON

absent (see Fig. 24). Clarke (1965, pp. 348–9) suggested that the relative uniformity of Zechstein microfloras that he obtained from Cumberland and the East Midlands indicated the existence of a relatively uniform vegetation during the Upper Permian. He also considered, in view of the dominance of the assemblages by bisaccate miospores and the fact that similar assemblages were obtained at different localities and from different lithologies, that the parent vegetation was some distance away.

Small numbers of microplankton (*Baltisphaeridium debilispinum* and *Veryhachium reductum*; Plate XIII) have been recorded in the assemblages from 902 and 1065 ft. The specimens are comparable with forms described by Wall and Downie (1963) from the Lower Permian Marl of Sykehouse and Conisbrough, Yorkshire. The small proportion of microplankton in the assemblage from the lower part of the formation at Sutton (0·4 per cent from 1065 ft) may indicate the proximity of the area to a shoreline at the time of deposition (cf. Wall and Downie 1963), whilst the increase in the abundance of microplankton in the higher part of the formation (3·3 per cent from 902 ft) is suggestive of deeper water or a greater distance from the shore-line. G.W.

MIDDLE PERMIAN MARL

The Middle Permian Marl crops out in a strip up to 2 miles wide near the western margin of the district. The outcrop forms low ground in the north, bounded by the dip-slope of the Lower Magnesian Limestone on the west and the scarp of the Upper Magnesian Limestone on the east. In the south, between Wigthorpe and Welbeck Abbey, where the Upper Magnesian Limestone is absent and where there is a substantial thickness of sandstone at the top of the Middle Permian Marl, the upper boundary is much less distinct. At outcrop the formation thickens southwards from about 80 to over 140 ft, but eastwards there are wide thickness variations (Fig. 25) with maxima of over 180 ft around Apleyhead and Babworth, in the Bevercotes–Markham Moor area, and in the extreme north-east of the district. Wilson (1926b, p. 181), commenting on the thicknesses of the Middle Permian Marl and Lower Magnesian Limestone, noted their complementary nature and attributed this to the Marl filling hollows on the Limestone surface.

The Middle Permian Marl consists essentially of red mudstone with green bands and mottling, which contains, in addition to a thick development of sandstone at the top and bottom in the south-western area (see below), thin bands of sandstone, dolomitic sandstone and siltstone. The thin sandstones may be white, grey, green, red, brown or mottled and, in contrast to the thick sandstone at Worksop (see below), are generally fine-grained and, in some instances, micaceous.

Bands, normally thin, of conglomerate and breccia (including clay-flake breccia) occur widely but intermittently at several horizons, generally in the upper part of the Marl. The pebbles and angular fragments they contain are up to about 4 inches in diameter and are composed of (in approximate order of abundance) quartzite, sandstone, igneous rocks, chert, conglomerate, mudstone, limestone, dolomite and gypsum. The matrix may be sandy, marly or gypsiferous. The most notable conglomeratic horizon is that of the 'Harworth Breccia', $6\frac{1}{2}$ ft thick and 4 ft below the top of the Marl in Harworth Colliery No. 1 Shaft (Gibson 1924, pp. 57–8; Versey 1925, pp. 219–20). This horizon is tentatively identified at a number of other localities, and may be recognizable beyond the boundaries of this district (Edwards 1967, p. 125).

Thin bands of dolomite and dolomitic limestone are found in the Marl at some localities, and to the north and east of a line from Firbeck to Bevercotes

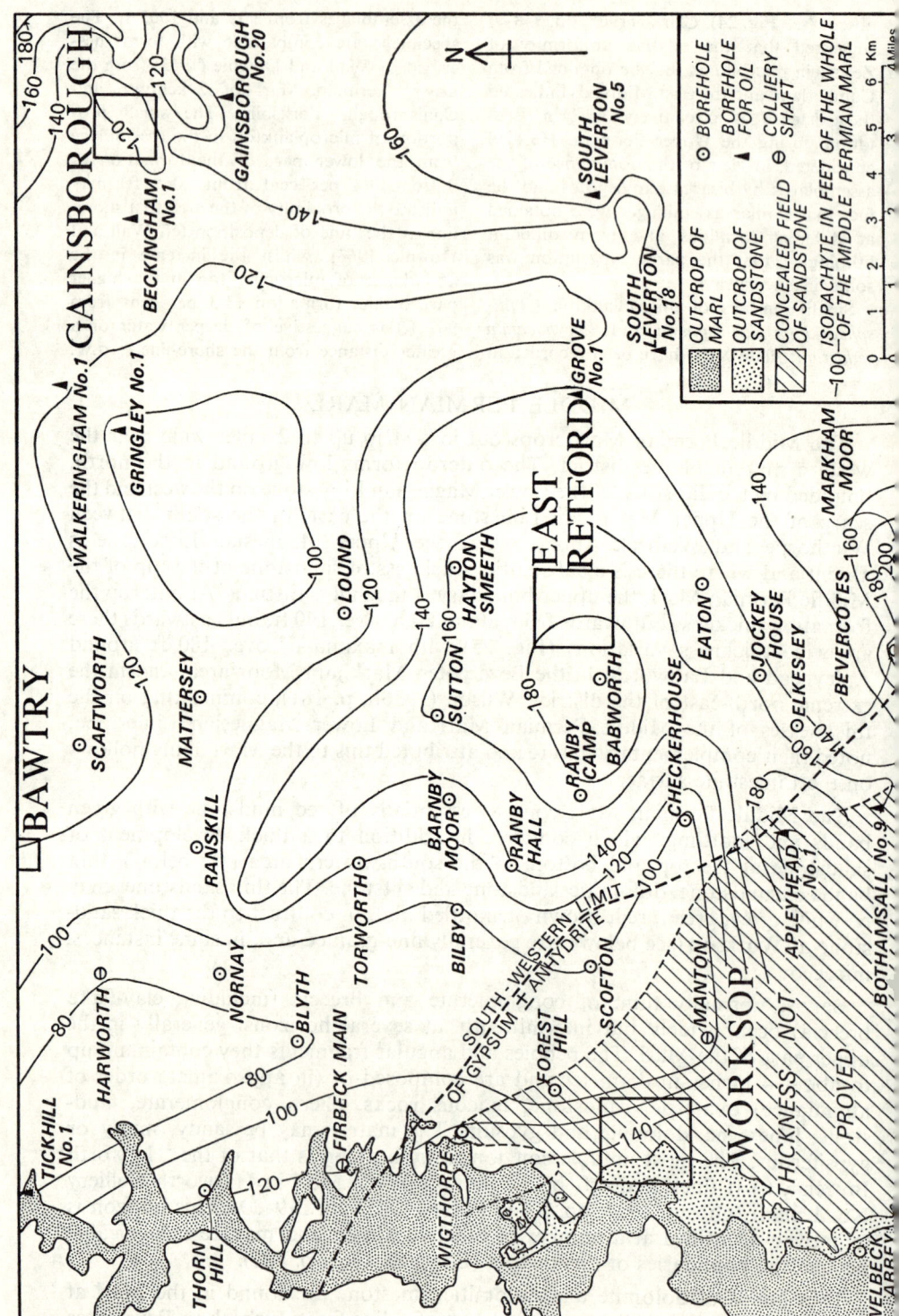

(Fig. 25) gypsum, occurring as nodules and veins as well as in thin bands, is ubiquitous. Anhydrite is found with, or in place of, gypsum in many of the boreholes in the north and east of the district, the most notable occurrence being the band 12 ft 8 in thick at the base of the Marl in Harworth Colliery No. 1 Shaft.

Reference has already been made to the local interbanding of dolomite and red mudstone at the junction of the Lower Magnesian Limestone and Middle Permian Marl, suggestive of a passage between the two formations (p. 122), and to the possibility of considerable thicknesses of sandy dolomite and dolomitic sandstone at the top of the Lower Magnesian Limestone in the Worksop–Checkerhouse area being of Middle Permian Marl age (p. 122 and Fig. 21).

Thick sandstones in the Middle Permian Marl. In the Worksop area the upper part of the Middle Permian Marl consists of sandstone (Fig. 25), which is worked at outcrop for building sand. The sandstone is apparently 20 to 30 ft thick over much of its field, but up to almost 90 ft are recorded in the Bothamsall Oilfield. It is red, grey, greyish green or yellow in colour, variably consolidated, and fine to coarse, but largely medium-to coarse-grained. Outcrop sections show it to be typically strongly false-bedded, and aeolian as well as fluviatile bedding may be present. It is separated from the Lower Mottled Sandstone in some sections by a few feet of mudstone containing bands of sandstone and thin breccia, which, although included for convenience with the Middle Permian Marl, may be the lateral equivalent of the Upper Permian Marl or Upper Magnesian Limestone.   E.G.S.

TABLE 3. *Mechanical analyses of some sandstones from the Middle Permian Marl*

Sample No. MR			Per cent quartiles (mm)			Sorting Coefficient
			25 (Φ_1)	50 (Φ_2)	75 (Φ_3)	$\sqrt{\Phi_1/\Phi_3}$
	(1)	*Sandstone unit (including building sand) from near top of Middle Permian Marl*				
27758–60		Mean	0·41	0·30	0·23	1·35
		Range	0·10	0·10	0·11	0·21
27761			0·40	0·28	0·14	1·39
27749–51		Mean	0·32	0·27	0·24	1·21
		Range	0·31	0·29	0·29	0·31
27753–4		Mean	0·25	0·19	0·15	1·28
		Range	0·03	0·02	0·00	0·08
27767			0·28	0·22	0·17	1·28
27769			0·25	0·21	0·19	1·15
	Mean of above 11 samples		0·33	0·26	0·20	1·28
	Range of above 11 samples		0·31	0·29	0·29	0·37
	(2)	*Sandstones from lower part of Middle Permian Marl*				
29139–41		Mean	0·17	0·14	0·11	1·25
		Range	0·05	0·04	0·04	0·14
	Mean of all above samples		0·29	0·23	0·18	1·27
	Range of all above samples		0·36	0·34	0·35	0·37

Channel and spot samples of this sandstone, chiefly from exposures, have been studied petrographically, and mechanical, modal and heavy mineral analyses are given in Tables 3 and 4 and Fig. 28 respectively. Medium-grade sand predominates in the samples and the degree of sorting is not generally high. Clastic grains, particularly those of the coarser grades, are commonly sub-rounded to well rounded. They comprise: quartz (mainly igneous), which is predominant; a range of rock fragments including chert, pure and argillaceous quartzites, and metasediments; minor feldspars, principally orthoclase with scattered fine micas; illite and other clay minerals; and sparse heavy minerals. In the ferruginous samples, ferric oxide or ferric hydroxide commonly forms irregular matricial impregnations, and pellicles around the clastic grains. Dolomite, a minor constituent, occurs as fine disseminations in the matrix. Calcite is sporadically present, in places forming spheroidal concretions.

TABLE 4. *Summary of modal analyses of some sandstones from the Middle Permian Marl*

Sample No. MR		Approximate proportions (by volume per cent)					
		Quartz, cherty silica	Feldspars	Composite particles, rocks, clay, ferric oxide, hydroxide	Muscovite, illite, biotite, chlorite	Carbonate	Minor constituents (including heavy minerals)
	(1) *Sandstone unit (including building sand) from near top of Middle Permian Marl*						
27758; 60	Mean	76	6	17	1	0	tr
	Range	4	1	4	1	0	0
27761		84	3	12	1	0	tr
27749–51	Mean	82	6	8	1	3	tr
	Range	4	2	2	tr	3	0
27753–4	Mean	86	6	7	tr	1	tr
	Range	2	0	2	0	0	0
27767		79	7	12	1	1	tr
27769		83	2	14	1	0	tr
	Mean of above 10 samples	82	5	11	1	1	tr
	Range of above 10 samples	13	5	13	1	4	0
	(2) *Sandstones from lower part of Middle Permian Marl*						
29139–40	Mean	80	3	11	1	5	tr
	Range	0	2	6	2	1	0
	Mean of all above samples	81	5	11	1	2	tr
	Range of all above samples	13	5	13	2	5	0

Detrital heavy minerals comprise a typical granitic suite, with traces only of metamorphic minerals: predominant zircon (mainly colourless and subhedral to euhedral, rarely purplish, rounded, polycyclic grains); subordinate tourmaline (green, brown, rarely blue) of igneous or pegmatitic origin (Krynine 1946, p. 68); rutile and leucoxene; garnet; opaque minerals—principally ilmenite and magnetite; apatite; traces of monazite; staurolite; anatase.   R.K.H.

Over most of its field in the present district this upper sandstone of the Middle Permian Marl is distinguishable from the overlying Lower Mottled Sandstone by its petrographical features (including the relative coarseness and roundness of its component grains) (see p. 165), by its bedding and in some instances by its colour. South of the district, however, the two sandstones become more alike and are difficult to distinguish by superficial examination: in the Chesterfield (112) district, for instance, sandstones of Middle Permian Marl age are included in the Lower Mottled Sandstone, only the upper part of which, therefore, is the equivalent of the Lower Mottled Sandstone of the East Retford district (Smith and others 1967, pp. 209–10, fig. 33).

Up to 40 ft of sandstone, in many instances grey or light-coloured, are recorded at the base of the Middle Permian Marl in boreholes from Bothamsall to South Leverton. Unlike the sandstone at the top of the Marl, however, it is generally fine-grained. It is probably the equivalent of the dolomitic sand and sandy dolomite assigned to the top of the Lower Magnesian Limestone farther west (see above and p. 122).

The Middle Permian Marl marks the silting up of the first phase of the Zechstein Sea. The only chemical deposition in the Middle Permian Marl of the district is represented by thin beds of dolomite, gypsum and anhydrite which become more conspicuous north-eastwards away from the land. The bulk of the formation consists of land-derived mud and silt with, near the shore in the south-west, thick deposits of sand. The latter were probably partly river-borne, but may also have originated by wind action, accumulating in water or as dunes at the water's edge.

No fossils have been recorded from the Middle Permian Marl.   E.G.S.

DETAILS

The only exposures north of Firbeck are shallow sections of stiff red clay seen in ditches and ponds and around the bases of uprooted trees. 'Gypseous marls' have been dug out "at the bottom of the hill half a mile west of Oldcoates on the Firbeck road" (Sedgwick 1835, p. 102), though the exact position of this, one of only two localities in the district where gypsum is recorded in the Middle Permian Marl at outcrop, is uncertain. Red and grey clay of the Middle Permian Marl is said to have been seen below the Upper Magnesian Limestone in Whalejaw Quarry [5843 8876] near Oldcoates.

At Letwell, the Middle Permian Marl was formerly exposed in Lamb Lane and by the roadside [559 870] near St Peter's Church (Aveline 1880, pp. 19–20). Both localities showed marl with bands of sandstone underlying the Upper Magnesian Limestone, but in the roadside section there were, in addition, pebble beds at two horizons. The higher of the pebble beds was 6 in thick, consisting of yellow sandstone with quartz pebbles, and the lower was about 4 ft thick, consisting of coarse red sand with pebbles which were most numerous at the base. This section was visited by Versey (1925, pp. 217–8) in 1914 when it was already overgrown, though the harder bands were still visible. He notes the occurrence of concretionary masses cemented by dolomite and baryte with a peripheral arrangement of pebbles. Among the pebbles he records devitrified rhyolite and a much-altered ?trachyte. The most detailed sections in the Letwell area, however, were provided by pipeline trenches in the nineteen-thirties and were described by Mackintosh (1939, pp. 282–94). In the red and green marls, gypsiferous in the upper part in his Letwell Lane section and with thin bands of dolomitic skerry and hard dolomitic sandstone in Carlton Lane, Firbeck, he recorded up to four conglomerate horizons, 8 to 25 ft below the Upper Magnesian Limestone. The pebbles in the conglomerate bands varied in size from minute to about 4 in diameter, were mostly rounded, though a few were angular, and were contained in either a sandy or marly matrix. A selection of pebbles examined by Mackintosh (*ibid.*, pp. 288–90) and Dr. J. Phemister (E 17186–95) consisted of sandstone, quartzitic sandstone, conglomerate, chert and decomposed igneous rocks.

In the railway cutting [5874 8694] south-east of Langold the topmost beds of the Middle Permian Marl are exposed in the following section:

	ft
Dolomite (*Upper Magnesian Limestone*): see p. 155	8
Marl, grey sandy laminated .. 2 to	3
Marl, red, with bands (up to 15 in) of red and grey mottled dolomitic sandstone containing geodes up to	9

In 1946 a trench section [575 853] in Middle Permian Marl was visible alongside Rotherham Baulk, 1 mile E of Gildingwells. This showed red, with patches of grey, marl containing bands, 2 to 6 in thick, of grey and mottled dolomitic siltstone. A specimen of the latter shows (E 22574) angular quartz grains up to 0·1 mm with minor amounts of chert, micas and heavy minerals (zircon, leucoxene and ilmenite) in a matrix of dolomite, in places granular, with some calcite. Brown streaks are produced by limonite staining.

Half a mile farther east along Rotherham Baulk, on the south side of the road, there is a large disused brick-pit [582 854] in the Middle Permian Marl. This is now overgrown, but when Aveline (1880, p. 19) surveyed the ground a section of red marl with bands of hard, thinly bedded, fine-grained, white sandstone was visible.

There are no exposures in Carlton in Lindrick at the present time, but temporary sections visible in pipe trenches in 1936 have been described in detail by Mackintosh (1939, pp. 295–7, fig. 2). They showed marl with numerous thin bands of sandstone and siltstone (from a fraction of an inch to almost a foot in thickness), largely green in colour, but also red and mottled. No pebble beds were found at this locality, but Mackintosh compares the sandy bands with the matrix of the pebble beds in the Letwell–Firbeck area (see above) and at Fox Covert (see below).

The basal 6 ft of the Middle Permian Marl are seen in the old quarry [556 826] east of Lindrick Common; they consist of red mudstone which rests abruptly on the Lower Magnesian Limestone. In the railway cutting [5594 8100] north-east of Shireoaks Colliery up to 18 ft of Middle Permian Marl were formerly visible. The section exposed in 1967 comprised about 8 ft of red mudstone with green bands, interbanded with red siltstone and red, fine-grained sandstone, and containing thin seams of greenish grey coarser sand (Smith *in* Neves and Downie 1967). R.F.G., E.G.S.

Four samples (MR 29139–42)[1] were taken through this section, at the following distances from the top: 0 ft, 4 ft, 5 ft, 8 ft. The topmost sample is friable reddish brown fine sand composed of subrounded to rounded grains, predominantly of quartz with subordinate composite rock particles, minor potash feldspars and micas. There is interstitial dolomite and ferric oxide and hydroxide staining. The second sample is friable pale greyish green fine sand (sorting coefficient 1·26) composed of moderately rounded quartz, minor rock particles, feldspars, micas and intergranular carbonate (Tables 3 and 4). The third sample differs in being a consolidated, reddish brown, silty fine sandstone, with fine micas on bedding planes, but containing conspicuous finely granular (0·01 mm) interstitial carbonate. The lowest sample is a

---

[1] Numbers prefixed MR refer to specimens in the Museum Reserve collection of the Institute of Geological Sciences.

friable pale reddish brown fine sand of similar composition to the highest sample. Heavy mineral suites (Fig. 28) from the three higher samples are similar, though proportions differ. There is a conspicuous garnet content in these suites.   R.K.H.

Up to 7 ft of reddish brown marly sand can be seen overlying the Lower Magnesian Limestone in Lady Lee Quarries [5615 7976] where, 250 yd SW of Haggonfields, there is a small outlier of Middle Permian Marl.

The old brick-pits [573 800] in the Middle Permian Marl between the canal and railway on the north-western outskirts of Worksop were referred to by Aveline (1880, pp. 18–19), who described the 12-ft section then visible as consisting of red marl with bands of reddish brown and white sandstone which were up to 6 in thick in the upper part. The pits have since been extended and deepened, and in 1946 some 20 ft of red marl with a little mottling and containing bands of grey and mottled sandstone and dolomitic sandstone up to 12 in thick (E 22575) were visible in one part. The pits are now either overgrown or infilled and partly built over.
          R.F.G., E.G.S.

The sandstone that forms the upper part of the Middle Permian Marl at outcrop southwards from the Gateford area has been seen in a number of large sand-pits between Fox Covert, north-north-west of Gateford Common, and Worksop. Several still provide good sections.

The face of the old pit [562 822] north-north-west of Fox Covert, formerly known as Payling's or Lindrick Sand Quarry and worked for building sand, is now obscured except for the top 9 or 10 ft, but in 1941 the following section was visible:

	ft
Mudstone and clay, reddish brown, with thin bands and lenses of grey and yellow sand	5
Interbanded reddish brown mudstone and coarse yellow sand	3
Sand, soft red, with some mottling, medium- to coarse- grained, false-bedded	8
Sand, grey, false-bedded	16
Conglomerate, pale greenish grey, with silty sand matrix; pebbles up to 2½ in diameter, seen to	0¼

The conglomerate, 1 to 2 ft thick, was first recorded and described by Mackintosh (1938, p. 204; 1939, pp. 290–2); it was green throughout with a matrix comparable to that of the Letwell conglomerates (see above). Pebbles of dolomite, volcanic rocks including rhyolite, quartzite, quartz, chert and calcareous sandstones were recorded. Near the base of the conglomerate "a layer of dolomitic stones in irregularly shaped pieces together with much fine quartz and marl" was seen. Mackintosh (1939, p. 291) gives the section below the conglomerate visible in 1937:

	ft	in
Conglomerate .. ..1 ft to	2	0
Clay, blue		4
Sandstone, hard red calcareous ..		2
Marl, red	3	6
Skerry (grey stone)	1	3
Marl	—	—

          E.G.S.

The old sand-pit in Fox Covert itself has been tipped in, but a section [5629 8227] visible in 1959 was similar to that in Payling's Quarry, viz: red marl 8 ft, on false bedded red sandstone with green mottling 20 ft.

The large sand-pit [564 823] north-east of Fox Covert was worked until recently for building sand (sandstone in Middle Permian Marl) and moulding sand (Lower Mottled Sandstone), but is now being filled with refuse by Worksop Corporation. However, a good section is still available on the eastern side of the pit:

	ft
LOWER MOTTLED SANDSTONE	
Sandstone, red fine-grained, with coarse-grained false-bedded bands, and level-bedded layers and lenses of red and green mudstone and sandy mudstone; 2-in pebble band near top	20
Sandstone, yellow, with coarse rounded grains; hard and cemented at base	1½
MIDDLE PERMIAN MARL	
Breccia, grey clay-flake (see below), normally 2 in thick but locally expanding to 6 in where bed cuts down into underlying mudstone 2 in to	0½
Mudstone, reddish brown, with layers and inclusions of fine to coarse green sand	8

	ft
Sandstone, red fine- to coarse-grained, with green patches; false-bedded in part	23
Sandstone, greenish grey coarse-grained, seen to	6

The 8 ft of mudstone with overlying thin breccia are included in the Middle Permian Marl, but it could be that they are the lateral equivalents of the Upper Magnesian Limestone. This mudstone band is the one exposed at the top of the two sand-pits previously described. South-eastwards from the Fox Covert area it can be traced at outcrop by the topographical slack it produces for a few hundred yards to the vicinity of California [5667 8177]. Beyond this point it is presumably too thin to produce a feature, and a further ⅔ mile to the south-east is represented by one or both of the thin mudstone bands below the Lower Mottled Sandstone in the sand-pit at Gateford (see below).   R.A.E., E.G.S.

The breccia noted above (E 33535) is composed of consolidated, roughly bedded, pale buff to greyish green poorly sorted, fine to medium sandstone, with prominent, bedded, flaky, pale greyish green mudstone galls. A clay gall (3 cm across) is pale pinkish brown with a greyish green selvedge. The sandstone consists of angular clastic grains (averaging 0·15 mm) scattered in a calcite cement (Plate XIV, fig. 3). The angularity is apparent and has resulted from extensive replacement of the clastics by calcite, as shown by ghost grains. The clastics comprise: clear igneous quartz; polygranular quartz (quartzite, chert); siltstone; orthoclase (with authigenic overgrowths) and microcline. The coarse calcite cement forms interlocking plates about 0·8 to 1·0 mm across. Some dolomite is also present.

Combined spot samples (MR 27753) of the lowermost 15 ft of sandstone exposed in this sand-pit consist of poorly sorted fine to medium sand-grade, rounded to well rounded clastics (Table 4): quartz, potash feldspars, composite rock particles, minor micas, ferric oxide and hydroxide, and fine (0·03 mm) dolomite rhombs (Plate XIV, fig. 1). Heavy minerals (Fig. 28) comprise zircon (including polycyclic purplish grains), tourmaline, opaque ores, rutile, apatite, pale pink to colourless garnet, leucoxene and traces of monazite.

A sample (MR 27754) of the red sandstone immediately underlying the 8 ft of red mudstone is ferruginous and dolomitic, composed of poorly sorted, moderately packed, sub-rounded to well rounded clastics. The last-named comprise igneous quartz, rock particles (chert, pure and argillaceous quartzites, and siltstone), potash feldspars, opaque oxides, leucoxene and sparse heavy minerals (Fig. 28). Ferric oxide and hydroxide form pellicle-like coatings around clastic grains. Coarsely crystalline dolomite, which noticeably replaces clastic grains, occurs mainly as an incoherent cement.   R.K.H.

In the sand-pit [576 813] at Gateford (Plate I), the Lower Mottled Sandstone and the sandstone in the Middle Permian Marl are worked for moulding sand and building sand respectively[1]. Known for many years as Turner's Sand-pit, it is now operated by Messrs. Alan S. Denniff Limited. A section measured here is very similar to that previously described:

	ft
LOWER MOTTLED SANDSTONE	
Sandstone, red, with a few thin grey bands; fine-grained, micaceous in parts; darker red argillaceous bands and clay galls; not markedly false-bedded	30
Sandstone, brown with coarse rounded grains (6 to 18 in thick)	1
MIDDLE PERMIAN MARL	
Mudstone and siltstone, red, with grey bands	0¾
Sandstone, red argillaceous and very micaceous	3¼
Breccia (½ to 1½ in thick) (see below)	thin
Mudstone, red, with bands of sand which contain impersistent breccias like that above	2
Sandstone, grey and red, becoming predominantly grey and greenish grey in bottom half; medium to coarse with many rounded grains; false-bedded, particularly in top half, seen to	23

---

[1] A recent extension of the pit to the north-west has proved a hitherto unknown fault throwing down north and bringing in a considerable thickness of Bunter Pebble Beds. There must be a further fault of similar displacement but opposite throw between here and Gateford Hill, so that the Pebble Beds occupy a narrow trough.

There are said to be a further 1 or 2 ft of sandstone below this section, and, below that, red mudstone.

Mr. D. B. Smith has recently studied the false bedding in the sandstone of the Middle Permian Marl in this pit (Plate XVA). He notes that, as in the Lower Mottled Sandstone above, there is foreset lamination and bedding towards the south in the top 3 to 4 ft. The underlying 8 ft show conspicuous foreset laminae dipping south and south-west, and the bottom part of the sandstone is cross-laminated with foresets dipping north and has cross bedding units up to 4 ft thick.

E.G.S.

A channel sample (MR 27749) through the top 10 ft of predominantly reddish brown sandstone underlying the 2-ft marl band is compact, fine- to medium-grained and poorly sorted (Table 3), with sporadic bleached spheroidal concretions. Clastic grains are mainly subrounded, but the coarsest grains are well rounded. Quartz predominates, with minor feldspar, rock particles, stained by ferric oxide and hydroxide, and fine (0·05 mm) interstitial dolomite rhombohedra (Table 4). Heavy minerals comprise predominant zircon (including rounded, polycyclic, purple grains), rutile, subordinate tourmaline (O = dark green, E = pinkish brown; O = dark brown, E = pale brown), rounded apatite, minor garnet, and opaque ores.

A channel sample (MR 27750) of the underlying 3 ft of reddish brown sandstone is similar. The component clastic grains are subrounded to well rounded, coated with ferric oxide, but rather better sorted than in the first sample (Table 3). Modal compositions (Table 4) are closely similar, as are heavy minerals (Fig. 28).

A channel sample (MR 27751) of the mainly grey, medium sandstone forming the lowest part of the section consists of friable to unconsolidated, well sorted (Table 3), rounded to well rounded clastics: quartz, composite rock particles, minor potash feldspar, dolomite and ferric oxide and hydroxide (Table 4). Small (3 to 4 mm diameter) concretions, oblate to spheroidal, scattered through this sand, are hard, indurated, and composed of clastic grains cemented by calcite and subordinate dolomite. It is noteworthy that calcite is rare in the enclosing sandstone, and the concretions appear to have formed through the migration

of calcite to these discrete centres, though the physico-chemical process involved is unknown.

A specimen of the sequence containing the breccia consists of: reddish brown, laminated marl grading up into siltstone (1 cm); followed by gravelly and sandy breccia (4 cm) in which phenoclasts are, in places, commonly orientated and roughly graded; succeeded by reddish brown sandstone (1 cm); a pale greyish green sandstone band (0·5 cm); and, topmost, a graded, reddish brown fine to medium sandstone band, 1·5 cm thick.

In the breccia (E 36106), phenoclasts are markedly angular (less commonly rounded), and range from irregular to platy. Platy phenoclasts towards the top of the band are in parallel orientation. Phenoclasts and fine gravel, which is rounded, consist of the following lithologies, in order of relative abundance: (1) calcirudite including microcrystalline dolomite (8 to 16 microns), with minor silt, reddish brown ferruginous and argillaceous dolomite (Plate XIV, fig. 5), and oolitic dolomite (ooliths average 0 08 to 0·1 mm across); (2) quartz individuals and polygranular quartz, including igneous quartz, quartzite, sheared quartzite; (3) sparse polygranular quartz-feldspar aggregates of granophyric and microgranitic appearance. Coarse sand grains are well rounded and include the lithologies noted above. The fine, intergranular matrix comprises fine sand, silt, clay and dolomite, with ferruginous staining (Plate XIV, fig. 4). Ferric oxide also forms pellicle-like coatings around clastic grains.

A specimen (MR 27748) from the 9 in of mudstone and siltstone $3\frac{1}{4}$ ft above the breccia is a bedded, reddish brown siltstone with conspicuous fine micas on bedding planes, and thin reddish brown clay partings. The siltstone consists of fine silty resistates stained with ferric oxide, ferruginous clay, and a trace of carbonate. X-ray powder photographs (NEX 257, 840) showed quartz, illite, feldspar, minor chlorite and possible kaolinite.   R.K.H.

Good sections of the Middle Permian Marl sandstone were formerly seen in the large sand-pits west [580 798] and south-west [581 794] of Worksop station. The sandstone here is soft, medium- to coarse-grained with many well rounded grains, and false-bedded throughout. At the former locality there are 12 ft of red, on 4 ft of yellow, on 8 ft of grey sandstone, and at the latter 20 ft of red sandstone with yellow and grey bands, especially in the lower part. In the pits west of the station striking evidence of the post-depositional nature of the staining of the sandstone was seen. At the junction of the yellow and grey beds there is a band containing black staining, evidently due to the presence of a manganese mineral. The black staining preferentially affects the coarser bands of sandstone and generally follows the false bedding, but in several places it rims irregular pockets of yellow sand which extend downwards into the grey sand, and through which the bedding passes undisturbed. The bedding is also unaffected where 'dykes', 6 to 24 in wide, of soft grey sand cut through the yellow sandstone.   R.F.G.

In the pit west of Worksop station a channel sample (MR 27758) through 8 ft of the red sandstone consists of poorly sorted clastics predominantly of medium sand grade (Table 3). The clastics consist chiefly of subrounded to rounded quartz and polygranular silica, with minor potash feldspars, conspicuous rock particles and much intergranular ferric oxide and hydroxide, and a trace of micas (Table 4). Heavy minerals comprise (Fig. 28) subangular to subhedral, colourless zircon, well rounded tourmaline (O = very dark brown, E = colourless; O = dark green, E = colourless), conspicuous reddish brown and yellowish brown rutile, minor colourless garnet, rounded leucoxene aggregates, opaque ores (ilmenite and magnetite), and traces of staurolite and monazite. A sandstone (MR 27759) occurring below this sample, is better sorted (sorting coefficient = 1·24), but heavily impregnated with a black manganiferous cement (see above).

A channel sample (MR 27760) through 8 ft of the yellow and grey sandstone near the base of the section, is friable, thinly bedded and variably iron-stained. The component clastic grains are poorly sorted (sorting coefficient = 1·37), the coarser grains being well rounded, and comprise predominant quartz, polygranular composite grains, ferric oxide and hydroxide (commonly coating the grains) and minor potash feldspars and micas.

Combined channel samples (MR 27761) taken through 18 ft of red sandstone in the

pit south-west of the station consist of poorly sorted (sorting coefficient = 1·45) rounded to well rounded clastic grains—predominantly quartz, with subordinate composite particles, intergranular ferric oxide, and very minor feldspars (Table 4). Heavy minerals (Fig. 28) include the varieties described from the previous locality, but with prominent apatite.
                                                R.K.H.

Castle Hill [5831 7884], Worksop, is formed of Middle Permian Marl sandstone, and about 20 ft of fine- to coarse-grained red sandstone with thin greyish brown bands are exposed on the north side. A spot sample (MR 27767) consists of well graded, fine to medium, rounded to subrounded clastics—quartz, rock particles, minor potash feldspars and micas—with intergranular ferric oxide (Tables 3, 4). The heavy mineral suite (Fig. 28) is similar to that from the pit west of Worksop station (see above).

South-west of Worksop a few feet of soft, red, medium- to coarse-grained sandstone with rounded and subrounded grains can be seen in small disused sand-pits north-north-west of Worksop Manor [5682 7833], west-south-west of Castle Farm [5706 7677] and west of Sloswicks Farm [5610 7537]. A spot sample (MR 27769) from the pit near Castle Farm shows that the clastics are moderately sorted and comprise predominant quartz, subordinate polygranular particles, ferric oxide and hydroxide, and very minor feldspars and micas (Table 4). Heavy minerals (Fig. 28) comprise the varieties described from the pit west of Worksop station (see above).

The 3 ft of brown, false-bedded sandstone exposed in the road [5497 7686] near Birks Cottages, 475 yd NW of Welbeck Bar, lies in the lower part of the Middle Permian Marl. This exposure may be the one referred to by Aveline (1880, p. 18), who saw overlying red marl.   R.F.G., R.K.H., E.G.S.

Colliery sections and cored boreholes through the Middle Permian Marl are as follows:

Bevercotes Colliery No. 1 Shaft (194 ft 2 in at 956 ft 5 in). The Middle Permian Marl consists of red mudstone with green banding and mottling, and contains bands of dolomite, one of which, 4 ft 10 in thick at 894 ft 7 in, is described as a largely dark grey argillaceous and dolomitic oolite. There are nodules and veins of gypsum and, at the base, 11¼ ft of pale green, fine-grained sandstone. In No. 2 Shaft the formation is recorded as red mudstone with bands of pale green or white sandstone and of gypsum, but its upper limit is not determinable.

Checkerhouse Borehole (99 ft 8 in at 646 ft). The Middle Permian Marl is made up of 44 ft 4 in of red marl with bands of blue sandy marl and thin limestone bands on 55 ft 4 in of red and blue marl with gypsum. Immediately below are 79¾ ft of red and grey sandy limestone (see p. 131) which have been included in the Middle Permian Marl by Edwards (1951, pp. 101, 152).

Firbeck Main Colliery (118¼ ft at 123 ft 7 in). The formation consists of red marl with beds of grey sandstone up to a recorded thickness of 3 ft 11 in; thin gypsum bands occur in the bottom 38 ft 5 in.

Harworth Colliery No. 1 Shaft (81 ft at 512 ft 10 in). The succession is: stone-bind 4 ft 2 in, conglomerate 6½ ft, blue marl 10 in, red marl with gypsum bands up to 5 in thick, 56 ft 10 in, on anhydrite 12 ft 8 in. The conglomerate is the well known 'Harworth Breccia' and has been described by Gibson (1924, pp. 57–8) and Versey (1925, pp. 219–20). It consists of rounded to subangular pebbles set in a matrix of gypsum, and there are also ramifying veins of gypsum through the rock. It is greenish when fresh, turning red on exposure, and comprises alternating medium-grained (average pebble size 7 mm) and coarse-grained (pebbles up to 10 cm) beds. The pebbles are composed of lithic and crystal tuffs, devitrified rhyolites, soda-syenite, basalts, andesites, quartzites, sandstones, feldspathic grits, greywackes, limestones (including one specimen containing a trilobite), red and purple mudstones and detrital gypsum; and commonly have a thin coating of dolomite. Apparently they are largely of Charnian or Uriconian origin.

Hayton Smeeth Borehole (166 ft at 1143 ft). The Middle Permian Marl consists of red and blue marl with gypsum bands and veins, and bands of grey sandstone up to 3 ft thick (for details see Wilson 1926a, p. 199).

Loxdale Cottages Borehole, Oldcoates (88½ ft at 158½ ft).

Manton Borehole (91¼ ft at 337¾ ft). The succession is: red and grey sandstone with marl bands 21 ft 3 in, on red and blue marl and sandy marl with a 3-in limestone band (at 304 ft) 70 ft. The Middle Permian Marl rests on red and grey sandstone and limy sandstone 56 ft 10 in, underlain by limestone 26 ft 2 in, on red and grey limy sandstone 16 ft 2 in, the whole of which is assigned to the Lower Magnesian Limestone (p. 131). The section compares closely with those of Manton Colliery shafts (see below).

Manton Colliery (86 ft 11 in at 324½ ft in No. 2 Shaft; 86½ ft at 352½ ft in No. 4 Shaft). The sections consist of red and grey marl with sandstone bands in No. 2 Shaft, and red marl with sandstone and limestone bands in No. 4 Shaft, respectively resting on 130 ft 7 in of red and blue 'limy sandstone', and on 131 ft of red and grey 'limestone and limy sandstone'. The sandstone and associated limestone here referred to the upper part of the Lower Magnesian Limestone have been regarded as part of the Middle Permian Marl by Edwards (1951, pp. 101 and 208).

Scofton Borehole (103½ ft at 405½ ft). The Middle Permian Marl consists of red marl with grey bands and bands of grey and red sandstone (one 9 ft in thickness).

Shireoaks Colliery (incomplete 47½ ft at 64¾ ft). The section is recorded as red marl with bands of light-coloured sandstone.

South Leverton Borehole (Ford 1920, p. 100) (145¾ ft at 1664 ft 2 in). The section is red and grey marls with gypsum 120 ft 8 in, on brownish red stone with anhydrite 2 ft 7 in, on skerries, red marl and gypsum 22½ ft.

Sutton Borehole (160 ft 4 in at 838 ft). The formation consists of brown and grey sandy marl with gypsum throughout, occurring in beds (up to 3 in thick), veins and 'balls'. In the lowest 62 ft there are bands of grey sandstone, usually containing gypsum and up to 4¾ ft thick.

Thorn Hill Borehole [5762 8903] provides a section of the upper 101 ft 10 in of the formation, consisting largely of red mudstone with green bands. The mudstone is sandy in parts, and there is a 4½-ft band of fairly coarse, friable sandstone with well rounded grains and sporadic small pebbles 13½ ft from the top and therefore roughly at the horizon of the Harworth Breccia. Also present are

K

thin bands of sandstone with a total thickness of about 4 ft, and thin gypsum bands totalling less than 2 ft.

Wallingbrook Wood Borehole. The 21 ft 7 in of cores seen represented the basal part of the formation and consisted of red mudstone and sandy mudstone with bands (up to 2 ft 5 in thick) of greenish grey, grey and red, fine-grained sandstone with rounded and subangular grains and inclusions of greenish grey and red mudstone.

Welbeck Abbey Borehole. 42 ft of core representing the lowest $92\frac{1}{4}$ ft of the Middle Permian Marl were recovered. These consisted of red mudstone and sandy mudstone with grey bands, 'fish-eyes', micaceous partings, bands of soft, pale grey, micaceous sandstone up to 12 in thick, and bands of red sandstone with clay pellets.

Wigthorpe Borehole (133 ft at 143 ft). Here all but the top 12 ft of the Middle Permian Marl were cored. The cored section consists of $20\frac{1}{2}$ ft of grey and red sandstone with thin bands of red marl, on $112\frac{1}{2}$ ft of red marl with bands of fine to coarse, mostly grey, sandstone. The top 25 ft of the Lower Magnesian Limestone consists of $1\frac{1}{2}$ ft of pale fawn 'limestone' on $23\frac{1}{2}$ ft of reddish grey calcareous sandstone, and by comparison with Checkerhouse Borehole and Manton Colliery shafts (see p. 149) it may be that these measures should be included with the Middle Permian Marl (see pp. 130–1).

Boreholes at Worksop include the two at the waterworks on the north side of the town. No. 1 Borehole showed 134 ft of red clay with sand and stone beds (for details see Lamplugh and Smith 1914, p. 126) on limestone, though at least half of the top 10 ft of the latter is red and may possibly belong to the Middle Permian Marl. No. 2 Borehole, the cores of which were examined by Mr. R. F. Goossens ended at 315 ft after penetrating 100 ft of Middle Permian Marl. Here the top $\frac{1}{2}$ in of the Marl is a breccia which overlies 5 ft of red and grey marl with bands of red sandstone, which rest on 17 ft of measures consisting largely of fine-grained, grey-banded, red sandstone with coarse bands. The remaining $77\frac{1}{2}$ ft of the beds are red mudstones with grey and mottled bands, and sporadic bands of sandstone, some of which are dolomitic. A specimen of sandstone from $221\frac{1}{2}$ ft, $6\frac{1}{2}$ ft below the top of the Marl, examined microscopically by Mrs. J. E.

Morey (E 22588), consisted of quartz grains up to 0·25 mm diameter, but averaging 0·1 mm, mostly angular in outline, together with fragments of feldspar, chert, granulite, slate, chlorite, mica, ferric oxides and hydroxides and sparse heavy minerals (see below), loosely cemented by finely granular dolomite and calcite. A further sliced specimen (E 22589), 6 in lower, shows a passage between fine-grained dolomitic sandstone and what is described as arenaceous dolomite, the latter consisting of angular quartz grains with accessory fragments similar to those listed above, embedded in a compact dolomitic and limonitic cement. Possible solution cavities at the junction of the dolomite and sandstone contained well formed calcite crystals up to 2 mm in diameter.

Other specimens of sandstones from between 224 and 245 ft have also been examined in thin section (E 22590–4). In general they are similar to the specimens from $221\frac{1}{2}$ ft described above, the quartz grains varying in size from 0·1 to 0·5 mm, and the carbonate matrix being fine to coarsely crystalline and mostly consisting of dolomite.   E.G.S.

Heavy minerals separated from the sliced specimens of sandstone comprise the following: angular to subhedral colourless and rounded purplish zircon, dark reddish brown prismatic rutile, subangular tourmaline (O = dark brown, E = colourless; O = dark green, E = pale brown), well rounded apatite, colourless to pink, step-etched garnet, ilmenite, magnetite, leucoxene and sparse anatase. Approximate proportions are indicated in Fig. 28.

Modal analyses of sandstone samples (E 22588, E 22590–4) from this borehole are not directly comparable with those of the outcrop samples of Middle Permian Marl sandstone since the former contain considerable proportions of carbonate, principally dolomite, which, in places, has replaced clastic grains extensively.   R.K.H.

At Worksop Brewery No. 2 Borehole the Middle Permian Marl succession is as follows:

	Thickness ft	Depth ft
Upper Magnesian Lime- *stone*		$34\frac{3}{4}$
Mudstone, red; sandy and micaceous at base	$0\frac{1}{2}$	$35\frac{1}{4}$

	Thickness ft	Depth ft
Sandstone, red, with rounded grains; sporadic small pebbles; clay pellets	0¾	36
Mudstone, red ..	0¾	36¾
Sandstone, yellow, dolomitic, with bands of red-mottled grey mudstone	2¼	39
Mudstone, red, with bands of pale grey, fine-grained sandstone, some of which are slightly micaceous (*about 25 ft of core missing— ?sand*) ..	59	98
Sandstone, pink except in top and bottom 12 in where pale grey; rounded and subangular grains; pellets of red mudstone ..	12	110
Mudstone, red, with bands of hard, pale grey, fine-grained dolomitic sandstone (*about 6 ft of core missing*) ..	27	137
Sandstone, pale greenish grey, with rounded and subangular grains; grey argillaceous partings at top	2¼	139¼
Sandstone, pink, with rounded grains; a few partings of red mudstone	1½	140¾
Mudstone, red, with sandy micaceous bands and thin bands of pale grey, fine-grained sandstone (*9½ ft of core missing*)	13¾	154½
Dolomite, pale grey, crystalline, with a few vughs in lower part ..	2½	157
Mudstone, hard grey, with red mottling ..	0¼	157¼
Mudstone, red (*11¾ ft of core missing*)	20¾	178
Lower Magnesian Limestone	—	—
		E.G.S.

A specimen (E 33363) of dolomitic sandstone from 38½ ft is well bedded, rather soft and incoherent with alternating finer and coarser bands, averaging 1 cm thick. In section abundant clastic grains—mainly subangular quartz (0·08 to 0·2 mm) with subordinate feldspar and sparse grains of zircon, apatite and tourmaline—are scattered in a finely granular dolomite cement ($\omega = 1·684 \pm 0·002$). The dolomite marginally replaces the resistates. Differences in the proportion of cement to clastics are reflected in the coarser and finer bands noted macroscopically. In the finer bands, the proportion (by volume) of dolomite to clastic grains is estimated at about 45:55, and in the coarser, 35:65. After solution of a finely crushed portion of the whole rock in hot concentrated HCl, the total insoluble residue was determined as 57 per cent.

Pale grey, fine-grained dolomitic sandstone (E 33364) from 125 ft is thinly bedded and contains darker streaks and segregations, indicating some degree of replacement. The bulk of the rock is an aggregate of silt- and fine sand-grade clastic grains (0·01 to 0·2 mm), with low to moderate sphericity and showing poor grading. These consist mainly of igneous quartz with lesser metamorphic quartz, potash feldspars, cherty silica and intergranular micas—illite, muscovite and probable glauconite. There are scattered opaque grains of leucoxene, zircon and tourmaline (O = deep yellowish brown; E = colourless). The abundant cement consists of dolomite ($\omega = 1·684 \pm 0·002$) which occurs in two forms: as tiny (0·01 mm) porphyroblastic rhombs mainly bordering on, or replacing, the clastic grains, and as larger (0·2 mm) grains of similar outline and size to the clastic grains. There is some evidence to suggest that the second form of dolomite represents completely replaced clastic grains. The approximate proportion (by volume) of carbonate to resistates is estimated visually as 45:55.   R.K.H.

Immediately below the Middle Permian Marl, from 178 to 217½ ft, are dolomitic sandstone and sandy dolomite which have been included with the Lower Magnesian Limestone (p. 132) but which may (like similar developments at Manton Colliery and Manton, Checkerhouse and Wigthorpe boreholes described above) be of Middle Permian Marl age (see p. 122).

A nearby water borehole [5900 7938], off East Gate, Worksop, records about 21 ft of sand on 71 ft of red clay with thin bands of stone, on 53 ft of stone with red clay at 193 ft. However, it may well be that the bottom 53 ft of this succession is partly or wholly sandstone or sandy dolomite, which, by comparison with for example Worksop Brewery No. 2 Borehole, should be placed in the Lower Magnesian Limestone.

The N.C.B. boreholes that were open-holed through the Middle Permian Marl (see Appendix I and Fig. 25) all record the formation as red or brown mudstone, usually with bands of grey or green mudstone and hard bands. Except at Babworth, Forest Hill and Jockey House, gypsum is also recorded. The driller's log of Lound Borehole places the top of the Middle Permian Marl at 924 ft 1 in and, although this gives an anomalous thick section for the Upper Magnesian Lime-stone (see Fig. 26), it is accepted because a sample from 925 ft is strongly suggestive of the Harworth Breccia. It is described as consisting of 75 per cent grey subangular pebbles and coarse sand, and 25 per cent light red, medium-grained sand.

Chipping samples show the Middle Permian Marl in oil bores to consist essentially of red or reddish brown mudstone, silty mud-stone and siltstone with green or grey bands (for thicknesses see Fig. 25 and Appendix I). *Tickhill* No. 1 Oil Bore proved the basal 50 ft of the Marl, which contains tabular and fibrous gypsum and, in the bottom 30 ft, much fine-grained sandstone. In *Ranskill* No. 1 Oil Bore the Marl consists of about 150 ft of red and subordinate grey silty mudstone with sandy partings and numerous thin bands of gypsum. Anhydrite is associated with gypsum in a thin band at or near the base, and there is a band of sandy dolomitic limestone, about 5 ft thick, 100 ft higher. In *Gringley* No. 1 Oil Bore a thin band of micaceous sandstone is recorded, together with gypsum, in the 100 to 130 ft of Marl. In *Walkeringham* No. 1 Oil Bore the formation is about 125 ft thick; about 10 ft of calcareous siltstone occur at the top, and there is a thin band of pale grey oolitic limestone in the lower part. Gypsum is recorded in both oil bores at Walkeringham, and anhydrite in No. 2. In *Morton* No. 1 Oil Bore anhydrite occurs at

three horizons in the 130 ft of red mudstones —in indistinct bands just over 20 and just over 80 ft from the top, respectively about 3 and 10 ft thick, and in the basal few feet. At the highest and lowest horizons the anhydrite is associated with gypsum. At *Beckingham* the Middle Permian Marl averages about 125 ft in thickness. Thin bands of limestone are recorded in the lower part in three of the four oil bores, and all contain gypsum, in-cluding, in No. 5 Bore, a massive fibrous band about 5 ft thick some 25 ft from the top of the formation. A basal band of anhydrite occurs in Nos. 1 and 2 bores.

At *Gainsborough* the Marl ranges in thick-ness from 100 to about 165 ft. Gypsum or anhydrite, or both these minerals, have been noted in all but three of the oil bores, and there is some evidence that they occur in bands up to 10 ft thick. Grey calcareous siltstone is recorded from the top of the Marl in several bores, at which horizon pyrite is found in a number of instances. Thin bands and streaks of limestone, in some cases argillaceous or dolomitic, occur in several sections, and streaks and thin bands of fine-grained sandstone are found in others.

At *South Leverton* the Marl varies in thickness between about 125 and 170 ft. Gypsum is recorded in all the oil bores except No. 18; in No. 2 Bore anhydrite has also been noted. Sandstone occurs in nearly half the sections, including about 20 ft of fine-grained sandstone mixed with siltstone at the top of the Marl in No. 3 Bore, and about 10 ft of pale green, fine-grained sand-stone and about 35 ft of grey, fine-grained sandstone at the base of the Marl in Nos. 9 and 15 bores respectively. At *Grove* the formation is 145 ft thick in No. 1 Oil Bore and 140 ft in No. 2. Gypsum is recorded in both, together with anhydrite in No. 1, where there is a red gypsiferous sandstone at the top of the Marl and about 20 ft of white fine-grained sandstone near the base. In *West Drayton* No. 2 Oil Bore the Marl, 160 ft thick, contains about 20 ft of reddish brown and grey sandstone near the top; near its base a similar thickness of largely brown marly sandstone is separated by 15 ft of sandy marl from the uppermost beds of the Lower Magnesian Limestone which here consist of dolomite and sandstone (see p. 135). A little gypsum is recorded at about the middle of the Marl.

At *Apleyhead*, where the Middle Permian Marl is about 200 ft thick, gypsum is recorded only as a trace at No. 2 Oil Bore, but sandstone occurs in all three bores. In No. 1 Bore there are about 20 ft of reddish brown sandstone at the top of the Marl and rather less than this thickness of white fine-grained sandstone at the base; in Nos. 2 and 3 bores bands of sandstone occur throughout but in No. 2 are most conspicuous at the top and bottom. There is a band of dark red, medium-grained dolomitic limestone in the lower part of the section in No. 2 Bore, and a band of light grey slightly dolomitic limestone near the middle in No. 3.

At *Bothamsall* the formation is up to about 200 ft thick. Sandstone is a conspicuous feature, much of it occurring at the top—there are 60 ft in No. 9 Oil Bore, and most of the top 90 ft in No. 20 consists of sandstone —but there are also substantial thicknesses in the middle (45 ft in No. 10) and at the base (40 ft in No. 19). The sandstone varies in grain from fine to coarse and in colour from white, grey or pale green to orange, brown or red. Pebbles are recorded in three of the bores: in No. 4 sporadic pebbles occur in the fine- to medium-grained sand composing the top 20 ft or so of the formation, and there is also a pebble band about 50 ft from the top; in No. 10 bands of pebbly sandstone are recorded in the bottom 25 ft; and in No. 19 there are two conglomeratic horizons within 30 ft of, and about 60 ft from, the top, the former including pebbles of quartz and igneous rocks. There is a considerable proportion of dolomitic limestone interbedded with mudstone in the middle part of the Marl in No. 8 Bore and limestone bands also occur in the top 55 ft of the formation in No. 11 Bore.   E.G.S.

UPPER MAGNESIAN LIMESTONE

Variations in thickness of the Upper Magnesian Limestone are shown on Fig. 26, from which it can be seen that this formation is absent in the south-west around Worksop and thickens irregularly towards the north-east, attaining about 130 ft at one locality in the Gainsborough Oilfield. It exceeds 40 ft in thickness over more than half of the district, but only locally, in the Gainsborough area and at Lound, does it exceed 100 ft. At outcrop its dip-slope forms a belt of well drained arable land up to two miles wide from Tickhill to Gateford; it is 30 to 35 ft thick in the north and thins gradually southwards to Gateford, where it disappears. The Limestone is normally white, pale grey, cream or buff in colour, but red and pink varieties are fairly common, and yellow, brown, mauve, light green and mottled rocks have been noted. It varies in composition from almost pure dolomite to almost pure limestone, the upper part being generally more dolomitic than the lower. Much of the rock is distinctly crystalline, though oolites are common, and finely granular bands, which have a porcellanous appearance in hand specimen, occur. Microscopic examination of the crystalline rocks often indicates that they are recrystallized oolites or organic limestones. A pisolite has been recorded at one locality. The bulk of the Limestone is compact and fine-grained, but saccharoidal bands are not uncommon and coarse-grained dolomite occurs rarely.

The Limestone is generally thinly bedded or flaggy, with bedding planes, which may be irregular and anastomosing to enclose lenses of rock, less than 3 in apart. Beds rarely exceed 9 in, and have not been seen to be more than 18 in; they are commonly separated by films of grey, red or purple clay. False bedding, large- and small-scale, is common in places. Bands of mudstone, silty mudstone and siltstone are found in many sections, particularly near the top and bottom of the formation, where they indicate passage into the Upper Permian Marl and Middle Permian Marl respectively. They are chiefly red or reddish brown in colour, but grey and greyish green mudstones and siltstones

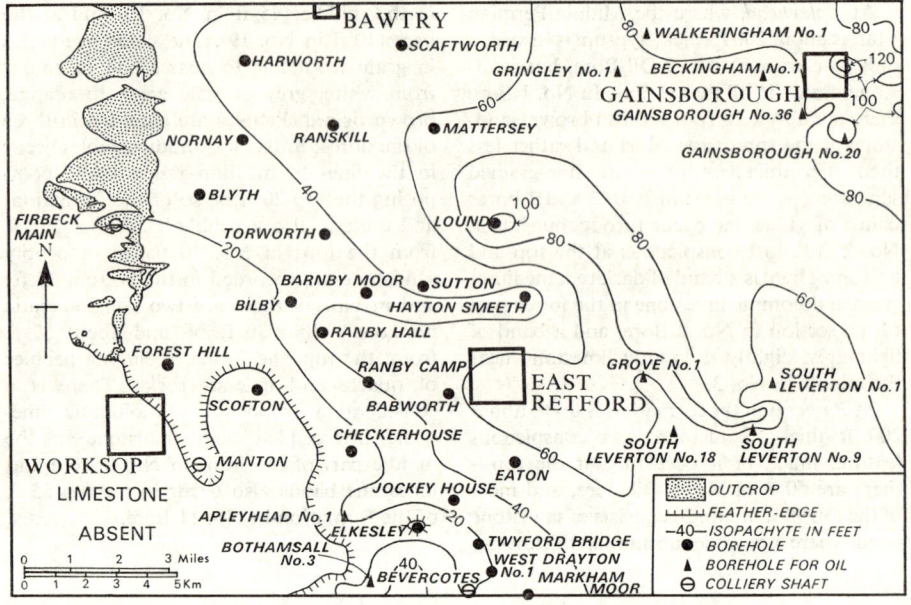

FIG. 26.   *Isopachyte map of the Upper Magnesian Limestone*

also occur. In addition the basal part of the Limestone is argillaceous or silty at some localities.

The Limestone is locally pyritic, and small amounts of limonite are commonly present. Anhydrite has been recorded in a number of underground sections in the east of the district, and traces of galena have been noted at one outcrop locality. Dendrites, probably formed of a manganese compound, are common, and black spheres, possibly hydrocarbons, have been noted in several boreholes.

Lithologically comparable with the Lower Subdivision of the Lower Magnesian Limestone, the Upper Magnesian Limestone was evidently formed under similar shallow-water conditions, though its more restricted fauna (see below) suggests a higher salinity.   E.G.S.

Palaeontology. The fauna of the Upper Magnesian Limestone is fairly uniform from north Yorkshire to Nottinghamshire and is similar to that of the equivalent Seaham Beds of County Durham (Smith 1970). The commonest fossils in the East Retford district, as elsewhere, are the alga *Tubulites permianus* and the bivalves *Liebea* and *Schizodus*. Conditions in the Upper Magnesian Limestone sea, while inimical to most of the organisms common in the Lower Magnesian Limestone, must have been very favourable for these stenohaline forms, for they are locally abundant. The effects of salinity on the distribution of bivalve genera in the Upper Permian rocks of northern England are discussed by Logan (1967).

The only other fossils recorded from the Upper Magnesian Limestone of the district are algal remains comprising stromatolite and carbonaceous debris, a few, largely indeterminate, gastropods, and ostracods. Lists of fossils collected from exposures are given under the details of localities on pp. 155–6: the only underground localities from which fossils are recorded are Manton Colliery No. 4 Shaft and Worksop Waterworks Borehole (p. 156).   J.P.

DETAILS

Some 16 ft of 'limestone'[1] are exposed in the road cutting [5839 9303] ½ mile W of St Mary's Church, Tickhill; *Tubulites permianus* has been collected here, and *Liebea squamosa* and *Schizodus obscurus* are common near the top of the section. The same three species were collected from 5 ft of hard oolite forming the bottom part of a 15-ft section of 'limestone' in the cutting [5850 9283] near Friary Farm. On the outskirts of Tickhill 12 ft of 'limestone' are exposed in quarries at Friars' Hill [584 929], and up to 14 ft in quarries near St Mary's Bridge [5882 9283]. 'Limestone' is exposed in several old quarries on the roadside between Tickhill and Woolthwaite, including up to 10 ft in those at Limestone Hill. Sedgwick (1835, pp. 105–6) gives a detailed section of the 17½ ft of rocks formerly visible in one of the latter quarries. South-west and south-south-west of Tickhill, 'limestone', up to 18 ft thick, is seen over almost a mile of the railway cutting between Friars' Lane [5830 9269] and Harworth Junction [5905 9120]. Geodes are seen in places, and at one point [5870 9164] there is a small anticline with an east-north-easterly axis, in the otherwise regularly dipping strata. This fold brings up the basal beds of the formation, which consist of interbedded 'limestone' and mottled mudstone and are exposed at the bottom of the cutting for 400 yd to the south. In another cutting [5958 9000], farther south on the same railway line and 1 mile NNE of Oldcoates, up to 20 ft of 'limestone', buff and more thinly bedded in the upper part and grey below, are exposed. Shells including indeterminate gastropods, *L. squamosa*, and *S. obscurus* are found throughout but are more abundant in the upper part, where *T. permianus* also occurs. At the southern end of this cutting [5964 8983] red clay of the Upper Permian Marl rests on the 'limestone' (p. 159).

Several quarries at Oldcoates show 10 to about 15 ft of flaggy 'limestone' in which individual beds are generally ⅛ to 4 in thick, though one is up to 18 inches in places. The latter bed, as well as some of the thinner ones, is coarser in grain than usual. Wedge bedding is common in this area, and small-scale false bedding is also seen in places. Specks of galena are recorded from the Oldcoates area

(Sedgwick 1835, p. 106). At Whalejaw Quarry [5843 8876] 30 ft of 'limestone', probably the full thickness at this point, are said to have been seen resting on red and grey clay of the Middle Permian Marl. At the old quarry [5869 8860] 100 yd S of St Mark's Church, Oldcoates, *T. permianus*, indeterminate gastropods, *L. squamosa* and *S. obscurus* were collected from the top 6 ft of the exposed 'limestone'. At the old quarry [5875 8835] 350 yd SSE of the church and on the south side of Goldthorpe Mill, the following fossils were collected from the 13-ft section of 'limestone', being especially abundant near the top: indeterminate carbonaceous plant remains, *T. permianus*, *L. squamosa*, *Schizodus*?. The thick bed, 12 to 18 in thick, which occurs here is said to be the *raison d'être* of the quarry as a source of building stone. The bottom 7 ft of the section consists of oolitic 'limestone'. *Naticopsis sp.* was collected in 1860 from the Upper Magnesian Limestone of the Oldcoates area by Mr. R. Gibbs, but the exact locality is unknown.

In the railway cutting [5874 8694] on the south-eastern side of Langold up to 8 ft of Upper Magnesian Limestone, consisting of grey, thinly bedded dolomite, can be seen resting on Middle Permian Marl (see p. 144). The base of the Limestone is hereabouts a well defined line.

The best exposure in the Carlton in Lindrick area is provided by the old quarry [5943 8443] at Jerusalem Farm, where there are 6 ft of flaggy dolomite containing *T. permianus*, *L. squamosa* and *S. obscurus*. Rock specimens from these beds have been examined in thin section (E 34599–601) by Mr. K. S. Siddiqui who reports that they consist of ooliths and comminuted organic debris with some silt-grade quartz, set in a microcrystalline dolomite cement. One specimen shows scattered mica flakes and ferric oxide particles, and another a little secondary calcite. The ooliths are spheroidal to ellipsoidal, 0·3 mm in mean diameter, and are composed of concentric growths of carbonate with, in some cases, a quartz or carbonate nucleus, and in others an infilling of dusty dolomite.

[1]See footnote on p. 123.

An analysis of a specimen of Upper Magnesian Limestone from Carlton by H. L. Belbin (Mackintosh 1939, pp. 297–8), showed 92·42 per cent dolomite with 4·78 per cent calcite (SiO_2 1·6 per cent, Fe_2O_3 + Al_2O_3 1·5 per cent, $CaCO_3$ 55 per cent, $MgCO_3$ 42·2 per cent).

There are no exposures between Carlton and Gateford, where the formation dies out. The larger of the two small outliers north of Gateford Hill was formerly quarried (Aveline 1880, p. 21), and fragments of 'limestone' in the soil on the north side of the outlier have yielded T. permianus, L. squamosa and S. obscurus. The same three species have been identified from fragments of 'limestone' picked up in a field [5825 8186] on the main outcrop about ⅓ mile NNE of Raymoth Farm. Quarrying has taken place in this vicinity [5810 8131] (op. cit.), though no traces now remain.   R.F.G., E.G.S.

Most of the boreholes cored through the Upper Magnesian Limestone are near the outcrop, and only scanty records, in many cases not made by a geologist, exist for the majority. The following is a list of boreholes and shafts at which cores or sinking samples of the Upper Magnesian Limestone have been seen; where the formation is penetrated from top to bottom the thickness and depth is given in brackets: Bevercotes No. 1 Shaft (13¾ ft at 762¼ ft), Checkerhouse (8 ft 8 in at 546 ft 4 in), Cowlishaw Plantation, Elkesley Pumping Station. Firbeck Main Colliery, Harworth Colliery (41 ft 4 in at 431 ft 10 in), Hayton Smeeth (23¾ ft at 977 ft or 25¼ ft at 978½ ft[1]), Jockey House (29 ft 8 in at 706 ft, or 39 ft 8 in at 716 ft[2]), Loxdale Cottages (20 ft at 70 ft), Manton (5 ft 2 in at 246½ ft), Manton Colliery (No. 2 Shaft 6 ft 10 in at 254 ft 10 in, No. 4 Shaft 11 ft at 266 ft[3]), Rockley (30 ft 8 in at 871 ft 2 in), South Leverton (54 ft 2 in at 1518 ft 5 in[4]), Sutton (45 ft at 677 ft 8 in), Thorn Hill, West Drayton No. 1 (20 ft at 810 ft), Worksop Brewery New Borehole (10 in at 34¾ ft), Worksop Waterworks (old boring ? 9 in at 201 ft 5 in, new boring 7½ ft at 215 ft).

In Cowlishaw Plantation Borehole only the top 2¾ ft of the Upper Magnesian Limestone were penetrated; it is described as white or greenish 'limestone' becoming pale pink in the lower part. It is thinly bedded, the bedding planes, many of which are irregular, being up to 3 in apart. Shelly layers and small-scale false bedding were observed. In Elkesley Pumping Station Borehole the Upper Magnesian Limestone is described as 'a hard, pinkish or yellowish mottled rock, riddled with minute cavities' (Lamplugh and Smith 1914, p. 69). At Harworth Colliery the 'limestone' is mostly brownish grey and has a 3-in clay band 5 ft 10 in from the top. Rockley Borehole shows light grey 'limestone' with bands, up to 1 ft thick, of 'red and blue marl'; a detailed section is given by Lamplugh and Smith 1914, p. 73. The details of the section at Sutton Borehole, quoted by Wilson (1927, p. 138), are:

	ft	in
Sandstone, hard reddish limy, with indeterminate shells	7	5
Marl, strong blue and red ..	2	0
Sandstone, hard reddish limy	4	5
Limestone, hard grey, with indeterminate shells in lower part ..	31	2

In Thorn Hill Borehole [5762 8903] 13 ft 1 in of grey and white, locally yellowish, 'limestone' with sporadic dendrites were seen at the top of the borehole. At Worksop Brewery 10 in of greyish yellow dolomitic limestone is underlain by 2 ft of red mudstone and sandstone on 2¼ ft of yellow dolomitic sandstone in which rounded quartz grains make up about 60 per cent of the rock. These lower measures are assigned to the Middle Permian Marl, but it may be that the dolomitic sandstone could be more correctly regarded as a leaf of the Upper Magnesian Limestone. The 7½ ft of Upper Magnesian Limestone in the new borehole at Worksop Waterworks is buff-coloured except for the basal 1 in which is red and also sandy, and there are thin bands of dolomitic mudstone, found chiefly towards the base. T. permianus and S. obscurus have been collected from the upper part. Rock samples (E 22586–7) from

---

[1] According to whether or not 1½ ft of 'grey sandstone' at the base are included.
[2] According to whether or not 10 ft of 'limestone with marl bands' at the base are included.
[3] Schizodus obscurus was collected here.
[4] This section is at variance with that given by Wilcockson (1950, p. 462), viz. 40 ft 10 in at 1519 ft 2 in.

the upper part, examined by Mrs. Morey, are of dolomite with a little calcite and ankerite; small amounts of limonite occur in patches and very finely disseminated through the rock. A band of porcellanous limestone in one specimen consists entirely of finely granular dolomite; generally, however, the rock is distinctly crystalline, slightly coarser where organic debris is replaced. Outlines of ooliths can be seen in places.

Knowledge of the Upper Magnesian Limestone in other boreholes in the district is derived from examination of chipping samples, which, in the case of oil bores, have been described in detail by British Petroleum Co.'s geologists. These samples show dolomitic and calcitic limestone; where both occur, as is generally the case, the dolomite is usually in the upper part of the formation. The rock is, in general, white, grey, cream or buff in colour, fine-grained and crystalline or oolitic. Locally, however, the colour may be red or pink (as is commonly the case in the Bothamsall oil bores), mauve, yellow or brown, and the rock may be saccharoidal or even coarse-grained (as in several of the South Leverton oil bores), and in places porcellanous. Pisolithic limestone was noted at Gainsborough No. 9 Oil Bore. Bands of mudstone, silty mudstone and siltstone are fairly common, especially in the Gainsborough area, where, at No. 12 Bore, carbonaceous partings are also recorded. Pyrite, usually in the form of small pellets and streaks, occurs locally, again especially at Gainsborough. Anhydritic limestone is recorded at a few oil bores in the north-east, and there are indications that it occurs elsewhere. The presence of fossils has been noted at Gringley No. 1 Oil Bore.   E.G.S.

UPPER PERMIAN MARL

The Upper Permian Marl is present at outcrop north of Gateford, forming a narrow belt of country which is largely drift-free, but in which exposures are practically non-existent. It is absent, or in the absence of the Upper Magnesian Limestone cannot be distinguished from the Middle Permian Marl, over a wide area in the south-west of the district and around West Drayton in the south (Fig. 27); elsewhere it varies irregularly in thickness up to about 190 ft in the extreme north-east.

The Upper Permian Marl consists of red with subordinate greyish green mudstones and silty mudstones, locally micaceous, and contains bands and lenses of red and grey sand of Lower Mottled Sandstone type (see pp. 161–5). It passes laterally into Lower Mottled Sandstone, and in many sections in which there is an interbanding of mudstone and sand the two divisions cannot be separately defined. In Fig. 27 they have therefore been treated as one unit for isopachyte purposes, though mudstone predominates over sandstone in the east and sandstone over mudstone in the west. In several sections there is a thin bed of coarse wind-rounded red sand at the base of the Marl, and in some others a thin development of dolomite breccia or of conglomerate occurs in this position. Breccia and coarse sand are also recorded above the Upper Magnesian Limestone in some sections where the Upper Permian Marl is absent, and are there included in the Lower Mottled Sandstone (see pp. 162–5).

Gypsum occurs in the Upper Permian Marl in the north and east of the district (Fig. 27), and the oil bores in the north-east also show anhydrite. The latter occurs principally in a bed up to about 30 ft thick and on average about 40 ft from the base. This is taken to be the Upper Anhydrite of County Durham, which has already been traced southwards to the Scunthorpe area (Smith and Francis 1967, p. 93). A thinner bed of anhydrite, associated with anhydritic marl and gypsum, occurs immediately above the Upper Magnesian Limestone in a few of the Gainsborough oil bores; this may well be the equivalent of the Billingham Main Anhydrite of Durham.

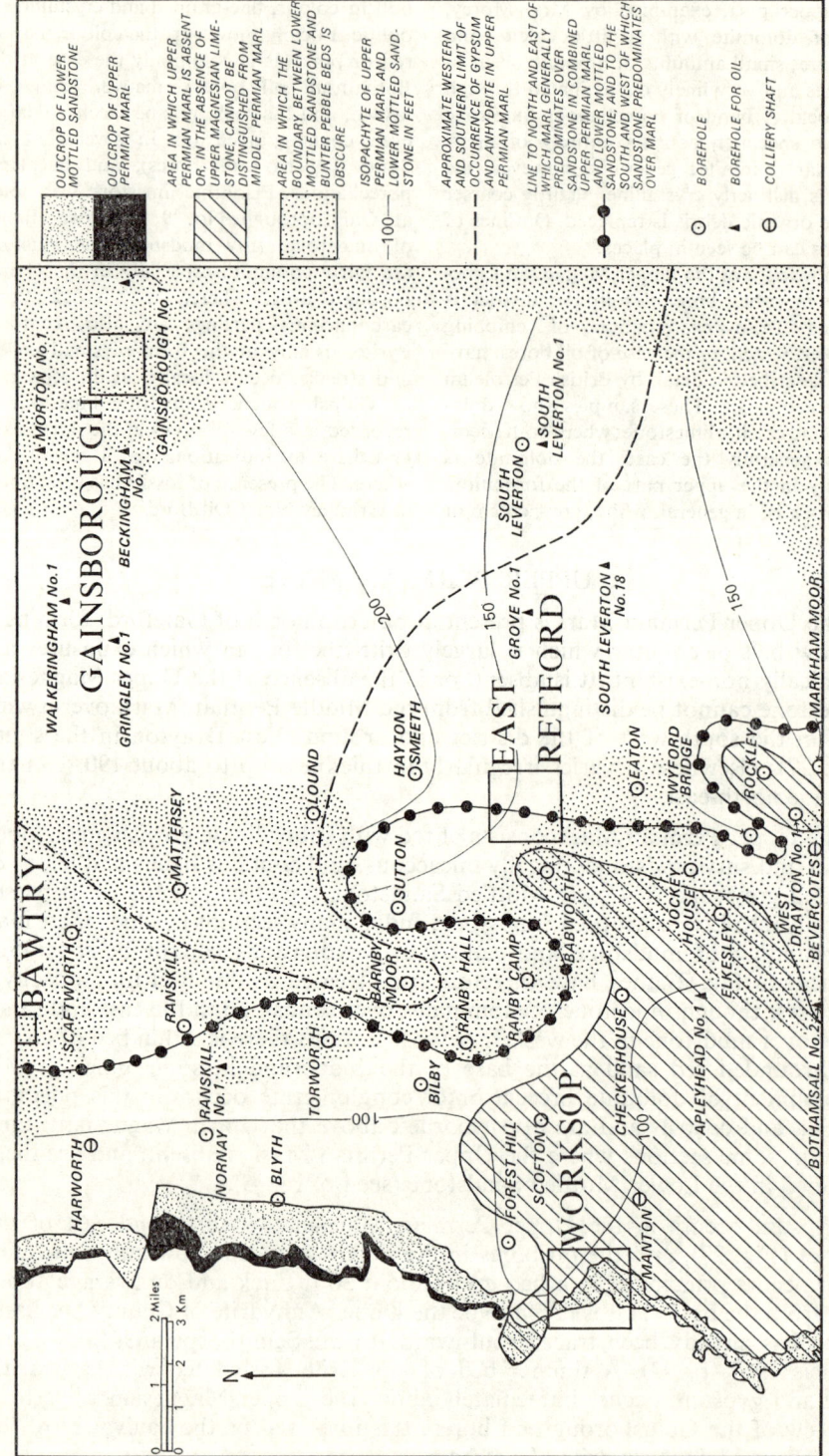

Fig. 27. *Sketch-plan of the Upper Permian Marl and Lower Mottled Sandstone*

Thin bands of dolomite occur locally in the bottom part of the Marl, in places in sufficient number to suggest a passage by alternation into the Upper Magnesian Limestone.

Like the closely comparable Middle Permian Marl, the Upper Permian Marl consists of terrigenous mud and silt which give way shorewards to sand, and which thicken and contain an increasing quantity of evaporites away from the land. It represents the final silting up of the Zechstein basin of deposition.

No fossils have been recorded from the Upper Permian Marl.   E.G.S.

DETAILS

Red clay can be seen overlying Upper Magnesian Limestone in the railway cutting [5964 8983] 1 mile NNE of Oldcoates (p. 155), and is visible at the bottom of the cutting [598 885] of the same railway line through the basal part of the Lower Mottled Sandstone, ½ mile E of Oldcoates (p. 166). 'Red and variegated marls, interstratified with thin shaly beds' were formerly visible in the old brickyard [6023 8937] at Holme Farm between Styrrup and Oldcoates (Aveline 1880, p. 22), but the sections are now overgrown, a mere 2 ft of red clay being visible at one point in 1946. Sections through the Upper Permian Marl were seen in pipe trenches in North Carlton [presumably around 5945 8465] in 1936 (Mackintosh 1939, p. 298). Up to about 10 ft thick, the red mudstones and clays with green bands were sandy and micaceous and contained lenses of sand. These beds showed a passage into the overlying Lower Mottled Sandstone.

South of Carlton the Upper Permian Marl has been seen at outcrop only in auger samples. Red sand, assigned to the Upper Permian Marl because of its coarse grain which contrasts with the fine texture of the Lower Mottled Sandstone, can be augered above the Upper Magnesian Limestone on the larger of the limestone outliers north of Gateford Hill and on the main outcrop near Raymoth Farm [5810 8132]. The sand is associated with red sandy clay at the latter locality.   R.F.G., E.G.S.

In No. 1 Shaft of Harworth Colliery 37 ft 2 in of red marl, red sandstone and red and greenish grey sandy marl succeed the Upper Magnesian Limestone at 390½ ft. At Manton Colliery the Upper Permian Marl is not distinct, though in No. 4 Shaft 4 ft 7 in of red and mottled marl with beds of grey and red sandstone occur (at 239 ft 4 in) above 15 ft 8 in of red and grey sandstone which

rest on the Upper Magnesian Limestone. In Manton No. 2 Shaft there are 1½ ft of red and grey marl (at 214 ft 7 in) on 18¾ ft of red and reddish grey sandstone, coarse-grained in the bottom 3 ft, on Upper Magnesian Limestone. Samples from No. 1 Shaft of Bevercotes Colliery give the following section for the Upper Permian Marl: red and green silty marl 28 ft, coarse red silty marl 40¼ ft, red and pale green sandstone (unsorted fine to coarse grains, with a layer of very coarse sandstone) 10¼ ft at 748½ ft on Upper Magnesian Limestone; in No. 2 Shaft the Upper Permian Marl cannot, in the absence of the Upper Magnesian Limestone, be separated from the Middle Permian Marl, but it evidently consists largely of silty marl with micaceous partings.

Very few boreholes have been cored through the Upper Permian Marl. Loxdale Cottages Borehole, Oldcoates, showed 44 ft at 50 ft depth, consisting of sandy marl 8 ft, red marl 22 ft, on sandy marl 14 ft. In Cowlishaw Plantation Borehole there are only 4 ft 2½ in of reddish brown and greenish grey laminated silty mudstone with disturbed bedding at 174 ft 8½ in between Lower Mottled Sandstone and Upper Magnesian Limestone. The bottom 2½ in are a breccia of dolomitic limestone fragments in a matrix of greenish grey silty mudstone containing coarse sand grains. In Hayton Smeeth Borehole, 40 ft of red marl and sandstone on 88¾ ft of red and grey marl and red sandy marl at 953¼ ft represent the Upper Permian Marl, though W. Gibson, who saw the cores, included in this formation the succeeding 41½ ft of red sandstone and marl (Edwards 1951, p. 185 fn.). The most reliable of several versions of the Upper Permian Marl succession in South Leverton Borehole is probably that of Ford (1920, p. 100), which gives: variegated marl with gypsum beds 23 ft 4 in, chocolate-red hard rock 6 in, on dark grey shale (soapstone) 2 ft 10 in at 1464¼ ft. In Rockley Borehole, Lamplugh and Smith (1914, p. 73) allocate the measures between 773 and 840½ ft to the Upper Permian Marl. These measures contain a considerable proportion of red and blue or grey sandy marl, but the greater part is of sandstone (see p. 168).

Information from other coal and water boreholes through the Upper Permian Marl is obtainable only from the drillers' logs and the study of chipping samples. All the records show red marl or sandy marl, usually with grey or green bands and commonly with sandstone bands; gypsum occurs at Mattersey Borehole and probably also at Barnby Moor, Bilby and Lound. The thickest sections are 129 ft at Barnby Moor, 141 ft 5 in at Lound and 140¼ ft at Scaftworth. The formation cannot be separated from the Lower Mottled Sandstone in Ranskill Borehole, and is absent in Babworth, Checkerhouse, Elkesley, Elkesley Pumping Station, and Scofton boreholes.

In *Ranskill* No. 1 Oil Bore the only signs of Upper Permian Marl are the reddish brown and greyish green micaceous mudstones interbedded with the 30 ft of red and white, fine- to medium-grained, micaceous sandstones which succeed the Upper Magnesian Limestone. These measures have been assigned to the Lower Mottled Sandstone (p. 169). At *Gringley* No. 1 Oil Bore approximately 50 ft of dark brown and bluish grey silty mudstone with a thin band of sandy limestone were recorded. The *Walkeringham* oil bores show about 140 ft of red mudstone with grey bands, and with bands of fine-grained sandstone in the upper part, one of which at No. 2 Bore is 45 ft thick and could be alternatively regarded as a leaf of Lower Mottled Sandstone: both sections have gypsum in the lower part, and at No. 2 Bore there is a distinct band of red-stained anhydrite and white gypsum a few feet from the base. In *Morton* No. 1 Oil Bore 85 ft of reddish brown mudstone are succeeded by about 40 ft of interbanded red mudstone and silty sandstone which may alternatively be regarded as Lower Mottled Sandstone. Anhydrite and gypsum are recorded in the bottom 40 ft of the Upper Permian Marl, with a bed of anhydrite about 6 ft thick 30 ft or so from the base.

Three of the *Beckingham* oil bores show 150 to 165 ft of Upper Permian Marl, which includes a high proportion of siltstone and fine-grained sandstone in the middle and upper parts. In the fourth, No. 2 Bore, there are only about 50 ft of mudstone, corresponding to the lower part of the other sections; the higher measures are essentially sandstones and have been included in the Lower Mottled Sandstone (see p. 169). Gypsum is recorded in the mudstones, and in Nos. 1 and 2 bores there is a band of anhydrite, about 15 ft thick in the former, and respectively about 45 and about 30 ft from the base of the Upper Permian Marl.

In the *Gainsborough* Oilfield the Upper Permian Marl varies in thickness between 45 and 190 ft, but over half the bores show between 120 and 170 ft. It is recorded as red or reddish brown mudstone or silty mudstone with grey or green bands, slightly calcareous in parts, and with bands of siltstone and fine-grained sandstone. Gypsum and anhydrite are recorded in nearly every bore, most commonly in the bottom part, and the principal evaporite horizon is a band of anhydrite up to about 30 ft thick, a few feet to 60 ft but generally between 30 and 50 ft from the base. This is generally white or pale grey, but mauve and bluish varieties are also recorded, and the texture varies from finely crystalline to granular. Anhydritic mudstone with anhydrite and gypsum is also prominent at the base of the Upper Permian Marl, immediately overlying the Upper Magnesian Limestone, in a few bores. Thin dolomite bands are recorded in four bores, in two cases immediately below the upper anhydrite band, and in one case, near the base, showing passage into the underlying Upper Magnesian Limestone.

In many of the *South Leverton* oil bores it is impracticable to separate the Upper Permian Marl and Lower Mottled Sandstone, together up to about 140 ft thick. Some record no mudstone and others no sandstone, though this is probably due to sampling deficiencies. In general there appears to be interbanding of mudstone, siltstone and sandstone, with the first two predominating at the bottom and the last at the top. The mudstones and siltstones are chiefly red with subordinate green and grey bands; mica is common, particularly in the silty measures, and traces of gypsum are recorded in several bores. In

No. 13 Bore a thin band of limestone was noted 25 to 30 ft above the top of the Upper Magnesian Limestone.

In the *Grove* oil bores the Upper Permian Marl and Lower Mottled Sandstone are together 110 to 130 ft thick, and consist of interbanded red silty mudstone and fine- to coarse-grained red sandstone, micaceous in No. 2 Bore. In No. 2 Bore sand predominates in the top 70 ft, which part could be fairly assigned to the Lower Mottled Sandstone, but in No. 1 Bore it is apparently concentrated in the middle portion.

No Upper Permian Marl is present in *West Drayton* No. 2 Oil Bore.

In the *Bothamsall* Oilfield the Upper Permian Marl is represented by up to about 30 ft of mudstone and siltstone in three of the bores; in two others mudstone beds occur in the sandstone or interbanded sandstone and dolomite which succeed the Upper Magnesian Limestone. In the remaining bores at Bothamsall and in the *Apleyhead* oil bores it is absent. In several of the Bothamsall bores and in Apleyhead No. 3 Bore, there is a thin bed of conglomerate or breccia above the Upper Magnesian Limestone, which may be compared with a similar deposit at the base of the Upper Permian Marl elsewhere.

E.G.S.

LOWER MOTTLED SANDSTONE

The Lower Mottled Sandstone has an outcrop, up to a mile wide, which extends across the district from Tickhill in the north to Welbeck Park in the south, and on which much of the town of Worksop is built. Near Worksop it has been worked fairly extensively for moulding sand. It consists generally of red or pink sandstone, locally containing grey bands, streaks and patches. It is largely fine-grained, though medium-grained sandstone is not uncommon; it is micaceous in part and pebbles are absent except at a few localities, where they occur sparsely. The sandstone is friable to compact, and there is only one record (p. 169) of it in a cemented state. Argillaceous partings are common and there are

TABLE 5. *Mechanical analyses of outcrop samples from the Lower Mottled Sandstone*

Sample No. MR		Per cent quartiles (mm)			Sorting Coefficient
		25 (Φ_1)	50 (Φ_2)	75 (Φ_3)	$\sqrt{\Phi_1/\Phi_3}$
22728–9	Mean	0·24	0·19	0·17	1·18
	Range	*0·05*	*0·02*	*0·00*	*0·13*
27731		0·17	0·15	0·13	1·14
27732–5	Mean	0·20	0·17	0·14	1·23
	Range	*0·12*	*0·08*	*0·08*	*0·06*
27741		0·32	0·21	0·15	1·46
27764–5	Mean	0·22	0·18	0·15	1·21
	Range	*0·02*	*0·01*	*0·00*	*0·07*
27758		0·23	0·18	0·13	1·33
	Mean of above 11 samples	0·22	0·18	0·14	1·24
	Range of above 11 samples	*0·17*	*0·08*	*0·08*	*0·35*
	Moulding sands				
27742–4 ⎱ 27746–7 ⎰	Mean	0·18	0·15	0·12	1·24
	Range	*0·13*	*0·12*	*0·11*	*0·15*
27755–6	Mean	0·15	0·13	0·11	1·19
	Range	*0·00*	*0·02*	*0·03*	*0·14*
	Mean of all above samples	0·20	0·17	0·13	1·23
	Range of all above samples	*0·13*	*0·12*	*0·12*	*0·35*

many records of bands and lenses of red, or less usually, green mudstone and sandy mudstone. The sandstone may be false-bedded or level-bedded, and in the latter cases the fine-grained, more argillaceous examples sometimes show lamination.

The Lower Mottled Sandstone varies widely in thickness between about 30 and about 150 ft, but most recorded thicknesses mean little because the formation passes laterally into the Upper Permian Marl (p. 157), and, over a large part of the district, the boundary between it and the Bunter Pebble Beds is obscure (Fig. 27) and probably also diachronous. Generally speaking the Lower Mottled Sandstone of the present district is regarded as a littoral facies of the Upper Permian Marl, and in Fig. 27 the two formations have been regarded as one unit for isopachyte purposes. It can be seen, however, that sandstone predominates over mudstone in the west, and only sandstone is present in the southwest of the district and around West Drayton in the south.

Where the Upper Permian Marl is absent, there is in places a thin conglomerate or breccia or a variable, but usually thin, development of coarse-grained

TABLE 6. *Summary of modal analyses of samples from the Lower Mottled Sandstone*

Sample No. MR		Approximate proportions (by volume per cent)					
		Quartz, cherty silica	Feldspars, authigenic overgrowths	Composite particles, clay, ferric oxide and hydroxide	Muscovite, illite, biotite, chlorite	Carbonate	Minor constituents including heavy minerals
27728–9	Mean	77	10	11	tr	2	tr
	Range	*1*	*1*	*0*	*0*	*2*	*0*
27731		63	14	16	6	1	tr
27732–5	Mean	67	15	14	3	0	0·5
	Range	*12*	*4*	*10*	*2*	*0*	*0·5*
27741		66	18	15	1	0	tr
27764–5	Mean	75	11	12	0·5	1	tr
	Range	*7*	*3*	*5*	*0*	*1*	*0*
27768		62	15	16	6·5	0	tr
	Mean of above 11 samples	69	14	14	2	1	tr
	Range of above 11 samples	*16*	*9*	*10*	*6·5*	*3*	*0·5*
	Moulding sands						
27742–4 27746–7	Mean	74	10	14	1	tr	tr
	Range	*9*	*6*	*11*	*0·5*	*0*	*0*
27755–6	Mean	71	12	16	1	0	tr
	Range	*12*	*1*	*10*	*0·5*	*0*	*0*
	Mean of all above samples	71	12	14	2	0·5	tr
	Range of all above samples	*17*	*12*	*13*	*4·5*	*3*	*0·4*

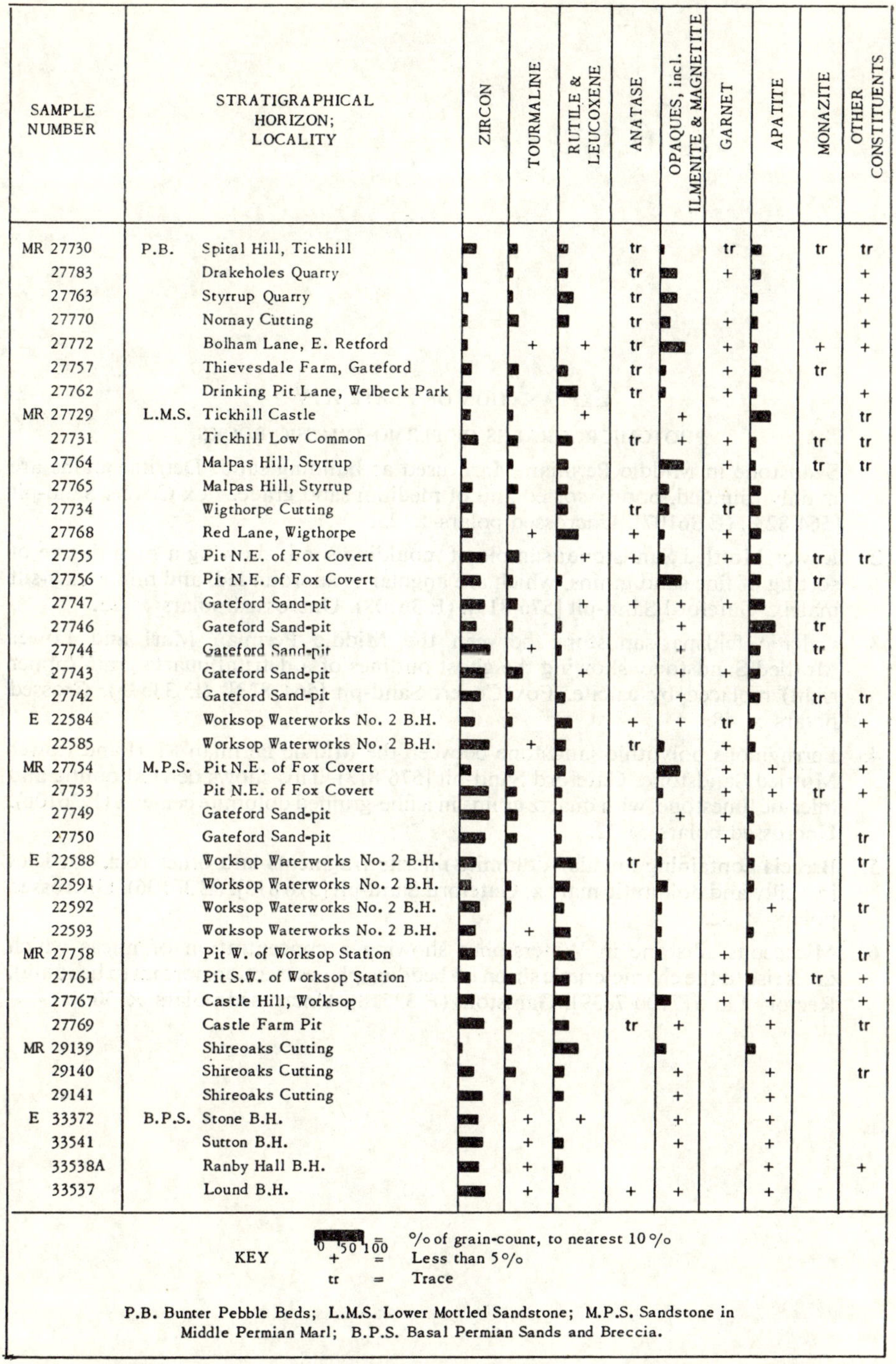

SAMPLE NUMBER	STRATIGRAPHICAL HORIZON; LOCALITY	ZIRCON	TOURMALINE	RUTILE & LEUCOXENE	ANATASE	OPAQUES, incl. ILMENITE & MAGNETITE	GARNET	APATITE	MONAZITE	OTHER CONSTITUENTS
MR 27730	P.B. Spital Hill, Tickhill				tr		tr		tr	tr
27783	Drakeholes Quarry				tr		+			+
27763	Styrrup Quarry				tr					+
27770	Nornay Cutting				tr		+			+
27772	Bolham Lane, E. Retford		+	+	tr		+		+	+
27757	Thievesdale Farm, Gateford				tr		+		tr	
27762	Drinking Pit Lane, Welbeck Park				tr		+			+
MR 27729	L.M.S. Tickhill Castle			+		+				tr
27731	Tickhill Low Common				tr	+	+		tr	tr
27764	Malpas Hill, Styrrup						+		tr	tr
27765	Malpas Hill, Styrrup									
27734	Wigthorpe Cutting				tr		tr			
27768	Red Lane, Wigthorpe		+		+		+			
27755	Pit N.E. of Fox Covert			+	+		+		tr	tr
27756	Pit N.E. of Fox Covert								tr	
27747	Gateford Sand-pit			+	+	+	+			
27746	Gateford Sand-pit					+			tr	
27744	Gateford Sand-pit		+			+			tr	
27743	Gateford Sand-pit					+				
27742	Gateford Sand-pit			+		+			tr	tr
E 22584	Worksop Waterworks No. 2 B.H.				+	+	tr			+
22585	Worksop Waterworks No. 2 B.H.		+		tr		+			
MR 27754	M.P.S. Pit N.E. of Fox Covert						+	+		+
27753	Pit N.E. of Fox Covert						+	+	tr	+
27749	Gateford Sand-pit					+	+			
27750	Gateford Sand-pit						+			tr
E 22588	Worksop Waterworks No. 2 B.H.				tr					tr
22591	Worksop Waterworks No. 2 B.H.									
22592	Worksop Waterworks No. 2 B.H.							+		tr
22593	Worksop Waterworks No. 2 B.H.		+							
MR 27758	Pit W. of Worksop Station						+			tr
27761	Pit S.W. of Worksop Station						+		tr	+
27767	Castle Hill, Worksop						+	+		+
27769	Castle Farm Pit				tr	+		+		tr
MR 29139	Shireoaks Cutting									
29140	Shireoaks Cutting					+		+		tr
29141	Shireoaks Cutting					+		+		
E 33372	B.P.S. Stone B.H.		+	+		+		+		
33541	Sutton B.H.		+			+		+		
33538A	Ranby Hall B.H.		+					+		+
33537	Lound B.H.		+		+	+		+		

KEY

0 50 100	% of grain-count, to nearest 10 %
+ =	Less than 5%
tr =	Trace

P.B. Bunter Pebble Beds; L.M.S. Lower Mottled Sandstone; M.P.S. Sandstone in Middle Permian Marl; B.P.S. Basal Permian Sands and Breccia.

FIG. 28. *Heavy detrital minerals in the Permo-Triassic sandstones and associated rocks*

EXPLANATION OF PLATE XIV

PHOTOMICROGRAPHS OF PERMO-TRIASSIC ROCKS

1. Sandstone in Middle Permian Marl, used as building sand. Detrital grains are mainly rounded, poorly sorted and of medium sand grade. Fox Covert Sand-pit [564 823]. (E 36107). Uncrossed polars × 25.

2. Lower Mottled Sandstone: sample of moulding sand, showing a high degree of sorting of fine sand grains, which are angular to subrounded, and minor clay-silt matrix. Gateford Sand-pit [576 813]. (E 36108). Uncrossed polars × 25.

3. Calcitic feldspar-sandstone between the Middle Permian Marl and Lower Mottled Sandstone, showing the ghost outlines of a detrital quartz grain (upper right) replaced by calcite. Fox Covert Sand-pit [564 823]. (E 33535). Crossed polars × 48.

4. Ferruginous polylithic sandstone between the Middle Permian Marl and Lower Mottled Sandstone, Gateford Sand-pit [576 813]. This shows detrital oolitic and micritic limestone, with quartz grains in a fine-grained dolomite cement. (E 36106). Uncrossed polars × 32.

5. Breccia containing angular dolomite-micrite fragments and other rock particles in a silty and dolomitic matrix. Gateford Sand-pit [576 813]. (E 36106). Uncrossed polars × 32.

6. Micaceous siltstone in Waterstones showing a concentration of micas which gives rise to the characteristic sheen on bedding planes (section normal to bedding). Rectory Lane [7100 7639], Gamston. (E 33518). Uncrossed polars × 30.

PLATE XIV

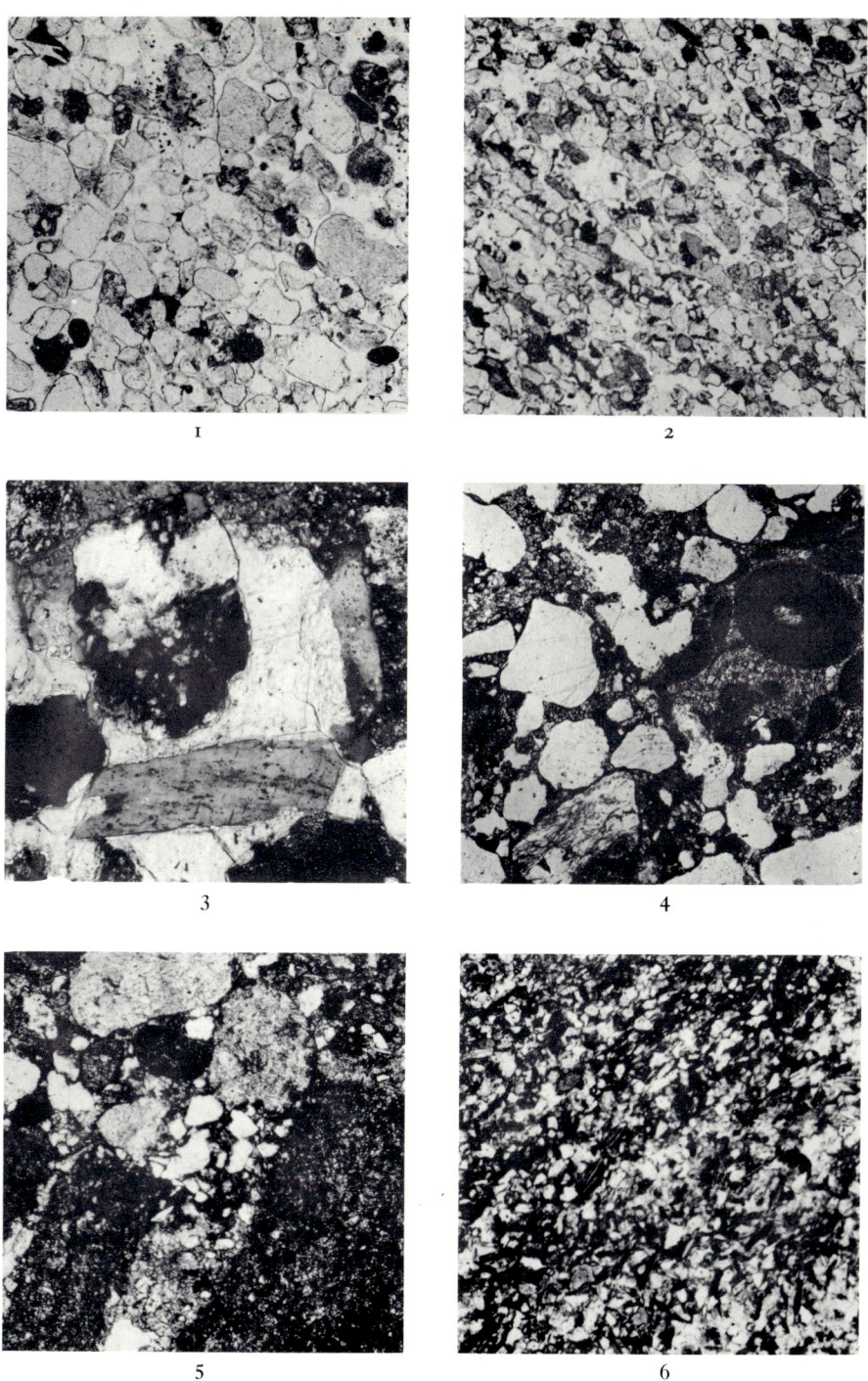

sand at the base of the Lower Mottled Sandstone. A similar development occurs at the base of the Upper Permian Marl elsewhere (p. 157).   E.G.S.

Channel and serial spot samples from exposures at outcrop have been analysed petrographically, and mechanical, modal and heavy mineral analyses are presented on Tables 5 and 6 and Fig. 28 respectively.

Fine sand predominates in the samples, with a small silt-clay content. The degree of sorting is generally high, particularly for the moulding sands. Fine sand grains are commonly angular to subangular while coarser grains range from subrounded to subangular. In thinly bedded samples, elongated clastics are orientated along the bedding. Authigenic overgrowths, especially of feldspar, present highly ragged outlines. Quartz (mainly from igneous sources) predominates, with subordinate potash feldspars, rock particles (chert, pure, argillaceous or sheared quartzites, slate, siltstone), minor micas (which are in places concentrated on bedding planes), clay minerals, ferric oxide and hydroxide (commonly occurring as pellicle coatings around clastic grains or impregnating the matrix), traces of carbonate and heavy minerals. The last-named (Fig. 28) comprise predominant zircon, subordinate tourmaline, rutile and leucoxene, ilmenite, magnetite, conspicuous apatite, minor to rare garnet, traces of staurolite, anatase and monazite. The suite is closely similar to that described from the Middle Permian Marl (p. 143), stemming principally from igneous sources or earlier sedimentary rocks carrying similar heavy minerals.

Comparison of the petrography of the Lower Mottled Sandstone and the sandstone in the Middle Permian Marl. The following comparisons are based on the analysed samples of the upper sandstone unit in the Middle Permian Marl (p. 142) and those of the Lower Mottled Sandstone, including moulding sands (see above). In general, and taking into account the range of values found (Tables 3 and 5), the Middle Permian Marl sandstones are of coarser overall grain (medium sand grade) and are less well sorted than the Lower Mottled Sandstone except for two samples (E 27741, 27768). The moulding sands in particular, though showing a similar mean coefficient of sorting to the other Lower Mottled Sandstone samples, have a considerably lower range of coefficients. Clastic grains in the Middle Permian samples are generally more rounded, grade for grade, than are those in the Lower Mottled Sandstone, though overall comparison is difficult since coarser sand grades, such as those forming the bulk of the Middle Permian sandstones, are on average more rounded than the finer grades.

The modal analyses presented (Table 4) indicate that within their respective ranges and means, the Middle Permian samples have a higher clastic quartz to feldspar ratio than those from the Lower Mottled Sandstone. Carbonate (principally dolomite) is more abundant in the Middle Permian samples. Heavy mineral suites (Fig. 28) indicate rather higher apatite and lower garnet proportions in the Lower Mottled Sandstone samples. However, many analyses would be required to confirm these differences.   R.K.H.

L

DETAILS

The northernmost exposures of Lower Mottled Sandstone in the district are seen in caves [5938 9287 to 5940 9283] on the east side of Castle Hill in the grounds of Tickhill Castle. They show red, fine- to medium-grained sandstone with argillaceous partings in parts and clay galls in others. R.F.G., E.G.S.

Random samples (MR 27728) from a small cave [5939 9285] consist largely of subrounded quartz (averaging 0·25 mm), potash feldspars and micas with calcite matrix, and sparse heavy minerals including zircon and apatite. The clastic grains are moderately well sorted, with sorting coefficient 1·24. Random spot samples (MR 27729) from a large cave [5939 9285] are composed of well sorted clastics (sorting coefficient = 1·11): quartz, potash feldspars, composite rock particles, with minor micas, clay minerals, dolomite and disseminated ferric oxide. Heavy minerals include subangular to prismatic, colourless to pink zircon, rounded apatite, tourmaline (with pleochroism O = olive green, E = colourless; O = yellowish brown, E = pale yellow; O = dark bluish green, E = colourless), reddish brown rutile, leucoxene, opaque ores (ilmenite, magnetite), anatase and a trace of staurolite. Estimated relative abundances are shown in Fig. 28. R.K.H.

About ¾ mile SSE of Tickhill Castle there are old shallow sand-pits [599 917] in the Lower Mottled Sandstone on the low drift-free ridge between Tickhill Low Common and Styrrup Carr. The pits are now overgrown, but 2 to 4 ft of soft fine-grained red sand and sandstone can be seen in banks of nearby ditches. R.F.G., E.G.S.

A random spot sample (MR 27731) from one of the ditches [5999 9170] is well graded (sorting coefficient = 1·14) and consists of quartz, conspicuous potash feldspars, composite rock particles, opaque ferric oxide, micas (illite, muscovite), clay minerals—particularly chlorite—and minor carbonate. Heavy minerals (Fig. 28) are similar to those from Tickhill Castle save for the presence at this locality of minor garnet and a trace of monazite. R.K.H.

Sixteen feet of pink to red and pale yellow sandstone, somewhat disturbed, can be seen in the old sand-pit [5922 9023] on the south side of Styrrup Lane, about 1 mile WSW of Styrrup and near Malpas Hill. Here the Lower Mottled Sandstone is faulted against the Upper Magnesian Limestone, which is exposed close by. This section is evidently the one referred to by Aveline (1880, p. 20) and Mackintosh (1939, p. 287) who both recorded pebbles and placed the sand in the Middle Permian Marl. Versey (1925, p. 218), however, placed it in the Bunter. E.G.S.

Samples were collected from the top 4 ft and bottom 5 ft of the section. The higher sample (MR 27764) is pale pinkish brown, moderately consolidated fine- to medium-grained (sorting coefficient = 1·24) with dark reddish brown mud flakes and galls. It is composed of predominant quartz, subordinate potash feldspars, composite rock particles, a little ferric oxide, carbonate and mica. Heavy minerals comprise zircon, tourmaline (green and brown), apatite, opaques (ilmenite, hematite, magnetite), leucoxene, minor garnet, and a trace of monazite (Fig. 28). The lower sample (MR 27765) is very friable and pale yellow, and composed of well sorted (sorting coefficient = 1·17) fine sand: quartz, potash feldspars, composite rock particles, very minor micas, carbonate and ferric oxide. Heavy minerals include the above species except, apparently, garnet, and the proportions of apatite, zircon and tourmaline are a little higher in this sample. R.K.H.

The junction between the Lower Mottled Sandstone and the Upper Permian Marl was formerly exposed in the railway cutting [598 885] ½ mile E of Oldcoates, but at present all that is visible is red sand at the top of the overgrown section and red clay at the bottom.

On the side of the main Tickhill–Worksop road [5914 8315] at Wigthorpe, up to 12 ft of false-bedded, pebble-free sandstone are exposed. The sandstone is generally red, but has grey streaks and contains clay galls and partings, especially towards the top. R.F.G., E.G.S.

Serial spot samples (MR 27732–5) were collected at the base, and at 3, 6 and 9 ft respectively above the base of the exposure. Sorting varies between samples (Table 5); the mean granularity is about 0·17 mm, and mean sorting coefficient is 1·23. The main clastic component is quartz in all samples, with conspicuous potash feldspars, rock particles, ferric oxide and hydroxide, and with minor micas and discrete clay minerals. Clear, ragged, secondary authigenic overgrowths occur around dusty orthoclase. Heavy minerals are similar to those described from Tickhill, and estimated abundances in one sample (MR 27734) are shown in Fig. 28.

R.K.H.

From the above locality for $\frac{1}{2}$ mile to the south, further sections in up to 6 ft of soft red false-bedded, fine- to medium-grained sandstone can be seen on the main road and in Red Lane. At the southernmost section [5890 8234], the highest in the succession, rare small pebbles were noted in the sandstone, which here has a slightly mottled appearance.

R.F.G., E.G.S.

A spot sample (MR 27741) of sandstone from this last locality is poorly graded (sorting coefficient = 1·46) and consists of subrounded to subangular quartz, conspicuous potash feldspars, minor polygranular rock particles, interstitial ferric oxide and hydroxide, illite, muscovite and a trace of heavy minerals. A spot sample of reddish brown bedded fine to medium sandstone from an exposure [5922 8292] in Red Lane is moderately sorted and consists of subrounded to subangular clastic grains (median quartile 0·18 mm) of quartz, potash feldspars with angular secondary overgrowths, and composite rock particles. Other constituents are ferric oxide and hydroxide, and minor micas (muscovite, illite, biotite, chlorite), with a trace of heavy minerals: apatite, zircon, tourmaline, garnet, rutile, anatase and leucoxene (Fig. 28).

R.K.H.

Good sections of the bottom part of the Lower Mottled Sandstone are provided by the two large sand-pits in the Gateford area, the one [564 823] north-west of Gateford Hill (Plate I), and the other [576 813] south of the village. Details of their respective 21 ft and 31 ft of Lower Mottled Sandstone, worked for moulding sand, are given on pp. 145–6. It should be noted, however, that the bottom 1 ft or so of the beds assigned to the Lower Mottled Sandstone in each case is yellow or brown in colour and contains coarse rounded grains. This is thought to be the equivalent of the coarse sand that occurs at the base of the Upper Permian Marl in some other sections (pp. 157 and 159).

E.G.S.

A spot sample (MR 27755) of greyish buff sandstone from the pit north-west of Gateford

Hill is friable and composed largely of fine sand (median quartile 0·14 mm; Table 5) which is well graded (sorting coefficient = 1·12), with a small proportion of clay and silt. Reddish brown flakes are also present. The major and trace-mineral components (Table 6 and Fig. 28 respectively) are closely similar to those from the moulding sands of the other Gateford pit (see below); no dolomite was detected. A second spot sample (MR 27756) of overlying reddish brown, fine sandstone is similar (colour apart) to the first, except for a higher content of ferric oxide and a slightly higher coefficient of sorting (1·26).

At the pit immediately south of Gateford five serial spot samples (MR 27742–4; 46–7) were taken at heights of 3, 9 to 12, 13, 16 and 18 ft above the base of the moulding sand. The samples are fine to medium sandstones, moderately to poorly consolidated and reddish brown, except for one sample (MR 27746) which is pale grey. Mechanical analyses (p. 161) show median quartiles ranging from 0·09 to 0·21 mm, and sorting coefficients from 1·18 to 1·33. The samples consist predominantly of well sorted, moderately packed, fine sandy clastics, which are angular to subrounded, with minor interstitial silt and clay (Plate XIV, fig. 2). The clastic grains comprise clear igneous quartz, conspicuous rock particles (chert, pure and argillaceous quartzites, slate, siltstone), potash feldspars with authigenic overgrowths, discrete opaque iron oxides, sparse micas (illite, biotite, kaolinite, muscovite, chlorite) and heavy minerals (Fig. 28). The last-named include conspicuous zircon (colourless euhedra and rare polycyclic purplish, rounded grains), tourmaline, rounded apatite, deep reddish brown to yellow, subrounded prisms and euhedra of rutile, opaque ores (ilmenite, magnetite), sparse garnet, leucoxene, and traces of staurolite and monazite. Ferric oxide forms pellicle-like coatings around clastic grains. Sparse dolomite occurs. Reddish brown, thinly bedded mudstone (MR 27745), 16 ft from the base of the exposure of Lower Mottled Sandstone, consists of chlorite, illite, hematite and quartz (X-ray powder photographs NEX 256, 841).

R.K.H.

The top 6 ft of the Lower Mottled Sandstone, overlain by Pebble Beds, and consisting of red and green-mottled, false-bedded, fine-grained sandstone with mud-stone bands and clay galls are seen in the sand-pit [591 796] $\frac{1}{8}$ mile ESE of Worksop station (see also p. 175). A similar section was formerly visible in the now-filled sand-pit [586 799] 200 yd N of the station.

South of Worksop the only exposure of Lower Mottled Sandstone within the district consists of a few feet of red sand at the bottom of a 6-ft cutting [5695 7570] on a bridle path in Manor Hills woods $\frac{1}{2}$ mile ENE of Sloswicks Farm.   R.F.G., E.G.S.

In the majority of the boreholes of the district there is difficulty in distinguishing the Lower Mottled Sandstone from the lower part of the Bunter Pebble Beds (see p. 170), and in some cases also in fixing a boundary between the Lower Mottled Sandstone and Upper Permian Marl (see p. 157). Generally speaking, where there is a sufficiently detailed log the upper boundary of the formation is drawn below the lowest occurrence of pebbles, to include the lower, fine-grained, marly portion of the Bunter, and the lower boundary is drawn where the proportion of sandstone is exceeded by that of mudstone.

Few boreholes have been cored through the formation. At Blyth Pumping Station most of the 124 ft of sandstone succeeding the thin Upper Permian Marl at 198 ft in No. 2 Well can be assigned to the Lower Mottled Sandstone; apart from the bottom 1 ft, which is greenish grey, it is red with marl 'nodules'. In Checkerhouse Borehole the 116$\frac{3}{4}$ ft of red and grey sandstone with marl bands resting on the Upper Magnesian Limestone at 537 ft 8 in apparently represent this formation. The bottom 8 in consist of coarse grey sandstone. In Cowlishaw Plant-ation Borehole the Lower Mottled Sandstone consists of reddish brown sandstone with grey patches and mudstone bands, with its base at 170$\frac{1}{2}$ ft from the surface. Despite careful examination of the cores by Mr. C. G. Godwin, it was not found possible to demarcate the top of the division other than to say it is probably somewhere between 80 and 104 ft from the surface, giving a thickness of about 80 ft. In Elkesley Pumping Station boreholes the lowest 151 ft of the Bunter contain bands of red, fine-grained, micaceous sandstone with mudstone partings, but pebbles occur to within 80 ft of the base

at 581 ft accordine to Lamplugh and Smith (1914, p. 69), who noted that "no clear distinction could be drawn here between Pebble Beds and Lower Mottled Sandstone". At 142 ft from the base of the Bunter they record (*idem*) a hard greyish layer of sandstone containing "curious cylindrical markings in high relief, like the casts of the burrows of some aquatic animal". In Rockley Borehole the Lower Mottled Sandstone is, according to Lamplugh and Smith (1914, p. 73), repre-sented by the 108 ft of red sandstone (with 'marl joints and balls' and bands of red sandy marl and marl) at 773 ft. There is, however, little evidence for the top at 665 ft (the lowest recorded pebbles are at 625 ft 9 in), and about 60 per cent of the beds between 773 and 840$\frac{1}{2}$ ft are red sandstone, the rest being red and blue sandy marl. In South Leverton Borehole, Lamplugh and Smith (1914, p. 87) place 112 ft 7 in of red sandstone, sandy marl and mottled marl with base at 1437 ft 7 inches in the Lower Mottled Sandstone, but the bottom 87 ft contain as much marl as sandstone and could equally well be placed in the Upper Permian Marl Sutton Borehole showed 96$\frac{1}{2}$ ft of red sand-stone and marly sandstone with thin bands of red and grey marl and sandy marl above Upper Permian Marl at 597$\frac{3}{4}$ ft. In the water borehole at Upper Morton Grange the lowest 141 ft 2 in of the Bunter, described as red marly sandstone with two thin bands of hard sandstone and not bottomed at 468 ft 8 in, probably represent the bulk of the Lower Mottled Sandstone.

In the Worksop area cores of the Lower Mottled Sandstone were obtained from boreholes at the brewery and at the water-works. The brewery boreholes commenced in the Lower Mottled Sandstone 34 to 36 ft from its base, and there is a detailed log only of No. 2, which was cored in the basal 9 ft 11 in of the formation. It consisted of 8 ft 8 in of red, bedded, fine-grained sandstone, soft and friable in parts, with subangular grains and sporadic argillaceous and mica-ceous bands, on 1$\frac{1}{4}$ ft of grey, coarse-grained sandstone with rounded grains and scattered small pebbles. In Worksop Waterworks No. 2 Borehole the Lower Mottled Sandstone may be as much as 137 ft thick; the base is at 207$\frac{1}{4}$ ft and pebbles are recorded above 70 ft. Only the bottom 42$\frac{1}{4}$ ft were cored, however, and these consist of 40 ft of red sandstone with some green mottling, chiefly fine-grained but medium-grained in parts with clay galls and, at intervals, marl bands, on 2 ft of coarse-grained grey sandstone (see pp. 162 and 165), on 3 in of breccia composed of marl, sand and dolomite. Coarse sandstone (E 22584) from 206 ft examined by Mrs. Morey consists chiefly of angular and rounded quartz fragments (up to 0·5 mm), but there are also grains of feldspar, cherty silica, micaceous siltstone, quartzite, mica and possibly of basalt. Distinct micaceous and gritty bands bring out the bedding. The rock is generally poorly compacted, but some patches have a carbonate or siliceous cement. The basal breccia (E 22585) shows large fragments of finely granular dolomite in a dolomite/calcite/ankerite matrix. The rock is much-fractured, and fissures are filled with quartz grains and fragments of chert, feldspar and siliceous ash cemented by dolomite with mica.

Samples from Bevercotes Colliery shafts do not enable the Lower Mottled Sandstone to be distinguished from the Bunter Pebble Beds, but in Harworth Colliery No. 1 Shaft it may be represented by the 79 ft 4 in of red, grey and yellow sandstone, marly with marl bands in the top 23 ft 10 in, at 353 ft 4 in.

Samples from the shaft sinkings at Manton Colliery indicate that the Lower Mottled Sandstone is probably 71 ft 4 in thick (at 213 ft 4 in) in No. 2 Shaft, and 66½ ft (at 255 ft) in No. 4 Shaft. The former show red sandstone and marly sandstone with mica, and the latter, red marly sandstone with bands of marl.

Uncored boreholes for coal at which tentative attempts can be made to delineate the Lower Mottled Sandstone are as follows: Bilby (104 ft of marly sandstone at 310 ft), Blyth (63 ft of red sandstone at 78 ft), Forest Hill (95 ft of red marly sandstone with pockets of red marl at 280 ft), Hayton Smeeth (84 ft of red marly sandstone at 824½ ft) and Manton (94 ft of red sandstone and marl at 173½ ft).

In *Ranskill* No. 1 Oil Bore the lowest 30 ft of the 'Bunter', resting on the Upper Magnesian Limestone, consist of red and white, fine- to medium-grained micaceous sandstones with rounded grains. These contain partings of red and green mudstone (see p. 160). In *Gringley* No. 1 Oil Bore the lowest 100 to 200 ft of the 'Bunter' is described as soft, fine-grained, dark brown sandstone with mudstone partings, and is suggestive of the Lower Mottled Sandstone. In *Walkeringham* No. 1 Oil Bore the Lower Mottled Sandstone comprises 70 ft of fine- to medium-grained sandstone, and bands of similar rock occur in the underlying top 30 ft of the Upper Permian Marl (see p. 160). In Walkeringham No. 2 Bore no pebbles were recorded in the bottom 130 or so feet of the 'Bunter'; these strata consist largely of fine-grained sandstone with thin silty and mudstone bands; sandstone also occurs in beds arbitrarily assigned to the Upper Permian Marl, one such bed being 45 ft thick (see also p. 160). In the log of *Morton* No. 1 Oil Bore the Lower Mottled Sandstone cannot be distinguished from the lower part of the Bunter Pebble Beds: the top 40 ft of the measures classed as Upper Permian Marl (see p. 160) contain as much sandstone as mudstone and could alternatively have been regarded as Lower Mottled Sandstone. In the *Beckingham* oil bores the Upper Permian Marl contains a high proportion of siltstone and fine-grained sandstone of Lower Mottled Sandstone type: in No. 2 Bore 100 ft of measures classed as Upper Permian Marl in the other bores consist of fine-grained, largely red and pink sandstone, which is silty or argillaceous in places and contains bands of siltstone. It is micaceous and much of it has a calcareous cement. The upper limit of the Lower Mottled Sandstone is ill defined at Beckingham; no pebbles are recorded in the bottom 90 ft of sandstones which succeed the above-described 100 ft in No. 2 Bore, nor in the bottom 90 to 130 ft of sandstones succeeding the strata classed as Upper Permian Marl in Nos. 1 and 5 bores. In No. 4 Bore, however, rare pebbles are recorded in the basal Bunter.

In many of the records of the *Gainsborough* oil bores it is not possible to distinguish the Lower Mottled Sandstone from the Pebble Beds, but some show that up to about 120 ft of sandstone at the base of the Bunter is fine-grained, pebble-free and contains bands of mudstone and siltstone. Bands of similar sandstone are common in the upper part of the strata classed as Upper Permian Marl (p. 160).

In the *South Leverton* and *Grove* oil bores it is also impracticable to separate the Lower Mottled Sandstone from the Upper Permian Marl, the combined thickness of which is up to about 140 ft at South Leverton and 110 to 130 ft at Grove. Fine- to medium-grained, largely red and commonly micaceous sandstone is interbanded with mudstone and siltstone, the sandstone generally tending to predominate in the upper part.

In *West Drayton* No. 2 Oil Bore the Lower Mottled Sandstone cannot be distinguished from the lower part of the Pebble Beds, the lowest 400 ft of the 'Bunter' being recorded as 'coarse sand'.

Overlying the Upper Magnesian Limestone in *Apleyhead* No. 1 Oil Bore are at least 90 ft of red and white, fine- to medium-grained, pebble-free sandstone with beds of red siltstone and silty mudstone. In this position in No. 3 Bore there are 80 ft of similar rocks, but pebbles are recorded from the upper half

and there is a conglomerate at the base (see also p. 161). In No. 2 Bore pebbles are recorded throughout the Bunter, although the bottom 100 ft contain bands of brick-red silt suggestive of the Lower Mottled Sandstone. It is not therefore clear whether the Lower Mottled Sandstone of the Apleyhead area contains pebbles locally, or whether samples from No. 2 Bore have been contaminated by pebbles from higher horizons.

In the logs of the *Bothamsall* oil bores the Lower Mottled Sandstone cannot be satisfactorily separated from the Pebble Beds. A few bores show up to 90 ft of fine- to medium-grained, pebble-free sandstone at the base of the Bunter, but in others up to 50 ft of coarse-grained sandstone, overlain by a widely varying thickness of fine-grained sandstone, occur in this position. The coarse-grained sandstone may be the equivalent of that noted at the base of the Upper Permian Marl elsewhere (p. 157).   E.G.S.

BUNTER PEBBLE BEDS

The outcrop of this formation, covering some 80 square miles of the district, extends in a belt, 6 to 7 miles wide, from Bawtry in the north to Clumber Park in the south. The basal beds, where not concealed by drift, form a fairly prominent escarpment overlooking the lower ground of the Lower Mottled Sandstone to the west. The outcrop generally gives land of indifferent agricultural quality, especially in its west-central and southern parts, where there is little drift cover. It was formerly forested, and much woodland remains in the south.

Complete sections in boreholes east of the outcrop show a general thickening from less than 600 ft in the south of the district to about 1000 ft around Gainsborough in the north-east. In many sections, however, it is not possible to distinguish the lower part of the Pebble Beds from the underlying Lower Mottled Sandstone (Fig. 27), and there is evidently a passage, probably diachronous, between the two formations (p. 162).

The Pebble Beds consist of sandstone, locally pebbly, which is generally markedly false-bedded, but also level-bedded or massive in parts. The colour, reflecting the amount of iron staining, is for the most part red or pink, but substantial thicknesses of brown, yellow and grey sandstone also occur, and mottling is fairly common. In places the topmost beds are noticeably light-coloured, or even white in some sections, but this feature is apparently less common than in the adjacent ground to the south (Edwards 1967, pp. 129, 132). The Pebble Beds are mainly soft, but can be quite hard where cemented with dolomite or calcite.

The sand is chiefly medium- to coarse-grained, though fine bands also occur, and it varies from poorly to well sorted. The grains are subangular, rarely angular, to rounded and consist mainly of quartz with a substantial proportion of feldspars. Minor constituents are rock particles, clay minerals and iron oxides and hydroxides. Mica (muscovite, biotite and illite) is common on some bedding planes and particularly notable at a few localities at the top of the Pebble Beds.   E.G.S.

Heavy detrital minerals separated from seven samples from scattered exposures of the Pebble Beds (Fig. 28) comprise a suite typified by that from Thievesdale Farm (see p. 174). Relative proportions of heavy minerals determined by grain counts show a general similarity to those from the Lower Mottled Sandstone except that the proportion of zircon seems to be considerably lower in the Pebble Beds.   R.K.H.

Pebbles or galls and irregular fragmentary inclusions of red or green mudstone are common in the sandstone, occurring both in bands and scattered through the rock. Lenses and discontinuous partings and bands, up to 3 ft thick, of red, greyish green or banded red and green mudstone, and also of silty mudstone and siltstone, are found from place to place. Some of the oil bores in the north-east of the district show that the top 30 ft or so of strata assigned to

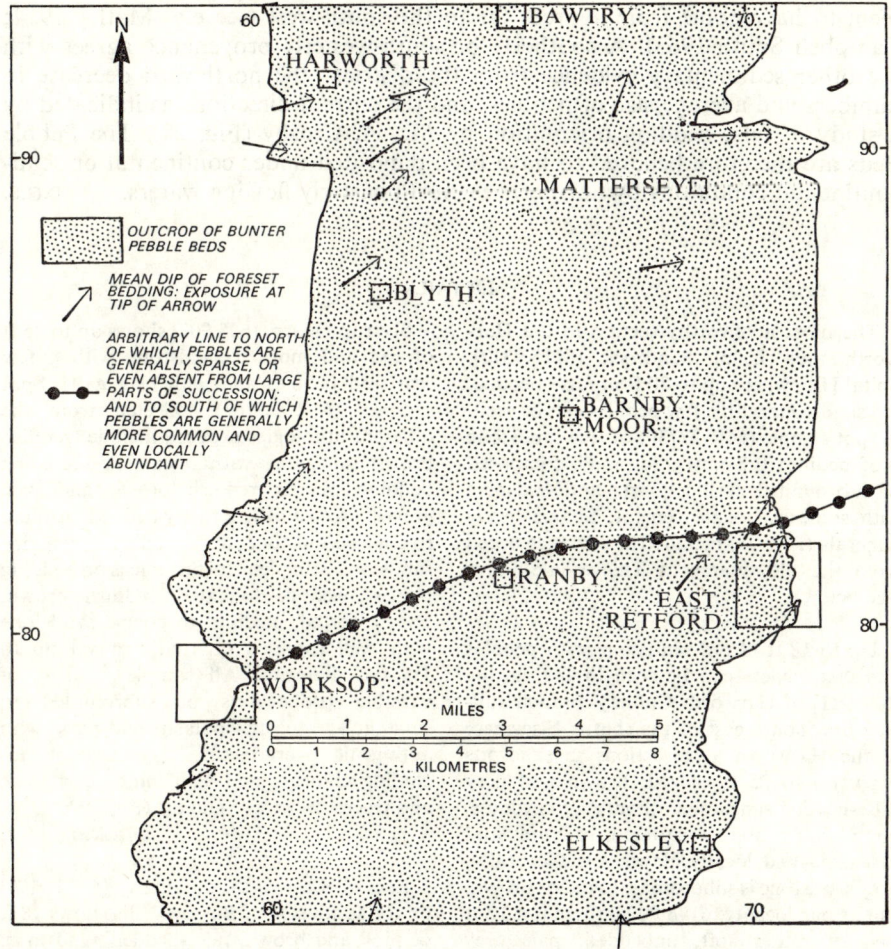

FIG. 29.   *Sketch-map of the Bunter Pebble Beds outcrop*

the Pebble Beds consist of interbanded mudstone, siltstone and sandstone, and it may be, as Smith and Warrington (1971) have suggested, that these beds should be included with the Green Beds (p. 178).

Some parts of the Pebble Beds, especially in the north of the district, are pebble-free, but pebbles have been recorded from the majority of exposures and underground sections. They occur scattered through the sandstone, but also in lenses and bands, where they are occasionally sufficiently concentrated to form conglomerate. They are of all sizes up to 6 inches diameter, but large pebbles are rare. In addition to the northward decrease in numbers (Fig. 29), the pebbles also become, on average, smaller in that direction. A few dreikanter occur, but most of the pebbles are rounded and the commonest shape is oblate. Their surfaces may be smooth or pitted. The bulk of the pebbles are formed of quartzite, which is found in a wide variety of colours, but some consist of vein-quartz, chert, siltstone, feldspathic sandstone, schist and igneous rocks, including porphyry and kaolinized granite. Some of the igneous pebbles may have been derived from the Charnwood Forest region (see p. 174), but for the most part the pebbles, in common with those from the Bunter Pebble Beds of the Midlands, seem to have come from sources far to the south-west (see e.g. Matley 1914; Campbell Smith 1963). A southerly or south-westerly provenance agrees with the other sedimentary evidence from the district—the northward decrease in numbers and average size of pebbles, and the current directions as indicated by a study of false-bedding orientation by Mr. A. Crosby (Fig. 29). The Pebble Beds are therefore believed to have been deposited under continental or semi-continental conditions by northerly or north-easterly flowing waters.   E.G.S.

DETAILS

The most noteworthy section in the Harworth area is at Sandrock [6115 9312], Spital Hill, where up to 30 ft of false-bedded, red sandstone with a few pebbles can be seen. A spot sample (MR 27730) of this sandstone is of medium (0·3 mm) grade and is composed of subrounded clastic quartz and feldspars with some carbonate cement. Sparse heavy minerals (Fig. 28) include the varieties noted from the old quarry at Thievesdale Farm (see below).   R.K.H., E.G.S.

Up to 18 ft of red sandstone with pebbles and clay pellets can be seen in the cutting [629 911] of Harworth Colliery branch railway line, south-east of the shafts. Elsewhere in the Harworth area sections in sand-pits show up to 25 ft of red, brown and grey false-bedded sandstone, of varying hardness, containing pebbles—scattered, in bands and in lenses—and, locally, clay pellets. Generally, grey sandstone is subordinate to red or brown, but at one locality [6198 9004] there is a 12-ft bed of rather soft, unbedded, pale grey sandstone.   E.G.S.

Styrrup Quarry [605 902] shows up to 25 ft of red and mottled sandstone with a few scattered pebbles and with clay galls. Spot samples (MR 27763) collected from the different lithologies exposed include: reddish brown sandstone with mica flakes coating bedding planes; reddish brown and pale greyish green mottled, thinly bedded, medium to coarse sandstone; dark reddish brown, fine sandstone with plentiful fine micas on bedding planes; well consolidated, reddish brown, poorly sorted, medium to coarse sandstone with well rounded quartzitic gravel up to 1·5 cm diameter. All samples consist of variable proportions of subrounded to subangular quartz, potash feldspars with authigenic overgrowths, clay minerals including chlorite, and also minor carbonate and traces of heavy minerals (Fig. 28).   R.K.H., E.G.S.

East of Scaftworth, the quarry [6763 9173] on the south side of Barrow Hills shows 18 ft of pink and brown, false-bedded sandstone, which contains a few pebbles at several horizons and in which pellets of green mud-stone occur, both scattered throughout the rock and concentrated in bands. South of Stone Hill there are natural sections [674 905] in up to 10 ft of pink and yellow sandstone, again with pebbles and mudstone pellets. Yellow is the predominant colour of the sandstone seen in several small sections be-tween Stone Hill and Drakeholes, none of which shows pebbles. At Drakeholes [706 904] exposures in the canal cutting and the old quarry show up to 12 ft of red and yellow, friable, false-bedded sandstone near the top of the Pebble Beds (see p. 186). Pebbles are localized and not abundant, and pellets of green mudstone occur. E.G.S.

A channel sample (MR 27783) through 12 ft of sandstone in the above quarry is composed of mainly medium sand clastics: rounded to subrounded quartz, potash feldspars, micas (muscovite, biotite, illite), dolomite matrix, rock particles, ferric oxide and hydroxide, and sparse heavy detrital minerals (Fig. 28). R.K.H.

North-east of Drakeholes, diggings on the south side of the Chesterfield Canal, near Scott's Wood, show brown sand, which is taken as evidence of an inlier of Bunter Pebble Beds on the upthrow side of the fault mapped along the canal at this point.

Exposures in the area around Blyth occur chiefly in road cuttings and show red or brown, false-bedded sandstone with sporadic pebbles and clay galls; the thickest section is that seen [6247 8793] flanking the main road immediately north of Nornay. E.G.S.

A channel sample (MR 27770) through 15 ft of sandstone in this cutting is moderately consolidated, reddish brown, and pre-dominantly of medium to coarse sand grade. Clastics comprise quartz, orthoclase (with authigenic overgrowths), composite rock particles, clay minerals (including illite and chlorite), ferric oxide staining and sparse heavy minerals (Fig. 28). Samples (MR 27771) of fine to coarse gravel were collected at random from the sandstone here. The gravel, which ranges from 2 to 4 cm mean diameter, is generally rounded, but has a range of sphericities. Roundness, estimated visually (Krumbein 1941) ranges from 0·5 to 0·8, and sphericity, also estimated visually (Ritten-house 1942), ranges from 0·5 to 0·8. The predominant rock-type is quartzite, tough,

exceedingly hard and indurated, and of a range of colours—purple, pale grey, white, buff. Quartzite pebbles range from sandstone to siltstone in grade. Dreikanter occur, though the quartzite gravel is more commonly ellipsoidal, elongated or oblate in shape. Sandstone and siltstone pebbles are quart-zitic, sporadically hematitic, and olive-green, buff, grey and reddish brown in colour; they are aligned along bedding planes, and dreikanter are only rarely developed. Pebbles of granite and porphyry, up to 3 cm diameter, are sparse, the former being highly kaolinized and of a pinkish white colour. R.K.H.

There are several exposures in the Mattersey–Ranskill–Lound area, including those seen in the bottoms of gravel pits, but the only thick sections are provided by the sand-pits in Ellis Plantation [684 879] south-south-east of Mattersey Hill. These show up to 15 ft of soft, brown, false-bedded sand (superficially similar to glacial sand) in which pebbles are small and scattered and which contain pellets of both red and green mudstone together with red, clayey, micaceous bands. At one place [6861 8778] there is a lenticular band of red mudstone with green bands, said to be up to 3 ft in thickness, in the sand.

East of Mattersey and Lound the only exposures are in the vicinity of Wiseton Home Farm and in Messrs. Lound Aggregate's gravel pits. Details of the former, which are very near the top of the Bunter Pebble Beds, are as follows: (a) [7140 8926], 300 yd SE of the farm, 8 ft of soft red micaceous sandstone with bands of coarser yellow sand; (b) [7137 8929], a small exposure 250 yd SE of the farm, soft yellow sandstone with red mud-stone pellets; (c) [7118 8939], a silage pit immediately south of the farm, 5 ft of soft red micaceous sandstone.

At Thievesdale Farm, near Gateford, an old quarry [5904 8124], now filled in, showed, at the time of survey, about 20 ft of soft, generally red, but in parts brownish grey, false-bedded sandstone with scattered small pebbles, largely of quartzite. Numerous clay galls were seen at several horizons. E.G.S.

Samples (MR 27757) from 15 ft of these beds are composed of moderately sorted clastics averaging 0·2 mm. Quartz (rounded to subrounded and mainly igneous in origin), potash feldspars (subangular, and where authigenic overgrowths occur, angular) and polygranular rock and clay particles stained with hematite are the main constituents, with minor green chlorite, reddish brown biotite, muscovite and illite. Sparse heavy minerals include the following: colourless, elongated prisms of zircon; rounded apatite, prisms and rounded grains of tourmaline, reddish brown and yellow rutile, anatase, pale pink garnet, yellow monazite, leucoxene, and opaque iron ores including ilmenite and magnetite.

Gravel (E 34598) ranges from 5 to 56 mm mean diameter, and is mostly rounded, but includes faceted dreikanter with smooth to pitted faces. Hard, quartzose gravel pre-dominates, one sliced pebble consisting of sheared, mylonitized quartz aggregate, with strained, interlocking quartz individuals associated with strings of quartz granules, stemming from a high-rank metamorphic provenance. Arenites are subordinate and mostly fine-grained including purplish brown equigranular coarse siltstone formed of igneous quartz, chert, microcline, micas, chlorite and secondary hematite. Igneous rocks are less conspicuous, but include a quartz-feldspar-porphyry with phenocrysts of orthoclase, altered amphibole, pyroxene and quartz, set in a microcrystalline groundmass of quartz, feldspar, leucoxene, ferric oxide and clay minerals. This rock may have come from the Charnwood Forest area.   R.K.H.

In the Carlton Forest area, sections of Bunter Pebble Beds can be seen in sand and gravel pits (see p. 224). The best sections [6089 8354, 6013 8218] show respectively: 18 ft of hard red false-bedded sandstone with bands of small pebbles and, at one horizon, mudstone pellets; and 30 ft of false-bedded sandstone, red except for greenish grey streaks and patches, containing a few pebbles (Plate XVB).

The only sections of note between Carlton Forest and the River Idle are: (a) a small disused sand-pit [6275 8085] 600 yd NNW of Scofton church, 8 ft of soft red sandstone with scattered pebbles; (b) the road cutting [6856 8080] between Babworth Hall and Babworth church, 5 to 10 ft of red sandstone with a few pebbles; (c) the old sand-pit [6960 8013] north of West Hill, Retford, 20 ft of red sandstone in which the bedding becomes indistinct and contorted where the rock protrudes into the overlying Head (see p. 225); (d) the railway cutting [6909 7988] west-north-west of West Hill, 25 ft of soft red massive level-bedded sandstone with rare pebbles.

Sections [e.g. 7016 8297] along the east bank of the River Idle north-west of Bolham show soft red massive micaceous sandstone with scattered small pebbles. At Bolham itself [7050 8267] 20 ft of Bunter Pebble Beds, exposed beneath glacial sand and gravel and boulder clay, consist of soft, bright red, rather coarse-grained sandstone with marked false bedding; mica occurs on some bedding planes, and pebbles are small and scarce. South of Bolham the top 10 ft or so of the Pebble Beds can be seen [7073 8207] along Bolham Lane. They consist of pink, strongly false-bedded sandstone with greyish green patches, containing small pebbles, clay galls and bands of red, argillaceous, micaceous, pebble-free sandstone.   E.G.S.

A channel sample (MR 27772) taken through these beds is composed of moderately sorted quartz with subordinate potash feldspars, composite rock particles and ferric oxides and hydroxides, and conspicuous micas (muscovite, illite, chlorite and biotite). Sparse heavy minerals (Fig. 28) include monazite.

The junction between the Pebble Beds and Green Beds is exposed in the railway cutting [710 805] east-north-east of East Retford station and north of Thrumpton. The Pebble Beds here consist of yellow and pink, false-bedded, medium-grained, pebbly sandstone. A channel sample (MR 27774) through

---

EXPLANATION OF PLATE XV

A. Both the red and grey banded sandstone above, and the grey sandstone with conspicuous foreset bedding below, form part of the Middle Permian Marl. They are worked for building sand. (See also Plate I).

B. The Bunter Pebble Beds, here worked for building and constructional purposes, consist of red sandstone with greenish grey bands and patches. Small pebbles, mostly occurring in bands, and false bedding can be seen.

(L 567)

A. Sandstone in Middle Permian Marl at Gateford Sand-pit

Plate XV

B. Bunter Pebble Beds in Carlton Forest Quarry, near Worksop

(L 595)

9 ft of the sandstone, is composed principally of medium (0·3 mm) sand clastics: quartz, subordinate potash feldspars with clear authigenic overgrowths, micas (muscovite, biotite), composite mineral particles with intergranular ferric oxide and hydroxide, leucoxenic dust, dolomite, and sparse detrital heavy minerals.

The top of the Pebble Beds with overlying Green Beds was formerly seen at two other localities in East Retford—off Moorgate [7087 8166] and off Spital Hill [7088 8139]. At the latter locality 4 ft of soft, pink, massive, coarse-grained sandstone with scattered small pebbles were seen, and Dr. V. A. Eyles, who described the exposure, mentions traces of old pillar and stall workings, evidence of the rock having been mined.   R.K.H., E.G.S.

The basal part of the Bunter Pebble Beds, resting on Lower Mottled Sandstone, is exposed in the sand-pit [591 796] ⅓ mile ESE of Worksop station. The section here is:

		ft
DRIFT		
Soil, pebbly sand and gravel	3 to	7
BUNTER PEBBLE BEDS		
Red, false-bedded, medium-grained sandstone, green-mottled in places, with scattered pebbles and clay galls and mudstone bands	20 to	30
LOWER MOTTLED SANDSTONE		
Red and green-mottled, false-bedded (on a smaller scale than Bunter Pebble Beds), fine-grained sandstone, with mudstone bands and clay galls		6

A similar section was visible in 1946 in the sand-pit [586 799], now filled, 200 yd N of the station.

There are several exposures of Pebble Beds in Welbeck Park, the thickest being in Drinking Pit Lane where a cutting [5722 7539] through the escarpment reveals up to 12 ft of red false-bedded sandstone. Here pebbles are abundant in places, and there are a few clay galls.   R.F.G.

Combined spot samples (MR 27762) taken through 10 ft of sandstone at this exposure consist mainly of medium, well sorted sand composed of quartz, orthoclase with authigenic overgrowths, composite particles, clay minerals, dispersed ferric oxide, and sparse

heavy minerals (Fig. 28) with a trace of deep blue sapphire and yellowish brown staurolite.   R.K.H.

Some 5 ft of red, false-bedded sandstone with pebbles and clay galls are exposed in the canal bank [6274 7977] at Osberton Hall, and 8 ft of red, pebbly sandstone can be seen in the bank of the River Ryton [6052 7895], 800 yd NNW of Manton Colliery. Numerous shallow sections visible in Manton Wood in 1946 showed a few feet of red, pebbly sandstone similar to that exposed in the railway cuttings immediately north of the wood.

In Clumber Park the best sections are provided by the old sand-pit [6437 7508] ¼ mile SE of Hardwick Grange, and the path cutting [6229 7442] 300 yd WSW of Clumber House. The former shows 10 to 12 ft of soft, red, false-bedded sandstone with pebbles, which are mostly small but up to 5 inches in diameter, scattered and in bands; and the latter shows 11 ft of red, false-bedded, pebbly sandstone with clay galls.

There are no important sections in the area east of Clumber Park between East Retford and Elkesley. At Elkesley 15 ft of friable, false-bedded, coarse-grained sandstone, brownish grey at the top, pink and red below, are exposed on the east bank of the River Poulter [6933 7530]. There are several discontinuous mudstone partings in the sandstone. West-south-west of the village, at Pepperly Hill [6819 7525], 6 ft of soft, dark red, massive sandstone are exposed. The rock is coarse-grained and the grains vary from rounded to angular; pebbles are small and scarce.

On the southern edge of the district, 1½ miles SW of Elkesley, a quarry [6725 7388] adjacent to Redhill Lane, Bothamsall, shows up to 25 ft of friable false-bedded sandstone, deep red except for the top 2 to 3 ft which are pink. Dr. Eyles, who examined the section, reports that the constituent grains are angular or subangular, and pebbles, mostly small but up to 4 inches in diameter, are sparse; he recorded irregular fragments of red mudstone in the sandstone. This section is close below the Green Beds, which crop out a few yards to the south-east of the quarry.

In underground sections there is often difficulty in drawing a boundary between the Bunter Pebble Beds and the Lower Mottled

Sandstone, even in cored boreholes. Apart from oil bores (see below) the only complete sections of the Pebble Beds are as follows (thickness and depth to base in brackets): Bevercotes Colliery (including Lower Mottled Sandstone, 620 ft 2 in at 670 ft in No. 1 Shaft and 621 ft 10 in at 668 ft 11 inches in No. 2 Shaft), Eaton Borehole (about 595 ft at 607 ft), Hayton Smeeth Borehole (640½ ft at 740½ ft), Rockley Borehole (564 ft 4 in at 665 ft), South Leverton Borehole (778 ft at 1325 ft), Twyford Bridge Borehole (including Lower Mottled Sandstone, about 600 ft at about 625 ft), West Drayton No. 1 Borehole (including Lower Mottled Sandstone, 740 ft).

Details of the Bunter Pebble Beds in cored boreholes are given below. At Carr Lane Borehole, Gainsborough, the top 377½ ft were described as fine red sandstone with bands of red marl totalling 40 ft; pebbles are only recorded near the top. In Checkerhouse Borehole the lower part of the Pebble Beds, consisting of red sandstone with pebbles and with bands and nodules of red marl, was proved resting on Lower Mottled Sandstone at 420 ft 11 in. In Cottam Borehole the Pebble Beds were not bottomed at 1458 ft, but 757 ft 8 in were proved if the 41 ft 8 in of red marly and micaceous sandstone at 742 ft, excluded from the formation by Lamplugh and Smith (1914, p. 66), are included. The remaining 716 ft consist of red sandstone, coarse-grained at the top, with pebbles, and, at 1277 ft, a 1-ft conglomerate. Pebbles consisting of 'pellets of red marl and nodules of baryta?' are recorded towards the bottom of the section (*idem*). The lowest 90 ft or so of the Pebble Beds were cored in Cowlishaw Plantation Borehole, but the lower boundary was obscure (see p. 168). Here the formation is recorded as reddish brown, fine- to coarse-grained sandstone with grey patches and bands, containing scattered pebbles as well as a few pebble bands; the pebbles, up to 2 inches in maximum diameter, are chiefly of quartz or quartzite. The boreholes at Dunham Bridge Pumping Station show up to 619½ ft of undivided Bunter, not bottomed, which is composed of sandstone with marl bands, and with pebbles chiefly in the upper part. Water boreholes into the Pebble Beds at East Retford include the four at the waterworks, which, starting near the top of the formation, show 600 ft of red sandstone with pebbles and

patches of red marl, the pebbles being both scattered and concentrated in bands. In the Elkesley Pumping Station boreholes Bunter rocks extend to 581 ft with the possible base of the Pebble Beds at 430 ft (see p. 168). Pinkish grey to dull red, coarse-grained, cross-bedded sandstone is recorded with widely scattered small pebbles (1 to 2 inches in diameter) and more abundant, larger pebbles (up to 3 inches in diameter) which occur in bands between 265 and 275 ft. A well at Markham Moor House shows the top 90 ft of the Pebble Beds, consisting of red sandstone with a 3-ft band of red clay 45 ft from the top. Rampton Hospital Borehole ended in Pebble Beds at 1005 ft, having proved 467 ft of pink coarse-grained sandstone with scattered pebbles, numerous clay galls and bands and partings of micaceous mudstone (Lamplugh and Smith 1914, p. 106). At Rockley Borehole the Pebble Beds extend from 100 ft 8 in to 665 ft (*ibid.*, pp. 72–3) and consist of red sandstone, coarse-grained in the upper part, with pebbles and bands, up to 3 ft thick, of red and grey or blue marl. At South Leverton Borehole there are 728 ft of red sandstone with pebbles and, in the lower part, marl bands. At Upper Morton Grange Borehole (Lamplugh and Smith 1914, pp. 92–3) Pebble Beds extend from the surface to 327½ ft, and possibly to the bottom of the hole at 468 ft 8 in, though no pebbles are recorded below the former depth. They are described as red marly sandstones with pebbles, and with bands of marl which are generally red, but occasionally grey near the top of the formation.

Samples of the Pebble Beds from Bevercotes Colliery shafts show sandstone, which is red except at the top of No. 2 Shaft, where it is yellow. It is medium-grained with sub-angular to rounded grains, and with pebbles up to 6 in long. Marl pellets occur and there is a 21-in band of red and green banded silty marl 87 ft 5 in from the top in No. 1 Shaft. At Harworth Colliery red sandstone with gravel pockets is recorded to 274 ft. At Manton Colliery, where the base is doubtfully at 149 ft in No. 2 Shaft, and 188½ ft in No. 4 Shaft, the Pebble Beds consist of red sandstone with pebbles, scattered and in bands, and pockets and bands of red marl.

Chippings from uncored coal and water boreholes give few details of the Bunter Pebble Beds, showing, in general, red, pink or brown, fine- to coarse-grained sandstone with rounded or subangular grains and containing pebbles and mudstone pellets and bands.

Samples from oil bores, examined by the geologists of the BP Company, give more information, but in many instances this is insufficient to determine the position of the Pebble Beds–Lower Mottled Sandstone junction. *Ranskill* No. 1 Oil Bore shows about 280 ft of undivided Bunter, consisting of fine- to coarse-grained sandstone with subangular to subrounded grains, quartz pebbles and a few silty partings. Some 775 ft of Bunter, probably an under-estimate, are recorded in *Gringley* No. 1 Oil Bore, including between 100 and 200 ft of Lower Mottled Sandstone (see p. 169) at the base. The Pebble Beds portion consists of reddish brown sandstone with pebbles, which are apparently concentrated in bands in the lower part.

Complete sections of the Pebble Beds are provided by the *Walkeringham, Morton, Beckingham* and *Gainsborough* oil bores, with thicknesses ranging from about 800 to almost 1000 ft. They consist generally of red, reddish brown, or occasionally white or grey, fine to medium and less commonly coarse-grained sandstone, with scattered pebbles and sporadic bands and partings of mudstone, silty mudstone and siltstone. The sandstone varies from poorly sorted to well sorted, and the grains range generally from subangular to rounded, though angular examples have been noted; some horizons are mica-rich. Pebbles consisting of quartzite, quartz, chert, schist, igneous rocks and feldspathic sandstone have been recorded; generally they seem to be scattered, though conglomerate bands have occasionally been noted, and they appear to be more common in the lower part than the upper. Apparently the rock is in places poorly consolidated, but in others cemented by calcite. Some bores indicate that the top 20 to 40 ft of the sandstone is interbedded with mudstone and siltstone (see pp. 171–2 and 178), and in a few others the uppermost beds of sandstone are white.

The *South Leverton* and *Grove* oil bores, in which the Pebble Beds are respectively between 750 and 850 ft and up to about 770 ft thick, present a similar picture. At *West Drayton* No. 2 Oil Bore there are a little over 700 ft of Bunter rocks, the lowest 400 ft of which are described as coarse sand and include the Lower Mottled Sandstone. At the top there are about 30 ft of fine- to medium-grained white and brown, partly micaceous, sandstone, which rest on about 275 ft of brown medium-grained, pebbly sandstone. The *Apleyhead* and *Bothamsall* oil bores penetrated the lowest 400 to 600 ft of the Pebble Beds, which consist of mostly red, pebbly sandstone with streaks and thin bands of red mudstone and siltstone. Traces of bitumen were recorded near the top of Nos. 2 and 4 bores at Bothamsall.   E.G.S.

GREEN BEDS

These beds were first recognized by Aveline (1861, p. 18) to the south of the East Retford district, and have since been described in detail in the neighbouring Ollerton district (Lamplugh and others 1911; Edwards 1967). Smith (1912, pp. 252–7) summarized all information then available. The beds have previously been considered as a subdivision of the Waterstones, but in the East Retford district they are treated as a separate division and are shown with a distinctive colour and symbol on the one-inch and six-inch maps, where they are named Keuper Green Beds. They are believed to be present at outcrop throughout the district, but, because of extensive drift cover in the north, have only been mapped south of the Lound–Clayworth area. They are apparently present east of the outcrop except in the north, and where absent are believed to have passed laterally into the Keuper Marl (Smith and Warrington 1971).

The Green Beds have been shown to be diachronous (*ibid.*), and while of late Scythian age throughout this district are regarded as Anisian farther south (Fig. 22). The average thickness of the Green Beds in the East Retford district is 30 to 50 ft, compared with 20 to 30 ft in the Ollerton country to the south

(Edwards 1967, p. 132), and the thickest proved section is 70 ft in Cottam Borehole.

The Green Beds consist essentially of green, grey, or greyish green mudstones, silty mudstones and siltstones with, in some sections, red or brown bands. Smith and Warrington (1971, pp. 203–6) have shown that to the north and east of a line drawn from East Retford through the South Leverton Oilfield and midway between Cottam and Dunham it is commonly found that the bottom few feet of the Green Beds are mainly or entirely red. They suggest (*idem*) that, although the basal red measures may be locally absent, their apparent absence in some sections could be due to deficiencies of the logs, and in other sections the beds may be represented by red mudstone and siltstone intercalated with sandstone and included in the uppermost part of the Bunter Pebble Beds. They use the term Green Beds *sensu lato* for the whole of the formation, including the basal red measures, and the term Green Beds *sensu stricto* for the dominantly green measures.

The basal red measures are thin, probably not exceeding 15 ft, in the East Retford district, but it has been shown (*ibid.*, pp. 204–5) that they thicken towards the north and east, becoming, beyond the boundaries of the district, as thick or even thicker than the overlying green measures.

Thin beds of fine- or medium-grained sandstone are found in the Green Beds, and these may be of almost any colour (green, red, brown, yellowish brown, white and mottled); in them, as well as in the silty beds, mica is common. Thin gypsum bands are described in several underground sections of the Green Beds, and anhydrite is recorded in Cottam Borehole, in Beckingham No. 4 Oil Bore (where there may be as much as 10 ft), and in several of the Gainsborough oil bores. Salt pseudomorphs have been found at one exposure (p. 179). The impersistent conglomerate which has been noted to the south of the district has not been seen at outcrop, but is recorded in a well at Gamston, in No. 1 Shaft of Bevercotes Colliery and in two of the Dunham Bridge Pumping Station boreholes. Mostly the conglomerate appears to be thin, but at Dunham Bridge No. 1 Borehole it is 10 ft thick. The thin basal development of sand with a single pebble at Big Lane Borehole (p. 179) represents the only known occurrence of the conglomerate within the district north of Gamston.

No animal fossils have been recorded from the Green Beds, but the abundant miospores and locally common carbonized plant debris which they contain have led Smith and Warrington (*ibid.*, p. 221) to believe that the green coloration of the formation is primary and that the Beds represent fluvial deposits.

From East Retford southwards abandoned brick and tile pits on the outcrop show that the Green Beds were formerly worked extensively.   E.G.S.

DETAILS

The Green Beds have not been traced at outcrop to the north of the Lound–Clayworth area, but Aveline (1880, pp. 24–5) records 'white or blue' clay in the now-overgrown clay pit [698 909] east of Everton. 'White clays and red marls' which he saw (*idem*) in the dyke alongside Blackbank Road, Everton Carr, were probably also an exposure in the Green Beds.

The basal 2 to 3 ft of the Green Beds were seen by Dr. V. A. Eyles in 1947 at Bolham [7050 8267]. They consisted of red mudstone with yellow and greenish grey bands, resting on Bunter Pebble Beds and overlain by thin boulder clay. The most northerly exposure visible at the present time is in Bolham Lane [7073 8207], between Bolham and East Retford, where 5 ft of banded green, brown and red mudstone and silty mudstone with rare thin bands (up to 1 in thick) of brown-mottled, but largely green, soft micaceous sandstone, can be seen resting on Bunter Pebble Beds. The Green Beds were formerly worked at the old Brick and Tile Works [7086 8195] adjacent to this exposure, at least 15 ft being visible in the face at one time.   E.G.S.

An X-ray powder photograph (NEX 253) of dark red flaky mudstone from the basal 2 ft of the Green Beds here showed major illite and minor chlorite and quartz, and an unidentifiable ferric oxide mineral. A specimen (E 33514) of an intercalated mottled greenish grey, wavy-bedded sandy bed, contains flakes or galls of purple, micaceous, silty mudstone. The sandy constituents are poorly sorted, but well rounded with a 'frosted' appearance. Fine micas (biotite, muscovite and chlorite) are concentrated on bedding planes of the silty portions. Igneous quartz predominates with conspicuous rock particles (quartzite, argillaceous quartzite, chert, volcanics), and potash feldspars with clear authigenic overgrowths. The matrix consists of silty clastics, micas and clay minerals.   R.K.H.

The only exposure now visible in East Retford is in the railway cutting [710 805] north-east of the station. At the south-western end of the cutting 6 ft of red and green mudstone with bands of brown silty micaceous sandstone can be seen resting on Bunter Pebble Beds, and at the north-eastern end 10 ft of mudstone are exposed. The latter comprise 2½ ft of red mudstone overlain by green mudstone striped with red, and containing bands of green micaceous siltstone and silty mudstone and rarer thin bands of yellowish brown sandstone.   E.G.S.

A channel sample (MR 27775) through the basal 2½ ft contains calcite, distributed mainly along bedding planes and fractures. An X-ray powder photograph (NEX 275) showed major illite, minor chlorite and quartz. The overlying 2 to 3 ft are largely greyish green and are coarser grained in general, with salt pseudomorphs and fine white micas sprinkled on bedding planes. The coarser clastics are of fine sand grade (averaging 0·09 mm) and comprise subangular quartz, rock particles, feldspars, with finer muscovite, biotite and chlorite, all partly cemented by calcite. There are scattered opaque ores and traces of zircon and tour-maline. An X-ray powder photograph (NEX 276) of the finer portion showed major quartz, with subordinate illite and chlorite.
                                        R.K.H.

In 1947 there were two further exposures of Green Beds in East Retford, both in old pits since filled in, one [7087 8166] off Moorgate and the other [7088 8139] off Spital Hill. Dr. Eyles described the sections respectively as: 6 to 8 ft of bedded sandy marl, red at top; and 5 to 7 ft of light greenish grey and dark red mottled mudstone with thin bands of harder sandy mudstone. Stiff bluish grey clay with faint pink mottling seen by Dr. Eyles in the old pit [7135 8055] at Newton, East Retford, may have been Green Beds or possibly boulder clay (see p. 221).

There are no exposures of the Green Beds south of East Retford, but they produce, where not covered by drift, a grey, or in places red, clay soil with, locally, slabs of grey sandstone. There are overgrown clay pits at Apple Pie Plantation [6918 7745], in the vicinity of Brick Yard Farm [6961 7693] between Eaton and Elkesley, and just beyond the district boundary near Bothamsall. In 1947 shallow sections of grey and greyish brown clay were visible in the vicinity of Brick Yard Farm, up to a maximum thickness of 6 ft at one locality [6969 7696].

The Green Beds have been proved in a number of boreholes for coal and water in the central and southern parts of the district, as well as in Bevercotes Colliery shafts. A detailed section is provided by Big Lane Borehole, Clarborough, where an examination of the cores was carried out by Mr. R. A. Eden. Here the beds are 42 ft 2 in thick at 146 ft, and consist of green, with subordinate red, mudstone and silty mudstone, micaceous in parts; approximately 50 per cent of the bottom 7 ft are red. There are sandy beds and laminae and thin bands of fine- or medium-grained green sandstone. A few gypsum bands, up to 2 inches in thickness, occur in the upper part, and there are thin bands of reworked Bunter sand in the bottom 3 in, with a single ½-in diameter quartzite pebble recorded from the base. In Hayton Smeeth Borehole, a mile to the west, 91 ft of grey sand and marl are recorded at 100 ft, but it is unlikely that all these measures represent the Green Beds.

In South Leverton Borehole 42½ ft of grey and red marl at 547 ft (for details see Lamplugh and Smith 1914, p. 87) evidently comprise the Green Beds, and in Cottam Borehole they are thought to be represented by the 70 ft 4 in of strata at 700 ft 4 in. The latter are described as grey, green and red marls with 2 ft of anhydrite 13 ft 8 in from the base, and with spar joints (gypsum veins) below this horizon. The red marly and micaceous sandstone between 700 ft 4 in and 742 ft, placed in the Keuper by Lamplugh and Smith (1914, p. 66), is considered to be the top part of the Bunter Pebble Beds. In Rampton Hospital Borehole the Green Beds, described as greenish grey marl and sandstone, are 29 ft thick at 538 ft.

Green Beds are recorded in all four of the wells at Grove Pumping Station: No. 1 shows 44 ft at 455 ft, composed, apart from 10 ft of red and green marl, of blue marl and stone; No. 2 shows 50 ft at 465 ft, consisting of 43 ft of sticky green marl with gypsum in the lower part on 7 ft of red silty clay; No. 3, none of which was cored, shows 49 ft at 453 ft, consisting of 38 ft of greenish grey marl, mottled in the top 18 ft, on 11 ft of red marl; No. 4 shows 52½ ft, consisting of 41 ft of greyish green marl, mottled in the top 16 ft, on 11½ ft of red marl. In Grove Hall Water Well there are 51 ft of Green Beds (42 ft of soft blue clay with stone bands on 9 ft of red silty clay) at 436 ft (for details see Lamplugh and Smith 1914, p. 75).

A well [7092 7644] at Gamston, presumed to start in the Waterstones, but not cored in the top 18½ ft, shows that the Green Beds probably extend to 50½ ft, consist of grey shale with some red clay, and have conglomerate at the base. Some 120 yd to the south there are at least 28 ft of Green Beds, logged as grey shale, in another well [7092 7633], which also commenced in Waterstones. Rockley Borehole, nearby, shows 38¼ ft of blue and red marl (at 100 ft 8 in), and in the well at Markham Moor House there are 30 ft of blue clay on 4 ft of blue stone resting on Pebble Beds at 132 ft. Green Beds are recorded in both shafts at Bevercotes Colliery —at 49 ft 10 inches in No. 1 and at ?47 ft 1 inch in No. 2. In No. 1 Shaft they consist of just over 30 ft of pale green and red silty marl with 10 in of white medium-grained sandstone 15 in from the base, where there is a thin conglomerate of ¼-in quartzite pebbles in a well cemented sandstone matrix. In No. 2 Shaft there are 28½ ft of green silty marl with red bands, containing a 9-in band of green, medium-grained sandstone near the top; similar sandstone is recorded in samples immediately below these beds, and this may also be part of the Green Beds.

At Dunham Bridge Pumping Station, Nos. 1 and 2 boreholes provide similar sections of the Green Beds, viz. 33 ft and 32 ft of blue and bluish green marl and clay on 10 ft and 3 ft of conglomerate at 760 ft and 795 ft respectively. In No. 3 Borehole 22 ft of green and grey marl, with red bands in the upper part, rest on 17 ft of red marl with thin sandstone bands, which, if included in the Green Beds, give a total thickness of 39 ft at 757 ft.

The Green Beds are apparently absent in Carr Lane Borehole, Gainsborough, and in the Gainsborough waterworks boreholes. They were not recorded at Eaton, Twyford Bridge and West Drayton No. 1 boreholes, but other sections show that they are present in this area.

In *Gringley* No. 1 Oil Bore green marl occurs between the Pebble Beds and Keuper Marl, but no thickness has been recorded. In *Walkeringham* No. 2 Oil Bore the Pebble Beds are overlain by about 15 ft of reddish brown silty mudstone, succeeded in turn by about 30 ft of light grey mudstone and siltstone with gypsum. Green Beds are not recognizable in *Morton* No. 1 Oil Bore, nor in three of the four oil bores at *Beckingham*. In Beckingham No. 4 Bore, however, about 30 ft of light greyish green silty mudstone rest on about 10 ft of anhydrite overlying the Pebble Beds. In the *Gainsborough* Oilfield, Green Beds are not recorded in over half of the bores, but 10 to 60 ft of grey or green mudstones, silty mudstones and siltstones, commonly gypsiferous or anhydritic, are recorded in the remainder. Red bands are included in the Green Beds in several bores, with 10 ft of red mudstone at the base in No. 58 Bore. Thin fine-grained sandstones occur, and mica is noted in a number of sections.

All but four of the *South Leverton* oil bores show Green Beds, which are more easily recognizable than in sections farther north because of the presence of overlying Waterstones (see p. 183). Here the beds, varying in thickness from about 20 to about 60 ft, consist of greenish grey, grey or pale green mudstones, silty mudstones and siltstones, with red, brown or pink banding or mottling in some sections. Sandstone, which may be white, red or light green, is recorded in four bores, and gypsum in three. Brown mudstone is recorded at the base of the Green Beds in No. 13 Bore, and red mudstone and siltstone, possibly part of the Green Beds *s.l.* (see p. 178), is interbedded with sandstone at the top of the Pebble Beds in several bores.

Green Beds cannot be identified in *Grove* No. 2 Oil Bore, but in No. 1 Bore 50 ft or so of largely greyish green or green siltstones and silty mudstones overlie the Pebble Beds.

The base of the Green Beds is at about 55 ft in *West Drayton* No. 2 Oil Bore, but samples were taken only from the bottom 10 ft. These show green marls resting on about 5 ft of red marl with traces of sandstone. This is an apparently isolated occurrence of red measures below the Green Beds *s.s.* outside the area in which Smith and Warrington (1971, pl. 12) have shown them to occur.

E.G.S.

WATERSTONES

This formation (designated Keuper Waterstones on the one-inch and six-inch maps) has previously (see for example Lamplugh and others 1911) included the Green Beds, but the latter are distinguished as a separate formation in the East Retford district (see p. 177). The Waterstones in this restricted sense are only present to the south of a line drawn from the northern limit of their outcrop in the Lound area through Grove, South Leverton and Cottam, and are thus less extensive than the Green Beds. Like the Green Beds, they pass laterally into the Keuper Marl towards the north and east (Figs. 22 and 30). Smith and Warrington (1971, pp. 218–20) have shown that the Waterstones are probably partly late Scythian and partly Anisian in age in the East Retford district, but farther south they are evidently Anisian or younger.

At outcrop, which is to a large extent concealed by drift deposits, the Waterstones are estimated to be about 70 to 100 ft thick, thinnest in the north. Sections to the east show more variation, owing partly to the difficulty of defining the top of the formation (see below). Generally, even when allowance has been made for the exclusion of the Green Beds from the formation, the Waterstones are considerably thinner than in the Ollerton district to the south (Lamplugh and others 1911, p. 26; Edwards 1967, p. 133), though several sections show upwards of 135 ft, and 180 ft are recorded at Dunham Bridge Pumping Station. Also they are probably, on the whole, less well defined than they are farther south.

The Waterstones consist of interbanded and interlaminated mudstone, silty mudstone, siltstone, sandy mudstone and sandstone, which are generally red or brown, but bands of grey or green mudstone and siltstone occur, and the sandstones may be almost any shade of red, brown, grey, green or yellow. The sandstones are normally fine-grained and thinly bedded. They have micaceous bedding planes, and mica is common also in the siltstones and sandy mudstones. The resemblance of the micaceous surfaces to watered silk is, according to Woodward (1887, p. 227), the origin of the term 'Waterstones'[1]. The only record in the district of pebbles in this formation is at Big Lane Borehole, where they occur in one thin band (see p. 183). Gypsum, not seen at outcrop, is common in underground sections, where it occurs as thin beds and veins, nodules and as an impregnation of sandstones. Salt pseudomorphs are also found. Indications

---

[1]The term is also applied colloquially in this district, particularly by farmers, to skerries in the Keuper Marl, probably referring, in this case, to their water-bearing capacity.

M

of the shallow-water or partly supra-tidal origin of these beds are provided by ripple-marks and supposed sun-cracks and rain-pits.

The basic lithological difference between the Waterstones and the Keuper Marl is the greater proportion of sandy material in the former. Individual sandy beds are, however, impersistent, and there appears to be in many places a gradual upward passage of the Waterstones into the Keuper Marl. The junction is not exposed, and has been cored in few recent boreholes. On the limited evidence available, therefore, it is impossible to draw more than an arbitrary and admittedly inconsistent boundary between the formations. Farther south, Elliott (1961, p. 216), working with detailed sections, has been able to define the upper boundary of the Waterstones by sedimentary structures, placing it at the top of the dominantly current-deposited, quartzose micaceous rocks.

In the northern part of the East Retford district, where Waterstones are not recognized there is, in some sections—as for example in some of the Gains-borough oil bores—an above-average amount of sandy material in the lower part of the Keuper Marl.

No fossils have been found in the Waterstones of the East Retford district, but *Lingula tenuissima* occurs farther south near Eakring (Rose and Kent 1955) and rare plants, possible worm-tracks, fish and tetrapod footprints are recorded in the Nottingham district (Lamplugh and others 1908, pp. 38–9; Lamplugh and others 1911, p. 27).   E.G.S.

DETAILS

The only exposures of these beds are in the Gamston area in the south of the district. A section is exposed [7100 7639] in Rectory Lane, south of Gamston House, and has been described by Dr. V. A. Eyles as 15 to 20 ft of alternating beds of reddish brown mudstone, sandy mudstone and thinly bedded red, brown or yellow sandstone. The sandstone and sandy mudstone have very micaceous bedding planes with large flakes of muscovite.
                                        E.G.S.

Serial specimens (E 33515–20; MR 27777–8) were examined of the arenaceous members in 3 ft of this section. The bulk of the speci-mens consist of well sorted, closely packed subangular to angular clastic grains, aver-aging 0·05 to 0·09 mm. Coarser, rounded sand grains up to 0·5 mm are sporadically scattered among the finer clastics. Quartz, chiefly of igneous derivation, predominates and exhibits secondary overgrowths in places. Feldspars are subordinate and com-prise mainly orthoclase, microcline, and sodic plagioclase. The potash feldspars in particular are turbid through alteration, and are conspicuously enclosed in clear, euhedral secondary feldspar overgrowths. Though siliceous and feldspathic overgrowths are common, they do not appear to form a coherent sand cement, but they contribute to some extent to the relative compactness of these sandstones. Rock particles are of minor importance and include fine siltstone, mudstone and polygranular quartz. Detrital micas tend to be concentrated in ferruginous, fine-grained silt and mud layers. They are commonly orientated in the well bedded specimens. Fresh muscovite and degraded micas are common, together with a deep brown to reddish brown biotite variably altered to green chlorite. The intergranular matrix consists of very fine micas and clay minerals, with patchy hematite staining. Detrital heavy minerals are relatively con-spicuous in places, and include colourless zircon, rounded apatite, prismatic tourmaline, rutile, garnet, leucoxene and anatase. Opaque specks of ferric oxide occur throughout. Hematite also forms secondary coatings along joints, and in one specimen (E 33516) selenite occurs likewise.

The distribution of detrital mica is variable, for in one specimen (E 33520) mica is sparse, save in scattered clay galls, while in another (E 33518) mica-poor sandstone is intercalated with mica-rich sandstone (Plate XIV, fig. 6), the junctions between the bands being highly irregular, and indicating winnowing and re-working of the sediment.   R.K.H.

Also in Rectory Lane, 100 yd NW of St Peter's Church, 10 ft of interbanded red mudstone and greyish brown, micaceous siltstone and sandstone, the latter in layers up to 8 in thick, are exposed [7080 7609]. At Twyford Bridge, ⅔ mile SW of Gamston, an old pit [7034 7545] shows glacial sand and gravel resting on 3 ft of Waterstones, which consist of hard, reddish and brownish grey, flaggy, fine-grained sandstone interbedded with thin bands of red mudstone. In the sandstone, bedding planes are very micaceous, and there is a little mica distributed through the rock.

Several boreholes and water wells on and east of the outcrop give details of the Waterstones, but few cores have been examined by a geologist. The principal exception is Big Lane Borehole, Clarborough, where coring probably commenced in the Waterstones at 35½ ft and proved their base at 103 ft 10 in. Here they consist of red mudstone, silty mudstone and siltstone, laminated in parts, and containing numerous bands of sandstone, most of which are brown, and which commonly have micaceous bedding planes. At 62 ft 7 in from the base there is a ½-in layer of unsorted, subangular, quartzite pebbles up to ½ inch in diameter. Gypsum is common but not abundant, occurring in thin beds, nodules and as an impregnation of sandstone. Salt pseudomorphs are common in the upper part and sun-cracks occur throughout. Near the base, ripple-marks were observed at one horizon and possible rain-pits at another.

In Hayton Smeeth Borehole the upper part of the 91 ft of grey sand, marl and clay at 100 ft (see p. 179) probably belongs to the bottom part of the Waterstones. In South Leverton Borehole the 151¾ ft of red and grey marl with sandstone ribs (see Lamplugh and Smith 1914, p. 87) at 504½ ft are classed as Waterstones. According to Lamplugh and Smith (1914, p. 106) the beds are 165 ft thick at 509 ft in Rampton Hospital Borehole, where they differ from the overlying Keuper Marl only in containing a higher proportion of sandstone. Both formations consist of red and grey marls with subordinate sandstone or skerry bands and with 'streaks, veins and knots of gypsum'. Micaceous bedding planes are common, and in the sandy layers salt pseudomorphs, lustre-mottling of selenite and ripple-marks are abundant.

In No. 2 Well at Grove Pumping Station the 67 ft of strata succeeding the Green Beds consist of red marl with 'stone bands' 58 ft, 'very hard stone' 7 ft, on red marl 2 ft at 415 ft, and probably represent the Waterstones. Rockley Borehole, starting in the Waterstones below 8 ft of drift, shows 'red and grey sandy marl with mica' resting on Green Beds at 62 ft 5 in.

In Bevercotes Colliery shafts the lowest few feet of the Waterstones were seen below drift, which is probably thin. Samples from No. 1 Shaft were of red and green mudstone and soft, brown and yellow, micaceous sandstone; No. 2 Shaft showed red and green mudstone with a 16-in band of mottled sandstone. In the well at Markham Moor House there are 92 ft of Waterstones, described as 'silvery marl and hard stone', and Dunham Bridge Pumping Station Nos. 1 and 2 boreholes showed 179 and 180 ft of red marl, sandy marl and sandstone at 717 and 760 ft respectively.

The Waterstones are apparently absent at Carr Lane Borehole, Gainsborough, and in the Gainsborough waterworks boreholes, and cannot be distinguished from the Keuper Marl in the records of Cottam Borehole, Grove Nos. 1 and 3 water wells and Grove Hall, Eaton, West Drayton No. 1 and Dunham Bridge Pumping Station No. 3 boreholes.

The Waterstones are absent in the *Gringley, Walkeringham, Beckingham* and *Gainsborough* oil bores, though at Gainsborough a few sections show a greater than average amount of sandy beds in the bottom 20 to 70 ft of the Keuper Marl. Similarly in *Morton* No. 1 Oil Bore there is a higher proportion of silty material in the bottom 85 ft than in the rest of the Keuper Marl. The majority of the *South Leverton* oil bores show Waterstones, which vary in thickness from 20 to, exceptionally, about 160 ft. They consist of interbedded red, reddish brown and grey mudstones, silty mudstones and siltstones, some of which are micaceous, and red, grey and green sandstones, generally fine or fine- to medium-grained and only rarely coarse. Gypsum has been recorded in Nos. 12 and 13 bores. With the exception of a 10-ft bed of red, fine-grained, poorly sorted sandstone above the Green Beds in No. 1 Bore, the Waterstones are not recorded in the *Grove* oil bores. E.G.S.

KEUPER MARL

The Keuper Marl crops out over some 85 square miles in the east of the district, forming, away from the broad valleys of the Idle and Trent, undulating ground of pleasant aspect, and providing land generally of good agricultural quality.

The full succession is present only in the north-east of the district, where the easterly bores of the Gainsborough Oilfield show that the Keuper Marl is up to almost 900 ft thick. Incomplete sections to the south show that the overall thickness must have been considerably less there, but this is due almost entirely, as Smith and Warrington (1971, pp. 208–9, 221) have demonstrated, to a lateral passage of the lower part of the Marl into the Waterstones and Green Beds. There is, throughout the district, very little variation in thickness of the rocks between the Bunter Pebble Beds and the Clarborough Beds horizon (p. 185 and Fig. 30). Smith and Warrington (*ibid*. pp. 218–20) have shown that the basal beds of the Keuper Marl are of late Scythian age in the north, where the Waterstones are not developed, and probably of late Anisian age in the south, where they succeed the Waterstones (see Fig. 22).

The Keuper Marl consists essentially of mudstones with silty mudstones and siltstones, the whole predominantly red in colour but with green or greyish

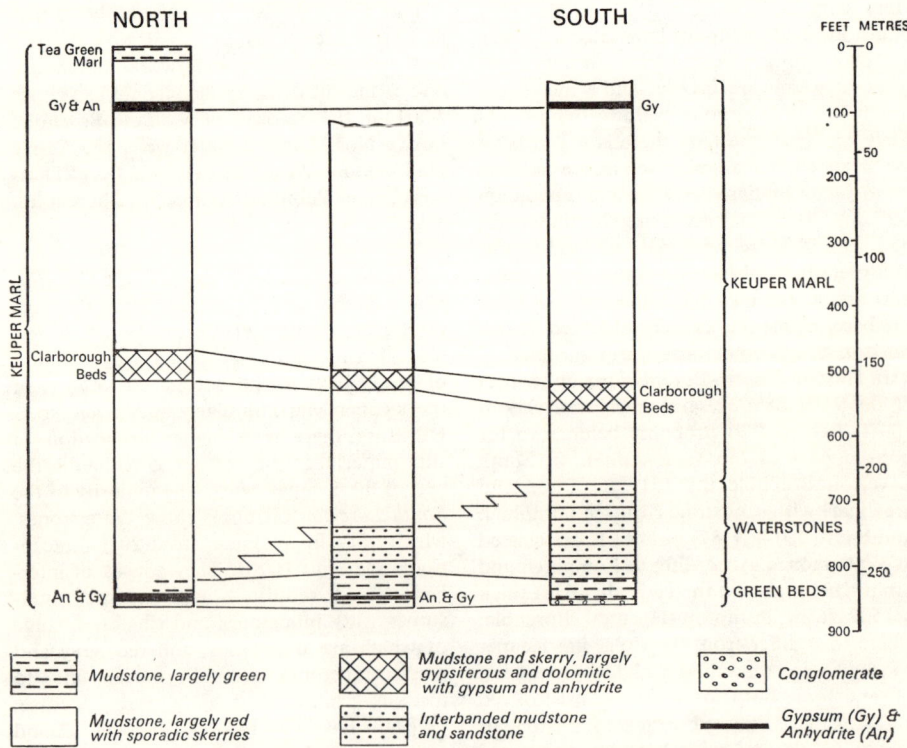

FIG. 30. *Generalized sections of the Green Beds, Waterstones and Keuper Marl in the northern, central and southern parts of the district*

These are based on borehole and outcrop information. Skerries and minor evaporite horizons are not shown

green bands and patches. It also contains thin bands of dolomitic siltstone and fine sandstone (skerries), anhydrite and gypsum, the last occurring in bands and irregular veins and as an impregnation of both marl and skerry. The lithology and sedimentary features of the Keuper Marl to the south of this district have been described in detail by Elliott (1961) who has used them to erect six formations between the Waterstones and Rhaetic. Such a subdivision of the Keuper Marl has not been possible in the East Retford district because of the poor exposure and lack of cored borehole sections.

Skerries are less prominent than in the country to the south, and individual bands can seldom be traced far. They vary from a few inches to a foot or two in thickness, but, being relatively hard, they often form marked topographic features quite disproportionate to their thickness, and are chiefly responsible for the undulating nature of the Keuper Marl outcrop. They are composed of varying proportions of dolomite and of detrital quartz and feldspars with conspicuous mica and heavy minerals. Much of the dolomite is secondary, impregnating the rock and replacing quartz and feldspar, but there is also evidence of detrital dolomite. The skerries are dominantly green or greyish green in colour, but also red or brown. Lamination and small-scale cross bedding are common, and are particularly well seen in thin section, where grading can sometimes be recognized. Ripple-marks also occur. The heavy minerals include tourmaline, leucoxene, zircon, apatite, rutile, garnet, anatase, baryte, brookite and opaque ores. They are commonly concentrated in certain laminae.

Gypsum occurs almost throughout the Keuper Marl, but there are only three important horizons, at or near the base, near the middle (the Clarborough Beds, see below) and within 100 ft of the top. At each of these three main horizons anhydrite is also recorded. The lowest evaporite horizon consists chiefly of anhydrite, and is continuous with that found in the Green Beds where that facies is present (see pp. 180–1); it has not been seen at outcrop. Gypsum of the two higher horizons at outcrop is either fibrous and displaces the country rock, or is present as an impregnation. Underground, these two horizons are known only from the logs of chipped boreholes, and at both anhydrite and gypsum are recorded, up to 15 ft thick at the higher level. Smith and Warrington (1971, p. 220) have shown that the lowest evaporite horizon is apparently of late Scythian age everywhere.

The Clarborough Beds are a distinctive division of the Keuper Marl, consisting of about 10 to 40 ft of mudstones and silty mudstones with much gypsum and a higher than normal proportion of skerries. The gypsum occurs in bands and veins and also as an impregnation of both mudstone and skerry. At depth it is associated with anhydrite. Because of the irregular development of the secondary gypsum and the lateral impersistence of individual skerries, the top and bottom of the beds are arbitrarily defined, and cannot be taken at precisely the same stratigraphical horizons from place to place. Generally, however, the beds as a whole maintain a constant stratigraphical position, and Smith and Warrington (1971, p. 220) have shown that they are apparently everywhere of the same age— late Ladinian or early Carnian (see also p. 196).

The Clarborough Beds lie about 300 to almost 400 ft above the top of the Bunter Pebble Beds, the interval increasing northwards (Fig. 30). This places them towards the bottom of the Keuper Marl in the south, where the Waterstones facies is developed, and towards the middle of the Keuper Marl in the north. At outcrop they are a valuable marker horizon, forming together with the overlying skerries a marked feature which can be traced throughout the district. At depth, however, they cannot always be recognized in the lithological logs of chipped boreholes, though Smith and Warrington (1971, p. 210) have suggested that they may be the cause of the regional marker recognized by Balchin and Ridd (1970) on sonic geophysical logs of oil bores to the east. Firman (1964, p. 190) has suggested that to the south, the East Bridgeford Gypsum, apparently occurring near the boundary of Elliott's (1961) Carlton and Harlequin formations, may be the correlative of the gypsum at the Clarborough Beds horizon.

The Tea Green Marl, up to about 25 ft thick, forms the topmost beds of the Keuper Marl. It consists of green or greyish green mudstones and silty mudstones with, locally, thin beds of grey fine-grained sandstone. It has not been delineated on the map.

The Keuper Marl was evidently deposited for the most part in a hypersaline sea at some distance from the shore-line. The only fossils recorded in the present district are miospores (pp. 195–6), but these are valuable chronostratigraphical indicators.   E.G.S.

DETAILS

Keuper Marl below the Clarborough Beds at outcrop. The only exposures worthy of note in the Walkeringham area in the extreme north are contained in abandoned brick-pits. The thickest section [7566 9230] in Walkeringham Brick-pit, north of Highfield Farm, shows 10 ft of red mudstone with green bands, close below the base of the Clarborough Beds; and the only sections visible in the extensive pits between Gringley Bridge and Leys Farm, which exploited the strata some 60 to 80 ft below the Clarborough Beds, are at Shaw Brick Works [738 920] near Dunstan Farm, and consist of up to 6 ft of weathered red mudstone.

There are several exposures in the bank of the Chesterfield Canal immediately north of the tunnel at Drakeholes, the section [7060 9063] nearest the tunnel showing 10 ft of red mudstone, striped with green, and containing bands of micaceous silty mudstone. These beds, dipping gently to the north, cannot be far above the Bunter Pebble Beds, which are exposed at the southern end of the 150-yd long tunnel, and point to the absence of the Waterstones facies in this northern area. This fact is borne out by the record of the strata in the now completely overgrown clay pit [698 909] east of Everton (see p. 178) and by exposures of red and green mudstone with thin skerries seen in road-widening operations between Drakeholes and Prospect Hill in

1964. There are no exposures in the vicinity of Everton at the present time, though red clay has recently been seen in a trench section; the two small outliers of Keuper Marl on the north side of the easterly trending fault south of the village have been mapped on soil and augering evidence.   E.G.S.

Exposures in the Wiseton–Clayworth–Hayton area are scattered, and most show no more than a foot or two of red clay or mudstone with green bands and sporadic thin skerries; the best are probably those in Wheatley Beck south-west of North Wheatley, which show about 8 ft of gently folded red and green mudstone with bands, up to 3 in thick, of greenish grey sandstone. The sandstones here are laminated and micaceous, and show current bedding and ripple-marking in places.

There are numerous small sections in the area between Hayton and Grove; the most notable are: (a) an exposure of a mapped skerry horizon [7312 8456] in Topyard Lane, Hayton, the section consisting of 6 ft of brown, green and grey silty mudstone with bands of green, fine-grained sandstone and brown, thinly bedded argillaceous sandstone; (b) part of the same skerry is seen in the roadside on Clarborough Hill [7320 8391], the section comprising 2 ft of interbedded grey sandstone and marl; (c) the best section [7310 8261] of several visible in the railway cutting north-east of East Retford is located due south of Clarborough Hall, and shows 5 ft of interbanded red and green mudstone (the green bands making up about 30 per cent of the whole), slightly silty in parts; (d) a gully [7338 8122] at Little Gringley exposes several feet of red and green mudstone containing a mapped 3-ft skerry, which consists of interbanded green mudstone and fine-grained sandstone in approximately equal proportions.

There are a number of persistent skerry bands in the area south of the latitude of East Retford. The outcrop of one has been traced for as much as 3 miles, from near Little Gringley to a point some 400 yd S of Gamston Wood. It is partially exposed [7275 7895] in Dog Kennel Lane, ¾ mile SW of Grove, where 3 ft of interbanded skerry and green marl rest on 2 ft of red mudstone.

R.A.E., E.G.S.

Clarborough Beds at outcrop. The small pit [756 928] on the north side of Cave's Lane, Walkeringham, shows the following section in the Clarborough Beds:

	ft
Mudstone, red jointed, with brown sandy bands; minor contortions ..	4
Skerry with gypsum bands up to 4 in thick; minor contortions	2
Mudstone, red, with thin gypsum bands, seen to	1

Exposures in the brick-pit on the south side of Cave's Lane (Walkeringham Brick-pit) show skerries with gypsum bands at two horizons in the Clarborough Beds, the upper [7561 9275] corresponding to that exposed on the north side of the lane (see above). This skerry thickens to the south, and can be traced around the western side of Highfield Farm and as far as the fault to the south of the farm; it is exposed [7567 9224] 250 yd NE of the farm, where 4 ft of grey siltstone with bands of silty mudstone and soft clay, but no gypsum, were seen. The lower skerry cannot be traced beyond Walkeringham Brick-pit, and gypsum associated with it is only seen within 150 yd of Cave's Lane. There are two exposures in this distance: one [7557 9269] shows a thin skerry with gypsum bands, and the other, an open trial pit [7560 9262], shows 7 ft of sandy mudstone with thin bands of gypsum and skerry. The skerry can be seen on the south side of the brick-pit [7566 9230], overlying mudstone

and sandy mudstone (containing thin bands of gypsum) which are presumed to be the basal part of the Clarborough Beds. E.G.S.

A specimen (E 37466) of skerry from this last locality is a very porous, greyish green argillaceous dolomitic siltstone riddled with small cavities (averaging ½ mm across), resulting in a relatively low density and biscuit-like texture reminiscent of the lower dolomite of the Permian Hampole Beds (Smith 1968, p. 467). In section, the present specimen consists of fine- and even-grained angular resistates (averaging 0·06 mm), which include igneous quartz, plagioclase and orthoclase, the last showing secondary overgrowths. Mica flakes are particularly common and include biotite, muscovite and chlorite. There are streaky aggregates of clay minerals associated with very fine dolomite (0·004 mm). Coarser dolomite grains (0·04 mm) impregnate much of the rock, sporadically replacing the resistate grains. Heavy detrital grains, including apatite, zircon, tourmaline (green and brown varieties), rutile, anatase and leucoxene, are commonly concentrated with granular dolomite along certain bedding planes. R.K.H.

The Clarborough Beds form a rounded scarp from Walkeringham to Gringley on the Hill, where they are concealed for a short distance by glacial deposits. The only section in this area is provided by an excavation [7369 9129] on the north side of Gringley village, where a skerry band surmounts a 12-ft face of slipped and weathered red mudstone: there is no evidence of gypsum, but the exposure falls within the feature lines which have here been taken as the boundaries of the Clarborough Beds. E.G.S.

Southwards from Gringley on the Hill the Clarborough Beds are the dominant control in the scarp which rises on the east of the Idle flats. There are few exposures between Gringley and Wheatley Field but the abundant slabs and fragments of skerry in the soil show that the beds contain a high proportion of skerry in this area. The soil also contains sporadic gypsum fragments. A large block of gypsiferous mudstone has been dug out [7448 8878] near Highfield Farm, Clayworth, and gypsiferous mudstone has been encountered in a cesspit [7454 8650] at Haughgate Hill and dug from a hole [7361 8559] near Hillside House, Hayton. Some 3 ft of red mudstone with hard silty bands is exposed [7443 8725] near Clayworth Field, and an excavation [7473 8553] at Wheatley Field shows fine-grained brown and grey sandstone, some of it laminated and ripple-marked.

The Clarborough Beds can be seen at a number of places in North Wheatley, chiefly in old pits and road cuttings. The most noteworthy exposure [7561 8602], approximately 675 yd WNW of St Peter's and St Paul's Church, shows about 15 ft of red and green mudstone, contorted in part, with gypsum and bands of hard flaggy sandstone. The gypsum occurs in bands (particularly in the green beds), in thin, anastomosing veins, and as irregular impregnations. The mudstone can be seen to thicken where it contains bands of crystalline gypsum, but not where the gypsum is present as an impregnation. A badly overgrown pit [7606 8583], 150 yd WSW of the church, reveals red and green mudstone with green sandstone; 2½ ft of gypsiferous mudstone with irregular veins of gypsum up to 2 in wide were visible at the base of the section in 1959.

Red and green gypsiferous mudstone showing gentle folding is exposed in an old excavation [7601 8547], 5 ft deep, at South Wheatley.

The best sections of the Clarborough Beds can be seen along the outcrop between Clarborough and Grove. An old gypsum pit [7393 8351], 400 yd SSW of Clarborough Hill Farm, shows the following section:

	ft	in
Silty clay soil with pebbles.. ..	1	10
Mudstone, red and green, with bands of fine-grained green sandstone; some contortion	4	8
Mudstone, red and green, with sandstone bands ¼ to 4 in thick; gypsum bands and veins up to 3½ in; local gypsum impregnation	14	0

A pit [7396 8336] (Plate XVIA) 150 yd SSE of this exposure, worked for gypsum until 1953 (see p. 239), shows the following section:

	ft
Silty clay soil	1½
Clay, red and green, with yellowish mottling	1¼
Mudstone, red and green, and green fine-grained sandstone; 1-ft gypsiferous lens	4½
Mudstone, red and green gypsiferous, with bands of fine-grained sandstone and fibrous gypsum	10

The gypsum beds, chiefly horizontal but also ramifying, are up to 6 in thick, and gypsum makes up perhaps 15 per cent of the lowest 10 ft. The sandstone bands show ripple-marking and salt pseudomorphs.

R.A.E., E.G.S.

Specimens of mottled reddish brown and green fairly hard marl (E 33531), and of overlying greyish green skerry (E 33532), from this exposure were examined in thin section. The lower specimen is a complex of small phenoclasts up to about 6·0 mm across, scattered in a gypsiferous silty matrix. The phenoclasts comprise rounded pellets of gypsum with disoriented anhydrite relicts (0·016 mm across), and turbid aggregates rich in leucoxene and secondary anhydrite. The bulk of the rock is a mesh of quartz silt (averaging 0·04 mm) with scattered feldspars and micas, with patches of intergranular fibrous gypsum, and streaks of finely granular dolomite. Fluxioned laths of secondary anhydrite are common, and average 0·06 by 0·01 mm. They are commonly associated with phenoclasts. Scattered heavy detrital grains include apatite and zircon.

The overlying skerry (E 33532) is a dense, fine-grained, roughly bedded aggregate of evenly sorted clastic quartz and feldspars (averaging 0·06 mm) scattered in a coarsely crystalline gypsum base, containing streaks of

EXPLANATION OF PLATE XVI

A. The section consists of weathered red and green gypsiferous mudstone with bands of fine-grained sandstone and irregular bands of fibrous gypsum. This pit was formerly worked for gypsum, and is the type locality for the Clarborough Beds.

B. This gypsiferous horizon occurs towards the top of the Keuper Marl, and is here of similar lithology to the Clarborough Beds in the accompanying photograph.

(L 596)

A. Clarborough Beds, formerly worked for gypsum, near Clarborough Hill Farm

PLATE XVI

B. Gypsum in Keuper Marl, east bank of River Trent near Dunham Bridge

(L 598)

laminated finely granular turbid dolomite. Micas also show preferred orientation, and include brown and green biotite, chlorite and muscovite. Heavy detrital minerals (yellowish brown tourmaline, zircon, TiO_2 polymorphs, leucoxene and opaque ores) are concentrated as placer-like laminae.   R.K.H.

The section visible in the railway cutting [7453 8266] west of Clarborough Tunnel probably represents nearly the whole of the Clarborough Beds in this area:

	ft
Skerry and mudstone, interbanded; some gypsum impregnation ..	3
Mudstone, red and green, impregnated with gypsum in parts; ramifying and horizontal bands of gypsum up to 2 in thick	8
Mudstone, red gypsiferous (poorly exposed)	8
Skerry, green	1
Mudstone, red with gypsum impregnation in parts; horizontal and ramifying bands of gypsum up to 2 in thick	8

Red gypsiferous mudstone with a 2-in gypsum band is exposed in the road cutting [7354 8155] 600 yd ESE of Welham Hall.

A section [7386 8117] in one of the old plaster pits east-north-east of Little Gringley shows:

	ft
Skerry and green mudstone, interbanded	2½
Mudstone, red (poorly exposed) ..	4
Mudstone, red and green gypsiferous, sandy in parts, with bands of gypsum up to 2 in thick	3

Nearby [7390 8124], but on the other side of a fault, 6 ft of green sandstone with partings of green mudstone can be seen. A skerry band within the Clarborough Beds is also exposed a little farther south, in the gully known as Deer Leap [7383 8077], north of Castle Hill Wood.

There are no exposures in the Grove area at the present time, but in 1947 sections near Grove Hall were examined by Dr. V. A. Eyles, who recorded red and green mottled mudstone with thin bands of pale grey, flaggy, silty sandstone. In the section of a water well at Grove (Lamplugh and Smith 1914, p. 75) 30 ft of strata with their base at 77 ft are assignable to the Clarborough Beds, viz:

	ft
Red clay and skerry	3
Hard gypsum	5
Blue and red stone with beds of gypsum	15

South of Grove a small unexposed outlier of Clarborough Beds forms the hill surmounted by Lodge Field Clump.

R.A.E., E.G.S.

Southwards from this area one or more mappable skerry bands make up a conspicuous part of the Clarborough Beds. 6 ft of green skerry interbanded with green and a little red mudstone are exposed [7490 7794] near the old brick-pit at Lady Well, north-north-east of Nether Headon. No gypsum was seen, but this section apparently forms the upper part of the Clarborough Beds, the lower part having been worked along with underlying mudstone in the brick-pit, now overgrown. There are several small exposures of skerry in the neighbourhood of Headon, south of which village the base of the Clarborough Beds is shifted a considerable distance to the west by two faults throwing down south. The villages of Upton and Askham are built on the Clarborough Beds, Askham on an outlier in which good sections are provided by the railway cutting west of the village. Here [7319 7488], on the south side of the road bridge over the railway, 4 ft of pale green sandstone and gypsiferous mudstone with ramifying, but mainly horizontal, beds of gypsum up to 2 in thick, are seen 8 ft from the top of the cutting. Eight feet below, there is an exposure of 8 ft of red silty mudstone with lenses of green silty mudstone and siltstone; these strata contain much gypsum, both as an impregnation and in bands up to 6 in thick.   E.G.S.

Specimens of the uppermost pale green sandstone (skerry), marl and gypsum, and of the lower 8 ft of gypsiferous mudstone, were examined in thin section. The topmost specimen (E 33523) consists of a pale buff aggregate of interlocking gypsum plates, charged with convoluted mosaics of fine dolomite grains which exhibit local laminations suggesting bedding. Fine silty clastic grains are scattered throughout. The fine and even granularity of the dolomite, its association with detrital resistates, and rough bedding, all suggest an origin penecontemporaneous with sedimentation. The gypsum, however, is crystallographically unrelated

EXPLANATION OF PLATE XVII
THIN SECTIONS OF TRIASSIC ROCKS

1. Photomicrograph of finely cross-laminated dolomite-siltstone from a skerry in Keuper Marl, Treswell [7810 7931]. The dark specks indicate laminae with concentrations of heavy detrital minerals. (E 33528). Uncrossed polars × 25.

2. Photomicrograph of gypsiferous dolomitic pseudo-breccia from the Clarborough Beds, in a railway cutting [7319 7488] near Askham. Coarse gypsum (pale grey) has patchily replaced dolomite-silt. (E 33524). Partly crossed polars × 17.

3. Macrophotograph of a thin section (E 33528) of the skerry detailed in (1) above. This shows festoons of finely cross-laminated dolomite, and quartz-mica-silt, with delicate graded bedding, flame structures, dislocations and a prominent convoluted siltstone dyke transgressing laminations and disturbing underlying sediment. Uncrossed polars × 4·2.

PLATE XVII

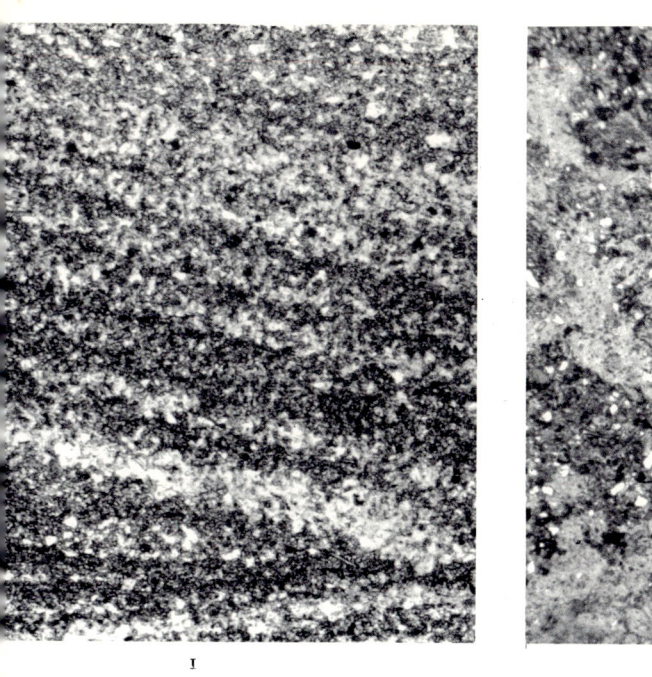

1

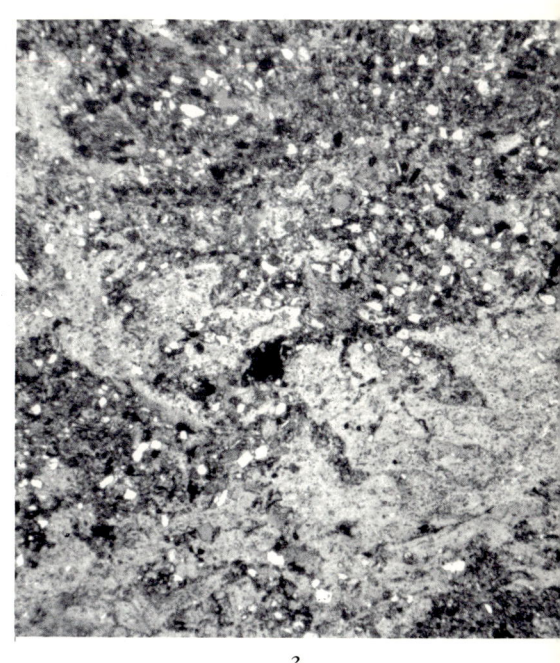

2

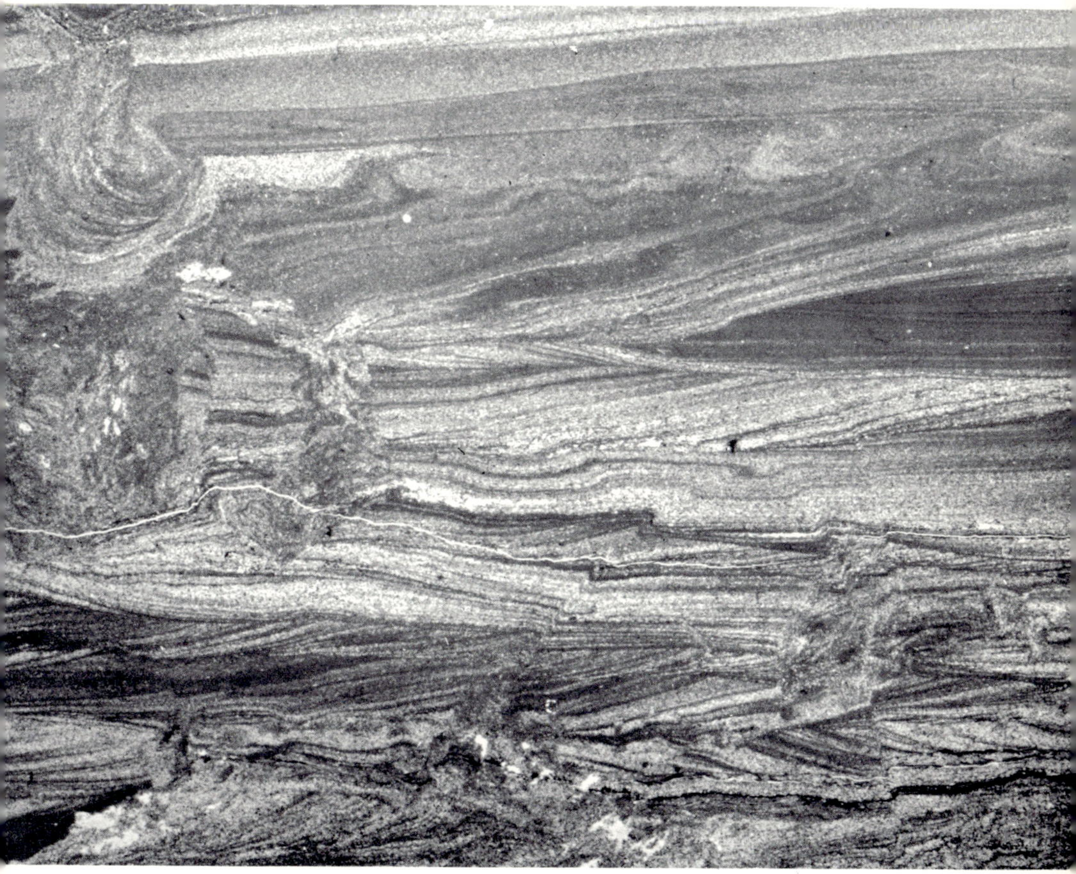

3

THIN SECTIONS OF TRIASSIC ROCKS

to any sedimentary feature, and is clearly secondary. A specimen (E 33521) of the underlying bed, a green siltstone, is dense, close-grained and shows traces of irregular bedding. In section, clastic grains are well sorted, subangular and average 0·04 mm; they consist of quartz, plagioclase and ortho-clase, with conspicuous mica flakes arranged in preferred orientation along the bedding in places. The micas include muscovite, reddish brown to green biotite (partly chloritized) and chlorite. Heavy detrital minerals include apatite, tourmaline (green and yellowish brown varieties), zircon, and leucoxene. Dolomite forms scattered interstitial rhombs averaging 0·04 mm, and these, with the clastic grains, are set in coarse-poikilitic plates of secondary gypsum. Selenite forms clear margins to fractures.

Both dolomite and the later gypsum replace quartz and other detrital constituents. The relative hardness of this rock is due to the interlocking gypsum matrix, which tightly cements the clastic grains and dolomite and is reinforced by them.

A pale greyish green, fine-grained gypsi-ferous dolomitic sandstone (E 33522), under-lying the previous specimen, contains a lower proportion of clastic grains (averaging 0·08 mm) which comprise quartz, plagioclase, orthoclase (with clear secondary overgrowths) and micas. There are streaks of micro-granular dolomite associated with con-centrates of heavy detrital minerals—zircon, leucoxene, apatite, rutile, garnet, anatase and tourmaline. The dolomite grains are com-monly zoned with pinkish rhombohedral cores and clear anhedral overgrowths. Clastics and dolomite are all enclosed in a clear, coarsely crystalline framework of secondary gypsum.

The lowest specimen (E 33524) (Plate XVII, fig. 2) from this section differs from the higher ones in that the residual, finely granular, dolomite, which forms laminae and con-voluted streaks with associated detrital grains and micas, is heavily stained with ferric oxide—presumably hematite. The coarse secondary gypsum plates in which these clastic and primary components are scattered are not so stained, and the gypsum clearly post-dates the iron staining. The reason for the highly convoluted state of the dolomite segregations is not clear. It may be due to soft-sediment movement, or to lateral creep, following or accompanying the emplacement of the secondary calcium sulphate (presumably anhydrite) and its subsequent change in volume through hydration to gypsum. In the emplacement process, large-scale replacement of primary dolomite and clastic material may well have occurred with resulting disturbance of any primary bedding structure.   R.K.H.

There is a further exposure of mudstone and gypsiferous mudstone with irregular, but generally horizontal, bands of gypsum in the cutting north of the bridge, and there are small exposures of skerry in the Clar-borough Beds in and around Askham village.
                                                           E.G.S.

A specimen (E 33525) of skerry from Top Street, Askham [7362 7484] is dense, very fine-grained, bedded, and pale grey with patchy iron staining. It consists of a laminated aggregate of very fine (0·016 mm), evenly grained dolomite, which is convoluted and disrupted in places. There are intercalated laminae of silty clastic grains (averaging 0·016 mm)—quartz, feldspars and micas (chlorite, green biotite, illite, muscovite)—oriented along the laminae. Dolomite commonly exhibits pale pinkish brown cores with clear overgrowths, and marginally replaces the detrital grains. The evenness of grain, and laminated fabric, further suggest that the dolomite was formed penecontem-poraneously with sedimentation. Opaque specks of leucoxene appear to be mainly detrital (after rutile). An X-ray powder photograph (NEX 346) of the whole rock showed major dolomite, quartz, chlorite and illite.   R.K.H.

South-west of Askham, and just beyond the district boundary, the Clarborough Beds are well exposed [7319 7488] where their scarp is cut through by the main Lincoln road. Dr. Eyles gives the following descrip-tion of the section in the cutting:

	ft
Soft and poorly bedded dark red mud-stone with greenish grey patches and partings	6½
Hard, grey, fine-grained, silty sand-stone; broken and of irregular thickness	0½
Soft and poorly bedded red mudstone, mottled and streaked with light grey mudstone; horizontal veins of gypsum at base	2½

	ft
Hard, grey, fine-grained sandstone on sandy mudstone; gypsum veins ($\frac{1}{8}$- to $\frac{1}{4}$-in) and impregnation in places 6 in to	$0\frac{3}{4}$
Soft, mainly light grey mudstone, locally veined with gypsum 0 to	1
Soft, dark red mudstone, locally veined with gypsum.. ..1 ft to	2
Reddish brown and light grey, flaggy sandstone, interbedded with mudstone .. .. .. about	$3\frac{1}{2}$
Mudstone and sandy mudstone on sandstone, the whole much-veined and impregnated with gypsum (veins up to 3 in thick); thickness uncertain owing to possible faulting about	10
Hard, grey, silty sandstone	$0\frac{3}{4}$
Red and green mudstone .. about	3
Grey, flaggy sandstone	$0\frac{3}{4}$
Red mudstone, seen to	2

Keuper Marl above the Clarborough Beds (including the Tea Green Marl) at outcrop. Exposed sections in the Walkeringham–Gringley–Beckingham area are few and shallow. For example, 3 ft of red mudstone with green bands and thin bands of green, flaggy siltstone can be seen below alluvium in the bank of the stream [7723 9156] east-south-east of Highfield House; and 45 yd NE of this locality another shallow exposure of red mudstone contains a 12-in skerry band. A temporary exposure [7515 9042] at Gringley Grange showed 4 ft of red clay containing abundant slabs of skerry, some of which were lying in a vertical position. This disruption, together with the presence of pebbles in the top 3 ft of the clay, is considered to be the result of frost-heaving.

Few skerry bands can be mapped in these beds in the extreme north of the district. The only one traceable east of the Trent can be followed for about $\frac{1}{2}$ mile across the falling ground north of Thonock Hall, and is calculated to be only 25 to 30 ft below the top of the Keuper Marl. It is not exposed, but forms a small feature, and there are numerous slabs of sandstone in the soil. The outcrop of a gypsum band has been shown on the map extending from near Thonock Hall to Gainsborough station. An exposure [8267 9233] 500 yd NW of Thonock Hall shows $4\frac{1}{2}$ ft of red and green mudstone with many bands and stringers of gypsum

which give the rock a nodular appearance. This exposure, calculated to lie some 50 to 75 ft below the top of the Keuper Marl, may however be at a slightly higher horizon than the gypsum band, which, as indicated by spoil heaps of red clay, was formerly worked extensively between Castle Hills and Gainsborough, and which is exposed [8191 9008] at Gainsborough station. Here 6 ft of red mudstone contain irregular bands and nodular masses of gypsum, together with bands of skerry and gypsiferous mudstone.

There are no exposures of Tea Green Marl north of Gainsborough, the supposed outcrop being largely covered by drift. Augering in the vicinity of Thonock Hall suggests that there may be about 20 ft of dominantly green clay at the top of the Keuper Marl in that area.

East of Clayworth 5 ft of red mudstone with green bands is visible in the ditch [7669 8783] east of Wheatley Wood Farm, and in the fields south-west of this locality material excavated from trenches just prior to the resurvey included many slabs of skerry and some gypsum.

A large number of skerry bands have been mapped in the Saundby–Bole area, their outcrops being broken by two WSW-trending faults. The most noteworthy exposures are at Saundby, where stream sections [7822 8865 and 7863 8790] show skerry interbanded with red mudstone.   E.G.S.

A specimen (E 33534) of bedded greyish green skerry from the latter locality is composed largely of very finely and evenly granular dolomite (0·01 to 0·016 mm) concentrated into laminae occurring roughly parallel with the bedding. There is less cross bedding of the laminae than in the specimens from Treswell and Newton-on-Trent (see below). There is much fine quartz silt, feldspars and micas (showing preferred orientation) intercalated with the dolomite laminae.   R.K.H.

A temporary exposure [7872 8847] in Saundby showed 10 ft of red mudstone with thin greyish green skerry bands, the top 3 ft being markedly contorted.

Red and green mudstone with thin, unmappable skerries and bands of gypsum up to 1 in thick are exposed at several places in Wheatley Beck east of North Wheatley.   E.G.S.

South-east of Gainsborough red mudstone with green bands and patches is visible in ditches and excavations; 7 ft being seen at one place [8205 8913]. The Keuper Marl in the railway cuttings in this area is now grassed over, but Burton (1867, p. 316) described the topmost beds (the Tea Green Marl) in the cutting at Lea [833 867] as 'blue marl'.

There are few exposures of note between Clarborough and the Sturton-le-Steeple–South Leverton area; up to 5 ft of mudstone are seen at several places in Oswald Beck, and 5 ft of skerry with mudstone bands are visible near the top of the cutting [7522 8260] east of Clarborough Tunnel. At Sturton-le-Steeple 6 ft of red and green mudstone are exposed [7853 8438] a few yards south-west of the Methodist Chapel in a small stream. The two streams that converge west of North Leverton, and the combined stream where it flows through the village, provide numerous sections, individually no more than 6 ft thick, in red and green mudstone with skerry bands up to 6 in. About 300 yd NW of Knaith 10 ft of Keuper Marl are exposed [8274 8487] in the bank of the Trent; the upper part consists of red mudstone, but the lower 4 ft are mainly grey and contain skerry partings and impersistent bands of gypsum and gypsiferous mudstone.

At Knaith there are excavations [8310 8486, 8315 8476], about 4 ft deep, into the Tea Green Marl; both are now grassed over but there is much green clay about. Tea Green Marl is exposed in the bank of a pond [8318 8454] 350 yd ESE of Knaith Hall, where greyish green mudstone can be seen close below Rhaetic shales, and in a ditch [8342 8366] 700 yd E of Red Hill, where 2 ft of green mudstone were visible at the time of the survey.

About 300 yd NW of Trent Port the downwash which covers much of the steep bank on the eastern side of the Trent flood-plain is patchy and Keuper Marl can be seen at the surface in places. Exposures [8324 8169] of red mudstone with green bands and patches may be seen at intervals over a vertical range of about 25 ft.

There are many exposures in Lee Beck, in and west of Treswell; the best [7757 7928], 630 yd WSW of the church, shows 7 ft of mudstone with skerry bands up to 2 in thick. Where the stream borders the churchyard [7810 7931] there are exposures of a 1½-ft skerry in red and green mudstone.   G.H.R.

The bottom part of this skerry (E 33527) consists of finely laminated cross-bedded units, in which bedding planes are disrupted and convoluted. The laminae consist mainly of finely and evenly granular (0·008 mm) dolomite, with dispersed, orientated mica flakes (chlorite, muscovite, illite, biotite) and intercalated or interspersed quartz silt (averaging 0·04 mm). Micas appear to be largely concentrated with the dolomite aggregates. Leucoxene is disseminated throughout. X-ray powder photographs (NEX 260–1) confirmed the main clay minerals and dolomite. A specimen (E 33528) from the middle of the skerry consists of finely cross-bedded laminations of dolomite and quartz silt (Plate XVII, figs. 1, 3), averaging 2 to 3 mm thick. In section the laminar structure is extremely delicate with weak grading, rupture, convolution and overfolds. Dolomite is evenly grained and averages 0·008 mm. There are laminae rich in oriented mica flakes (muscovite, chlorite, brown and green illite). An X-ray powder photograph (NEX 347) showed quartz, dolomite, feldspar, illite and kaolinite. The topmost part of the skerry (E 33529) is thinly bedded and finely cross-laminated, though containing a higher ratio of quartz to dolomite. Reddish brown marl from this exposure (E 33530) consists of chlorite, illite, quartz, dolomite and minor hematite (X-ray powder photograph NEX 890, by Mr. K. S. Siddiqui). R.K.H.

South-west of Treswell there are a few sections in North Beck [759 788], north of Bottom Woodbeck, which include an exposure of a 1-ft grey flaggy skerry.

South-east of Cottam 2 ft of rather broken red mudstone are seen [8354 7953] in the east bank of the Trent, 380 yd N of the railway viaduct over the river. Near Church Laneham, and 140 yd S of Trentfield House, 3 ft of red and green mudstone are exposed [8168 7730] in the west bank of the Trent.

On the southern edge of the district, exposures [8213 7377] in the bluff on the east bank of the Trent south of Dunham Bridge (Plate XVIB) show the following section:

	ft
Mudstone, red, with green bands ..	6
Skerry, with gypsum	1½
Mudstone, red, with rare grey bands..	10

10 yd to the south the two lower divisions have passed into:

	ft
Mudstone, red and·grey, with gypsum ribs	3
Mudstone, grey, with a few red partings	$4\frac{1}{2}$
Mudstone, red, with grey partings ..	6

These rest on:

	ft
Mudstone, grey, with red bands; impersistent band of interbanded gypsum and mudstone up to 1 ft thick at top .. .. .. 2 to	3
Mudstone, grey and red, with gypsum in ribs up to 3 in thick	3

G.H.R.

A specimen (E 33526) from the greenish grey skerry in this bluff, consists of delicately cross-laminated, festoon-bedded grains of dark, turbid dolomite, micas and conspicuous heavy minerals, averaging 0·04 mm across.

A heavy mineral separation was made of this specimen by grinding to < 80-mesh B.S.S., and mechanical panning. Mr. Siddiqui reports that the heavy crop obtained, ignoring ferric oxides, includes the following minerals in decreasing order of abundance: tourmaline, leucoxene, apatite, baryte, rutile, brookite, zircon, garnet, anatase. Though the proportion of dolomite is low, the specimen shows closely similar micro-sedimentation structures to that (E 33528) described from Treswell (see above). In both cases the bulk of the constituent minerals is almost certainly detrital. The laminae are intercalated with quartz silt (averaging 0·06 mm) containing, in addition to quartz, feldspars, chlorite, biotite and muscovite. In places the quartz grains are tightly interlocking.

An X-ray powder photograph (NEX 890) by Mr. Siddiqui of the red marl (MR 27780) above the skerry showed illite, chlorite, dolomite, hematite and quartz.   R.K.H.

Keuper Marl in boreholes. Of the few boreholes other than oil bores in the Keuper Marl, Carr Lane Borehole, Gainsborough, and Gainsborough Waterworks boreholes provide the thickest sections (about 700 ft, including the horizons of the Waterstones and Green Beds), and Rampton Hospital Borehole the most detailed record (see Lamplugh and Smith 1914, pp. 105–6, and also p. 183 of the present account). The

records of all of them show red and green or grey mudstone or marl with bands of skerry and gypsum, and in none are the Clarborough Beds distinguishable.

About 590 ft of Triassic rocks, probably an overestimate, were logged above the Pebble Beds in *Gringley* No. 1 Oil Bore, including an unknown thickness of Green Beds at the base (see p. 180). Anhydrite is recorded about 100 ft from the top of the succession and gypsum at several horizons in the top 225 and bottom 125 ft. *Walkeringham* Nos. 1 and 2 oil bores show 430 and 435 ft of Keuper Marl respectively, the former succession resting directly on Pebble Beds and the latter on 45 ft of Green Beds (see p. 180). Gypsum occurs throughout, with anhydrite at several horizons in No. 2 Bore. Chippings of red and green siltstone and silty mudstone with 20 per cent of gypsum and anhydrite recovered from a depth of about 50 ft in No. 2 Bore probably indicate the Clarborough Beds.

In *Morton* No. 1 Oil Bore about 525 ft of Keuper Marl are recorded. Traces of gypsum are recorded at many horizons, but notably in a band about 20 ft from the base, and associated with anhydrite in another band about 340 ft from the base. The latter occurs at the base of 30 to 40 ft of silty measures with gypsum, and the whole sequence probably represents the Clarborough Beds.

Beckingham oil bores show up to 565 ft of Triassic rocks above the Pebble Beds, which thickness, with the exception of the bottommost 30 ft (Green Beds) at No. 4 Bore, is all assigned to the Keuper Marl. Gypsum and anhydrite occur at various levels, but are particularly common at or near the base in Nos. 1, 2 and 5 bores, where they correspond with the anhydrite band in the Green Beds of No. 4 Bore (p. 180). In No. 2 Bore the anhydrite in hard brown siltstones 350 to 400 ft above the Pebble Beds may represent the Clarborough Beds.

Eleven of the *Gainsborough* oil bores give a complete section of the Keuper Marl, which, including a small thickness of Green Beds at the base in some cases (see p. 180), are between 825 and 890 ft thick. Several other bores start very near the top of the Keuper Marl, in some instances in the Tea Green Marl. The latter, up to about 25 ft thick (Fig. 31), is generally described as light green, grey or greyish green marl or silty mudstone, calcareous in some instances, and many contain bands of grey fine-grained sandstone. A pale greyish green dolomitic limestone is recorded at this horizon in No. 10 Bore.

Gypsum is present at many horizons in the Keuper Marl of the Gainsborough oil bores, but notably at three, where it is in each case associated with anhydrite. The first of these is at or near the base (sometimes the same band is within the Green Beds facies, see p. 180), where up to 10 ft of gypsum or anhydrite are noted. Salt pseudomorphs were seen at this horizon in No. 54D Bore. The second is 300 to 400 ft above the top of the Pebble Beds, where siltstone and gypsiferous mudstone are also recorded, apparently at the Clarborough Beds horizon. The highest is some 730 to 800 ft above the Pebble Beds, and includes an estimated 15 ft of gypsum and anhydrite in No. 46D Bore. This last horizon is evidently that exposed at Gainsborough station and worked at out-crop between Castle Hill Woods and the town (p. 239).

In the *South Leverton* Oilfield the bores show up to about 500 ft of Keuper Marl with variable thicknesses of Waterstones (p. 183) and Green Beds (p. 181) below. Gypsum is recorded at many horizons, but especially in Nos. 2, 4 and 13 bores at about 300 to about 400 ft above the Pebble Beds. This horizon shows a concentration of siltstone and sandstone bands, and is evidently the Clarborough Beds. Anhydrite was recovered from this level in No. 2 Bore.

In *Grove* Nos. 1 and 2 oil bores there are respectively 450 and about 480 ft of Triassic rocks above the Pebble Beds. Included in the former figure are about 60 ft of Green Beds and Waterstones, but these formations were not identified in No. 2 Bore and the whole thickness is assigned to the Keuper Marl. Gypsum, gypsiferous and anhydritic mudstone, and siltstone occur at several horizons, but the Clarborough Beds are not recognizable.   E.G.S

PALYNOLOGY OF THE GREEN BEDS AND KEUPER MARL

Material from the Green Beds and the Keuper Marl up to the horizon of the Clarborough Beds has been examined for miospores (Smith and Warrington 1971, pp. 212–8). The samples were obtained from three boreholes and seven surface localities; material from five of the latter proved barren. The Green Beds were sampled from Big Lane Borehole, Clarborough, at depths of 106 and 113 ft, Gainsborough No. 41 Oil Bore from 610 to 620 ft and Walkeringham No. 2 Oil Bore from 440 to 450 ft. Samples from the oil bores consisted of rock-bit cuttings.

The miospore assemblages obtained from the Green Beds were profuse and varied; the following miospores have been recorded from all the Green Beds assemblages:

Calamospora sp., Verrucosisporites jenensis, V. thuringiacus, Angustisulcites gorpii, A. klausii, Taeniaepollenites jonkeri, Alisporites grauvogeli, Colpectopollis ellipsoideus, Triadispora crassa and *Voltziaceaesporites heteromorpha.*

The following miospores have been observed in a number of Green Beds assemblages:

Punctatisporites triassicus, Scabratisporites scabratus, Apiculatasporites plicatus, Cyclotriletes oligogranifer, ?C. triassicus, Anaplanisporites protumulosus, Microreticulatisporites opacus, Cycloverrutriletes presselensis, Convolutispora cf. *wicheri, Rugulatisporites mesozoicus, Retitriletes jenensis, ?Lapposisporites armatus, ?Densoisporites caretteae,*

Lundbladispora nejburgii, Saturnisporites praevius, Podocarpeaepollenites thiergartii, Alisporites microreticulatus, Microcachryidites sittleri, Paravesicaspora planderovae, Sulcatisporites kraeuseli, Voltziaceaesporites nephrosaccus, Angustisulcites grandis, Taeniaepollenites hengeloensis, T. multiplex, cf. *Taeniaesporites novimundi, Lunatisporites puntii, Triadispora falcata, ?Jugasporites renalis, ?Illinites chitonoides, I. kosankei, ?I. trivisus, Cycadopites accerrimus* and *C. subgranulosus.*

The presence of the above miospores in assemblages from the Green Beds indicates that the Green Beds are of late Scythian age in the East Retford district.

A sample from the Keuper Marl at a depth of 370 to 380 ft in Walkeringham No. 2 Oil Bore yielded a sparse assemblage containing *Punctatisporites sp., Apiculatasporites plicatus, Verrucosisporites jenensis, Alisporites grauvo-*

geli, *?Colpectopollis ellipsoideus, Micro-cachryidites sittleri* and *Cycadopites sp.* Samples from the Keuper Marl at depths of 160 to 170 and 210 to 220 ft in the same bore yielded the following miospores:

Calamospora sp., Punctatisporites sp., Apiculatasporites plicatus, Cyclotriletes microgranifer, cf. *C. pustulatus, Verrucosisporites jenensis, Cyclogranisporites arenosus, Lapposisporites* cf. *loricatus, Microreticulatisporites sp., Lundbladispora?, Densoisporites sp., Saturnisporites praevius, Aratrisporites quadriiuga, A. saturni, Podocarpeaepollenites thiergartii, Alisporites grauvogeli, Voltziaceaesporites heteromorpha, Sulcatisporites kraeuseli, Chordasporites magnus, Colpectopollis ellipsoideus, Microcachryidites doubingeri, M. fastidiosus, M. sittleri, Triadispora crassa, T. falcata, T. plicata, Illinites chitonoides, I. kosankei, Angustisulcites gorpii, A. grandis, A. klausii, Tubantiapollenites balmei, Taeniaepollenites* cf. *hengeloensis,* and *Striatoabietites aytugii.* This assemblage is similar to those described from the Green Beds, but the absence of certain miospores occurring in that formation and the presence of a few additional forms are indicative of a younger (Anisian) age.

Material from an horizon in the Keuper Marl about 60 to 70 ft below the Clarborough Beds at Little Gringley [7338 8122] yielded the following miospores:

Leiotriletes?, Cyclotriletes margaritatus, C. cf. *microgranifer, Verrucosisporites morulae,*

Anapiculatisporites telephorus, Zonalasporites marginalis, ?Saturnisporites cf. *virgatus, ?Brachysaccus neomundanus, Illinites?, Angustisulcites* cf. *klausii, Alisporites parvus, A.* cf. *aequalis* and *Cycadopites accerrimus.*

The following assemblage was obtained from the Clarborough Beds at Askham railway cutting [732 748]:

Patinasporites cf. *funiculus, Podocarpeaepollenites thiergartii, Alisporites* cf. *aequalis, A.* cf. *grauvogeli, A. parvus,* cf. *Voltziaceaesporites sp., Scopulisporites minor, Brachysaccus sp., ?Klausipollenites devolvens, Platysaccus sp., Microcachryidites sittleri, Ovalipollis?, ?Illinites* cf. *chitonoides, Angustisulcites klausii, Triadispora crassa* and *T.* cf. *staplini.*

In Walkeringham No. 2 Oil Bore a sample from about 15 ft above the Clarborough Beds horizon yielded *Calamospora sp., Punctatisporites sp., Cyclotriletes?, ?Alisporites parvus, Minutosaccus potoniei* and *Lunatisporites kraeuseli.*

Many of the miospores recorded from around the horizon of the Clarborough Beds are known to occur in the Ladinian, and a number are also known from earlier deposits (of late Scythian and of Anisian age), but a small number (*Zonalasporites marginalis, Patinasporites* cf. *funiculus* and *Ovalipollis?*) are suggestive of a post-Ladinian age. As the last are very rare, however, the age of the Clarborough Beds is, at present, regarded as near the Ladinian–Carnian boundary.   G.W.

RHAETIC

A comprehensive account of the Rhaetic rocks in the East Midlands is given by Kent (1943, 1953, 1968, 1970). In the East Retford district they occur only in a small area east of the Trent to the north of Littleborough. Their outcrop is extensively obscured by drift, and information about them is virtually confined to a group of uncored oil bores at Gainsborough, and the railway-cutting section at Lea, now partly overgrown but described in detail by Burton in 1867. The Gainsborough oil bores show that the Rhaetic is from 37 to over 50 ft thick, comprising 20 to about 35 ft of Lower Rhaetic and 10 to more than 20 ft of Upper Rhaetic (Fig. 31). There is no discernible trend in the thickness variations in this small area.

The Lower Rhaetic, consisting largely of dark shales with *Rhaetavicula* [*Avicula*] *contorta,* is considered to be the approximate equivalent of the Westbury Beds, which form the major part of the Lower Rhaetic in southern England (Kent 1968, p. 174 and table 9). There is evidence of a stratigraphical break at the base of the *R. contorta* shales in Lea Cutting (p. 200), possibly representing the period during which the Sully Beds of Somerset were deposited. Kent (*ibid.,*

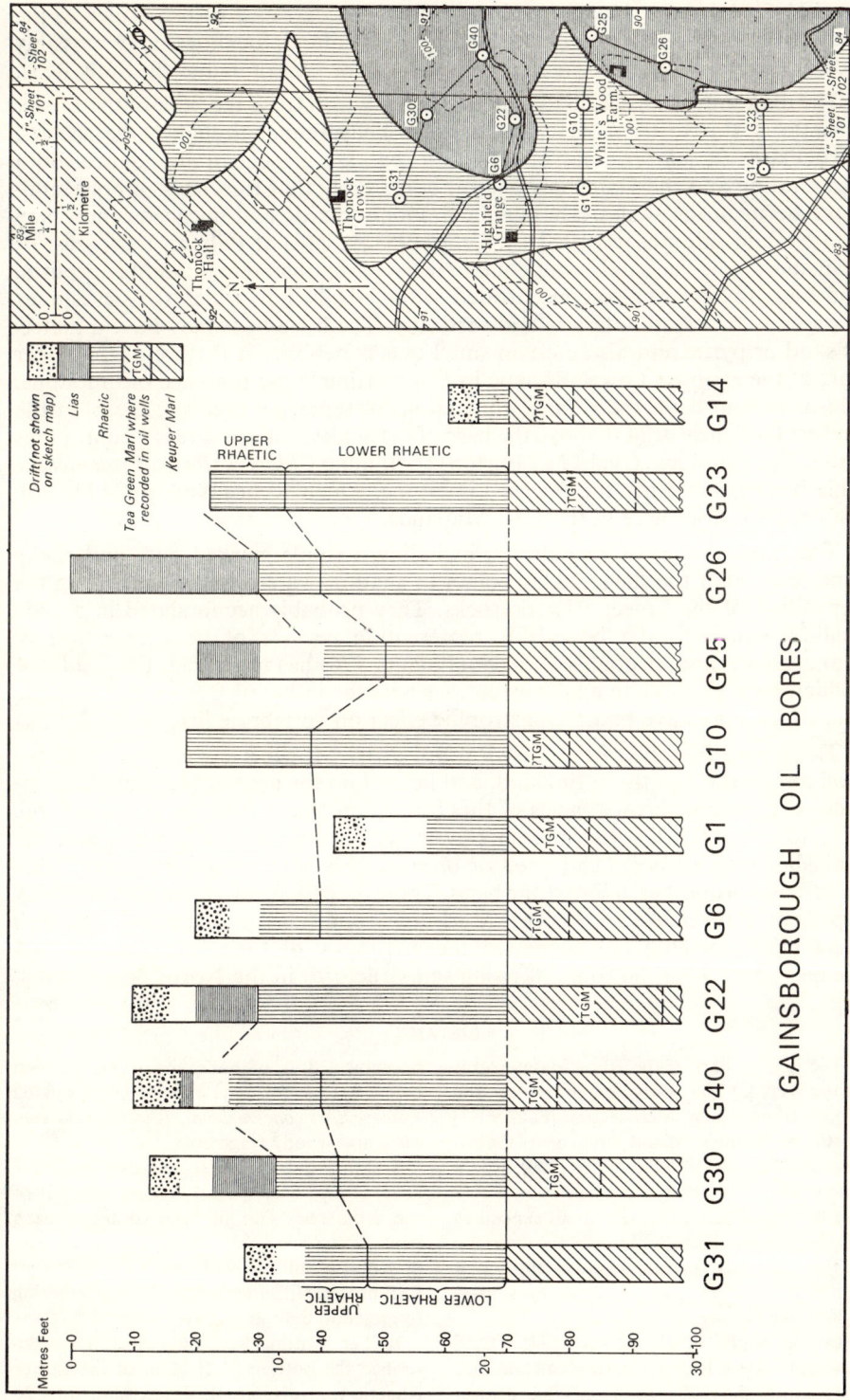

FIG. 31. *The Lias, Rhaetic and upper part of the Keuper Marl in the area east of Gainsborough*

pp. 174–5), however, has pointed out that the occurrence of fish scales in the Tea Green Marl at some more southerly Midland localities may possibly indicate an equivalence between this formation, or part of it, and the Sully Beds.

The dark grey to black, fissile, pyritic shales which form the bulk of the Lower Rhaetic in this district, have a restricted shelly fauna, though some species, particularly *R. contorta* and *Eotrapezium spp.*, are present in abundance. Both Lea Cutting and the Gainsborough oil bores show that thin beds of grey sandstone occur in the shales. These are commonly pyritic and micaceous and may also contain siderite, recorded by Kent (1953, p. 118) at Pilham, north-east of Gainsborough. The sandstones contain fish and reptilian remains and coprolites, and these fossils are also concentrated in bone-beds, into which the sandstones apparently pass both laterally and vertically. The bone-beds may have a matrix of sand or pyrite and also contain small quartz pebbles. A feature of the upper part of the exposed Lower Rhaetic in Lea Cutting is the presence of limestone. This occurs in the form of large concretions or septarian nodules, except at the highest level, nearly 24 ft above the base of the shales, where there is a continuous band, 1 to 3 in thick, (Bed 19 of Burton) containing *Chlamys [Pecten] valoniensis*. This band is evidently the *Pecten* Limestone, which Kent (1968, pp. 174, 176, 180) records elsewhere in the East Midlands.

The intercalation of sandstones, including a ripple-marked bed, and shelly limestone, with the dark shales points to shallow-water conditions during the deposition of the Lower Rhaetic rocks. They probably accumulated in a wide shallow sea, with the bone-beds representing periods of non-deposition of sediment or, more probably as Kent (1968, p. 176) has suggested, the results of sudden local changes in environment, such as the influx of fresh water into the sea, which may have had a catastrophic effect on vertebrate life.

The Upper Rhaetic of the East Retford district is generally equated with the Cotham Beds of southern England, and lies within the area of the local development of a reddish brown facies of this formation (Kent 1953, p. 119; 1968, pp. 178, 180, fig. 33). Greyish brown clay is seen in one poor exposure, and the Gainsborough oil bores show red or brown marls with, in one instance, sandy bands, and in another, a limestone band. These reddish brown measures probably represent a reversion to Keuper Marl-type conditions of deposition, though Kent (1970, p. 365) has suggested the possibility of their being redeposited Keuper Marl from the roofs of rising salt structures in the North Sea.   E.G.S.

DETAILS

The greater part of the Rhaetic outcrop is drift-covered. The only exposure of the Upper Rhaetic is a ditch section [8367 9031] north of White's Wood and near Gainsborough No. 10 Oil Bore (see p. 200), where a few feet of weathered greyish brown clay can be seen. There are a few small exposures of the Lower Rhaetic in addition to the section provided by the railway cutting at Lea which was described in detail by Burton (1867) (see below).

In the banks of the stream [8458 9219] flowing from the lake in Corringham Scroggs, little more than ½ mile E of the district boundary, 3 ft of dark shale can be seen below 8 ft of grey clay. The shale has yielded *Eotrapezium concentricum, Rhaetavicula contorta* and woody fragments.

Weathered dark grey shale is exposed [8371 8881] 800 yd S of White's Wood, and 2 ft of the same rock can be seen in the eastern bank of the pond [8318 8454] east-south-east of Knaith Hall. A ditch [8350 8347] 850 yd ESE of Red Hill, shows grey shale weathering to grey and light grey clay.

In Lea Cutting [834 866] Burton has described the bottom 25 ft 10 in of the Lower Rhaetic and divided the succession into 20 numbered beds. The cutting is now badly overgrown, and only the upper part of Bed 2 and Beds 3 to 9 are visible. Burton's section is given below, with details of Bed 1 and Beds 10 to 20 quoted direct from his account. Fossils collected during the survey have been identified by Dr. H. C. Ivimey-Cook who has also revised, where possible, Burton's palaeontological nomenclature. It has not been possible to interpret definitely *Schizodus cloacinus* (though it is probably *Eotrapezium concentricum*) and some of the vertebrate remains. Old generic names are given in square brackets.

Lea Cutting, near Gainsborough
(after Burton 1867)

Bed No.		ft	in
20	Shale, black fissile; *Rhaetavicula* [*Avicula*] *contorta*, '*Schizodus cloacinus*'	2	0
19	Stone (?limestone), dark grey, with veins of black fibrous gypsum [*sic*][1]; *Chlamys* [*Pecten*] *valoniensis, R.* [*A.*] *contorta, Modiolus minimus*, '*S. cloacinus*' and ? worm tracks 1 in to		3
18	Shale, black fissile, with nests of pyrite, horizontal veins of black fibrous gypsum [*sic*][1] and large septarian limestone nodules; abundant *R.* [*A.*] *contorta* and '*S. cloacinus*'	3	0
17	Sandstone, dark rubbly..	0	2
10 to 16	Shale, black fissile, with 0½-in pyritic sandstone 3 ft 2 in from base and 2-in bands of dark sandstone 4 in and 2 ft from base; *R.* [*A.*] *contorta* and '*S. cloacinus*'	4	8½
9	Sandstone, hard light grey laminated micaceous pyritic, with ripple-marks; *Eotrapezium* cf. *concentricum* and fish fragments including *Sphaerodus?*. Burton also records *M.*		

Bed No.		ft	in
	minimus, Perna?, Pullastra arenicola, R. [*A.*] *contorta*, teeth, bones, coprolites, ?worm tracks and drift wood	1	5
8	Shale, black fissile pyritic, with large flat limestone concretions; shale contains *E. concentricum, E. sp., Lyriomyophoria postera, M. minimus, 'Modiolus' sodburiensis, Protocardia rhaetica, R.* [*A.*] *contorta, Actaeonina oviformis?, 'Natica' oppelii*, fish fragments: concretions have yielded *E. concentricum?, E. sp.* and *R.* [*A.*] *contorta*	2	4
7	Sandstone, hard fine-grained micaceous and highly pyritic; *Gyrolepis alberti*, fish fragments and coprolites; Burton also records *R.* [*A.*] *contorta*	0	6
6	Shale, black fissile; *R.* [*A.*] *contorta* according to Burton	2	0
5	Bone-bed, soft, with sand matrix; coprolites ..	0	0½
4	Sandstone, soft grey micaceous; *Rhaetavicula?, Hybodus minor*, fish fragments. Burton also records *M. minimus, Pullastra arenicola, Nemacanthus sp.* and coprolites ..	0	4
3	Bone-bed, hard, with pyrite matrix and small quartz pebbles; *R.* [*A.*] *contorta, Acrodus minimus, Birgeria acuminatus, G. alberti, H. minor, Nemacanthus monilifer, Saurichthys* cf. *apicalis*, coprolites and fish fragments; Burton also records *Lepidotus* (*giebeli?*), *H. plicatilis*,		

[1]Almost certainly calcite "beef".

Bed No.		ft	in
	Sargodon tomicus, *Termatosaurus alberti* and *Ichthyosaurus sp*...	0	1
2	Shale, black fissile, with discontinuous layers of grey pyritic stone and nests of pyrite; Burton records: *R.* [*A.*] *contorta*, '*Schizodus cloacinus*' and coprolites ..	8	0
1	Sandstone, soft greenish grey, micaceous; Burton records *R.* [*A.*] *contorta*, *Pliosaurus?*, unidentified bones and teeth, coprolites. This bed fills hollows in the underlying surface of the Tea Green Marl average	1	0
	Mudstone, grey (Tea Green Marl)	—	—

Separation of Beds 3 to 5 is difficult, and, in view of the abundance of vertebrate remains and coprolites in Bed 4, the three together may perhaps be considered a single bone-bed.   G.H.R., F.G.S.

A sample of sandstone from Bed 9, examined in thin section (E 33533), is composed of poorly sorted clastics (0·03 to 0·6 mm) comprising quartz (chiefly igneous), microcline, orthoclase and micas, with abundant vertebrate remains (teeth, spines and scales) preserved in opaque to translucent collophane. Granular pyrite rims clastic grains and impregnates the sandstone. The cement is formed of fine-grained calcite, which has extensively replaced the clastic grains.

An X-ray powder photograph (NEX 894) of a specimen of black shale from Bed 8 taken by Mr. Siddiqui showed major chlorite and subordinate illite and quartz.   R.K.H.

Details of the Rhaetic rocks proved in the Gainsborough oil bores, the results of the examination of chippings by BP Exploration Co.'s geologists, are as follows (see also Fig. 31). *Gainsborough No. 1 Oil Bore* evidently started in drift overlying Lower Rhaetic, but there is no information about the top 15 ft. From 15 to 28 ft the bore passed through grey to black, slightly silty, pyritic shale

containing shell and fish fragments and with thin bands of greyish green, fine-grained, pyritic sandstone. These beds rest on Tea Green Marl. *Gainsborough No. 6 Oil Bore* shows a full section of Lower Rhaetic, 30 ft thick and consisting of dark silty shale and siltstone with thin bands of micaceous sandstone. The top 20 ft of the bore is partly in drift, but probably mostly in Upper Rhaetic, recorded as reddish brown, slightly micaceous marl, sandy or silty in places. *Gainsborough No. 10 Oil Bore* proved an incomplete section of the Upper Rhaetic, 20 ft thick, on 36 ft of Lower Rhaetic. The cellar of the bore, examined during the survey, showed 5 ft of greyish brown, slightly silty mudstone, the only other recorded part of the Upper Rhaetic being a thin basal greenish grey silty limestone recognized from chippings. Chippings from the Lower Rhaetic indicated dark grey to black, micaceous and pyritic, silty shale and siltstone with thin bands of pale grey, fine-grained calcareous sandstone. There is no record of the Rhaetic rocks in *Gainsborough No. 14 Oil Bore:* probably the basal 5 ft or so were penetrated beneath a similar thickness of drift. *Gainsborough No. 22 Oil Bore* gives a complete section of the Rhaetic, here about 40 ft, although no Upper Rhaetic rocks were recognized. The chippings showed only black silty shale with a thin band of fine-grained, silty sandstone at the base. The top 10 ft of *Gainsborough No. 23 Oil Bore* are presumed to be in Upper Rhaetic, though the chippings are atypically described as dark grey calcareous shale. The Lower Rhaetic consists of soft black pyritic shales in which traces of fossils were noted, and fine-grained white sandstones. The position of the base of the Rhaetic in this bore is in doubt. If taken at 65 ft as in the log the succession would be abnormally thick for the area; and it is thought more likely to lie between 45 and 50 ft. In *Gainsborough No. 25 Oil Bore* the Rhaetic is probably about 35 ft thick, it being impossible to fix the position of the Lias–Rhaetic boundary within 10 ft. The Upper Rhaetic was described as red to orange calcareous shale, resting on 20 ft of Lower Rhaetic recorded as dark grey to black pyritic siltstones. In *Gainsborough No. 26 Oil Bore* the Lias rests on about 10 ft of orange-brown shales, classed as Upper Rhaetic, which in turn rest on about 30 ft of Lower Rhaetic rocks similar to those described in No. 25 Bore, but with the addition in the lower part of pale grey, pyritic sandstone. In *Gainsborough No. 30 Oil Bore* the Lias is underlain by 10 ft of Upper Rhaetic, consisting of chocolate brown mudstones, which rest on almost 30 ft of Lower Rhaetic, consisting of black pyritic shales containing sporadic greenish layers. In *Gainsborough No. 31 Oil Bore* less than 20 ft of Upper Rhaetic, described as orange calcareous mudstone, were found below the drift and resting on 20 ft of Lower Rhaetic, recorded as dark grey to black finely micaceous siltstone. A complete succession of the Rhaetic must be present in *Gainsborough No. 40 Oil Bore*, but the upper boundary cannot be fixed. The Upper Rhaetic, probably less than 20 ft thick, is described as chocolate-coloured calcareous clay, and the Lower Rhaetic, up to 30 ft thick, chiefly as dark grey to black micaceous shales. It is not clear whether the light grey, fine-grained sandstone at the bottom of the succession should be included, or whether it is part of the Tea Green Marl. E.G.S.

REFERENCES

ANDERSON, F. W. 1964. *Aschemonella longicaudata* sp. nov. from the Permian of Derbyshire, England. *Geol. Mag.*, **101**, 44–7.

AVELINE, W. T. 1861. Geology of the country around Nottingham. *Mem. geol. Surv. Gt Br.*

—— 1880. The geology of parts of Nottinghamshire, Yorkshire and Derbyshire (Explanation of Quarter Sheet No. 82 NE). 2nd edit. *Mem. geol. Surv. Gt Br.*

BALCHIN, D. A. and RIDD, M. F. 1970. Correlation of the younger Triassic rocks across eastern England. *Q. Jnl geol. Soc. Lond.*, **126**, 91–101.

BUBNOFF, S. VON. 1935. *Geologie von Europa*, Bd 2, 717–92. Gebrüder Borntraeger, Berlin.

BURTON, F. M. 1867. On the Rhaetic Beds near Gainsborough. *Q. Jnl geol. Soc. Lond.*, **23**, 315–22.

CLARKE, R. F. A. 1963. British Permian and Triassic miospores. *Ph.D. Thesis*, University of London.

—— 1965. British Permian saccate and monosulcate miospores. *Palaeontology*, **8**, 322–54.

DEANS, T. 1950. The Kupferschiefer and associated lead-zinc mineralization in the Permian of Silesia, Germany and England. *Rep. XVIII Int. geol. Congr.*, (7), 340–52.

DUNHAM, K. C. 1960. Syngenetic and diagenetic mineralization in Yorkshire. *Proc. Yorks. geol. Soc.*, **32**, 229–84.

—— 1961. Black shale, oil and sulphide ore. *Advmt Sci. Lond.*, **18**, 284–99.

EDEN, R. A., STEVENSON, I. P. and EDWARDS, W. 1957. Geology of the country around Sheffield. *Mem. geol. Surv. Gt Br.*

EDWARDS, W. 1951. The concealed coalfield of Yorkshire and Nottinghamshire. 3rd edit. *Mem. geol. Surv. Gt Br.*

—— 1967. Geology of the country around Ollerton. 2nd edit. *Mem. geol. Surv. Gt Br.*

—— WRAY, D. A. and MITCHELL, G. H. 1940. Geology of the country around Wakefield. *Mem. geol. Surv. Gt Br.*

—— MITCHELL, G. H. and WHITEHEAD, T. H. 1950. Geology of the district north and east of Leeds. *Mem. geol. Surv. Gt Br.*

EISEL, R. 1909. Über die Varianten des *Productus horridus* Sowerby bei Gera. *Jb. Ges. V. Freund Naturwiss. Gera.*, 33–7.

ELLIOTT, R. E. 1961. The stratigraphy of the Keuper series in southern Nottinghamshire. *Proc. Yorks. geol. Soc.*, **33**, 197–234.

FIRMAN, R. J. 1964. Gypsum in Nottinghamshire. *Bull. Peak Distr. Mines hist. Soc.*, **2**, 189–203.

FORD, J. 1920. Record of the deep borings at Kelham and South Leverton. *Trans. Instn min. Engrs*, **58**, 94–107.

GEIGER, M. E. and HOPPING, C. A. 1968. Triassic stratigraphy of the southern North Sea Basin. *Phil. Trans. R. Soc.* (B), **254**, 1–36.

GIBSON, W. 1924. In *Summ. Prog. geol. Surv. Gt Br. for 1923*.

GRABAU, A. W. 1932. *Principles of Stratigraphy*. 3rd edit. New York.

GREBE, H. 1957. Zur Mikroflora des niederrheinischen Zechsteins. *Geol. Jahrb.*, **73**, 51–74.

—— and SCHWEITZER, H. J. 1962. Die Sporae Dispersae des neiderrheinischen Zechsteins. *Fortschr. Geol. Rheinld Westf.*, **12**, 201–24.

HIRST, D. M. and DUNHAM, K. C. 1963. Chemistry and petrography of the Marl Slate of S.E. Durham. *Econ. Geol.*, **58**, 912–40.

KENT, P. E. 1943. Upper Rhaetic Beds in North Lincolnshire. *Trans. Lincs. Nat. Un. for 1942*, 130–2.

—— 1953. The Rhaetic beds of the north-east Midlands. *Proc. Yorks. geol. Soc.*, **29**, 117–39.

—— 1968. The Rhaetic beds *in* Sylvester-Bradley, P. C. and Ford, T. D. (Editors) *The Geology of the East Midlands*. Leics. Univ. Press.

—— 1970. Problems of the Rhaetic in the East Midlands. *Mercian Geol.*, **3**, 361–73.

KIRKBY, J. W. 1861. On the Permian rocks of south Yorkshire; and on their palaeontological relations. *Q. Jnl geol. Soc. Lond.*, **17**, 287–325.

KLAUS, W. 1955. Über die Sporendiagnose des deutschen Zechsteinsalzes und des alpinen Salzgebirges. *Z. dt. geol. Ges.*, **105**, 776–88.

—— 1963. Sporen aus dem südalpinen Perm. *Jb. geol. Bundesanst.*, Wien, **106**, 229–363.

—— 1965. Zur Einstufung alpiner Salztone mittels Sporen. *Z. dt. geol. Ges.*, **116**, 544–8.

KRAUSKOPF, K. B. 1955. Sedimentary deposits of rare metals. *Econ. Geol.* Fiftieth Ann. Vol., Pt. I., 411–63.

KRUMBEIN, W. C. 1941. Measurement and geological significance of shape and roundness of sedimentary particles. *J. sedim. Petrol.*, **11**, 64–72.

KRYNINE, P. D. 1946. The tourmaline group in sediments. *J. Geol.*, **54**, 65–87.

LAMPLUGH, G. W., GIBSON, W., SHERLOCK, R. L. and WRIGHT, W. B. 1908. Geology of the country between Newark and Nottingham. *Mem. geol. Surv. Gt Br.*

—— HILL, J. B., GIBSON, W., SHERLOCK, R. L. and SMITH, B. 1911. The Geology of the country around Ollerton. *Mem. geol. Surv. Gt Br.*

—— and SMITH, B. 1914. The Water Supply of Nottinghamshire. *Mem. geol. Surv. Gt Br.*

LESCHIK, G. 1956. Sporen aus dem Salzton des Zechsteins von Neuhof (bei Fulda). *Palaeontographica*, **B100**, 122–42.

LOGAN, A. 1967. The Permian Bivalvia of Northern England. *Palaeontogr. Soc.* [*Monogr.*].

LOVE, L. G. 1962. Biogenic primary sulphide of the Permian Kupferschiefer and Marl Slate. *Econ. Geol.*, **57**, 350–66.

—— and AMSTUTZ, G. C. 1966. Review of microscopic pyrite from the Devonian Chattanooga Shale and Rammelsberg Banderz. *Fortschr. Miner.*, **43**, 273–309.

MACKINTOSH, A. H. G. 1938. *In* Report of field meetings for 1937. *Proc. Yorks. geol. Soc.*, **23**, 203–4.

—— 1939. Notable features in the Middle Permian Marls of South Yorkshire and North Nottinghamshire. *Proc. Yorks. geol. Soc.*, **23**, 282–301.

MAGRAW, D., CLARKE, A. M. and SMITH, D. B. 1963. The stratigraphy and structure of part of the south-east Durham coalfield. *Proc. Yorks. geol. Soc.*, **34**, 153–208.

MALZAHN, E. 1957. Neue Fossilfunde und vertikale Verbreitung der niederrheinischen Zechsteinfauna in den Bohrungen Kamp 4 und Friedrich Heinrich 57 bei Kamp-Lintfort. *Geol. Jb.*, **73**, 91–126.

MATLEY, C. A. 1914. Note on the source of the pebbles of the Bunter Pebble-Beds of the English Midlands. *Geol. Mag.*, Dec. VI, **1**, 211–5.

MITCHELL, G. H. 1934. Notes on the Permian rocks of the Doncaster district. *Proc. Yorks. geol. Soc.*, **22**, 133–41.

—— STEPHENS, J. V., BROMEHEAD, C. E. N. and WRAY, D. A. 1947. Geology of the country around Barnsley. *Mem. geol. Surv. Gt Br.*

NEWELL, N. D. and RIGBY, J. K. 1964. Geological studies on the Great Bahama Bank. *Spec. Publ. Soc. Econ. Pal. and Min.* No. 5., 15–79.

POTONIÉ, R. and KLAUS, W. 1954. Einige Sporengattungen des alpinen Salzgebirges. *Geol. Jb.*, **68**, 517–46.

RANKAMA, K. and SAHAMA, TH. G. 1949. *Geochemistry*. Chicago.

RICOUR, J. 1963. Problèmes stratigraphiques et caractères du Trias français. *Mem. Bur. Rech. Géol. Min.*, **15**, 19–28.

RITTENHOUSE, G. 1942. A visual method of estimating two-dimensional sphericity. *J. sedim. Petrol.*, **13**, 79–81.

ROSE, G. N. and KENT, P. E. 1955. A *Lingula*-bed in the Keuper of Nottinghamshire. *Geol. Mag.*, **92**, 476–80.

SCHAARSCHMIDT, F. 1963. Sporen und Hystrichosphaerideen aus dem Zechstein von Büdingen in der Wetterau. *Palaeontographica*, **B113**, 38–91.

SCHNEIDERHÖHN, H. 1923. Chalkographische Untersuchung des Mansfelder Kupfer-schiefers. *Neues Jb. f. Min.*, **47**, 1–38.

SCHOUTEN, C. 1946. The role of sulphur bacteria in the formation of the so-called sedimentary copper ores and pyrite ore bodies. *Econ. Geol.*, **41**, 517–38.

SCHUCHERT, C. 1915. The conditions of black shale deposition as illustrated by the Kupferschiefer and Lias of Germany. *Proc. Am. Phil. Soc.*, **54**, 259–69.

SCHWEITZER, H. J. 1962. Makroflora des niederrheinischen Zechsteins. *Fortschr. Geol. Rheinld Westf.*, **6**, 331–76.

SEDGWICK, A. 1835. On the geological relations and internal structure of the Magnesian Limestone, and the lower portions of the New Red Sandstone in their range through Nottinghamshire, Derbyshire, Yorkshire and Durham to the southern extremity of Northumberland. *Trans. geol. Soc. Lond.*, **3**, 37–124.

SHERLOCK, R. L. 1911. The relationship of the Permian to the Trias in Nottinghamshire. *Q. Jnl geol. Soc. Lond.*, **67**, 75–117.

—— 1926. A correlation of the British Permo-Triassic rocks. *Proc. Geol. Ass.*, **37**, 1–72.

—— 1947. *The Permo-Triassic Formation*. London.

SMITH, B. 1912. The green Keuper Basement beds in Nottinghamshire and Lincoln-shire. *Geol. Mag.*, **9**, 252–7.

SMITH, D. B. 1968. The Hampole Beds—a significant marker in the Lower Magnesian Limestone of Yorkshire, Derbyshire and Nottinghamshire. *Proc. Yorks. geol. Soc.*, **36**, 463–77.

SMITH, D. B. 1970 Permian and Trias *in* JOHNSON, G A. L. (Compiler). The geology of Durham County, pp. 66–91 in *Trans. nat. Hist. Soc. Northumb.*, **41**.

—— and FRANCIS, E. A. 1967. Geology of the country between Durham and West Hartlepool. *Mem. geol. Surv. Gt Br.*

SMITH, E. G. The Permo-Triassic rocks of the Worksop area *in* NEVES, R. and DOWNIE, C. (Editors) 1967. *Geological excursions in the Sheffield Region*. Sheffield.

—— RHYS, G. H. and EDEN, R. A. 1967. Geology of the country around Chesterfield, Matlock and Mansfield. *Mem. geol. Surv. Gt Br.*

—— and WARRINGTON, G. 1971. The age and relationships of the Triassic rocks assigned to the lower part of the Keuper in north Nottinghamshire, north-west Lincolnshire and south Yorkshire. *Proc. Yorks. geol. Soc.*, **38**, 201–27.

SMITH, W. CAMPBELL. 1963. Description of the igneous rocks represented among pebbles from the Bunter Pebble Beds of the Midlands of England. *Bull. Br. Mus. nat. Hist.* Mineralogy, **2**, No. 1.

SOHN, I. G. 1960. Palaeozoic species of *Bairdia* and related genera. *Prof. Pap. U.S. geol. Surv.*, 330A.

STONELEY, HILDA M. M. 1958. The Upper Permian flora of England. *Bull. Br. Mus. nat. Hist.* Geology, **3**, 295–337.

SWINNERTON, H. H. and KENT, P. E. 1949. *The Geology of Lincolnshire*. Lincoln.

TRECHMANN, C. T. 1925. The Permian Formation in Durham. *Proc. Geol. Ass.*, **36**, 135–45.

TRUSHEIM, F. 1963. Zur Gliederung des Buntsandsteins. *Erdöl-Z. Bohr-u. Fördertech.*, **79**, 277–92.

VERSEY, H. C. 1925. The Middle Permian rocks of Yorkshire. *Proc. Yorks. geol. Soc.*, **20**, 215–25.

WALL, D. and DOWNIE, C. 1963. Permian hystrichospheres from Britain. *Palaeontology*, **5**, 770–84.

WALTON, J. 1928. On the structure of a Palaeozoic cone-scale and the evidence it furnishes of the primitive nature of the double cone-scale in the conifers. *Mem. Proc. Manchr lit. phil. Soc.*, **73**, 1–6.

WARRINGTON, G. 1967. Correlation of the Keuper by miospores. *Nature, Lond.*, **214**, 1323–4.

—— 1970. The stratigraphy and palaeontology of the 'Keuper' Series of the central Midlands of England. *Q. Jnl geol. Soc. Lond.*, **126**, 183–223.

WESTOLL, T. S. 1941. The Permian fishes *Dorypterus* and *Lekanichthys*. *Proc. zool. Soc. Lond.*, **111**, 39–58.

WILCOCKSON, W. H. 1950. Sections of strata of the Coal Measures of Yorkshire. 3rd edit. *Midland Inst. min. Engrs.*

WILLS, L. J. 1910. On the fossiliferous Lower Keuper rocks of Worcestershire. *Proc. Geol. Ass.*, **21**, 249–331.

—— 1956. *Concealed Coalfields*. London.

WILSON, G. V. 1926a. The concealed coalfield of Yorkshire and Nottinghamshire. 2nd edit. *Mem. geol. Surv. Gt Br.*

—— 1926b. The thickness and distribution of certain members of the Permian System in Yorkshire and Nottinghamshire. *Summ. Prog. geol. Surv. Gt Br. for 1925*, 178–82.

—— 1927. The eastern boundary of the concealed coalfield of Yorkshire and Nottinghamshire. *Summ. Prog. geol. Surv. Gt Br. for 1926*, App. VII, 138–46.

WOODWARD, H. B. 1887. *The geology of England and Wales*. 2nd edit. London.

WOOLACOTT, D. 1919. Borings at Cotefield Close and Sheraton, Durham. *Geol. Mag.*, **6**, 163–70.

Chapter VI

JURASSIC

JURASSIC ROCKS are represented by two small areas of basal Lias, together totalling only some 60 acres, on the eastern margin of the district east of Gainsborough. There are no exposures within the district, for the greater part of the outcrop is covered by boulder clay, and information about these rocks is available only from the records and chippings of uncored oil bores and from exposures to the east.

Up to 30 ft of beds (the thickness proved in Gainsborough No. 26 Oil Bore, just east of the district boundary—see Fig. 31) are believed to be present, and these consist of hard, grey or white, fossiliferous flaggy limestone, apparently interbedded with grey clay. Gainsborough No. 30 Oil Bore showed only limestone, but other borehole records in the area suggest that clay may predominate in places. These beds are of 'Hydraulic Limestones' facies and probably belong entirely to the *Psiloceras planorbis* Zone (Wilson 1948, pp. 20–1; Swinnerton and Kent 1949, p. 40; Hallam 1968, p. 189).

Slabs of rock dug from ditches on or close below the dip-slope of the top of the limestones a few hundred yards east of the district boundary, near the northern end of White's Wood, have yielded a fauna including the following (identifications by Dr. H. C. Ivimey-Cook): echinoid spines and plates, *Liostrea irregularis*, *L.* cf. *irregularis* transitional to *L. hisingeri*, *Pteromya tatei*, *P. tatei altior*, *Gyrolepis?* (scale), *Saurichthys?* (tooth), but no ammonites were found.

Again a short distance east of the district boundary, but farther south near Marton, a collection of fossils from the Lower Lias was made by R. Tate in about 1875. Dr. Ivimey-Cook has re-examined the collection, which includes *Cardinia listeri*, *C. ovalis*, *Chlamys subulatus*, *Lucina limbata* and *Pseudolimea pectinoides*. This fauna suggests a slightly higher horizon in the Lower Lias—the *Alsatites liasicus* or *Schlotheimia angulata* zones—and probably came from beds overlying the Hydraulic Limestones.

The Hydraulic Limestones of this district are apparently very similar, both as regards lithology and faunal content, to beds of the same age in the Midlands and south-western Britain (Hallam 1968, p. 192). They were evidently deposited in shallow water under normal marine conditions. Whether the limestone bands were primary, or the products of diagenesis, or were partly primary and partly secondary has been the subject of some discussion (Kent 1936, 1957; Hallam 1957, 1960, 1964; Simpson 1957).   E.G.S.

REFERENCES

HALLAM, A. 1957. Primary origin of the limestone-shale rhythm in the British Lower Lias. *Geol. Mag.*, **94**, 175–6, 512–3.

—— 1960. A sedimentary and faunal study of the Blue Lias of Dorset and Glamorgan. *Phil. Trans. R. Soc.*, B, **243**, 1–44.

HALLAM, A. 1964. Origin of the limestone-shale rhythm in the Blue Lias of England: a composite theory. *J. Geol.*, **72**, 157–69.

—— 1968. The Lias *in* SYLVESTER-BRADLEY, P. C. and FORD, T. D. (Editors). *The geology of the East Midlands*. Leicester University Press.

KENT, P. E. 1936. The formation of the Hydraulic Limestones of the Lower Lias. *Geol. Mag.*, **73**, 476–8.

—— 1957. The limestone-shale rhythm in the British Lower Lias. *Geol. Mag.*, **94**, 429–30.

SIMPSON, S. 1957. On the trace-fossil *Chondrites*. *Q. Jnl geol. Soc. Lond.*, **112**, 475–95.

SWINNERTON, H. H. and KENT, P. E. 1949. *The geology of Lincolnshire*. Lincoln.

WILSON, V. 1948. East Yorkshire and Lincolnshire. *Br. reg. Geol. Inst. geol. Sci.*

Chapter VII

STRUCTURE

IN ADDITION to movements contemporaneous with deposition, the rocks of the district display evidence of tectonic activity at two principal periods. The earlier was the main phase of the Hercynian Orogeny during late Carboniferous times, and the later was of Tertiary (probably Miocene) date. Thus the Carboniferous rocks show Hercynian folding, superimposed upon which is the gentle easterly tilt of Tertiary origin (Figs. 32 and 33), and are affected by faults of both Hercynian and Tertiary ages; and the Permo-Triassic rocks show only Tertiary tilting with its associated minor warping and faulting (Fig. 34).

The movements contemporaneous with sedimentation during both Carboniferous and Permo-Triassic times apparently resulted in uneven subsidence, and are inferred from variations in the thickness of the deposits. In general, the amount of subsidence in the East Retford district increased towards the west and north during Carboniferous times, but, superimposed on this regional pattern, there was very marked local subsidence in the area of the south-easterly trending Gainsborough Trough, the effects of which are particularly evident in the Millstone Grit rocks (see Kent 1966, figs. 3–6; Howitt and Brunstrom 1966, figs. 4–5, pl. 32). During Permo-Triassic times there was an increase of subsidence towards the north-east (see e.g. Figs. 23 and 26).

It is apparent from a comparison of Figs. 33 and 34 that, even after allowance has been made for the original north-easterly slope of the Permo-Triassic basin, removal of the Tertiary tilt results in the regional dip of the Carboniferous rocks being to the west and north-west. This has been demonstrated by Kent (1966, pl. 20), who constructed a Permian palaeo-contour map of the whole of the East Midlands Coalfield. As Edwards (1967, p. 157) has suggested, it may well be that the oil now trapped in anticlinal areas of Carboniferous rocks in Nottinghamshire and Lincolnshire migrated eastwards up-dip from the deeper parts of the Carboniferous basin in pre-Tertiary times.

STRUCTURE OF THE CARBONIFEROUS ROCKS

Folding. In terms of the structure of the Carboniferous rocks, Edwards (1951, fig. 39) assigns practically the whole of the East Retford district to the south-western limb of a broad trough—the Maltby–Lincoln Trough—which lies between the major uplifted areas of the Askern–Spital Fold Belt to the north and the Eakring Anticline to the south. Recent exploration for coal and oil, however, has shown that superimposed on the general north-easterly dip are a number of appreciable folds, the axes of which are shown on Fig. 32. The major axes—the Maltby–Tickhill, Walkeringham–Gainsborough, Kiveton–Bothamsall and Whitwell anticlines—trend south-eastwards or eastwards, an alignment which is familiar in other parts of the East Midlands, while a number of minor axes have a north-easterly trend.

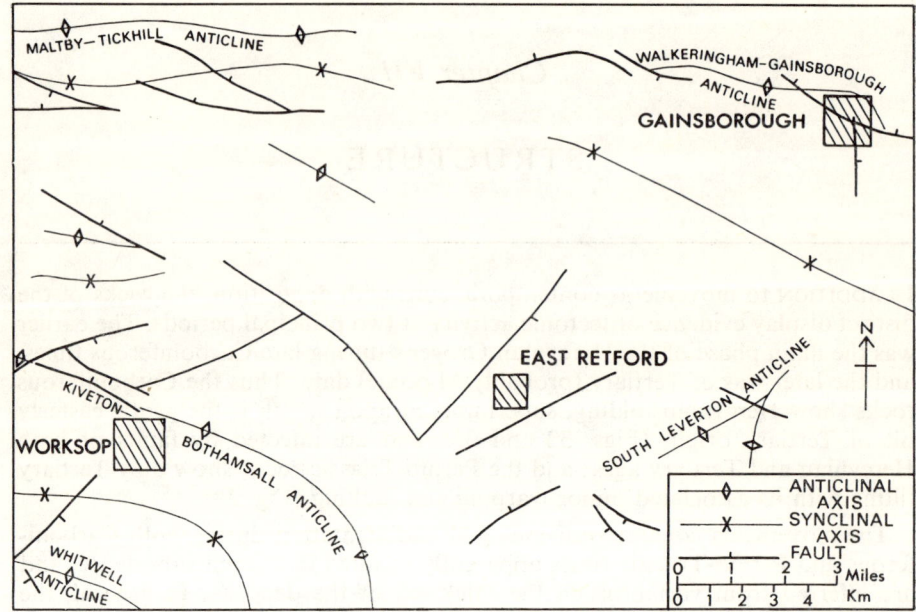

FIG. 32. *Structural sketch-map of the Carboniferous rocks, showing fold-axes and*
principal known faults

West of East Retford the details of the structure of the Carboniferous rocks
are illustrated by the contour map of the Top Hard or Barnsley Coal (Fig. 33).
In the north-west the anticline running west–east through Maltby Main Colliery
and Tickhill is a continuation of the structure through Maltby in the adjacent
Sheffield (100) district (Eden and others 1957, p. 150 and pl. vi), and is therefore
referred to as the Maltby–Tickhill Anticline. It appears to plunge out near
Bawtry, but its alignment with the Walkeringham–Gainsborough Anticline
(Fig. 32) suggests a link between the two structures.

The south-easterly plunging Kiveton Anticline, continued from the Sheffield
country (*idem*), runs through Shireoaks, Sunnyside and Rayton Farm, with
Worksop on its southern limb. The south-easterly plunge dies out in the vicinity
of Apleyhead, and from here the crest of the anticline rises again and curves on
to an almost southerly course through Normanton Hill and on into the Ollerton
(113) district. This southern culmination, although part of the same structure—
the Kiveton–Bothamsall Anticline—can be conveniently referred to as a
separate entity—the Bothamsall Anticline. It is important as an oil reservoir
(see p. 236). At Apleyhead itself there is a minor culmination on the axis, which
results in a small closure containing oil (see Brunstrom 1963, p. 5, fig. 4).

The dip of the Coal Measures on the north-eastern limb of the Kiveton–
Bothamsall structure is of the order of 7° for about a mile from the axis; it
then decreases suddenly to between a fraction of a degree and 3°. This gentle
north-easterly dip is maintained northwards to the Tickhill Anticline and
eastwards at least as far as East Retford, with slight easterly and north-easterly
flexures around Firbeck Main Colliery and Ranskill and in the area between
Scofton and Babworth. At Ranskill there is a low, closed, north-westerly
trending anticline which has been named the Ranskill Anticline by Howitt and

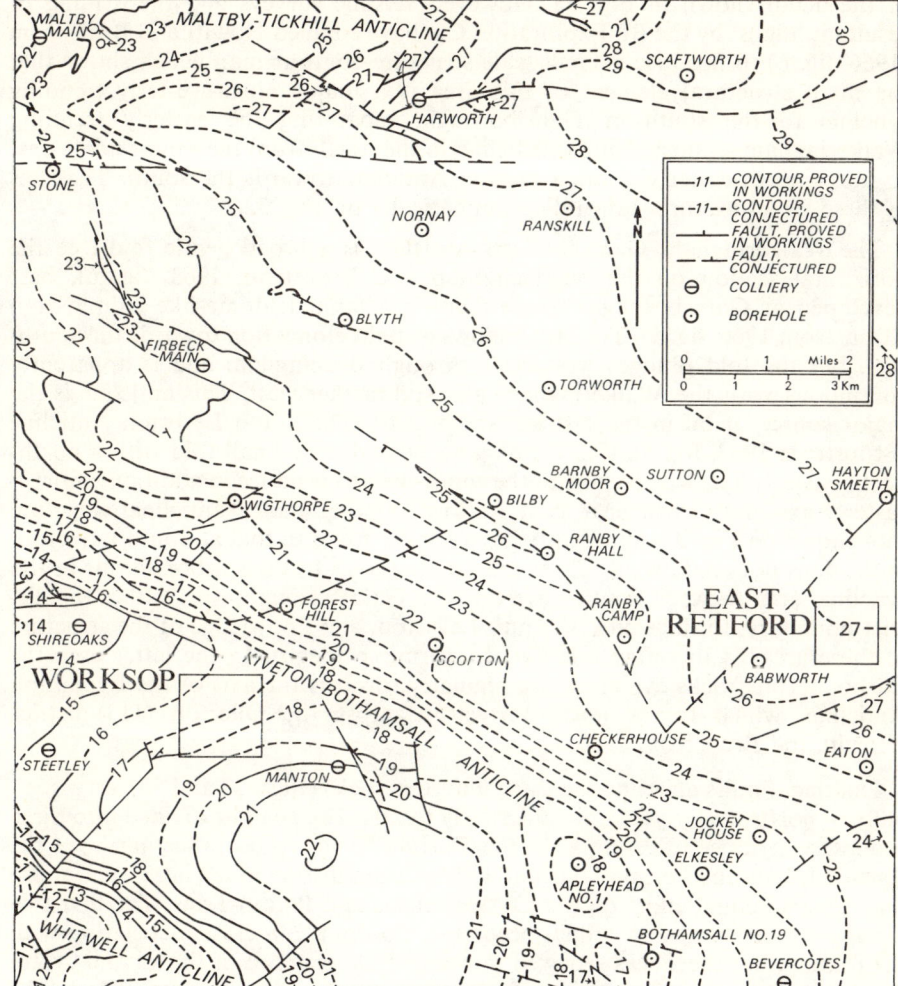

FIG. 33.   *Contour map of the Top Hard (Barnsley) Coal in the western part of the district*

Contours are in hundreds of feet below O.D.

Brunstrom (1966, p. 561, fig. 6). At Scofton there is a slight easterly trending anticlinal fold which swings north-eastwards towards Hayton Smeeth. South and south-east of this anticline is a gentle downfold through East Retford and a parallel low anticline through Eaton and Grove (see below).

To the south and west of the Kiveton–Bothamsall structure there is a shallow basin centred midway between Worksop and Clumber. The strata rise out of this depression over the axis of the south-easterly plunging Whitwell Anticline. Only a small part of the latter structure, named in the Sheffield (100) country to the west (Eden and others 1957, p. 150) lies in the district described here.

East of East Retford the absence of mine workings and exploratory boreholes for coal means that there is insufficient information to construct a contour map of the detail shown in Fig. 33. However, seismic surveys and the drilling of resultant 'highs' by the BP Exploration Co. have enabled Howitt and Brunstrom (1966, fig. 6) to produce a small-scale structure-contour map which shows that the main structural element of this area is a major north-westerly trending syncline to the south of Gainsborough, separating the easterly trending Walkeringham–Gainsborough Anticline in the north from the minor north-east to north-north-easterly South Leverton Anticline towards the south. The axes of these structures are shown diagrammatically on Fig. 32.

The Walkeringham–Gainsborough Anticline is a broad gentle fold (see the horizontal section on the one-inch map and Brunstrom 1963, fig. 6), best developed at Gainsborough, where there is a faulted, dome-like culmination (Brunstrom 1966, fig. 5). The latter shows a slight elongation towards the south-east, but the fold plunges westwards through Beckingham and is apparently continuous with the Walkeringham axis still farther west. This anticline is the major source of oil in the district (see p. 236). The South Leverton Anticline (Brunstrom 1963, fig. 5), also forming an oilfield, is a small fold with a north-north-easterly axis which splits to the south-west to produce in addition a north-easterly axis. At Grove near East Retford another minor anticlinal structure (not shown on Fig. 32) apparently has a similar trend to that at South Leverton, but it does not contain oil. There are no boreholes to the Carboniferous in the synclinal area between South Leverton and Gainsborough, nor, within the district boundary, to the south of South Leverton, and knowledge of the structure of these areas is therefore confined to seismic evidence. In the latter area the Carboniferous rocks evidently rise regularly towards the axis of the Egmanton Anticline, which trends north-westwards across the Ollerton (113) district (Howitt and Brunstrom 1966, fig. 6).

Faulting. Faults affecting the Carboniferous rocks (Figs. 32 and 33) are principally of north-westerly and north-easterly trends. They can be divided into three groups, viz: those affecting only the Carboniferous, those affecting both the Carboniferous and Permo-Trias but with reduced throws in the latter, and those that have an equal effect on the Carboniferous and Permo-Trias. The first two groups are of Hercynian origin, occurring, no doubt, in close association with the folding, but there was renewed movement along those of the second group during the Tertiary, to which date faults of the third group are restricted. All the faults are apparently normal, and they have small to modest throws. The biggest known faults, which as might be expected, are of Hercynian origin, are the two forming a trough north of Stone Borehole, those parallel to and crossing the Kiveton Anticline north of Worksop, the north-easterly fault crossing the Whitwell Anticline south-west of Worksop, and the north-westerly fault associated with the Gainsborough Anticline. The southern of the two 150-ft faults north of Stone is apparently the Thurcroft Fault of the Sheffield district (Eden and others 1957, pl. 6); it intersects Stone Borehole at a depth of about 2000 ft, cutting out the measures containing the Haughton Marine Band and Clown Coal (Plate VI). This fault has not been detected in the Permian rocks at the surface, but its northern neighbour has. The latter evidently splits above the horizon of the Top Hard Coal into two faults of opposite and approximately equal throw, forming a trough in which the Permian rocks are thrown down about 30 ft. The fault crossing the axis of the Kiveton Anticline throws a maximum of about 250 ft down on the south side, while the two faults forming a trough parallel to the axis have throws of up to 200 ft. In the Permo-Triassic rocks the former fault causes a displacement of up to 100 ft, and the two latter have throws of about 50 ft (Fig. 34). A fault of about 275 ft throw occurs in the Coal Measures in Ranby Hall Borehole, cutting out most of the strata between the Houghton Thin and Mainbright coals (Plates VI and VIII). It is assumed to trend south-eastwards as shown on Fig. 33, but has not been detected in the Bunter Pebble Beds at surface. The large fault affecting the Whitwell Anticline, throwing down to the south-east, has a maximum displacement of just over 200 ft; it does not appear to have affected the Permo-Triassic rocks. The northerly downthrowing Gainsborough fault has an estimated maximum displacement of more than 150 ft, apparently increasing north-westwards (Brunstrom 1966, fig. 5); this also has not been traced in the Permo-Trias.

Fig. 32 suggests that the intensity of both faulting and folding of the Carboniferous rocks diminishes eastwards. This, particularly as regards faulting, is in part a function of the limited information available in the east, but it has been pointed out by Kent (1966, p. 346) that structural complexity generally decreases eastwards across the East Midlands. He attributes this regional transition, not to an eastward reduction of orogenic stresses, but to a passage from unstable to stable shelf conditions.

STRUCTURE OF THE PERMO-TRIASSIC ROCKS

Contours on the base of the Permian rocks (Fig. 34) show a regional easterly or east-north-easterly dip of between 1 and 2 degrees. Contours on the top of the Bunter Pebble Beds (Fig. 34) run more nearly north–south and are more widely spaced than the Permian-base contours. These differences are a result of the north-easterly increase in thickness of the Permian rocks and the Pebble Beds. Slight sinuosity of the Permian-base contours in several places on Fig. 34, most noticeably around Ranby, could be indicative of faulting, for which, however, there is at present insufficient evidence. There appears to be some relationship between the sinuosity and spacing of the Permian-base contours and the more marked deviations of the Coal Measures contours (Fig. 33), e.g. the gentle swing of the former over the Kiveton Anticline north of Worksop, but there is otherwise little evidence of posthumous movement on Hercynian fold axes, a phenomenon that is more marked to the west (Eden and others 1957, p. 151, fig. 28, pl. 6; Mitchell and others 1947, p. 130, figs. 29 and 30) and south (Edwards 1967, p. 157).

Dips that are anomalous in direction or amount are seen from place to place in exposures of the Permo-Triassic rocks; some of these may be drag-effects from nearby faults and others may have been caused by superficial movements, of the cambering and valley-bulging type, of recent date. The most striking of the minor structures in the Permo-Trias, however, is the sharp fold which affects the Lower Magnesian Limestone and Middle Permian Marl south-west of Cotterhill Woods Farm, near Woodsetts. Here an old quarry [5567 8261] shows about 30 ft of bedded dolomite which is overlain by Middle Permian Marl (see p. 128) and dips at 20 to 30° to the north-north-west, representing one limb of an anticlinal fold or, possibly, a monocline (Plate XII). The quarry is on a marked ridge (up to 20 ft above the surrounding ground level) which continues west-south-westwards through the club house of Lindrick Common Golf Club and through Stubbings Lathe, and corresponds to the axis of the fold. This fold is therefore very similar in form and topographical expression to that described by Gibson and others (1908, pp. 108, 145) near Hucknall Torkard in the Derby

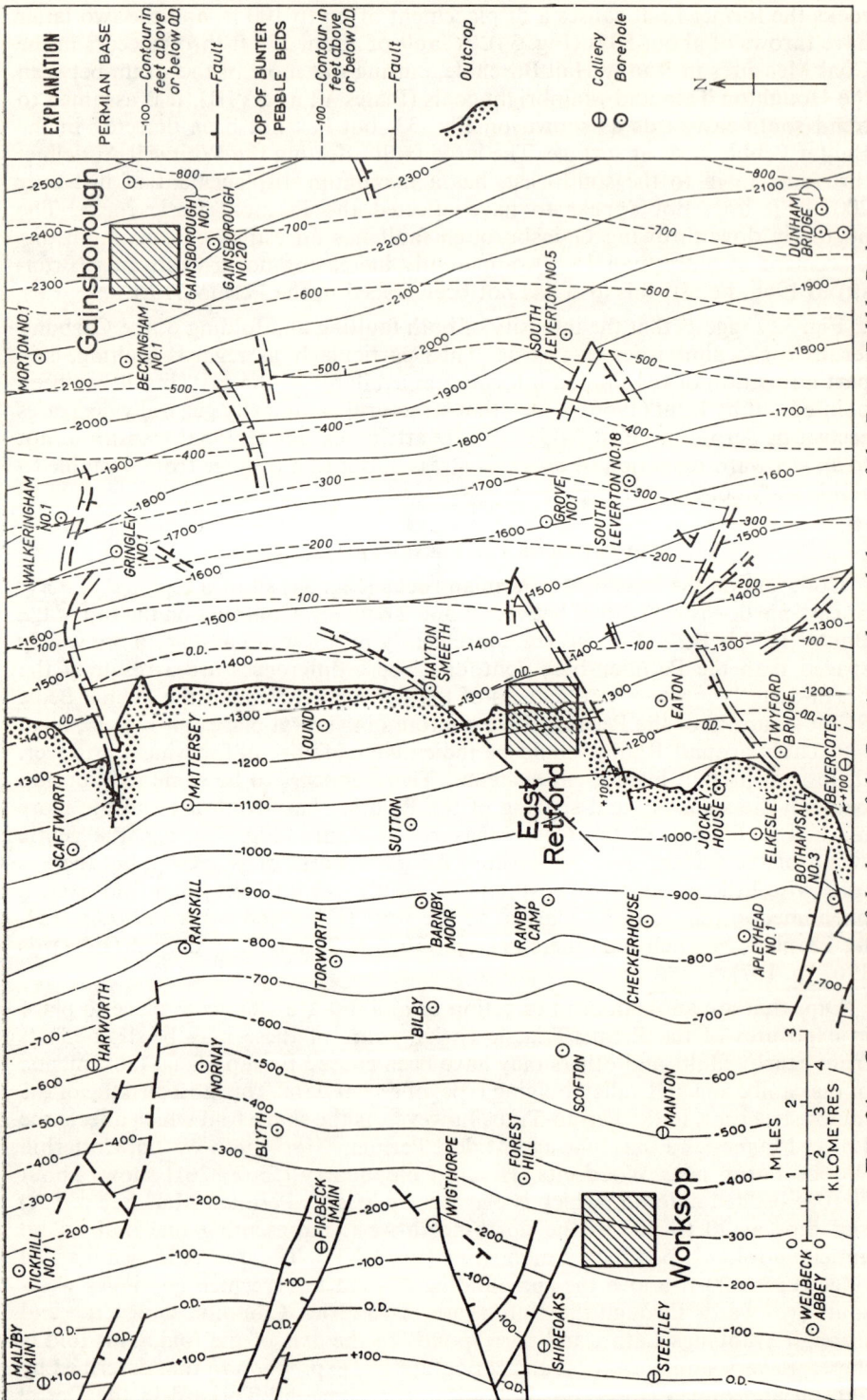

Fig. 34. *Contour map of the base of the Permian rocks and the top of the Bunter Pebble Beds*

(125) district, though the latter is directly related to the faulting in the Coal Measures below. In the Cotterhill Woods Quarry a south-hading fault was formerly visible near the axis of the fold, but this can only be a small fracture because it does not occur in the Top Hard Coal worked beneath. The nearest fault in the Top Hard projects to the surface some 200 yd to the south of the quarry and cannot be directly connected with the fold. This fault, however, is parallel or sub-parallel to the fold axis, and it seems likely that the stresses responsible for it also produced the fold. Small, ridge-forming anticlines of the same type occur to the north and south of the one described, both continuing westwards beyond the district boundary (Edwards 1961). The axis of the former runs from Sockeage Hill [5510 8319], where Lower Magnesian Limestone dipping gently to the north and south is exposed, through Lofties Plantation. That of the latter runs west-south-westwards through Moses Seat and Fan Field, and can be seen affecting the Lower Magnesian Limestone in the railway cutting [5365 8189] near Fan Field Farm.

Faulting. Because of the difficulty of recognizing faults at the surface on the wide outcrop of the Bunter Pebble Beds, or on the Keuper Marl where skerry bands are scarce, and the impossibility of detecting them beneath the mantle of drift in the Idle and Trent valleys, Fig. 34 provides only a partial picture of the faults affecting the Permo-Trias. It is believed, nevertheless, that the faults depicted are representative in that they are all normal, they have two general trends—north-easterly and north-westerly—and their throws nowhere exceed 100 ft and are usually 50 ft or less.

The mapping of surface faults is most accurate in the west because of the varied nature of the Permian succession, the exploration of the underlying Coal Measures and the general sparsity of drift there. A comparison of Figs. 33 and 34 for this area shows that faults affecting the Permo-Trias are also found in the Top Hard Coal. The two parallel faults in the Permo-Trias at Sandbeck constitute an apparent exception to this, but they throw down and hade towards each other and are inferred to have met in the Coal Measures above the Top Hard, which is affected only by the dominant northern one. A comparison of these text-figures also shows which faults are entirely of Tertiary age, and which were initiated in Hercynian times but were subject to posthumous Tertiary movement. Notable examples of the former group, affecting Permo-Triassic and Carboniferous rocks equally, are the faults near Harworth and Firbeck Main collieries, and of the latter group, which have much greater throws in the Carboniferous than in the Permo-Trias, the faults affecting the Kiveton–Bothamsall Anticline north-west of Worksop. Both groups of faults follow the same pattern.

Several faults form scarps, particularly in the north-west where Lower and Upper Magnesian Limestone are thrown against the Middle Permian Marl, but also in the north where the fault running along the Chesterfield Canal east of Everton forms a scarp of Keuper Marl overlooking the low-lying drift-covered Keuper Marl of Gringley Carr. Elsewhere faults have been detected by their displacement of mapped horizons, as is the case with the three north-easterly trending faults that conspicuously affect the outcrops of the Green Beds and Waterstones in the East Retford area.   E.G.S.

O

REFERENCES

BRUNSTROM, R. G. W. 1963. Recently discovered oilfields in Britain. *Wld Petrol. Congr.*, Frankfurt, Section 1, 1–10.

—— 1966. Indigenous petroleum and natural gas in Britain. *Proc. Joint Meeting of Inst. Pet. and Deutsche Gesellschaft für Mineralölwissenschaft und Kohlechemie e.V.*, 1965, 5–27.

EDEN, R. A., STEVENSON, I. P. and EDWARDS, W. 1957. Geology of the country around Sheffield. *Mem. geol. Surv. Gt Br.*

EDWARDS, W. 1951. The Concealed Coalfield of Yorkshire and Nottinghamshire. 3rd edit. *Mem. geol. Surv. Gt Br.*

—— 1961. Six-inch geological sheet SK 58 SW. *Geol. Surv. Gt Br.*

—— 1967. Geology of the country around Ollerton. 2nd edit. *Mem. geol. Surv. Gt Br.*

GIBSON, W., POCOCK, T. I., WEDD, C. B. and SHERLOCK, R. L. 1908. The geology of the southern part of the Derbyshire and Nottinghamshire Coalfield. *Mem. geol. Surv. Gt Br.*

HOWITT, F. and BRUNSTROM, R. G. W. 1966. The continuation of the East Midlands Coal Measures into Lincolnshire. *Proc. Yorks. geol. Soc.*, 35, 549–64.

KENT, P. E. 1966. The structure of the concealed Carboniferous rocks of north-eastern England. *Proc. Yorks. geol. Soc.*, 35, 323–52.

MITCHELL, G. H., STEPHENS, J. V., BROMEHEAD, C. E. N. and WRAY, D. A. 1947. Geology of the country around Barnsley. *Mem. geol. Surv. Gt Br.*

Chapter VIII

PLEISTOCENE AND RECENT

GENERAL ACCOUNT

APART FROM the considerable stretches of alluvial and terrace deposits associated with the Rivers Idle and Trent, drift does not occur abundantly in the East Retford district (Plate XVIII). Boulder clay is found in patches as evident remnants of once widespread deposits; it occurs chiefly on the higher ground, and includes a sandy facies on the outcrop of the Bunter Pebble Beds, which has been mapped separately. Glacial Sand and Gravel has a similar but more extensive distribution. In addition to these deposits, erratic stones, mostly quartzite pebbles and presumably of a remanié nature, are widely scattered over the drift-free areas. In the west of the district some of these pebbles, together with others lying on the surface of the Glacial Sand and Gravel, have been converted into ventifacts, a situation reminiscent of areas both to the north (Edwards 1936a) and to the south (Swinnerton 1914). Small areas of Head occur, chiefly in the bottoms of minor valleys, but also on the lower parts of slopes. Terrace deposits, mostly consisting of sand and gravel, flank and underlie the alluvium of the Rivers Idle and Trent; alluvium also occurs along the tributaries of these rivers. Associated with the alluvium and terrace deposits are areas of peat and small patches of calcareous tufa. Blown sand, derived from the Bunter Pebble Beds and sandy drift deposits, occurs in several areas and has been mapped locally.

The earliest record of Pleistocene events in the district is provided by the boulder clay, which, lying beyond the accepted limits of the Weichselian Glaciation (see e.g. West 1963, fig. 4; Penny 1964, fig. 1) and resembling the deposits of adjacent areas to the west and south (Eden and others 1957, pp. 152–9; Edwards 1967, pp. 159–60), is, like the latter, considered to be Older Drift, probably of Saale age. It is possible, however, that the clay filling the supposed sub-glacial channel at Harworth (see p. 220) and underlying the sandy boulder clay, may date from Elster times. The bulk of the boulder clay is apparently of local origin, though widely scattered Carboniferous and Permian erratics indicate ice movement from the north-west and west. In the north-east, near Gainsborough, the boulder clay contains numerous flints, derived from the east, and it would seem that in the East Retford district ice flowing off the Pennines on the west met ice moving in from the North Sea on the east. The deposits mapped as Glacial Sand and Gravel may be of more than one age, but the bulk of the high-level gravels, at least, is thought to have originated in late Saale times. In several places these deposits can be seen to overlie or to be intimately associated with the boulder clay, and they are interpreted as outwash material, derived essentially from the south because the contained pebbles are chiefly Bunter quartzites, too numerous and in many cases too large to have come from the local Bunter.

Much of the erosion responsible for the removal of the bulk of the glacial deposits of the district must have taken place during the Eemian (Ipswichian) Interglacial, following the retreat of the Saale ice. The only deposit that can be referred confidently to the Eemian Stage, however, is the small patch of gravel at Bawtry[1], shown on the map as Older River Gravel. It is possible that other gravels in the district may be the correlatives of this deposit, particularly the low-lying terrace-like Glacial Sand and Gravel that occurs in the Idle valley southwards from East Retford, but also some of the River Terrace deposits.

During the Weichselian Stage the district experienced periglacial conditions with accompanying frost-heaving of the pre-existing drifts, and it was probably then that most of the Head deposits were formed. Also in Weichselian times water, probably impounded by ice, must have covered the district up to approximately the 100-ft O.D. contour, a situation inferred from the existence of a 100-ft, locally-derived, strand-line deposit along the western side of the Vale of York between Tadcaster and Doncaster (Edwards 1936b). No strand-line deposits have been identified in the East Retford district, though it may be that some of the gravels occurring at or near the 100-ft level (see Plate XVIII) are in reality reworked glacial deposits along the beach that must have crossed the district.

Somewhat later in Weichselian times a lake at a lower level in the Vale of York resulted in the deposition to the north of this district of the 25-ft Drift (Edwards and others 1940, pp. 149, 155–6; Mitchell and others 1947, pp. 132, 138; Edwards and others 1950, pp. 65–6). It is believed that the First Terrace of the River Idle and probably that of the Trent are the valley equivalents of this Drift—indeed, that part of the terrace fringing the carr land north of Everton and Gringley on the Hill could be designated 25-ft Drift. The terrace deposits of the Idle valley have been subject to at least two periods of down-cutting which have resulted in the formation of flats at lower levels, and it is possible that this down-cutting has in places exposed and cut benches in gravels of pre-25-ft-Drift age (see p. 219). Similarly gravels at depth in the Trent valley may be older than the formation of the First Terrace of the Trent (the Flood-plain Terrace of some authors).

It has been shown that the original Trent flowed eastwards to the Lincolnshire coast, first through the Ancaster Gap and later through the Lincoln Gap (Jukes-Browne 1883; Swinnerton 1938). Swinnerton (*ibid.* p. 152, and *in* Swinnerton and Kent 1949, p. 105) suggested that the present course through the East Retford district is due to capture of the east-flowing Trent by a minor tributary of the Humber some time after maximum Saale Glaciation. Posnansky (1960, p. 305) suggested that this capture took place in Weichselian times, but the occurrence of boulder clay low down near the present river shows that this part of the Trent valley was already well developed at the end of the Saale Glaciation. It would appear therefore that the valley of the lower Trent either contained the Trent in pre-Saale times, or was that of a minor pre-Saale stream subjected to deep scouring by the Saale ice. That the latter is a possibility is supported by the evidence of extensive ice-scouring in central Lincolnshire, the area south-east of Gainsborough and the Fen country (Straw 1958) and in the Vale of Belvoir (Kent 1939). In either event it would appear then that the Trent was using its present route to the Humber in Eemian times, probably adopting it as soon as the

[1]Similar gravels occurring farther north have been dated by studies of the peat they contain (Gaunt 1966, p. 51, and see p. 218).

Saale ice had retreated to the east of Gainsborough (Straw 1963, pp. 177–9; King 1966, p. 54). This being so it is surprising that no Eemian deposits have been identified in the Trent valley within the present district, though it may be, as mentioned above, that some of the gravels at depth below the flood-plain and First Terrace may be of this age. It is also possible that the patches of gravel shown on the map as Glacial Sand and Gravel in the Laneham area (see p. 225) may be remnants of an old river terrace.

Evidently the Trent reverted to the Lincoln Gap during early Weichselian times when ice blocked the Humber and the Vale of York was flooded up to the 100-ft level (see above). As soon as the Humber Gap became passable again, however, the Trent again assumed its present course. The later Weichselian flooding resulting in the deposition of the Vale of York Drift probably meant that the Humber drainage was again impeded (Straw 1961, 1963) though this could have been only partial, as Straw (1963, p. 176) points out and as the grading of the First Terrace, believed to have been deposited at this time, shows.

There have been no physiographic changes of consequence since Weichselian times, and recent deposits, with the exception of small areas of blown sand, are confined to the river valleys.

Boulder Clay. Small patches of boulder clay occur throughout the district, but are least common on the outcrop of the Bunter Pebble Beds, where their place is taken by sandy till, a deposit which is considered separately below. As in the case of other glacial drifts, they tend to occur on the higher ground, indicating that they are remnants of a once widespread and since much-eroded sheet. At Harworth 15 to 20 ft of boulder clay are recorded (p. 220), a deposit which is thought to be the fill of a sub-glacial channel. Elsewhere infrequent exposures show only a few feet of clay, which is generally red or brown (particularly where it rests on Keuper Marl), and less commonly grey or mottled. The bulk of the erratics in the boulder clay are quartzite pebbles derived from the Bunter Pebble Beds, but Carboniferous and Permian cobbles and boulders are also found with, in the north-east, the addition of flints.

Sandy Boulder Clay. The term 'sandy boulder clay' has been applied to glacial till which is essentially composed of sand. Its incidence is confined to the outcrop of the Bunter Pebble Beds, where it has a patchy distribution (Plate XVIII) and usually occurs on the higher ground. It is interpreted as a locally derived ground moraine, its chief constituents being red or brown sand and quartzite pebbles which, overall, are not more abundant than in the underlying Pebble Beds. Local admixture with clay results in clayey sand or in pockets and patches of clay. Erratic cobbles and boulders, the latter composed of Carboniferous and Permian rocks indicating a westerly or north-westerly derivation, are widely distributed and in places abundant.

Apart from a 25-ft section near Bircotes which may be the fill of a channel in Bunter Pebble Beds (p. 221), the sandy boulder clay has not been observed to be more than about 10 ft thick, but exposures are sparse and, with one or two exceptions, poor. The paucity of exposures has led to mapping difficulties, especially as the soil produced by sandy boulder clay differs from that of the Bunter Pebble Beds only in being prone to have clayey patches and in containing scattered cobbles and boulders. Even in exposures it is not always easy to distinguish sandy boulder clay from Pebble Beds, or from Glacial Sand and Gravel which has been subjected to permafrost or solifluxion.

Glacial Sand and Gravel. Deposits mapped as Glacial Sand and Gravel are widespread but patchy, occurring chiefly on the high ground of the Bunter Pebble Beds outcrop, along the Idle valley and east of the Trent. They may be of more than one age (see pp. 215–7), but everywhere consist largely of varying proportions of sand and pebbles. Pebbles are abundant in certain bands of the high-level deposits on the Pebble Beds outcrop and have been worked as gravel, but they occur less commonly elsewhere and are sparse east of the Trent. The bulk of the pebbles consists of quartzites patently derived from the Pebble Beds, though some are too large to be of local origin; others are composed of a variety of rocks including sandstone, Keuper Marl, and igneous and metamorphic rocks. In addition to pebbles, there are angular and subangular quartz fragments, sporadic cobbles and blocks, mostly of sandstone, and in the north-west and east flints.

The deposits are false-bedded in some places and apparently horizontally bedded in others, and the gravel bands vary from well graded to only roughly graded. Beds of clayey sand and thin bands of clay occur locally.

The high-level deposits are the thickest; 50 ft or more are recorded at Gringley on the Hill and in the Broom Covert area, up to 40 ft at Barrow Hills, and 25 ft have been seen at Scrooby Top. The deposits along the Idle valley, which have a terrace-like form, are apparently thin, probably less than 10 ft in most places. There are few sections in the Glacial Sand and Gravel of the Trent valley: it is apparently thin in some places, but exceeds 20 ft in others.

Older River Gravel. A small area at about 30 ft above O.D. has been mapped at Bawtry. There are no exposures but the deposit is part of an extensive low-lying spread of cryoturbated gravel which extends northwards to the Hatfield area, and which has been shown by studies of contained peat near Austerfield to be of Eemian age (Gaunt 1966, p. 51).

It is possible that in some other parts of the district, deposits mapped as River Terrace or Glacial Sand and Gravel may be the southern equivalents of the Older River Gravel (see pp. 216–7).

Head. Small patches of Head, of no great thickness and occurring in minor valleys and depressions or on the lower parts of slopes, are widely scattered over the district. Some of the valley-bottom deposits may be partly alluvial in origin. Composition is very variable, ranging from clay to sand and gravel, reflecting the lithology of the local rocks, solid or drift, from which the deposits were derived.

River Terrace Deposits. These are readily divisible into the deposits associated with the River Idle and its tributaries and those associated with the River Trent.

The Idle deposits, occurring chiefly on the outcrop of the Bunter Pebble Beds, largely consist of sand and gravel, but north of East Retford they consist of clay where they overlie younger argillaceous rocks. This suggests that the deposits are locally derived, coming essentially from the west and east rather than from upstream. The sand and gravel, varying in thickness from a few feet to a recorded maximum of 33 ft, consists of fine to coarse, generally well bedded sand with subangular to rounded grains, containing bands and lenses of gravel. The gravel is usually poorly graded and consists chiefly of quartzite pebbles, the bulk of which must have been derived from pre-existing gravel deposits. The pebbles vary in size, generally ranging up to 4 inches in diameter, though larger specimens and sporadic cobbles occur; they also include wind-faceted examples. Included in the sand and gravel are thin seams, lenses and pockets of clay and seams of clayey sand. In the extreme north of the district, where the river enters the plain of the Vale of York, the terrace deposits apparently contain very little gravel.

Where the terrace deposits consist of clay this is generally brown or red, resembling reworked Keuper Marl; it is sandy in some places and pebbly in others, and has shelly and calcareous bands.

The Idle deposits occur on at least three levels up to 10 ft apart, but sections show that these are bench features cut in gravels originally filling the valley up to the level of the highest terrace. Deposits corresponding in age to the down-cutting that produced the benches are restricted to a thin skin of clay or sand overlying the gravels. The gravels, or at least their upper layers, are believed to be the time-equivalents of the 25-ft Drift of the Vale of York (see p. 216), but the lower layers may be older, possibly of the same age as the Older River Gravel (see pp. 216–8).

The Trent deposits, resting everywhere on Keuper Marl, form a terrace on the east side of the river north of Gainsborough and on the west side south of West Burton. This terrace, over 2 miles wide near Littleborough, is 10 to 20 ft above O.D., and its deposits also underlie much of the alluvium. Here and there on the flood-plain these underlying deposits protrude through the alluvium as low sandy mounds. These deposits can only be arbitrarily separated from the alluvium in boreholes: generally it has been the practice to draw the boundary at the junction of clays and silts above with sands and gravels below. The Trent terrace deposits are not so well known as those of the Idle: whereas the latter have been worked in extensive pits, there has been no attempt within the confines of the district to exploit the Trent deposits. Information about them is therefore limited to poor exposures and scattered boreholes for water, engineering purposes and geophysical shot-firing. Apparently they consist predominantly of sand with varying proportions of gravel. They extend down to 50 ft below O.D. and thicknesses of up to 50 ft have been proved. In places boreholes are suffi-ciently closely spaced to indicate that the sand and gravel fills an old channel or channels cut into the Keuper Marl.

The Trent Terrace is thought to have been formed, like the Idle terrace deposits, at the same time as the 25-ft Drift of the Vale of York, but some of the deeper sands and gravels described above may be older, possibly dating back to Eemian times (see p. 216).

Recent Deposits. These comprise calcareous tufa, alluvium, peat and blown sand.

Calcareous tufa has been mapped in two small areas, near Ordsall and East Drayton (Plate XVIII), respectively overlying terrace deposits and alluvium; it is also found locally within alluvium and peat. Its incidence is confined to the outcrop of the Keuper Marl and it has evidently been deposited around plants in shallow lakes, probably by lime-rich waters issuing from skerry bands (Lamplugh and others 1911, pp. 38, 61).

There are wide spreads of alluvium in the Idle and Trent valleys and narrow belts along minor streams. The Idle alluvium consists chiefly of clay and sandy clay with local sand, pebbles and beds of peat; the thickest section seen measured 8 ft. The Trent alluvium, up to 40 ft thick, consists of clay, sandy clay and silt with beds of sand and peat in places. The flood-plains of both Trent and Idle are subject to floods, some of which have been of disastrous proportions.

Surface peat occurs chiefly in the Tickhill area and in the lower part of the Idle valley. Thin in some areas, it exceeds 4 ft in others, and about 9 ft have been proved on Tickhill Low Common.

Blown sand, derived from the Bunter Pebble Beds and sandy drift, is widespread as small patches, but it has been mapped only in two places, both on the Keuper Marl outcrop—north-north-west of East Retford and east of the Trent near Dunham.   E.G.S.

DETAILS

BOULDER CLAY

In the western half of the district boulder clay, as distinct from sandy boulder clay, occurs only in scattered small patches at Harworth, Scaftworth, Firbeck, Steetley and Elkesley. At Harworth there are two separate deposits, above and below the sandy boulder clay (see p. 221). The upper deposit, a thin layer of brown clay, is found in the Lords Wood area [628 908], but it has not been found practicable to map it separately from the sandy boulder clay. The lower deposit is not now exposed, but Aveline (1880, p. 26) records the section in an extensive old brick pit [620 914], since filled in. Below 15 ft of sand with pebbles, he observed 15 to 20 ft of brown clay with a purple tint; it was nearly homogeneous and free from pebbles or rock fragments. This deposit is presumably the one referred to by Kendall and Wroot (1924, pp. 480–1), which they described as laminated with much contortion and eyes of sand, and which they inferred to have been laid down under lacustrine conditions. The form of the deposit, its low-lying position and its lithology, suggest a comparison with the fill of buried channels, thought to be of sub-glacial origin, in the Doncaster area (Smith 1963, p. 41; Gaunt 1965, p. 52).

Shallow excavations [6634 9214; 6635 9210] at Scaftworth reveal red clay with pebbles, and there are cobbles lying about on the surface nearby. The patch of boulder clay at Firbeck and the two in the Steetley area are not exposed and have been mapped on the evidence of the soil—a pebbly clay. At Elkesley in 1947 Dr. V. A. Eyles saw grey clay in the banks of a pond [6852 7509] and sandy clay with pebbles in two nearby trenches.

In the eastern half of the district boulder clay covers the high ground formed by the Lias, Rhaetic and topmost Keuper Marl east of Morton and Gainsborough. Scattered sections show up to 4 ft of brown clay, sandy clay and clayey sand with pebbles, cobbles and boulders. The cobbles and boulders are of sandstone, quartzite, limestone, basalt, dolerite and flint; small flints are common in the soil.

Elsewhere in the eastern half of the district boulder clay has been preserved in widely scattered small patches. There are five of these along the northern margin of the district between Everton and Walkeringham. They form low mounds with pebbly clay soil, and exposures in one of them [7030 9209] show 4 ft of red sandy clay with pebbles. Boulder clay underlies the western part of the Glacial Sand and Gravel at Gringley on the Hill (p. 224), as well records in the village show. It emerges at the surface on the south side of the village and extends southwards for almost three-quarters of a mile, covering the raised ground on which Topley Farm stands. It produces a brown silty clay soil with abundant pebbles—some formed from skerry, but mostly of Bunter-type—and a few small cobbles; there are no exposures. The two patches of boulder clay at Pear Tree Hill, Beckingham, the patch at Wheatley Grange and that at Saundby are not exposed, all having been delineated on the evidence of cobbles and abundant pebbles in the red clay soil. At West Burton, trial bores and excavations have shown up to 5 ft of boulder clay consisting of red and brown, rarely grey, clay and sandy clay with pebbles, resting on Keuper Marl.

Boulder clay is present beneath much of the Glacial Sand and Gravel in the northern part of East Retford (p. 225). There are no exposures at the present time, but in 1947 Dr. Eyles noted a number of sections in boulder clay overlying Green Beds. One [7050 8267] at Bolham, showed 2 ft of sandy clay with pockets of sand and a few pebbles;

another [7087 8166], off Moorgate, showed 6 ft of mottled brown and grey clay with a few pebbles and pockets of pebbly sand. In an old sand-pit [7088 8139], now filled in, off Spital Hill, 4 to 6 ft of mottled pebbly clay with pockets of pebbly sand were seen resting on Green Beds. An old record of the section at the disused brick and tile works [7086 8195] between Bolham and East Retford shows 6 ft of clay on Green Beds. This is presumed to be boulder clay, but the stiff, stoneless and unbedded bluish grey clay with faint pink mottling formerly seen in the old pit [7135 8055] at Newtown, East Retford, may well be the weathered top of the Green Beds. E.G.S.

There are two small patches of boulder clay capping low hills at Cowsland and Grovemoor farms between East Retford and South Leverton. A ditch section [7555 8032] near Grovemoor Farm showed 2½ ft of sandy clay with pebbles up to 9 inches in length. A patch of boulder clay about 1 square mile in extent at Rampton is thought to be thin, draping over the hill on which the Manor House stands. There are no exposures, but pebbles are much in evidence in some fields and brown clay and sandy clay are present in others. The boulder clay covering a low ridge of Keuper Marl north and west of Ragnall is a thin deposit of sandy clay with abundant pebbles. G.H.R.

SANDY BOULDER CLAY

The large expanse of sandy boulder clay around Bircotes in the north-west of the district produces a sandy or, in places, a sandy clay soil, with numerous pebbles and sporadic cobbles. The boundary separating it from Glacial Sand and Gravel to the north is arbitrary, and there is also evidence that patches of Glacial Sand and Gravel occur within the sandy boulder clay boundary. A thin layer of brown clay overlies the sandy boulder clay in the Lords Wood area [628 908] and is exposed in Wadsworth's Sand-pit (see p. 220). For convenience it has been included on the maps with the sandy boulder clay. This sand-pit [626 907] provides the best exposure of sandy boulder clay in this area and illustrates the difficulty of distinguishing weathered Bunter Pebble Beds from an overlying drift deposit largely derived from it. Two sections at the western end of the pit are as follows: (a) [6263 9072] brown clayey sand with a few pebbles 1½ ft, a layer of Bunter-type pebbles (up to 3 in diameter) 3 in, brown clay 6 in to 1 ft, on soft reddish brown sand with a few small pebbles in the top half, of which 2 ft are seen; (b) [6262 9067] brown sand with a few pebbles 1¼ ft, structure-less unsorted gravel consisting of subangular fragments of Permian limestone and Bunter-type pebbles with a sand (in places reddish brown sandy clay) matrix 3½ ft, on soft unbedded pebble-free reddish brown sand with pale patches 3 ft. Bunter Pebble Beds are seen 4 ft below the latter section. At the eastern end of the pit [6271 9066] the sandy boulder clay suddenly increases in thickness by cutting down into the Pebble Beds (Plate XIXA). Here there are 25 ft of reddish brown structureless sand with vague greenish patches; occasional pebble- and boulder-rich lenses show marked contortion. The boulders, which are ill sorted, subangular, and up to 18 inches in diameter, are composed of Permian limestone, Carboniferous sandstone, quartzite and basalt. Nearly everywhere at the eastern end of the pit the sandy boulder clay is overlain by a few feet of brown clay (see above). About 300 yd SE of Wadsworth's Sand-pit, sandy boulder clay is poorly exposed [6289 9042] in Dobb's Quarry; it contains pockets of small, well rounded pebbles of Permian limestone as well as many subangular boulders (up to 2 ft diameter) of the same rock. There is much red clay here, apparently occurring as lenses in the basal part of the sandy boulder clay. North of Bircotes, sections of sandy boulder clay were seen in trenches on a building site [6301 9215] near Plumtree Farm; these showed 4 ft of brown and grey sand, very clayey in places, with pebbles.

The small patch of sandy boulder clay adjoining the Ryton alluvium at Penny Acre [6392 9106], between Bircotes and Scrooby, is exposed in a ditch parallel to the river; 5 ft of red clayey sand with pebbles can be seen. Gravel workings at Scrooby Top show, in one area [6560 8919], that the Glacial Sand and Gravel (see p. 223) overlies unbedded reddish brown clayey sand with pebbles (mostly small, but a few large) scattered throughout. This deposit, of which the top 2½ ft were seen, is identified as sandy boulder clay.

Several patches of sandy boulder clay have been mapped in the area south of Mattersey, the largest being the one that includes Mattersey Hill. Sand-pits working Bunter Pebble Beds (p. 173) in Ellis Plantation [684 879] south-south-east of Mattersey Hill, show up to 3 ft of yellowish brown clayey sand with numerous pebbles and sporadic Permian limestone cobbles. The sandy boulder clay of Blaco Hill [696 880] is not exposed, but material from a deep excavation at the farm consisted of brown and red sand with lumps of clayey sand together with pebbles, cobbles and boulders.

Several patches of sandy boulder clay have been mapped in the Nornay–Torworth–Bilby–Barnby Moor area, where they form or cap low hills on the undulating surface of the Bunter Pebble Beds; there are no exposures worthy of note.

The largest expanse of sandy boulder clay in the district, some $2\frac{1}{2}$ square miles, is found in the Ranby–Babworth area, extending southwards to Morton Hill Farm. It has a sandy, pebbly soil with clay patches; sandstone blocks and boulders of sandstone and igneous rocks are common, especially in the vicinity of Green Mile Farm. Exposures are poor and sparse, viz: (a) A lane-cutting [6600 8173] 500 yd SSW of Green Mile Farm shows $2\frac{1}{2}$ ft of yellow sand with pebbles resting on $2\frac{1}{2}$ ft of reddish brown sand which contains pebbles (sporadic except in a 6-in layer), irregular lumps of cemented sand, a red clay lens and a basalt boulder (not definitely *in situ*). (b) A 4-ft excavation [6648 8166] in the wood 250 yd S of Green Mile Farm shows reddish brown sand with pebbles similar to the lower part of section (a) above. (c) The cutting [6506 8222] of the Chesterfield Canal near Ranby Barracks shows 6 ft of unconsolidated red silty clay, clayey sand and sand on Bunter Pebble Beds. (d) A 10-ft excavation [6568 8120] in the wood near Ranby House is largely overgrown but reveals red clay, green-mottled in parts and

sandy in places, near the top. (e) About 10 ft of brown pebbly sand is poorly exposed in the railway cutting [6693 7940] north-west of Little Morton. (f) An old clay pit [659 791], some 15 to 20 ft deep, covering several acres of ground east-north-east of Morton Hill Farm. The excavation is overgrown, but a 12-ft section of smooth, red and purple, laminated clay, with pebbly sand at the base has been described here. In 1947 Dr. Eyles found small exposures of reddish brown, grey-mottled, silty clay containing small stones, and recorded a number of boulders (up to $2\frac{1}{2}$ ft in diameter) lying about in the bottom of the pit. The latter included specimens composed of Upper and Lower Magnesian Limestone and Carboniferous Limestone and sandstones. The clay in this pit indicates either that the sandy boulder clay is locally almost a pure clay or that it is thin and underlain at this point by other drift. The latter circumstance is the most likely, and the clay recorded is possibly the fill of a buried channel similar to the occurrence at Harworth (see p. 220).

A small patch of drift at West Retford has been mapped as sandy boulder clay. It is exposed in an old sand-pit [6920 8147] near the railway line, where there is a 6-ft section of unbedded red clayey sand with pockets of pebbles and red clay.

There are several patches of sandy boulder clay scattered over about 2 square miles of the Bunter Pebble Beds outcrop west of Elkesley. A trench section [6686 7695], 700 yd SW of Top Farm, showed brown sandy soil with pebbles 1 ft, yellow sand with a few pebbles 1 ft, red sand with irregular inclusions of clay and sporadic pebbles 4 ft, on red sandstone. An old sand-pit [663 761] 400 yd SSW of Forest Farm, Elkesley, shows 5 to 7 ft of brown pebbly sand on Bunter Pebble Beds; the pebbles are mostly of Bunter type but include limestone. Trenches [638 747] south-west of Tank Wood in Clumber Park, examined by Dr. Eyles in 1947, showed up to

---

EXPLANATION OF PLATE XIX

A. The sandy boulder clay consists of reddish brown structureless sand containing ill-sorted boulders and pebbles which, in some parts, occur in lenses. The deposit has been worked for building sand.

B. The pebbles consist largely of quartzite, producing a high-quality gravel widely worked in this area. Sporadic lenses of sand can be seen in the deposit, which rests on Bunter Pebble Beds (showing dark) a short distance above water level.

(L 529)

A. Sandy Boulder Clay on Bunter Pebble Beds in sand-pit at Lords Wood, Harworth

PLATE XIX

B. Terrace gravel of River Idle, near Sutton

(L 535)

5 ft of brown, unbedded sand and gravel with some clayey sand towards the bottom, on red false-bedded sandstone. Pebbles up to 6 inches in diameter were observed and there were also rounded and subangular boulders (up to 1 ft in diameter) of black olivine-basalt. Signs of old clay pits just beyond the district boundary, together with the name 'Claypit Wood' on the 6-in map, suggest that this patch of sandy boulder clay is locally very clayey. R.A.E., E.G.S.

GLACIAL SAND AND GRAVEL

The Glacial Sand and Gravel shown on the map north of Bircotes is not exposed within the confines of the district, but is well seen in extensive workings around High Common Farm [6216 9375] just beyond the district boundary, where parts of it are rich in flints. The small patch of sand and gravel north-north-west of Harworth and some of the patches in the Scrooby–Serlby area contain small abandoned sand-pits, but these are overgrown and exposures show no more than a foot or two of pebbly sand. Some 3 to 4 ft of roughly bedded sandy gravel are exposed in the road cutting [6258 8923] in a small patch of Glacial Sand and Gravel 1 mile N of Nornay.

There is a large expanse of sand and gravel between Scrooby and Ranskill, which has been exploited by pits at Scrooby Top. At the time of the survey the pit on the west side of the Great North Road showed extensive, but largely obscured, sections, chiefly in false-bedded sand at least 25 ft thick. Gravel was worked in the southern part of the pit and can be seen in places along the abandoned southern face, near the Scrooby Top–Serlby road. At one point [6525 8888] 4 ft of well bedded and graded gravel with a few thin bands of almost pebble-free sand were overlain by 1 to 4 in of banded and contorted red clay. Some 75 yd to the SW [6519 8885] 2 ft of brown sand with abundant pebbles, showing bedding in the lower part, overlie 2½ ft of coarse false-bedded sand with small pebbles and angular and subangular quartz fragments, which, in turn, rest on 5 ft of false-bedded brown sand with reddish partings. A further 85 yd to the SW [6513 8881], 2 ft of brown sand with fairly numerous pebbles overlie 8 ft of brown sand, horizontally striped with bands of red clayey sand; 6 ft from the top of the section a band, 1 to 2 in thick, of red clay contains abundant pebbles. East of the Great North Road at Scrooby Top, active workings show a similar degree of variability in the sand and gravel. One measured section [6554 8912]

showed 10 ft of brown false-bedded sand with thin bands of gravel. Another [6560 8919] 100 yd away showed 4 ft of roughly graded gravel, 6 to 15 in from the base, with an undulating band, 3 to 9 in thick, of yellowish brown sand which becomes ochreous in places. In this area the Glacial Sand and Gravel is overlain on the low ground by terrace gravels and underlain in places by sandy boulder clay, but the succession is complicated by downhill passage of the Glacial Sand and Gravel into a frost-heaved solifluxion gravel in which the original bedding has been destroyed and surface clay incorporated, and which is similar in appearance to sandy boulder clay.

East of Scaftworth the prominence known as Barrow Hills, rising to some 140 ft above O.D., is capped by sand and gravel. This deposit probably does not exceed 40 ft in thickness—a shot hole [6744 9204] on the top of the western end of the hill recorded 39 ft of sand and gravel on sandstone, and Bunter Pebble Beds were seen by Aveline (1880, p. 25) in the sand-pits at the eastern end—but it extends down the slopes to below the 50-ft contour on the west and south. A tongue of sand and gravel connects the deposit of Barrow Hills with that forming the low hill between Scaftworth Hall and Ling's Wood. The sides of the sand-pits on Barrow Hills were much overgrown and covered by slip at the time of the survey, only a few feet of interbedded sand and gravel being visible here and there. The gravel consisted wholly of Bunter-type quartzite pebbles, but Aveline (idem) records "fragments of Magnesian Limestone, Coal Measures and other rocks, very little rounded". It is possible that the Carboniferous and Permian rocks, cobbles of which were seen lying about in the bottoms of the pits, may belong to sandy boulder clay underlying the Glacial Sand and Gravel. A quarry [6763 9173] on the margin of the Barrow Hills deposit, near the Bawtry–Gainsborough road and ¾ mile E of Scaftworth, showed 2½ ft of brown sand with abundant pebbles on 6 ft of unbedded red sand with scattered pebbles, some of which were of large size, on Bunter Pebble Beds. The bottom 6 ft of drift at this locality may well be sandy boulder clay.

Patches of Glacial Sand and Gravel form small hills, including Stone Hill [674 907] and Pusto Hill [696 900], in the Mattersey Thorpe area, and some of these contain old sand-pits. Exposures are poor, but shallow sections and the nature of the soil suggest that they consist largely of sand.

There are no exposures in the three low-lying patches of sand and gravel between Wiseton and Clayworth. They produce a brown sandy, pebbly soil and the northern-most patch has been penetrated by a number of shot holes which record 15 to 20 ft of sand.

The hill on which the village of Gringley on the Hill stands, and which is over 250 ft above O.D. at its highest point, is capped by Glacial Sand and Gravel, exposed in the old sand-pit [746 756] east of the village, where up to 17 ft of brown sand with pebbles are seen. The pebbles, which tend to occur in bands, are mainly of quartzite and sandstone, but include examples composed of red silty mudstone, presumably from the Keuper Marl. Gringley No. 1 Oil Bore, sited in the western end of this pit, proved 40 ft of sand and gravel, showing that the deposit attains a thickness of at least 50 ft. E.G.S.

South-east of Gringley on the Hill, and near the western edge of the Trent alluvium, there is a small patch of Glacial Sand and Gravel, about ¼ mile W of Bole, and another about ¾ mile ENE of Sturton le Steeple. The former produces a sand, or clayey sand soil, with pebbles in places, and shallow diggings show about 2 ft of brown sand. The latter forms a slight rise and old diggings show up to 1½ ft of brown sand with a few pebbles. G.H.R., E.G.S.

A few small patches of Glacial Sand and Gravel occur on the outcrops of the Lower Magnesian Limestone and Middle Permian Marl in the extreme west of the district, and there are numerous small patches on the Bunter Pebble Beds outcrop south of Rans-kill, two of which, east of Ranby, rest on sandy boulder clay. Two of the patches of sand and gravel concentrated in the Hodsock–Worksop area contain working sand-pits.

The patch centred on Broom Covert is exploited by two pits [608 834 and 610 832] showing, respectively, 7 ft of sand and gravel on 6 in of coarse gravel on Bunter Pebble Beds, and 20 ft of brown or yellow sand and gravel on Pebble Beds. The sand at the latter pit (Toulson's Quarry) is well bedded, commonly horizontally, but steep wedge-bedding is seen in places. Ripple bedding is common. Pebbles occur sparsely in thin bands, or compose thick beds of gravel, and sporadic cobbles are found. The pebbles are chiefly of quartzite, but specimens of gneiss, rhyolite, tuff, quartzitic breccia and arkose have been collected. As much as 50 ft of sand and gravel were originally present at this locality. R.A.E., R.F.G., E.G.S.

A mechanical analysis of a channel-sample through a 12-ft section of sand at this locality indicates a fairly low degree of sorting with the following quartiles: Φ_1 (25%) 0·37 mm; Φ_2 (50%) 10·28 mm; Φ_3 (75%) 0·21 mm. The sorting coefficient $\sqrt{\Phi_3/\Phi_1} = 1·76$. The coarser grains of quartz are mainly rounded to well rounded, while the predominant grades are subrounded to subangular. Quartz predominates (a modal analysis gave 75 per cent by volume of quartz and polygranular rock particles, mainly quartzite), with sub-ordinate alkali feldspars, mainly kaolinized orthoclase (15 per cent), and minor clay aggregates (5 per cent), ferric oxides (4 per cent) and heavy minerals. The last-named include zircon, rutile and tourmaline. R.K.H.

The sand-pit [6013 8218] north-west of Carlton Forest essentially works Bunter Pebble Beds below the Glacial Sand and Gravel, but the latter has been used for con-crete making. The section here is: brown sandy soil with pebbles 1 ft; 1 to 4 ft of yellow structureless sand with a variable pebble content, the base of which cuts down sharply in places into the underlying deposit; on pinkish brown bedded sand with pebble bands, false-bedded in parts, 6 ft, on Bunter Pebble Beds 30 ft. R.A.E., E.G.S.

Glacial Sand and Gravel forms a hill to the north-west of Bolham, near East Retford. Sections measured by Dr. Eyles in the eastern bank of the River Idle in this area are as follows: (a) [7010 8309] 10 ft of gravel on red sandstone; (b) [7016 8297] 6 ft 10 in of sand and gravel resting on an irregular surface of red sandstone. To the south-east this deposit is continuous with the terrace-like spread of gravel which extends, almost without interruption, down the eastern side of the Idle alluvium to the southern boundary of the district, rising from about 70 ft O.D. in the East Retford area to a little over 100 ft in the south. The soil overlying the latter deposit and shallow sections show sand or pebbly sand, but exposures are sparse. The deposit is evidently nowhere very thick. Two sections in the East Retford area measured by Dr. Eyles are as follows: (a) [7050 8267] at Bolham where up to 4 ft of pebbly sand and gravel rest on boulder clay; (b) [7087 8166] off Moorgate, East Retford, where 1 ft of sandy soil overlies 2 to 3 ft of loose brown sand which rests on boulder clay. Farther south a section [7084 7629] in the bank of the River Idle at Gamston shows 6 ft of unbedded silty sand with a few pebbles, and at Twyford Bridge, $\frac{2}{3}$ mile SE of Gamston, an old sand-pit [7034 7545] shows up to 4 ft of sand with pebbles, resting on Waterstones. Several small exposures around West Drayton show up to 5 ft of gravel or pebbly sand.

E.G.S.

Between this terrace-like spread of gravel and the Trent, Glacial Sand and Gravel is confined to small patches in the Grove and Laneham areas. The three patches in the former area, around Hutchinson's Holt, east of Castle Hill Wood and east-north-east of Grove, produce a sandy or sandy clay soil with abundant Bunter-derived pebbles and a few subangular sandstone blocks. Indications are that the deposit is thin, scattered ditch sections and auger holes showing only 2 to 4 ft of sandy clay and sand on Keuper Marl. The three patches around Laneham, north of the village, extending northwards from the Manor House and at Church Laneham, form slightly higher ground than the surrounding country and give rise to a sandy or sandy loam soil with many pebbles. The only exposures [816 772] are in the bank of the Trent north-east of the Manor House, where up to 2 ft of sand with pebbles rests on Keuper Marl.   G.H.R., E.G.S.

East of the Trent and south of Gainsborough, sand with scattered pebbles and rare flints covers much of the Keuper Marl and Rhaetic outcrops. A shot hole [8364 8711] 900 yd ENE of Lea proved sand to 30 ft, and two further shot holes [8366 8605 and 8356 8545] near Norbury Hills proved up to 20 ft of sand. An exposure [8354 7836] in a steep bank, bordering the alluvial flat, 300 yd N of the Foss Dike, shows about 2 ft of fairly well indurated reddish brown sand with rare flints. In the extreme south-east of the district, No. 2 Well at Dunham Bridge Pumping Station showed 5 ft of soil and loamy sand on Keuper Marl.   G.H.R.

HEAD

Head occupies the headwater valleys of several small streams on the western edge of the district, is found fringing the alluvium of the River Ryton at a number of places east of Scofton, and occurs in two small depressions in Clumber Park and one at Barnby Moor. The only sections in the western half of the district, however, are in the banks of the River Ryton [554 821] 1 mile SSE of Woodsetts, where Mr. W. N. Edwards has noted up to 3 ft of reddish brown and mottled clay and gravelly clay.

The alluvium occupying the shallow valley immediately east of Everton is flanked on its western side by Head, proved in excavations [6947 9129] to consist of 3 ft of sand with a few pebbles, resting on Bunter Pebble Beds. A patch of Head covering the slopes below the scarp of the Clarborough Beds north-west of Highfield Farm, Walkeringham, is exposed in an abandoned brick-pit [7548 9240]; it consists of 3 ft of unstratified red clay with pebbles which are particularly abundant in the bottom 1 ft.   E.G.S.

A patch of Head north of West Hill, East Retford, is exposed in an old sand-pit [6960 8013] north of the railway. Up to 6 ft of poorly graded, frost-heaved gravel containing rare sandstone cobbles rests on about 20 ft of Pebble Beds which have a very irregular top. Upward protrusions of Pebble Beds into the Head show contortion, and the drift pebbles are oriented with their long axes parallel to the margins of the protrusions. Inclusions of sandstone, up to 1½ ft across, also occur.

A strip of Head flanks the Idle terrace, masking the lower slopes of the Keuper Marl scarp, between Clarborough and Ordsall Hill. This deposit, derived from the Keuper Marl, produces a red or brown, pebble-free, clay soil; there are no exposures. There is a similar, but smaller, deposit east of Gamston.

R.A.E., E.G.S.

Elsewhere on the Keuper Marl outcrop, from North Wheatley southwards, Head is fairly widely distributed, generally occurring in the bottoms of dry valleys and the head-water valleys of small streams draining into the Trent. Scattered sections show up to 5 ft of red or brown clay or sandy clay with, in some instances, pebbles, which may be scattered throughout the deposit or concentrated in a layer at the base.

G.H.R., E.G.S.

Head occupies the broad, flat-bottomed valley below the Lias escarpment east of Gainsborough and extends northwards up a narrow valley to Highfield Grange. Sections in ditches up to 5 ft deep around Park Springs Farm [8355 8885] south of White's Wood, show a variable deposit consisting of red, brown or grey clay, and brown or grey sand. There is passage, both laterally and vertically, between clay and sand, and varying quantities of pebbles and flints occur in both. A few cobbles have also been seen.

G.H.R.

RIVER TERRACE DEPOSITS

River Idle and tributaries. There are no sections north of the Ranskill area, and information is available only from shot holes in the Bawtry–Everton area and from a study of the soil. The poorly defined stretch of terrace material flanking the carr land north-east of Everton varies between 12 and 16 ft above O.D., and overlies Keuper Marl. Surface indications are that the deposit is thin and consists of brown or red clay with scattered pebbles. North and north-west of Everton, where the terrace deposit rests on Bunter Pebble Beds with perhaps, in places, intervening boulder clay, shot holes show up to 12 ft of sand. There are no records east of Pasture Farm [6814 9286], but between that locality and Bawtry most of the scattered shot holes are reputed to have passed through sand and gravel, the maximum recorded thickness being 33 ft in a hole [6682 9293] 325 yd N of Scaftworth Grange. A shot hole [6585 9127] between Scaftworth and Scrooby is reported to have found sand and gravel to 20 ft beneath thin alluvial clay.

There is a large spread of gravel at 25 to 50 ft above O.D. in the Ranskill–Torworth area, forming two terraces, though the higher, 2 to 6 ft above the lower, has a restricted distribution. Workings ½ mile N of Danes Hill [674 865] show that the gravels immediately underlying the higher level are younger than those underlying the lower level. Ignoring the thin surface skin of clay that is present locally, the terrace flats are therefore essentially erosional features, apparently in one gravel deposit. Extensive pits and provings in the Ranskill–Torworth area show the gravel to vary between 3 and 14 ft in thickness, and to be 6 to 9 ft over much of the area.

Sections exposed in 1958 in the face of Messrs North Road Sand and Gravel Company's pit, ½ mile E of Torworth, demonstrate the lateral variability of the deposit: (a) [6666 8702] 1½ ft of clayey sand with pebbles rest on 5 ft of sand and gravel, which in turn rest on red sand (Bunter Pebble Beds). Some bands of sand are pebble-free, but pebbles are generally abundant and some lenses of gravel have no sand between the pebbles. There is no grading of the pebbles, which, with the exception of scattered cobble-sized specimens, range up to 4 inches in long diameter. (b) [6654 8702] 125 yd to the W, the gravel is 7½ ft thick with no overlying clayey sand. Extensive uptilting of pebbles was observed at this locality, in contrast to locality (a) (which is at the same level and is apparently a continuation of the same deposit) where all the pebbles lay with their long axes horizontal. There is no uptilting at locality (c) [6642 8677] where regularly alternating beds of gravel and almost pebble-free sand were seen to a depth of 4 ft.

There are several large gravel pits south-east and east of Lound, which have been worked by Messrs Lound Aggregates. Those south-east of the village are abandoned, being partly filled with surplus sand and flooded to within 4 ft of the surface. They were said to show about 4 ft of soil and subsoil (overburden) on an average of 14 ft of sand and gravel in which pockets of yellowish clay were found. A section [6992 8561] measured in 1948 by Dr. Eyles was as follows:

	ft
Dark brown sandy pebbly soil　0½ to	0¾
Well bedded sand and fine to medium gravel; some disturbance　..　2½ to	3

	ft
Contorted blue clay, weathering brown; almost stone-free; numerous rootlets .. .. 0 to	1½
Well bedded sand and fine to medium gravel, the gravel beds being lenticular with a number of dreikanter seen	7

The pits east of Lound, working in 1959, show a thinning of the sand and gravel and a marked decrease in the proportion of gravel to sand from west to east towards the river. This is illustrated by sections (a) [7023 8623] in the western and (b) [7050 8608] in the eastern pits:

	ft
(a) Sandy soil with pebbles	1½
Yellowish brown sand with, in bottom part, large lenses of brown and green mottled clay and thin lenses of gravel 2½ to	3
Gravel, roughly graded, with lenses of clay as above, and thin bands of sand	3½
Brown sand with thin bands of gravel .. .. .. 1 to	1½
Gravel, roughly graded, with thin lenses of sand	2
Brown sand and gravel	0¾
Not seen	1½
BUNTER PEBBLE BEDS	

	ft
(b) Brown sandy clay with pebbles (Alluvium)	2½
Brown sand with pockets and thin seams of clay .. .. up to	2½
Red sand with pockets and bands of gravel	4½
BUNTER PEBBLE BEDS	

E.G.S.

Between Sutton and East Retford the surface of the Idle terrace deposits is between 40 and 70 ft above O.D., occurring on three levels, the height intervals being up to 6 ft between the upper and middle levels and up to 10 ft between the middle and lower levels. Boreholes suggest that the deposits are up to 17 ft thick, but sections in gravel pits show no more than 15 ft. There are several large pits in the Bell Moor Wood area belonging to the North Nottinghamshire Gravel Company (Plate XIXв). A section [6969 8509] in one of the pits being worked in 1959 showed:

	ft
Sandy clay soil .. .. up to	1¼
Yellow sand with pebbles (a proportion of the pebbles have their long axes aligned vertically) .. 1 to	1¼
Gravel	2¾
Fine-grained brown sand with coarse bands; rare small pebbles ..	0½
Gravel	2
Obscured to standing water level ..	2½

Another section [7000 8463] 600 yd to the south-east showed:

	ft
Brown clay with a few pebbles ..	1
Yellow sandy clay	2
Gravel	4½
BUNTER PEBBLE BEDS	

The gravel at the latter locality is locally false-bedded with dips of 5 to 10°.

A section [6879 8404] in the 1959 working face of Bell Moor Pit showed 14 ft of gravel below about 1 ft of soil, with evidence of Pebble Beds at the bottom of the face. The gravel was poorly sorted with pebbles ranging up to 4 inches in diameter, and the interstitial sand had subangular to rounded grains. Lenses of sand and gravelly sand, up to 1½ ft thick, were seen in the gravel, and there are said to be clayey seams. Ferruginous and tufaceous cementing occurred in places in the top 4 ft. The feature separating the two lower terrace levels (see above), hereabouts 3 to 6 ft high, crossed Bell Moor Pit and, as at Danes Hill (see above), the gravel forming the higher level rests on the gravel forming the lower.

East of the River Idle in the Sutton–Hayton–East Retford area, the deposits occur at two levels, but the higher, found at the edge of the terrace to the north of Welham, is bounded on the west by only a slight feature. In this general area the composition of the deposits changes from essentially sand and gravel to essentially clay roughly at the boundary between the Pebble Beds and higher rocks. Ditch sections of terrace material overlying the higher Triassic rocks are as follows:

(a) [7139 8433] 250 yd S of Tiln Holt, brown silty clay 4 ft, on roughly bedded gravel 1 ft; (b) [7141 8410] 500 yd S of Tiln Holt, dark brown sandy clay with sporadic small pebbles 4 ft, on light brown and grey sand with ditto 1½ ft, on greenish grey clay (? Green Beds);

(c) a number of sections 900 to 1200 yd SSE of Tiln Holt show 4 ft of reddish brown structureless silty clay; (d) at Little Lane [7298 8349], Clarborough, 1¾ ft of shelly and calcareous clay with 6 in of soil rest on Keuper Marl.

Southwards from East Retford, to the east of the Idle and Maun, the terrace forms one flat, rising quickly to about 90 ft above O.D. The soil here is generally a clay or loam, except in the south, where sand is more in evidence. The only section of note is 10 ft of sand seen in the bank of the Maun [7117 7514] ¼ mile N of West Drayton.

The deposits extending up the Meden and Poulter valleys have been divided into First and Second Terraces, respectively about 5 ft and 15 to 30 ft above the alluvium. There are fairly extensive areas of both where the rivers join to the west of West Drayton.

The surface of the Second Terrace west of West Drayton consists of pebbly sand, and there is a 15- to 20-ft bank [690 751], apparently all in sand with pebbles, overlooking the River Poulter 450 yd SSE of Elkesley church. Knowledge of the deposit at depth is otherwise confined to a few shallow excavations and temporary sections. An old flooded gravel pit [6989 7419] on the eastern edge of Lawn Covert shows 3 ft of rudely bedded, poorly sorted gravel with pebbles up to 3 inches in diameter and a sandy or clayey matrix; the pit is at least another 4 ft deep below water level. A shot hole [7012 7452] on West Drayton Avenue recorded sand and gravel to 15 ft on marl and sand (evidently Waterstones), and West Drayton No. 1 Bore, 150 yd to the ESE, showed 10 ft of clay with pebbles on Waterstones. Surface excavations for No. 1 Shaft of Bevercotes Colliery showed 11 ft of sand and gravel on 6 in of sand on Waterstones. The sand and gravel (about half and half) is brown, with greenish patches which are mainly along vertical fissures. The gravel is virtually unsorted, with pebbles up to 3 inches in diameter, and shows a little contorted bedding in the lower part. Sporadic clay lenses occur. The basal bed of sand is brownish grey and follows irregularities in the underlying Waterstones surface.

Old pits [680 746] in the Second Terrace gravels at Broom Hill reveal 6 to 8 ft of rudely bedded sand and fairly coarse gravel, and higher up the Poulter valley another old pit [6398 7580] shows 6 ft of sand and gravel

underlying 1½ ft of pebbly sand soil and resting on Bunter Pebble Beds. The sand, showing some cross-bedding, occurs in bands up to 1 ft thick, and the gravel, consisting of pebbles up to 4 inches in diameter, is in lenses up to 1½ ft thick.

A shot hole [7049 7456] on the First Terrace east of the River Meden recorded sand and gravel, probably bottomed at 15 ft, and 6 ft of apparently level-bedded gravel were seen by Dr. Eyles in 1947 in an overgrown gravel pit [7051 7396] 600 yd to the south. Apart from ditches revealing a few feet of sand and gravel, there are no other sections in the First Terrace deposits of the Meden and Poulter valleys. Everywhere the soil produced by the deposits is a pebbly sand.

R.A.E., E.G.S.

In the Ryton valley there are small unexposed patches of terrace deposits near Serlby, and in the Bilby–Ranby area. The former occur 2 to 3 ft above the alluvium and produce a brown sandy soil with pebbles; the latter, 2 to 5 ft above the alluvium, have a loamy or loamy sand soil with pebbles and cobbles. The valley-bottom fill of the small stream flowing from near Hundred Acre Wood to join the Ryton north-west of Bilby, is included with the terrace deposits.

River Trent. North of Gainsborough, terrace deposits at the surface are confined to the east side of the Trent. Here they form a spread of sand with rare tiny pebbles, generally 10 to 20 ft above O.D. Sections are virtually non-existent but shot holes show about 18 ft of sand in the Morton area, which thickens northwards to about 50 ft at the district boundary. The surface of the terrace has been moulded by wind action, resulting in hummocks and ridges of blown sand (see p. 233) with intervening shallow, alluvium- and peat-filled hollows (p. 232). There is also an old peat-filled watercourse cutting across the terrace north-eastwards from Morton (p. 232).

Sand and gravel, regarded as terrace deposits (see p. 219), have been proved below alluvial clay and silt in a shot hole [7983 9133] south of Point Farm on the west bank of the Trent and in a borehole [810 901, exact site uncertain] at Trent Works, Gainsborough. The former hole failed to bottom sand and fine gravel at 60 ft (about 50 ft below O.D.), and the latter showed 11 ft of sand, gravel and 'shingle' on Keuper Marl at 43½ ft (25 ft below O.D.). Up to 14 ft of running sand were proved below warp and alluvial clay and silt at depths of 20 to 35 ft from the surface in three boreholes [811 900] at Albion Works, south-east of Trent Works. Other boreholes in the Gainsborough area show alluvium resting directly on Keuper Marl, and suggest that the sand and gravel deposits fill an old channel, or channels, following approximately the present course of the river.

E.G.S.

Much of Gainsborough is built on river terrace, which extends to the southern outskirts of the town. The deposits have been proved in one of the Gainsborough Waterworks boreholes [8161 8891], where 9 ft of sand were encountered resting on Keuper Marl. On the west bank of the Trent the eroded terrace deposits that underlie much of the alluvium come to the surface near the old rifle range in a low mound [802 891], approximately 400 yd long, on which there are a few inches of clay overlying the sand. Other such mounds occur farther south, on the right bank, on and in the neighbourhood of Lea Marsh. The largest of these stands several feet above the general level of the alluvium, and a borehole here [8139 8656] proved 50 ft of sand and gravel without bottoming the deposit. This depth approximates to a level of 35 ft below O.D., and neighbouring boreholes [8105 8850, 8065 8640] on Lea Marsh, have found the base of the deposit slightly more than 40 ft below O.D. These holes, together with numerous bores on the west bank of the river, drilled between West Burton and Bole Ings, prove the sand and gravel to be infilling a north–south channel, lying roughly beneath the present river course.

From Burton Round terrace deposits extend southwards to the Rampton area, occupying the greater part of the valley and reaching a maximum width of over 2 miles between Littleborough and Habblesthorpe. They give rise to a sandy soil with scattered pebbles, and in places, rare flints. North of the Roman road to Littleborough they are overlain in places by a foot or so of alluvial clay (not mapped), the resulting soil being a sandy clay. Boreholes along Upper Ings Lane have proved sand and gravel to 40 ft, resting on Keuper Marl and overlain by as much as 5 ft of clay. The terrace is bounded on its eastern side by a fairly well defined feature, up to 6 ft high, as far as Cottam, south of which the feature is indistinct. East of the feature, at Out Ings, the terrace deposits protrude through the alluvium, forming small hillocks of sand. Similar sand hillocks are present south of Cottam on both sides of the river, and boreholes in this general area prove that the base of the sand and gravel deposits is up to 47 ft below the surface. The sand varies in grain size from fine to coarse, and gravel, though not as common as sand, is generally present towards the base of the deposit. Some of the boreholes encountered bands of clay towards the middle of the sand and gravel.

South of Church Laneham a low terrace, only a foot or so above the alluvium, extends through Dunham, joining with a terrace in the north-east of the Ollerton (113) district. The deposit gives rise to a sandy soil with numerous pebbles, and a borehole [8110 7401] about 700 yd SW of St Oswald's Church proved sand and gravel to 20 ft. Beneath the alluvium, deposits of sand and gravel have been proved to a depth of 29 ft in shallow boreholes and at Dunham Bridge Pumping Station.

G.H.R.

CALCAREOUS TUFA

Two small areas of calcareous tufa have been shown on the map. The larger one [767 732], south-west of East Drayton, is the continuation of a deposit shown on the Ollerton (113) Sheet. It has been laid down on an alluvial flat by lime-rich water considered to be issuing from skerry bands in the Keuper Marl (Lamplugh and others 1911, pp. 38, 61). The top of the deposit lies between 9 in and 2 ft below the ground surface, and fragments of buff or light grey tufa are commonly ploughed up through the overlying alluvial clay. The tufa appears to have been built up over a framework of plant fragments, mainly reeds, which have rotted away leaving only calcareous casts.

G.H.R.

The deposit [720 789] south-east of Ordsall overlies Idle terrace, but skerry-bearing Keuper Marl crops out nearby on three sides, and the theory of tufa formation given above can be invoked. Ditch sections show the deposit to be at least 4 ft thick, and in the east, towards Low Farm, there is an admixture of organic material.

P

In addition to the mapped deposits described above, thin layers of calcareous tufa occur in the Idle alluvium and terrace deposits between Clayworth and Hayton Castle, and lumps of soft, white, crumbly tufa containing plant remains are to be found in the peat north-east of Morton (see p. 232).   E.G.S.

ALLUVIUM

There is alluvium, to a large extent covered by peat, in the Tickhill Low Common and Styrrup Carr areas west of Harworth. This occupies the low-lying ground around the headwaters of the River Torne, and consists of evidently thin, grey or dark clay and pebbly sand. The Torne flows northwards towards Doncaster and eventually joins the Idle near Epworth. There is a small patch of alluvium north of Welbeck Abbey in the valley of Walling Brook which drains south-eastwards to join the Poulter beyond the district boundary.

Other deposits of alluvium are associated with the Idle or Trent or tributaries which join them within the district boundaries, and are most easily described under the two river systems.

River Idle and tributaries. Alluvium, usually consisting of grey, brown or yellow clay, but, in places, of sandy clay or sand, underlies most of the peat of the carr land north of Everton and Gringley on the Hill. It is found at the surface in small areas where peat is absent or has been burnt off and in areas of thin peat subjected to deep ploughing. It is also seen in drains cut through the peat, and has been encountered in auger holes. Included in the area mapped as alluvium is a patch of thin shell-marl on the northern boundary of the district, north-west of Dunstan Farm, Gringley on the Hill. Gastropods from a similar patch nearby, but just over the district boundary, have been examined by Mr. R. V. Melville, who recognizes the following fresh-water forms: *Bithynia tentaculata* (Linné), *Planorbis leucostoma* Millet and *Limnaea spp.* including *L. palustris* (Müller). Associated with these in the deposit is the terrestrial gastropod *Succinea putris* (Linné).

There are no sections in the Idle alluvium between Bawtry and Clayworth. Where not covered by peat, which occupies much of the surface of the flood plain in this area and extends as far south as Tiln Holt, the alluvium produces a brown clay, sandy clay, or peaty clay soil. Sections in the alluvium flanking Toft Dyke Drain, a small tributary of the Idle east of Clayworth, show up to 8 ft of red or brownish grey clay with pebble bands. West of the river in the Clayworth–Lound area, scattered exposures show the alluvium to consist of brown sand or sandy clay with pebbles. Between Clayworth and Hayton Castle ditch sections show up to 8 ft of red, grey or brown clay, in which gastropods are common and which contains thin layers of peat, pebbly sand and calcareous tufa. Southwards from the Lound–Hayton Castle area to East Retford, sections in the Idle alluvium show 2 to 4 ft of brown or grey clay and peaty clay with, in places, bands of peat and sandy clay. From East Retford to West Drayton, where the Idle originates at the confluence of the Poulter and Maun, brown or bluish grey clay is seen at the surface on the flood-plain in many places. There are old sand and gravel pits [707 798], now waterlogged, at Ordsall, and sand and gravel have been recorded in a shot hole [8070 7793], at Eaton, and by Dr. Eyles in a 6-ft section [700 757] between Gamston and Twyford Bridge. It is not clear, however, whether these gravels are of recent alluvial origin, or whether they are part of the terrace-like gravels (see p. 228) which flank the flood-plain on the east and must underlie much of it.

The Idle alluvium is subject to floods, the major of which have been recorded since 1683 (Nixon 1960, pp. 11, 15, 67–9) and can be correlated with flooding of the Trent valley (p. 231). Notable floods in this century occurred in 1910, when East Retford suffered badly and there was a vast lake extending from Hayton downstream to the carr land north of Everton, and in 1947 when parts of East Retford were again flooded, as were Everton Carr and the low-lying areas around Bawtry and Scrooby.

A small unnamed tributary of the Idle, rising in the Jockey House area, flows parallel to the main river, about a mile to the west, and joins it about $1\frac{1}{2}$ miles N of East Retford. It has a narrow ribbon of alluvium along the 3 to 4 miles of its length, but the deposit, which produces a dark peaty loam soil with some gravel, is not exposed.   E.G.S.

The Idle's principal tributary in the district is the River Ryton, which enters the district north-west of Worksop and winds its way across country to join the Idle at Scrooby. For most of its length it flows across the Bunter Pebble Beds, but its headwaters and small tributaries drain the Permian dip-slope along the western edge of the district. Outside Whitewater Common, the associated alluvium occurs in a narrow belt up to, but generally much less than, ¼ mile wide; it consists of clay, peaty clay, sandy clay and loam, but sections are few. Whitewater Common, south of Harworth and north-west of Blyth, is a flat-bottomed, roughly circular depression about 1 mile in diameter[1], covered by a spread of alluvium continuous with that of the Ryton. This alluvium produces a dark pebbly clay soil, which becomes sandy in the east and contains rounded dolomite fragments in the west. Ditch sections generally show up to 3 ft of clay, or sandy clay, dark at the top and greyish brown below, on a gravelly layer 6 in or so thick. This rests on red sand or sandy clay, which is probably the solid in most places. Whitewater Common is apparently the site of an old lake: the channel running from near Whitewater Gorse [617 893] to the Ryton north of Blyth Wood and now occupied by a drain, may have been the exit from it.

R.F.G., E.G.S.

River Trent and tributaries. The Trent, tidal and navigable throughout the length under consideration, meanders northwards along the eastern edge of the district, and has a flood-plain ½ to 1½ miles wide. The latter is underlain by alluvium, consisting of clay, silty clay and silt up to about 40 ft thick and containing, in places, beds of peat and sand. Where sand with a few pebbles is found at or near the surface, as in the old channel which runs from Old Trent Road, east of Beckingham, towards Point Farm, it is clearly alluvium, but where sands and gravels are proved at depth, it is not clear whether they should be included with the alluvium or assigned to the underlying gravels. Generally the practice has been to draw the boundary at the base of the argillaceous and silty deposits, and the sands and gravels have therefore been considered under 'River Terrace Deposits' on pp. 228–9.

North and east of Beckingham, all but one of the boreholes for water and seismic purposes, show up to 20 ft of clay resting directly on Keuper Marl: the exception [7983 9133], south of Point Farm, shows sand and fine gravel to underlie the clay to a depth of more than 60 ft. Boreholes for the Gainsborough Flood Protection Scheme, upstream from Gainsborough Bridge, show peat below the river and beneath part of Beckingham marshes (Potter *in* Nixon 1960, p. 51). One bore in the river bed revealed at least 9 ft of peat below 3 ft of silt. Boreholes in and near Gainsborough show up to 40 ft of clay and silt including a 4-ft bed of peat at one locality [8107 9002], and 16 to 21 ft of 'peat and bog' in two holes [815 886] on the south side of the town. Some of these boreholes show the alluvium to rest on sand and gravel (see pp. 228–9), as it does at West Burton Power Station [791 853], where the alluvium consists of up to 34 ft of grey silty clay with, in one case, 3½ ft of peat and clay at the base. A borehole [8194 8430] on Out Ings proved 15 ft of clay, and trial holes at Cottam up to 10 ft (averaging 6 ft) of clay, in all cases resting on gravel. Three of the Cottam holes showed peat up to 5½ ft thick at the base of the alluvium. Shot holes [8235 7460 and 8175 7443] showed 20 and 15 ft of clay respectively, on gravel, but at Dunham Bridge Pumping Station No. 1 Well there are 29 ft of alluvium resting directly on the Keuper Marl.

G.H.R., E.G.S.

The alluvium produces a loam or clay soil which is very fertile when well drained, though much of the flood-plain is occupied by water meadows. There is evidence of warping around Walkerith, and the largest of the patches of alluvium that occur on the terrace north of Morton is probably warp.

The Trent valley has suffered many floods, some caused by excessive flow, usually after a severe winter, some by tidal surges and others by a combination of these circumstances. During the worst floods, water extends far beyond the margins of the alluvium, on to the terrace deposits and the low-lying areas of solid rock, inundating villages and, frequently, parts of Gainsborough. Reports of flooding go back to mediaeval times, and major floods have been recorded for the last 300 years (Padley 1881; Nixon

---

It is possible that this depression owes its origin to the solution of evaporites in the under-
[1]lying Permian rocks.

1960), during which period the most severe, so far as the present district is concerned, were those of 1697, 1770, 1795, 1828, 1875, 1910, 1932 and 1947. The highest Trent flood recorded is that of 1795 and the worst in the twentieth century to date, that of 1947. There is an extensive and ever-growing system of flood defences, but these are still inadequate to contain floods equal to the worst of those recorded (Nixon 1960, p. 22). Some of the flood banks are of great antiquity, and in some cases may even predate the parish boundaries that run along them (Smith 1910, p. 576). There are now major and minor floodbanks, the function of the latter being to contain minor floods but to allow overspill into the washlands during a major flood. Much of the alluvium is designated as wash-land and a large area of this is subject to flooding almost annually (Nixon 1960, p. 8, plan no. 1).

The Trent is constantly modifying its course: old, silted-up channels can be seen in several places, and old maps, plans, documents and records tell of movement in recent historical times. The most celebrated changes of course in the present district are those which caused the amputation of the loops at Bole (No Man's Friend) and Burton (Burton Round). No Man's Friend was cut off in February 1792, according to Nixon (*ibid.* pp. 11, 20) during a flood, but according to Smith (1910, p. 575) 'by the river (possibly aided by the bore or "aegir")'. Burton Round, part of the river until 1797, in which year there was also a major flood, was described by Shakespeare (*King Henry IV, Part I*).

E.G.S.

Of the small tributaries flowing down the Keuper Marl dip-slope to the Trent, the only ones of note are Wheatley Beck, North Beck and the unnamed streams that rise at Beverley Spring and in the Upton–Askham area and empty into the Trent at Church Laneham. Sections in the alluvium of Wheatley Beck show up to 6 ft of brown or red clay, silty or sandy in places and pebbly in part, on up to 2 ft of sand or gravel. Gastropods are locally abundant in the clay, which also contains bands of peat and peaty clay up to 9 in thick. Sections in the alluvium of the southern tributaries show 2 to 6 ft of red, brown or grey clay, locally sandy in the lower reaches, with, in places, a layer of quartzite pebbles and skerry fragments at the base.

G.H.R., E.G.S.

PEAT

Peat covers most of the surface of Tickhill Low Common and Styrrup Carr south-east of Tickhill. Boreholes and excavations across the former (see p. 240 and Anon. 1958) showed up to $8\frac{3}{4}$ ft, the maximum thickness proved in the area.

In the Idle valley, peat occurs at or near the surface over much of the flood-plain north of Tiln Holt. In places, downstream from the Mattersey area, it covers the full width of the alluvium. It varies in thickness from a few inches to more than 4 ft, but there are few sections. Augering often shows that the peat passes at its margins through peaty clay into the alluvial clay or is interdigitated with clay. There is a patch of peat in the Ryton valley in the Brecks Wood area, and most of the flood-plain of the small tributary of the Idle south and west of Mattersey Wood is underlain by peat. Small patches of peat are also found overlying Idle terrace gravels at Lound Field and to the north-east of Barnby Moor.

Peat, shelly in places and mostly less than $2\frac{1}{2}$ ft thick, covers most of the carr land north of Everton and Gringley on the Hill: a specimen from Gringley Carr has been examined by Mr. C. O. Harvey (see p. 240). The patch of alluvium north-east of Tethering Lane Farm indicates an area in which the peat has been burnt off, and the small area of alluvium south-west of Ellicar Farm is one in which surface peat, originally 9 in thick, has been turned under by deep plough-ing.

There is an area of peat overlying Trent terrace deposits and evidently occupying an old stream channel, which extends from Morton north-eastwards to the district boundary. It is $1\frac{1}{2}$ to more than 4 ft thick and in places contains lumps of soft white tufa associated with plant remains. In the extreme north the peat is overlain by thin warp. Patches of peat, together with peaty sand, the latter not shown on the map, are found in hollows in the terrace to the north and north-west of Morton. The hollows and accompanying sand dunes (see below) are the result of wind action on the terrace sands.

Beds of peat are found in places at depth below the flood-plain of the River Trent; known details of these are given on p. 231.

E.G.S.

BLOWN SAND

Blown sand is shown on the map in only two areas, near Broom House, between Sutton and Bolham, and at Naylor's Hills, in the south-eastern corner of the district. Much of the surface of the Bunter Pebble Beds, Glacial Sand and Gravel and terrace sands is, however, subject to blowing, and loose sand drifted against hedges and other wind-breaks is widespread. The two patches of blown sand [6955 8370 and 6975 8360] resting on Pebble Beds near Broom House are aligned along hedges in this way, though the deposits here are more extensive and thicker (up to 10 ft) than elsewhere. Several small hummocks and ridges of blown sand occur on, and have been derived from, the Trent terrace north of Morton, but it has been found impracticable to delineate these deposits.

E.G.S.

The deposit at Naylor's Hills consists of loose yellow to fawn sand covering an area approximately 1 mile by ⅔ mile. An excavation [8330 7518] 150 yd W of the main road shows 6 ft of sand, and a further 4½ ft have been proved by augering. The sand rests partly on Keuper Marl, which at one place protudes through it, and partly on Glacial Sand and Gravel, with which it has an ill-defined boundary.

G.H.R.

REFERENCES

ANON. 1958. *Report of soils investigation: Doncaster Western By-pass Motorway.*

AVELINE, W. T. 1880. The geology of parts of Nottinghamshire, Yorkshire and Derbyshire (Explanation of Quarter Sheet No. 82 NE). 2nd Edit. *Mem. geol. Surv. Gt Br.*

EDEN, R. A., STEVENSON, I. P. and EDWARDS, W. 1957. Geology of the country around Sheffield. *Mem. geol. Surv. Gt Br.*

EDWARDS, W. 1936a. Pleistocene dreikanter in the Vale of York. *Summ. Prog. geol. Surv. Gt Br. for 1934*, Pt II, 8–18.

—— 1936b. A Pleistocene strand-line in the Vale of York. *Proc. Yorks. geol. Soc.*, **23**, 103–18.

—— 1967. Geology of the country around Ollerton. 2nd Edit. *Mem. geol. Surv. Gt Br.*

—— WRAY, D. A. and MITCHELL, G. H. 1940. Geology of the country around Wakefield. *Mem. geol. Surv. Gt Br.*

—— MITCHELL, G. H. and WHITEHEAD, T. H. 1950. Geology of the district north and east of Leeds. *Mem. geol. Surv. Gt Br.*

GAUNT, G. D. 1965. In *Summ. Prog. geol. Surv. Gt Br. for 1964*, 52.

—— 1966. In *A. Rep. Inst. geol. Sci. for 1965*. Pt 1: *Summ. Prog. geol. Surv. Gt Br.*, 51.

JUKES-BROWNE, A. J. 1883. On the relative ages of certain river valleys in Lincolnshire. *Q. Jnl geol. Soc. Lond.*, **39**, 596–610.

KENDAL, P. F. and WROOT, H. E. 1924. *Geology of Yorkshire*. Vienna.

KENT, P. E. 1939. Notes on the river systems in south Lincolnshire. *Proc. Geol. Ass.*, **50**, 164–7.

KING, C. A. M. *in* EDWARDS, K. C. (Editor), 1966. Nottingham and its region. British Association, Nottingham.

LAMPLUGH, G. W., HILL, J. B., GIBSON, W., SHERLOCK, R. L. and SMITH, B. 1911. Geology of the country around Ollerton. *Mem. geol. Surv. Gt Br.*

MITCHELL, G. H., STEPHENS, J. V., BROMEHEAD, C. E. N. and WRAY, D. A. 1947. Geology of the country around Barnsley. *Mem. geol. Surv. Gt Br.*

NIXON, M. 1960. Report on the Tidal Reach Improvement Scheme. Nottingham (Trent River Board).

PADLEY, J. S. 1881. *Fens and floods of mid-Lincolnshire*. Lincoln.

PENNY, L. F. 1964. A review of the last glaciation in Great Britain. *Proc. Yorks. geol. Soc.*, **34**, 387–411.

POSNANSKY, M. 1960. The Pleistocene succession in the middle Trent basin. *Proc. Geol. Ass.*, **71**, 285–311.

POTTER, H. R. *in* NIXON, M. 1960. *q.v.*, 39–55.

SMITH, B. 1910. Some recent changes in the course of the Trent. *Geogrl J.*, **35**, 568–78

SMITH, E. G. 1963. In *Summ. Prog. geol. Surv. Gt Br. for 1962*, 41.

STRAW, A. 1958. The glacial sequence in Lincolnshire. *E. Midld Geogr.*, No. 9, 29–40.

—— 1961. Drifts, meltwater channels and ice-margins in the Lincolnshire Wolds. *Trans. Inst. Br. Geogr.*, **29**, 115–28.

—— 1963. The Quaternary evolution of the lower and middle Trent. *E. Midld Geogr.*, No. 20, 171–89.

SWINNERTON, H. H. 1914. Periods of dreikanter formation in South Notts. *Geol. Mag.*, Dec. vi, **1**, 208–11.

—— 1938. The problem of the Lincoln Gap. *Trans. Lincs. Nat. Un.* for 1937, 145–53.

—— and KENT, P. E. 1949. The Geology of Lincolnshire. *Lincs. Nat. Un.*, Lincoln.

WEST, R. G. 1963. Problems of the British Quaternary. *Proc. Geol. Ass.*, **74**, 147–86.

Chapter IX

MINERAL PRODUCTS AND WATER

COAL

THE DISTRICT is everywhere underlain at depth by Coal Measures, and exploitation of the large reserves of coal that these contain began with the sinking of Shireoaks Colliery shafts in 1854–59. Firbeck Main Colliery, sunk in 1925, was closed in 1968. There are now six working collieries—Bevercotes, Harworth, Maltby Main, Manton, Shireoaks and Steetley—including, in Bevercotes, one of the newest and most highly mechanized in the Concealed Coalfield. Workings, however, are still confined to the west of the district; they extend eastwards for about seven miles in the north, about six in the south (excluding the new development at Bevercotes), and three to four in between. East of this worked belt the productive part of the Coal Measures has been systematically explored by deep boreholes, more than twenty in number, as far as a line running roughly north and south through East Retford. Beyond this, particulars of the Coal Measures are only imperfectly and locally known from oil bores.

The Top Hard (or Barnsley) Coal, locally joined to the Dunsil, is the principal —and until relatively recently was the only—seam worked. The Parkgate, High Hazles and Clown are now, however, becoming increasingly important. Details of the various seams are given in Chapter V, and general accounts of their physical and chemical properties were given by Mott (1945) and Dawe and Coles (1948).

There is, in the coalfield, a general increase of calorific value, caking properties and rank from south to north and with increasing depth of burial (Dawe and Coles 1948, p. 15). Over at least the working area of the East Retford district the 'brights' of the Top Hard are thus of coking quality (*ibid.*, p. 16), but only in the north-west can the seam as a whole be considered a coking coal (Mott 1945, p. 458), and only Maltby Main and Harworth collieries produce this commodity. These two collieries also produce gas coal. Steam coal, for which the Barnsley 'hards' are eminently suitable, is produced by all six collieries, as is house coal.

OIL

The first record of oil in the East Retford district was at South Leverton Borehole where, during a search for coal in 1914, 1 ft of brown bituminous oily shale was encountered in Coal Measures at 2563 ft 5 in, 603 ft below the sub-Permian unconformity (Ford 1920, p. 100). The precise stratigraphical horizon is uncertain, but this shale probably lies close below the Mansfield Marine Band.

In 1918 small quantities of oil were found in the Coal Measures in Hayton Smeeth Borehole and during the sinking of Harworth Colliery shafts. At the former locality about 2 gallons of oil were obtained from 2 ft 4 in of sandy shale with ironstone nodules at 2212 ft, some 77 ft above the Mansfield Marine Band,

and what is described as "a good green oil spring" was encountered in a thick sandstone at 2005½ ft, just over 200 ft below the presumed Top Hard Coal. At Harworth oil occurred in joints in the Mansfield Marine Band Cank at 2115 ft and in sandstone 15 to 43½ ft below the marine band. Subsequently oil was met in the Barnsley workings at Firbeck Main Colliery, evidently occupying a faulted wash-out. Details of these occurrences are given by Kent (1954, pp. 1707, 1709).

West Drayton No. 1 Borehole was drilled in 1940 to 3358 ft in search of coal, and the discovery of oil, which overflowed to the surface, in 16 ft of much-jointed siltstone and sandstone in the measures between the Kilburn and Black-shale coals, prompted the deepening of the hole to 3901 ft by the D'Arcy Exploration Co. Ltd. The only other showing of oil, however, was a slight trace 200 ft below the first. Details are given by Kent (1954, p. 1706).

The first borehole drilled specifically in search of oil was Gringley No. 1 Bore, put down in 1946 by the Anglo-American Oil Co. Ltd. The well was dry. West Drayton No. 2 Bore was drilled near No. 1 Bore by the D'Arcy Exploration Co. Ltd in 1953–54, but found only faint traces of oil in the Coal Measures and Millstone Grit.

Commencing in 1958, the BP Exploration Co. Ltd have drilled into a number of anticlinal structures in the Carboniferous rocks, proved by seismic surveys, and three of these—Bothamsall (1958), Gainsborough (1959) and South Leverton (1960)—have yielded oil in commercial quantities (Brunstrom 1963).

The Bothamsall Oilfield (ibid., p. 15, fig. 4), which straddles the boundary between the East Retford and Ollerton districts, is about ½ square mile in extent and is formed by a culmination on the Bothamsall Anticline (see p. 208), a continuation of the Kiveton Anticline. There are two oil-producing horizons, both near the junction of the Millstone Grit and Coal Measures, and these have been identified by Taylor and Howitt (1965, p. 198) as the Sub-Alton and Crawshaw sandstones, though the latter horizon is now (p. 49) thought to be the combined Crawshaw Sandstone and Rough Rock. A mile to the north of the Bothamsall field at Apleyhead a small oilfield occupying a separate anticlinal culmination was discovered in 1960 (Brunstrom 1963, p. 15, fig. 4); here oil is contained in the upper of the two Bothamsall sands, but the quantity is unlikely to be economic.

The Gainsborough Oilfield (ibid., pp. 16–7, fig. 6) is formed by a broad, faulted anticline in the Carboniferous rocks and is several square miles in extent, lying partly beneath the town of Gainsborough and the River Trent, in which areas development depends upon directional drilling. There are eleven horizons producing, or capable of producing, oil, in the Coal Measures below the Top Hard Coal and in the Millstone Grit (ibid., p. 16, fig. 2), the most important being just above and below the *Gastrioceras cancellatum* Marine Band. The gas : oil ratio is higher than average for the proved East Midlands oilfields, and some of the Gainsborough accumulations have gas caps. The original field has been extended by the discovery of oil in the nearby Beckingham area in 1964, where, however, the only productive horizon is a sandstone between the Swallow Wood–Haigh Moor group of coals and the Top Hard Coal (Brunstrom 1966, p. 17).

The South Leverton Oilfield (ibid., p. 16, fig. 5) covers about ½ square mile and is formed by a north-north-easterly trending anticline (Fig. 32). Oil occurs at two horizons in the Millstone Grit and one in the Coal Measures, but there is only one producing horizon, a sandstone now thought (p. 25) to be the Rough Rock. Cumulative production figures up to the end of 1964 (data kindly supplied by the BP Group) are as follows: Bothamsall, 86 573 tons; Gainsborough with Beckingham, 102 709 tons; South Leverton, 24 311 tons.

GRAVEL AND BUILDING SAND

The principal source of gravel in the district is the terrace deposits of the River Idle. These deposits (pp. 226–8, Plate XIXB) have been extensively exploited by several companies to the west of the river north of East Retford. They are variable in thickness up to about 15 ft, with a gravel : sand ratio of between 1 : 3 and 1 : 2 over wide areas. Generally the deposits thin and the proportion of gravel decreases eastwards. The gravel consists largely of quartzite pebbles of Bunter origin and is much in demand for making concrete, including high-quality pre-cast concrete, and also for the production of tarmacadam. Some of the sand fraction is sold as building sand, but much of it has been used for reclaiming the pits. South of East Retford there are shallow abandoned sand and gravel workings at Ordsall [707 797], and small abandoned excavations in the neighbourhood of West Drayton (p. 228).

The Bunter Pebble Beds and the sandstone in the Middle Permian Marl have been worked for building sand, the former in numerous pits (see pp. 172–5), now mostly abandoned, which are widely scattered over the outcrop, and the latter in a few pits in the Worksop area. Many of the patches of Glacial Sand and Gravel (pp. 223–5) have also been worked, chiefly for sand, but gravel has also been screened off. Other minor sources of sand and gravel are the sandy boulder clay (pp. 221–3 Plate XIXA) and certain deposits of Head derived from the Pebble Beds.

The principal sand and gravel workings active at the time of the survey, other than those in the Idle Terrace deposits, were at Gateford [576 813] in Middle Permian Marl sandstone (p. 145, Plates I and XVA), at Styrrup [605 902] and Mattersey [684 879], both in Pebble Beds (pp. 172 and 173), at Broom Covert [609 833] and Carlton Forest [601 821] in Bunter Pebble Beds and overlying Glacial Sand and Gravel (pp. 174 and 224, Plate XVB), and at Scrooby Top [652 888] in Glacial Sand and Gravel (p. 223).

The Bunter Pebble Beds and sandy drifts provide immense reserves of building sand, but the much more limited gravel reserves are rapidly being depleted. However, borings suggest that there are considerable quantities of gravel in places at depth in the terrace deposits of the River Trent, and these are, as yet, untouched.

BASIC REFRACTORIES

Until the late 1950's the Lower Magnesian Limestone was quarried by the Steetley Company Limited for processing into basic refractories at Steetley Quarries, but present-day workings are to the south-west, beyond the district boundary. Sections of the disused quarries are given on pp. 129–30. The Lower Magnesian Limestone of the Steetley area is an almost pure dolomite as the following analyses, reproduced from Thomas and others (1920, p. 83), show.

TABLE 7.   *Analyses of Dolomite from Steetley Quarries*

Lab. No. (and Slice No.)	560 (E 11737)	561 (E 11738)
	per cent	
Insoluble residue	·34	1·93
SiO_2	·09	·30
Al_2O_3	·76	1·84
Fe_2O_3	·01	·01
FeO	·39	·34
MnO	·05	·02
CaO	30·25	30·58
MgO	20·83	20·22
H_2O (at 105°C)	·03	·07
H_2O (above 105°C)	·26	·23
SO_3	·61	·44
C	·01	·01
CO_2	46·43	44·05
	100·06	100·04

BRICK CLAY

The clay formations of the district have been widely exploited for the manufacture of bricks, tiles, drainpipes etc., but there are no active workings at the present time. Old pits exist in the Middle and Upper Permian Marls, Green Beds, Keuper Marl and boulder clay, known details being given in Chapters V and VIII. Some of the pits are large, and of these the ones at Shaw Brick Works [738 920], abandoned about 1949, and Walkeringham Brick Works [754 928], both in Keuper Marl, are probably the most recently worked. The majority, however, are small and have long been disused.

BUILDING STONE

The Lower Magnesian Limestone has been worked extensively in the past for building stone, but there are no active quarries within the district boundaries at the present time. Apart from Steetley and Shireoaks quarries, the better-known localities in this region lie just west of the district boundary (Eden and others 1957, pp. 142–6, 174). Steetley stone was used to construct, among other buildings, St George's Church, Doncaster (Aveline 1880, p. 16), Shireoaks and Thorpe Salvin churches, and the original Sledmere House, near Great Driffield (G. H. Mitchell *in litt.*), the lowest beds in the old quarry providing the best building material. The building stone at Shireoaks, which came from the top of the quarries, was described by Aveline (*idem*) and noted by Sedgwick (1835, p. 84) and Howe (1920, p. 206).

The Upper Magnesian Limestone is flaggy and therefore of less value than the Lower Magnesian Limestone as a building stone. It has, however, been used on and near its outcrop, more particularly for walling.

Skerries in the Keuper Marl are the only other rock in the district to have been used as a building stone. Slabs can be seen incorporated in some of the churches and older buildings in the east, but their use was small-scale and local. An account of skerry as a building material is given by Lamplugh and others (1911, p. 68).

GLASS SAND

Worksop's glass bottle industry was founded on the grey sandstone occurring in the Middle Permian Marl in that area (pp. 145–8), but the high iron content of the sand meant that only low-quality, dark green and black bottles could be produced. Technical details of the sands from this locality, which have also been worked for refractory purposes and for building (see p. 237), are given by Boswell (1916, pp. 60 and 84; 1918, pp. 96 and 220). The present glass works at Worksop, operated by the Co-operative Wholesale Society Limited, imports all its raw materials from outside the district.

GYPSUM AND ANHYDRITE

As described in Chapter V, gypsum is widespread in the Permo-Triassic rocks, and in several formations it gives way to anhydrite at depth. The most important horizons are in the Middle Permian Marl (p. 139), Upper Permian Marl (p. 157), Green Beds (p. 178), Clarborough Beds (p. 185), and near the top of the Keuper Marl (p. 185). Sedgwick (1835, p. 102) mentions the digging of 'gypseous marls' from the Middle Permian Marl near Oldcoates, but otherwise only the last two of the above-listed horizons have been worked.

The gypsum in the Clarborough Beds has been worked at outcrop in pits at Walkeringham, Clayworth, North Wheatley, Clarborough, Little Gringley and Headon (see pp. 187–91 and W. H. Dalton *in* Ussher and others 1888, pp. 8–11). Most of the pits were abandoned by the later part of last century, but one [7396 8336] at Clarborough (Plate XVIA) was worked during four of the summers between 1947 and 1953, producing several hundreds of tons of gypsum per annum for agricultural purposes.

The gypsum horizon within about 100 ft of the top of the Keuper Marl, probably corresponding to the Newark Gypsum of southern districts, has been extensively worked for plaster in old pits between Castle Hills and Gainsborough, but there are no known records.

IRONSTONE

The vast amounts of ironstone contained in the Coal Measures are not an economic proposition. Two old shafts [5887 8113, 5589 8110] on the Middle Permian Marl outcrop near Coachroad Wood, north of Shireoaks Colliery, are, however, reputed to have been sunk into the Coal Measures for ironstone. A 16-in band of ironstone, with one or more thin bands in the shales above, was encountered at 234 ft from the surface in the nearby colliery shaft, but the hematite horizon occurring close below the Mexborough Rock in Manton No. 4 Shaft and at several localities farther west (Eden and others 1957, p. 172) is not recorded at Shireoaks, where the base of the sandstone is at 479 ft.

LIME AND MAGNESIA

Both the Lower and the Upper Magnesian Limestone were formerly extensively worked for burning to lime, and it is to this industry that many of the scattered small quarries owe their origin. The bulk of the output at Shireoaks Quarries was converted to lime, used for both building and agricultural purposes, and was said to be of excellent quality (Howe 1920, p. 206). Magnesia was produced at Steetley Quarry (*ibid.*, p. 205).

MOULDING SAND

The Lower Mottled Sandstone is a source of moulding sand for the iron and steel industry: it is quarried at Gateford (see p. 167) and was also formerly worked at Worksop (p. 168). Its petrography is discussed on pp. 165–8, and mechanical and modal analyses of sands from various localities, including Gateford, are given in Tables 5 and 6 (see also Boswell 1918, pp. 153 and 227). Quartz–feldspar ratios for the Lower Mottled Sandstone are included in Table 6.

PEAT

Surface peat occurs in small, low-lying areas south-east of Tickhill, along the northern edge of the district east of Bawtry, and in the Idle valley (see p. 232). In general the thickness has not been proved except where its base is within auger reach, but it is nowhere thought to be great; boreholes and excavations along the line of the Doncaster By-pass Motorway across Tickhill Low Common proved $5\frac{1}{2}$ to $8\frac{3}{4}$ ft.

A specimen of shelly peat from Gringley Carr examined by Mr. C. O. Harvey lost 63·2 per cent of volatile matter on ignition, but there are no known instances in the district of peat being exploited as a fuel. Its importance is agricultural and as a hazard in civil engineering projects, and in the latter connection buried peat in the alluvial deposits of the Trent (p. 231) should be remembered. Tests carried out on the peat along the line of the Doncaster By-pass Motorway by the West Riding County Council's Highways and Bridges Laboratory (Anon. 1958) show a moisture content of 100 to 650 per cent, mainly in the range 200 to 400 per cent, with a dry density of 8 to 12 lb per cu ft and cohesive strengths after consolidation of 100 to 200 lb per sq ft. The peat was removed from the motorway foundations.

ROAD-STONE

The Upper Magnesian Limestone and the harder beds of the Lower Magnesian Limestone are sources of road-stone, and are locally worked for this purpose. According to Dr. G. H. Mitchell (*in litt.*) quarries in the latter at Lindrick Common, working in 1930 when he visited them, produced a very good road-stone from 13 ft of compact, grey dolomite, much harder than the usual Lower Magnesian Limestone. The lower beds of the worked Lower Magnesian Limestone at Shireoaks Quarries are also unusually hard, owing to a silica content of up to 7 per cent, and were formerly worked for road-stone (Howe 1920, p. 206). Attrition tests on a sample of Lower Magnesian Limestone from Lady Lee Quarries, near Worksop, were described by Lovegrove (1929, pp. 10–1, 52).

The 17-in band of hard micaceous sandstone occurring in the Lower Rhaetic in Lea Cutting (Bed No. 9, p. 199) has, according to Burton (1867, p. 319), been used to construct the roadways at Gainsborough station.   E.G.S.

WATER

Except for a very small area in the north-east which drains into the River Witham, the district lies within the catchment of the River Trent. It has a rainfall varying from less than 23 inches in the north to rather more than 25 inches in the south-west, with an average annual potential evaporation of about 19 in.

There are no impounding reservoirs in the district and, apart from water abstracted from the rivers for agricultural and industrial purposes, public and private supplies are maintained from underground sources, with the 'Bunter' (see below) forming by far the most important aquifer. Detailed information about the Nottinghamshire and Lincolnshire parts of the district were given by Lamplugh and Smith (1914) and Woodward (1904), and recently Downing and others (1970) have reviewed the hydrogeology of the whole Trent Basin. Many of the old wells have been abandoned during this century owing to extensions of the piped public supply. The statutory undertakings responsible for the water supply of the district are the Central Nottinghamshire Water Board, the Lincoln and District Water Board and the Doncaster and District Joint Water Board. Small bulk supplies are imported from and exported to areas outside the district boundaries.

Spray irrigation, mainly of grass and green crops and potatoes, sugar beet and other roots, is becoming increasingly popular, especially on the sandy soils. Some ground water is used for this purpose, but 90 per cent is taken directly from rivers, streams, dykes etc. Licences were in force for the abstraction of nearly 400 million gallons for irrigation purposes in the district in 1966. Whereas most of the water abstracted for other purposes returns to the river system as effluent, nearly all the irrigation water is permanently lost to it.

The table below shows total abstractions of ground water in the East Retford district during the year 1969, based on figures supplied by the Trent River Authority and the National Coal Board.

TABLE 8.   *Water abstractions from underground sources in millions of gallons per year*

Aquifer	Wells and Boreholes	Collieries	Total
Drift (excluding that overlying Bunter Pebble Beds)	2	—	2
Keuper Marl	4	—	4
Bunter Pebble Beds and Lower Mottled Sandstone (including overlying Drift)	4280	666	4946
Permian (excluding Lower Mottled Sandstone)	120	805	925
Coal Measures	$< \frac{1}{2}$	169	169
Totals	4406	1640	6046

The above figures do not include: (1) certain unlicensed abstractions from wells, but these are unlikely to exceed 30 m.g.y. from the 'Bunter' and 5 m.g.y. from the Permian; (2) possible abstractions from terrace gravels of the River Idle in the Sutton and Lound areas, where licences exist for 280 m.g.y. The bulk of the water extracted from the Permo-Triassic rocks in collieries is from Manton, and the bulk of the water extracted from the Coal Measures is from Maltby Main.

Carboniferous Limestone Series. No water is abstracted from these deeply buried rocks, but Downing (1967) has described the chemistry of saline water samples from some of the oil bores. He shows that the total concentration of dissolved solids increases rapidly towards the north-east, from less than 30 000 milligrammes per litre (mg/l) at Apleyhead to almost 100 000 mg/l in the Corringham Oilfield near Gainsborough. Chlorides, particularly those of sodium and calcium, form a high proportion of these totals. Sulphates and carbonates are of little importance, and decrease in amount as the total concentration increases. These changes are part of the regional variation in Carboniferous Limestone waters of the East Midlands, from relatively soft, sodium sulphate waters in the south-west to strong, calcium chloride brines in the north-east.

Millstone Grit Series. Water samples have been taken at a number of horizons from the Millstone Grit in oil bores. The results of analyses have been examined by Downing and Howitt (1969), who found ionic contents and a regional trend similar to those displayed by the Carboniferous Limestone waters, although there is evidence of vertical variation in salinity in the Millstone Grit. Most of the samples come from sandstones in the Middle Grit Group (p. 16 and Plate II), and show that the total concentration of dissolved solids increases northwards from a little over 10 000 mg/l at Egmanton (one-inch sheet 113), to more than 100 000 mg/l at Blyton and Corringham (one-inch sheets 89 and 102). Strong brines appear to extend much farther south at lower horizons, for remarkably saline water (125 000 mg/l) was obtained from a sandstone evidently of pre-Marsdenian age at Apleyhead.

Coal Measures. The collieries and a few boreholes draw water from these rocks, but amounts are small. At least one well in the Steetley area reaches the uppermost Middle Coal Measures, which may contribute some water to its yield; while at Blyth Borehole (see p. 250) water, which was initially artesian, is obtained from the Upper Coal Measures. In both cases, however, the strata yielding water are close below the sub-Permian unconformity, and, since it is likely that hydraulic continuity exists between the Basal Permian Sands and the Coal Measures sandstones, this water may be derived from the Permian rocks.

At lower levels, abstractions from the Coal Measures form part of mine drainage yields. Two factors inflate the outflow from the collieries: (1) run-off from surface water supplies to the coal face, and (2) water from workings extending beyond the district. The latter factor is partly compensated by a small area of workings which drains to Dinnington Colliery in the Sheffield (100) district to the west. Because all the collieries in the East Retford district are more than 3 miles from the outcrop of the productive measures, their waters are hard and saline (Downing and others 1970): e.g. water from the Dunsil workings at Firbeck Main showed 48 000 mg/l of chloride.

Water samples taken in some oil bores indicate that concentrated brines appear farther down dip and also at lower horizons (Downing and Howitt 1969). Total dissolved solids reached 218 000 mg/l in the Sub-Alton Sandstone at Tickhill No. 1 Oil Bore. There is some evidence of a northward increase in concentration as in the Carboniferous Limestone and Millstone Grit.

Basal Permian Sands. These impersistent, generally thin, highly permeable and porous sandstones are tapped by a few wells west of Worksop, and by Blyth Borehole. The yield is almost always combined with that from the Lower Magnesian Limestone, but up to about 750 gallons per hour (g.p.h.) of hard water have been obtained from a 4-in borehole.

Lower Permian Marl. This formation is relatively impermeable and is not known to yield water in the district.

Lower Magnesian Limestone. This is the main Permian aquifer and the oldest formation cropping out within the district, although much of its catchment lies within the Sheffield (100) district to the west. Ground-water flow in this rock is in joints and fissures, and yields are therefore dependent on the local intensity of these fractures. A large diameter well is usually required to give more than 5000 g.p.h. By combining the water from the Lower Magnesian Limestone with that from the Basal Permian Sands in such a well, up to 9000 g.p.h. have been obtained. The largest single abstraction from the Lower Magnesian Limestone in the district is, however, that from Manton Colliery No. 1 Shaft—more than 80 000 g.p.h. in 1969. Some of the wells at Worksop Waterworks extract water from this formation as well as from the overlying rocks. Lower Magnesian Limestone water is hard, the few available analyses showing up to 1200 mg/l total solids including prominent carbonates. We are indebted to Mr. J. G. Hinchliffe for the analysis of water (pH value 7·2) from the Lower Magnesian Limestone in a recent well at Belmont Villa [5997 8842], Oldcoates:

	mg/l
Total Solids dried at 180°C	980·0
Chlorides as Cl	105·0
Nitrites	slight trace
Nitrates as N	5·95
Poisonous metals (Pb etc.)	less than 0·04
Total hardness (expressed as $CaCO_3$)	712·0
Temporary hardness (expressed as $CaCO_3$)	305·0
Permanent hardness (expressed as $CaCO_0$)	407·0

Middle Permian Marl. This formation is generally an aquiclude, but the sandstone contained in it near Worksop (p. 141) apparently contributes water to a few wells in that neighbourhood.

Upper Magnesian Limestone. This formation is hydrogeologically similar to the Lower Magnesian Limestone, but it is thin at outcrop in the north of the district and absent in the south, and is therefore relatively unimportant as an aquifer. Although a number of wells are thought to draw small amounts of water from the Upper Magnesian Limestone, little is known of the quality or quantity of the supply available.

Upper Permian Marl. This formation is impermeable, but it passes laterally into the Lower Mottled Sandstone (p. 157), which is part of the 'Bunter' aquifer (see below).

Bunter Pebble Beds with Lower Mottled Sandstone. The Bunter Pebble Beds form the principal aquifer in the district, and are supplemented by the Lower Mottled Sandstone, locally by the sandstone in the Middle Permian Marl and by overlying sandy drift. This composite aquifer has been studied in detail by Land (1966) as part of a wider survey of the nature and resources of 'Bunter' groundwater in Nottinghamshire. Theoretical resources are in excess of abstractions in the district, and the evidence suggests that, except locally, water levels have altered little in the last twenty years. Sites of the principal abstractions from these rocks are shown on Fig. 2. The highest recorded yield from a single source appears to be 75 000 g.p.h. from Manton Colliery No. 1 Shaft.

We are indebted to the Lincoln and District Water Board for the following analysis of a typical 'Bunter' groundwater from No. 2 Borehole at Ordsall Road Pumping Station, East Retford (p. 284), carried out in May 1971:

	mg/l
Calcium as Ca	52·1
Magnesium as Mg	5·6
Sodium as Na	8·5
Potassium as K	2·0
Bicarbonate as CO_3	68·4
Chloride as Cl	21·0
Fluoride as F	<0·1
Sulphate as SO_4	29·0
Phosphate as PO_4	0·05
Nitrate as NO_3	10·2
Iron as Fe	0·14
Silica as SiO_2	8·0
Total dissolved solids (at 180°C)	220
Carbonate hardness (as $CaCO_3$)	114
Non-carbonate hardness (as $CaCO_3$)	39
Total hardness (as $CaCO_3$)	153
pH	8·03

Triassic rocks above Bunter Pebble Beds. There are no records of water being obtained from the largely argillaceous Green Beds and Rhaetic rocks. Some water is present in the Waterstones, but any yield in wells sunk through that formation is masked by water pumped from the Pebble Beds. The bulk of the Keuper Marl is impermeable mudstone, but thin skerries were formerly the source of scores of small domestic and farm supplies. Keuper Marl water is generally very hard, and may be gypseous or saline. With the spread of piped supplies in recent years, many of the old wells have fallen into disuse.

Pleistocene and Recent. Sandy drifts lying on the outcrop of the Bunter Pebble Beds are in hydraulic continuity with that formation, and are included with it for hydrogeological purposes (see above). A few supplies are obtained from terrace gravels in the valleys of the Rivers Idle and Trent; these are likely (Downing and others 1970) to be in hydraulic continuity with the rivers, and extensive pumping would give rise to induced recharge from them. Since the Trent is highly polluted, such sources near to it are probably suitable only for industrial purposes. Up to 6000 g.p.h. have been extracted from terrace gravels.

C.G.G., E.G.S.

REFERENCES

ANON. 1958. *Report of soils investigation: Doncaster Western By-pass Motorway.* West Riding County Council, Highways and Bridges Department, Wakefield.

AVELINE, W. T. 1880. The Geology of parts of Nottinghamshire, Yorkshire and Derbyshire (Explanation of Quarter Sheet 82 N.E.) *Mem. geol. Surv. Gt Br.*

BOSWELL, P. G. H. 1916. *A memoir on British resources of sands suitable for glass-making.* London.

—— 1918. *A memoir on British resources of refractory sands.* Pt 1. London.

BRUNSTROM, R. G. W. 1963. Recently discovered oilfields in Britain. *Proc. 6th Wld Petrol. Congr.* Frankfurt. Section 1, 11–20.

—— 1966. Indigenous petroleum and natural gas in Britain. *Proc. Joint Meeting Inst. Petrol. and dt Ges. Minerw. Kohlechem. e.v.* Nov. 1965, 5–2.

BURTON, F. M. 1867. On the Rhaetic beds near Gainsborough. *Q. Jnl geol. Soc. Lond.,* **23**, 315–22.

DAWE, A. and COLES, G. 1948. The coal seams of Derbyshire, Nottinghamshire and Lincolnshire. *J. Inst. Fuel*, Oct. 1948, 12–23.

DOWNING, R. A. 1967. The geochemistry of ground-waters in the Carboniferous Limestone in Derbyshire and the East Midlands. *Bull. geol. Surv. Gt Br*. No. 27, 289–307.

—— and HOWITT, F. 1969. Saline ground-waters in the Carboniferous rocks of the English East Midlands in relation to the geology. *Q. Jnl Engng Geol.*, **1**, 241–69.

—— LAND, D. H., ALLENDER, R., LOVELOCK, P. E. R. and BRIDGE, L. R. 1970. The hydrogeology of the Trent River Basin. *Hydrogeological Rep. No. 5, Water Supply Pap. Inst. geol. Sci.*

EDEN, R. A., STEVENSON, I. P. and EDWARDS, W. 1957. Geology of the country around Sheffield. *Mem. geol. Surv. Gt Br*.

FORD, J. 1920. Record of the deep borings at Kelham and South Leverton. *Trans. Instn min. Engrs*, **58**, 94–107.

HOWE, J. A. (Editor) 1920. Special reports on the mineral resources of Great Britain, 6, Refractory materials. 2nd edit. *Mem. geol. Surv. Gt Br*.

KENT, P. E. 1954. Oil occurrences in Coal Measures of England. *Bull. Am. Ass. Petrol. Geol.*, **38**, 1699–713.

LAMPLUGH, G. W., HILL, J. B., GIBSON, W., SHERLOCK, R. L. and SMITH, B. 1911. The geology of the country around Ollerton. *Mem. geol. Surv. Gt Br*.

—— and SMITH, B. 1914. The water supply of Nottinghamshire from underground sources. *Mem. geol. Surv. Gt Br*.

LAND, D. H. 1966. Hydrogeology of the Bunter Sandstone in Nottinghamshire. *Hydrogeological Rep. No. 1, Water Supply Pap. geol. Surv. Gt Br*.

LOVEGROVE, E. J. 1929. Attrition tests of British road-stones. *Mem. geol. Surv. Gt Br*.

MOTT, R. A. 1945. The coking coal resources in the Yorkshire, Nottinghamshire and Derbyshire Coalfield. *Trans. Instn min. Engrs*, **104**, 446–68.

SEDGWICK, A. 1835. On the geological and internal structure of the Magnesian Limestone, and the lower portions of the New Red Sandstone in their range through Nottinghamshire, Derbyshire, Yorkshire and Durham, to the southern extremity of Northumberland. *Trans. geol. Soc. Lond.* **3**, 37–124.

TAYLOR, F. M. and HOWITT, F. 1965. Field meeting in the U.K. East Midlands oil-fields and associated outcrop areas. *Proc. Geol. Ass.*, **76**, 195–209.

THOMAS, H. H., HALLIMOND, A. F. and RADLEY, E. G. 1920. Special reports on the mineral resources of Great Britain. 16, Refractory materials: petrography and chemistry. *Mem. geol. Surv. Gt Br*.

USSHER, W. A. E., JUKES-BROWNE, A. J. and STRAHAN, A. 1888. The geology of the country around Lincoln. (Explanation of Sheet 83) *Mem. geol. Surv. Gt Br*.

WOODWARD, H. B. 1904. The water supply of Lincolnshire from underground sources. *Mem. geol. Surv. Gt Br*.

BOREHOLES AND SHAFTS

All the collieries, boreholes for coal and groups of oil bores in the East Retford district are listed below, together with the more important boreholes for water and other purposes. Their sites are indicated by National Grid References (in square brackets), all of which lie in 100-km square SK (or 43) and by bearings and distances from points on the one-inch map. They are also shown on Figs. 2 and 3.

A few sections have been published elsewhere, and in these cases reference is made to the appropriate works. The only logs quoted in any detail here are those of Eaton, Nornay, Scaftworth and Scofton boreholes, and a small part of Gainsborough No. 1 Oil Bore. In these logs all measures are grey and seatearths are argillaceous or silty unless otherwise stated, and the term 'striped beds' is used for interbedded or interlaminated sandstone or siltstone and silty mudstone or mudstone. Brief notes are provided for the other sections, but these include the depths of the main horizons and, in cored sections of the Coal Measures, details of the principal named coals. The following abbreviations have been used:

abt	about	KM	Keuper Marl
BPS	Basal Permian Sands and Breccia	LML	Lower Magnesian Limestone
c.	coal	LMS	Lower Mottled Sandstone
ca.	cannel	LPM	Lower Permian Marl
d.	dirt	M.B.	Marine Band
d.c.	dirty coal	MPM	Middle Permian Marl
Div.	Division (of Upper Coal Measures)	O.D.	Ordnance Datum
ft	feet	PB	Bunter Pebble Beds
GB	Green Beds	R.T.E.	Rotary Table Elevation
Ht	height	UML	Upper Magnesian Limestone
in	inches	UPM	Upper Permian Marl

Albert Paper Mills Borehole, East Retford

Ht above O.D. abt 55 ft.   6-in SK 78 SW.
Site [7033 8074] 490 yd at 20° from East Retford station. Drilled 1950 by C. Isler and Co. Ltd for Albert Paper Mills Ltd.

MADE GROUND to 5 ft.

DRIFT: Alluvium and ?Terrace sand and gravel to 14 ft.

PERMO-TRIASSIC: PB to bottom of borehole at 350 ft.

Apleyhead (BP) Oil Bores

There are three bores in this group. Notes on No. 1 are given below. No. 2 [6577 7663] was drilled to 3647 ft into the Millstone Grit Series abt 50 ft below the *G. cancellatum* M.B. and No. 3 [6559 7581] was drilled to approximately the same horizon at 3574 ft.

No. 1 Bore

R.T.E. above O.D. 144 ft.   6-in SK 67 NE.
Site [6551 7631] 510 yd at 175° from Apleyhead Farm, Elkesley.
Drilled 1960 by BP Exploration Co. Ltd. Largely open holed,
cored 3300–10 ft, 3350–449 ft, 3480–784 ft, 4800–12 ft. Cores
examined by C. G. Godwin.

PERMO-TRIASSIC: PB to abt 450 ft, LMS to 540 ft, UML to 560 ft, MPM to 780 ft, LML
to abt 973 ft, LPM to 986 ft.

MIDDLE COAL MEASURES: ?Top Hard at abt 1950 ft; Clay Cross M.B. at abt 2230 ft.

LOWER COAL MEASURES: ?Deep Soft at abt 2320 ft; Kilburn horizon at abt 3010 ft;
Alton M.B. at 3391 ft 4 in; Pot Clay M.B. horizon at abt 3500 ft.

MILLSTONE GRIT SERIES: *G. cancellatum* M.B. at abt 3561 ft; base at abt 4590 ft.

CARBONIFEROUS LIMESTONE SERIES to bottom of bore at 4813 ft.

Babworth Borehole

Ht above O.D. 106 ft.   6-in SK 68 SE.
Site [6895 8027] 680 yd at 151° from All Saints' Church, Babworth.
Drilled 1953 by Foraky Co. for N.C.B. Open holed to 1120 ft,
cored 1120 ft to bottom. Cores examined by R. A. Eden and
R. E. Elliott.

PERMO-TRIASSIC: PB and LMS to abt 717 ft, UML to abt 775 ft, MPM to abt 970 ft,
LML to 1183 ft, LPM to 1196 ft 10 in, BPS to 1197 ft.

UPPER COAL MEASURES: Hemsworth Div. to 1398 ft, Brierley Div. to 1508 ft (Blyth,
d.c. 16½ in at 1508 ft), Ackworth Div. to 1698 ft.

MIDDLE COAL MEASURES: Top M.B. at 1709 ft; Shafton M.B. at abt 1755¾ ft; ?position
of *Edmondia* Band at 1935¼ ft; Sharlston Top (c. 10½ in) at 1983 ft 7 in; ?Houghton Thin
(c. & d.c. abt 30 in) at 2108 ft 4 in; Mansfield M.B. at 2170¼ ft; Haughton M.B. at 2255 ft
2 in; Clown (c. 34 in on d.c. 1 in) at 2313 ft 2 in; Two-Foot M.B. at 2434 ft 10 in; Two-Foot
(ca. 3 in on c. 28 in) at 2437 ft 5 in; Kent's Thin (ca. & d.c. 23 in) at 2598 ft; Kent's
Thick (c. 10½ in, d. 2½ in, c. & d.c. abt 17½ in, d. 25½ in, d.c. 1½ in on c. 15½ in) at 2641
ft 2 in; Top Hard + Dunsil (c. with ¾ in d., 32 in, d.c. 4½ in, c. 23 in, d. 1¾ in on c. 11¼ in)
at 2760 ft 2 in (in diversion); Swallow Wood, Main Bed (c. 1 in) at 2852 ft 4 in; Haigh
Moor (ca. 2 in) at 2862 ft 2 in; 2nd Ell (c. 6½ in, d. 11 in on c. 6½ in) at 2973 ft 2 in; Clay
Cross M.B. at 3028 ft 1 in.

LOWER COAL MEASURES: Parkgate (c. 12 in) at 3259 ft 5 in; Low '*Estheria*' Band at
3452 ft 1 in; Low Silkstone (c. & d. 12 in) at 3453 ft 1 in.

Bottom of borehole 3501 ft.

Barnby Moor Borehole

Ht above O.D. 59 ft.   6-in SK 68 SE.
Site [6630 8364] 980 yd at 179° from Ye Olde Bell Hotel, Barnby
Moor. Drilled 1959–60 by Foraky Co. for N.C.B. Open holed to
900 ft, cored 900 ft to bottom. Cores examined by C. G. Godwin,
R. E. Elliott and M. Lock.

PERMO-TRIASSIC: PB + LMS to abt 400 ft, UPM to abt 529 ft, UML to abt 566 ft,
MPM to abt 700 ft, LML to 930 ft 8 in, LPM to 942½ ft, BPS to 944 ft 10 in.

UPPER COAL MEASURES: Badsworth Div. to 1031 ft 11 in, Hemsworth Div. to 1275 ft
5 in, Brierley Div. to 1413¼ ft (Blyth, c. 29 in at 1413¼ ft), Ackworth Div. to 1582¼ ft.

MIDDLE COAL MEASURES: Top M.B. at 1586½ ft; Shafton M.B. at 1659¼ ft; Sharlston
Low (c. 9 in) at 1964½ ft; Sharlston Yard (d.c. & d. 14 in) at 2004 ft 4 in; ?Houghton
Thin (d.c. & d. 11½ in) at 2010 ft; Mansfield M.B. at 2030 ft 2 in; Sutton M.B. at 2071

ft 7 in; Haughton M.B. at 2119¼ ft; Swinton Pottery (c. 10 in) at 2124 ft 10 in; Clown (c. 32 in) at 2175 ft 8 in; ?Mainbright (c., d.c. & d. 23 in) at 2256 ft 8 in; Two-Foot M.B. at 2292¼ ft; Two-Foot (d.c. 3½ in on c. 26 in) at 2294 ft 7 in; Abdy (c. 13½ in, d. 9 in, c. 2½ in on d.c. 3 in) at 2325 ft; High Hazles, Top Bed (c. 13½ in) at 2402 ft; Kent's Thin (c. 14 in) at 2447 ft 4 in; Kent's Thick, top leaf (c. 8 in on d.c. & d. 11½ in) at 2489 ft 8 in; bottom leaf (c. 10 in) at 2502 ft; Top Hard + Dunsil (c. with 13½ in d.c., 65 in) at 2653 ft 2 in; 1st Waterloo (d.c. 9 in) at 2695¼ ft; Swallow Wood, Main Bed (c. 8 in) at 2760 ft 2 in; Haigh Moor (c. 31 in, d. 1 in, d.c. 3 in, d. 24 in, c. 13 in on d. & c. 7 in) at 2771 ft 1 in; Clay Cross M.B. at 2952¼ ft.

LOWER COAL MEASURES: ?Deep Hard Roof Coal (c. 7 in on d.c. 8 in) at 3142¾ ft; Deep Hard (d.c. 1 in, d. 17 in, d.c. 5 in on c. 10½ in) at 3160½ ft; Parkgate (d.c. 10 in, c. 23 in on d.c. 5 in) at 3215 ft 7 in; Thorncliff (d.c. 13 in, d. 7 in, d.c. 5 in, measures 12 ft 7 in on d. & d.c. 15½ in) at 3273¼ ft; Threequarters (c. 7 in) at 3310 ft 2 in; Top Silkstone (d.c. & d. 21 in on c. 2½ in) at 3367 ft 1 in; Low 'Estheria' Band at 3401 ft; Low Silkstone (c. 5½ in on d.c. 11½ in) at 3402 ft 5 in.

Bottom of borehole 3415 ft 1 in.

Batchelor's Foods No. 1 Borehole, Worksop

Ht above O.D. abt 150 ft.   6-in SK 58 SE.
Site [5774 8018] 1770 yd at 272° from Worksop Infirmary. Drilled 1963 by F. Smith & Son Ltd for Messrs. Unilever Ltd.

MADE GROUND to 6 ft.

PERMO-TRIASSIC: MPM to 117 ft, LML to bottom of borehole at 251½ ft. No. 2 Borehole, 325 yd to the NW, was drilled to a similar depth.

Beckingham (BP) Oil Bores

There are four bores in this group. Notes on No. 1 are given below. No. 2 [7927 8996] was drilled to 3349 ft into Middle Coal Measures abt 170 ft below the Top Hard Coal. No. 4 [7912 9070] was drilled to 4324 ft into Lower Coal Measures abt 100 ft below the Kilburn horizon. No. 5 [7953 9056] was drilled to 3360 ft into Middle Coal Measures abt 220 ft below the Top Hard Coal.

No. 1 Bore

R.T.E. above O.D. 16 ft.   6-in SK 79 SE.
Site [7921 9036] 1440 yd at 88° from All Saints' Church, Beckingham. Drilled 1964 by BP Petroleum Development Ltd. Largely open holed, cored 4555–640 ft and 5504–12 ft. Cores examined by C. G. Godwin.

DRIFT: Alluvium etc. not recorded.

PERMO-TRIASSIC: KM to abt 560 ft, PB and LMS to abt 1450 ft, UPM to abt 1600 ft, UML to abt 1670 ft, MPM to abt 1795 ft, LML and LPM with thin BPS at base to abt 2150 ft.

UPPER COAL MEASURES: Ackworth Div. to abt 2200 ft.

MIDDLE COAL MEASURES: Top M.B. at abt 2215 ft; horizon of Mansfield M.B. at abt 2615 ft; Kent's Thick at abt 3010 ft; Top Hard at abt 3125 ft; Clay Cross M.B. at abt 3470 ft.

LOWER COAL MEASURES: Deep Soft at abt 3580 ft; Parkgate at abt 3685 ft; Top Silkstone at abt 3840 ft; Kilburn at abt 4175 ft; Pot Clay M.B. at 4559¼ ft.

MILLSTONE GRIT SERIES: G. cancellatum M.B. at abt 4640 ft.

Bottom of bore 5512 ft.

Bevercotes Colliery (No. 1 Shaft)

Ht above O.D. 98 ft.   6-in SK 67 SE.
Site [6948 7391] 1 mile WSW of West Drayton. Sunk 1955–8.
Samples examined by R. A. Eden.

DRIFT: Terrace deposits to abt 10 ft.

PERMO-TRIASSIC: Waterstones to abt 20 ft, GB to abt 50 ft, PB + LMS to 670 ft, UPM to 748½ ft, UML to 762¼ ft, MPM to 956½ ft, LML + LPM to 1181½ ft, BPS to 1182 ft 5 in.

UPPER COAL MEASURES: Ackworth Div. to 1296 ft.

MIDDLE COAL MEASURES: Top M.B. at 1303 ft 4 in; Sharlston Top (c. 20 in) at 1522¼ ft; Wales (c. 18 in) at 1624 ft 11 in; Mansfield M.B. at 1676 ft 10 in; Sutton M.B. at 1697 ft 11 in; Clown (c. 18 in) at 1813 ft 4 in; Mainbright (c. 37 in) at 1849¾ ft ; Two-Foot M.B. at 1873 ft 4 in; Two-Foot (c. 18 in) at 1874 ft 10 in; Abdy (c. 10 in) at 1897½ ft; High Hazles, Top Bed (c. 11 in) at 1972¾ ft; Kent's Thick (c. 37 in) at 2049¾ ft; Dunsil (c. 24 in) at 2224½ ft; 1st Waterloo (c. 11 in) at 2256 ft 7 in; Swallow Wood (c. 10 in, d. 14 in on c. 19 in) at 2321 ft 2 in; Haigh Moor (c. 7 in, d. 5 in on c. 20 in) at 2327¾ ft; Lidget (d.c. 3 in) at 2389 ft 11 in; 2nd Ell (c. 23 in) at 2444½ ft; Clay Cross M.B. at 2500 ft 7 in.

LOWER COAL MEASURES: Deep Soft (c. 49 in) at 2582 ft 5 in; Deep Hard (c. 7 in) at 2640 ft 8 in; Parkgate (c. 54 in, d. 54½ in on c. 6 in) at 2697 ft 7 in; Threequarters (d.c. of unrecorded thickness, measures 12 ft 8 in on c. 32 in) at 2797 ft 8 in.

Bottom of shaft 2828 ft 5½ in.

Big Lane Borehole, Clarborough

Ht above O.D. abt 60 ft.   6-in SK 78 SW.
Site [7303 8343] 500 yd at 292° from St John the Baptist's Church, Clarborough. Drilled 1957 by W. Bradley to test drilling equipment. Open holed to 35½ ft, cored 35½ ft to bottom. Cores examined by R. A. Eden.

DRIFT: Terrace deposits to between 1½ and 9½ ft.

PERMO-TRIASSIC: KM base above 35½ ft but not determined, Waterstones to 103 ft 10 in, GB to 146 ft, PB to bottom of borehole at 146½ ft.

Bilby Borehole

Ht above O.D. 66 ft.   6-in SK 68 SW.
Site [6385 8338] 320 yd at 21° from Bilby Farm, Barnby Moor. Drilled 1960 by Foraky Co. for N.C.B. Open holed to 700 ft, cored 700 ft to bottom. Cores examined by C. G. Godwin, M. Lock and R. E. Elliott.

PERMO-TRIASSIC: PB + LMS to abt 310 ft, UPM to abt 345 ft, UML to abt 369 ft, MPM to abt 470 ft, LML to abt 582 ft, LPM to abt 636 ft, BPS to 726 ft.

UPPER COAL MEASURES: Hemsworth Div. to 914 ft, Brierley Div. to 1087¾ ft (Blyth, c. 29 in at 1087¾ ft), Ackworth Div. to 1251¾ ft.

MIDDLE COAL MEASURES: Top M.B. at 1277¼ ft; Shafton M.B. at 1356½ ft; Sharlston Low (c. 6 in, d. 61½ in on c. 6½ in) at 1657 ft 8 in; ?Houghton Thin (c. 3½ in on d.c. & d. 15½ in) at 1716 ft 1 in; Mansfield M.B. at 1786 ft; Clown M.B. at 1996¼ ft; Clown (d.c. 3 in, c. 28 in on d.c. 1 in) at 1999¼ ft; Mainbright (d.c. 3 in, c. with ½ in d., 14 in, d.c. 2½ in, measures 9 ft 2 in, d.c. 3 in, c. 2½ in on d. & d.c. 47½ in) at 2062 ft 3½ in; Two-Foot M.B. at 2083 ft 7 in; Two-Foot (d.c. 3 in, d. 4½ in on c. 24½ in) at 2086 ft 7 in; Abdy (c. with 2½ in d, 18½ in, d. 2½ in, d.c. 6 in on c. 4 in) at 2115 ft 10 in; High Hazles, Top Bed (c. & d. 6 in) at 2212 ft 2 in; Kent's Thin (ca. 1 in on c. 17 in) at 2254 ft 4 in; Kent's Thick

(c. 2½ in, d. & d.c. 51¼ in, c. 2¾ in, d.c. & d. 6 in on c. 3½ in) at 2308 ft 7 in; Top Hard (d.c. 2½ in on c. with 1¼ in d.c., 34½ in) at 2472½ ft; Blidworth + Dunsil (c. 13½ in, d. 5½ in on c. 18 in) at 2497 ft 4 in; 1st Waterloo (d.c. 2 in on c. 14 in) at 2551½ ft; Swallow Wood, Main Bed (c. 7 in, d. 9 in, d.c. 5½ in on c. 5½ in) at 2621 ft 5 in; Haigh Moor (c. 26 in, d.c. 4 in, d. 54 in on c. 21 in) at 2662 ft 1 in; Lidget (d.c. 11 in) at 2745 ft 7 in; 2nd Ell (c. 5 in) at 2807 ft 1 in; Clay Cross M.B. at abt 2885½ ft.

LOWER COAL MEASURES: Joan (c. 13 in) at 2887 ft 1 in; Deep Soft (d.c. & c. 21 in, c. 14½ in, d.c. & c. 19 in, d. 2½ in, c. with 2 in d., 30 in on d. & d.c. 27 in) at 2999 ft 1 in; ?Deep Hard Roof Coal (c. 5 in) at 3029 ft 8 in; Deep Hard (d.c. 1½ in, c. 15½ in, d. 3½ in on c. 3½ in) at 3057¼ ft; Parkgate, main part (c. 24½ in, d. 2½ in, c. 14 in on d.c. 2 in) at 3131¾ ft; Thorncliff (c. 23 in on d. & d.c. 30 in) at 3194 ft 4 in; Threequarters (c. 10 in) at 3227¾ ft; Top Silkstone (inferior ca. 3½ in, ca. 21½ in, d. 1½ in, ca. 3 in on inferior ca. 3 in) at 3291 ft 7 in; Low '*Estheria*' Band at 3327 ft 4 in; Low Silkstone (c. 6 in, d.c. 6 in, c. 5 in on d.c. 5 in) at 3329 ft 5 in.

Bottom of borehole 3348 ft 7 in.

Blyth Borehole

Ht above O.D. 44 ft. 6-in SK 68 NW.

Site [6101 8693] 1600 yd at 264° from St Mary and St Martin's Church, Blyth. Drilled 1953–4 by Craelius Co. for N.C.B. Open holed to 403 ft, cored from 403 ft to bottom. Cores examined by R. F. Goossens. This borehole now functions as a well, water being extracted from the Upper Coal Measures.

DRIFT: Alluvium to 15 ft.

PERMO-TRIASSIC: LMS to 78 ft, UPM to 92 ft, UML to 118 ft, MPM to 197 ft, LML to 409 ft, LPM to 415¼ ft, BPS to 416 ft.

UPPER COAL MEASURES: Badsworth Div. to 615¼ ft, Hemsworth Div. to 894½ ft, Brierley Div. to 1082 ft 5 in (Blyth, c. 39 in at 1082 ft 5 in), Ackworth Div. to 1315¾ ft.

MIDDLE COAL MEASURES: Shafton M.B. at 1394 ft 10 in; Sharlston Top (d.c. 11 in) at 1618 ft 8 in; Sharlston Yard (d.c. 7 in) at 1747 ft; Mansfield M.B. at 1838 ft; Clown (c. with 2¼ in d.c. & d., 25 in) at 2031¼ ft; Two-Foot M.B. at 2169 ft 10 in; Two-Foot (ca. 3 in on c. 27 in) at 2172 ft 4 in; Abdy (c. 10½ in, d.c. 6 in, d. 3 in on c. 6½ in) at 2198¾ ft; Kilnhurst (c. & d. 29 in) at 2237 ft 10 in; High Hazles, Top Bed (c. 11 in on d.c. & d. 2½ in) at 2301¼ ft; Kent's Thin (c. 11¼ in, d.c. & d. 11½ in on d.c. 2¼ in) at 2326 ft 8 in; Top Hard (ca. & d.c. 1¼ in, c. 37¼ in, d. ½ in, c. 7½ in on d.c. & d. 1½ in) at 2573¾ ft; Blidworth + Dunsil (c. 13 in, d. 6 in, c. with ¾ in d., 24 in, d. 25½ in, d.c. & d. 5½ in, d.c. 8 in, d. 24 in, d.c. 5 in, d. 10 in on c. 4 in) at 2626 ft 2 in; 1st Waterloo (c. 3 in, d. 72 in, c. 10 in, d. 5 in on c. 3 in) at 2644¼ ft; Swallow Wood (c. with 4½ in d., 47 in, measures 32 ft 3½ in on ca. 25½ in) at 2760 ft 10 in; Haigh Moor (c. 27 in, d. 33½ in, c. 2½ in, d. 5½ in on c. 16 in) at 2779 ft 8 in; Lidget (d.c. 16 in) at 2842 ft 11 in; Clay Cross M.B. at 2980 ft 11 in.

LOWER COAL MEASURES: Flockton Thick (c. 28 in) at 3026 ft 4 in; Deep Soft (c. & d.c. 40 in, d. & d.c. 29 in, d.c. & c. 14 in, d. 13½ in, c. 32½ in, d. 9 in on d.c. 9 in) at 3092 ft; Parkgate, main part (d.c. 1¼ in, c. with 1 in d.c., 40¼ in, d.c. 2½ in, d. 50 in, c. 18 in on d.c. 2 in) at 3268 ft 5 in; Thorncliff (c. 38 in, d. ½ in on d.c. & d. 24½ in) at 3342 ft 8 in; Threequarters (c. 5 in, d. 20 in on c. 5 in) at 3389 ft 1 in; Top Silkstone (ca. 4½ in, d. 2 in, ca. 21½ in, c. 7½ in, d. 6½ in on d.c. 6 in) at 3466 ft; Low '*Estheria*' Band at 3493¼ ft; Low Silkstone (c. 14 in) at 3494 ft 8 in.

Bottom of borehole 3497 ft 2 in.

Blyth Pumping Station

There are two wells here, a few yards apart. No. 1 Well was sunk prior to 1909 to 200 ft (Lamplugh and Smith 1914, p. 54). The following notes apply to No. 2 Well.

Ht above O.D. abt 52 ft.   6-in SK 68 NW.
Site [6241 8714] 170 yd at 180° from St Mary and St Martin's Church, Blyth. Drilled 1951 by J. Thom Ltd for Worksop R.D.C.; now owned by Central Notts. Water Board. Open holed to 75 ft, cored 75 ft to bottom.

PERMO-TRIASSIC: PB to abt 74 ft, LMS to 198 ft, UPM to bottom of borehole at 200 ft.

Bothamsall (BP) Oil Bores

There are 21 bores in the Bothamsall Oilfield, of which 14 (Nos. 2 to 4, 8 to 13, 15, 16 and 19 to 21) lie within the area of 1-in Sheet 101 (see Fig. 3). Details of Nos. 1 and 5 bores (1-in Sheet 113), drilled into the Carboniferous Limestone Series, are given by Edwards (1967, pp. 271-4). Notes on No. 3 Bore are given below. Nos. 8, 9, 11, 12, 16, 19 and 21 bores were drilled into basal Coal Measures; Nos. 2, 4, 10, 13, 15 and 20 bores were drilled into the Millstone Grit Series abt 100 to 250 ft below the *G. cancellatum* M.B. horizon.

No. 3 Bore

R.T.E. above O.D. 113 ft.   6-in SK 67 SE.
Site [6632 7421] 920 yd at 83° from Normanton Larches Farm, Bothamsall. Drilled 1958-9 by BP Exploration Co. Ltd. Largely open holed, cored 3230–606 ft and 4525–41 ft. Cores examined by I. P. Stevenson, G. H. Rhys and G. D. Gaunt.

PERMO-TRIASSIC: PB + LMS to 570 ft, UPM to abt 580 ft, UML to abt 615 ft, MPM to abt 790 ft, LML + ?LPM to abt 1020 ft, BPS to 1026 ft.

MIDDLE COAL MEASURES: Top Hard at abt 1838 ft; Clay Cross M.B. at abt 2135 ft.

LOWER COAL MEASURES: Parkgate at abt 2356 ft; Kilburn horizon at abt 2865 ft; base of Coal Measures (Pot Clay M.B. presumed to be washed out) at abt 3380 ft.

MILLSTONE GRIT SERIES: *G. cancellatum* M.B. at abt 3414 ft; ?horizon of *R. superbilingue* M.B. at abt 3584 ft; igneous rocks 3590 ft 10 in to abt 3730 ft; base at abt 4405 ft.

CARBONIFEROUS LIMESTONE SERIES to bottom of bore at 4700 ft.

Bothamsall (BP) Water Borehole

Ht above O.D. 122 ft.   6-in SK 67 SE.
Site [6639 7419] 990 yd at 85° from Normanton Larches Farm, Bothamsall. Drilled 1961 by BP Exploration Co. Ltd.

PERMO-TRIASSIC: PB to bottom of borehole at 107 ft.

Carr Lane Borehole, Gainsborough

Ht above O.D. abt 15 ft.   6-in SK 88 NW.
Site [8154 8883] 1180 yd at 201° from Gainsborough station. Drilled 1961 by F. Smith & Son Ltd for Messrs. Sandars and Co. Ltd.

MADE GROUND to 2 ft.

DRIFT: Alluvium and Terrace deposits to abt 40 ft.

PERMO-TRIASSIC: KM to 724 ft, PB to bottom of borehole at 1101½ ft.

Checkerhouse Borehole

Ht above O.D. abt 110 ft.   6-in SK 67 NE.
Site [6583 7845] 770 yd at 303° from Morton Hill Farm, Babworth.
Drilled 1924–5 by Wigan Coal and Iron Co.

PERMO-TRIASSIC: PB to 420 ft 11 in, LMS to 537 ft 8 in, UML to 546 ft 4 in, MPM to 646 ft, LML to 954¾ ft, LPM to 965 ft 2 in, BPS to 965¾ ft.

UPPER COAL MEASURES: Brierley Div. to 1085¾ ft (Blyth, c. 14 in at 1085¾ ft), Ackworth Div. to 1283 ft 7 in.

MIDDLE COAL MEASURES: Top M.B. at 1290 ft; Mansfield M.B. at abt 1811 ft.

Bottom of borehole abt 1884 ft.

Note: depths below 1700 ft are not precise because of substantial deviation of the hole from the vertical.

Chequer House Pumping Station

Ht above O.D. 65 ft.   6-in SK 68 SW.
Site [6469 8149] 190 yd at 97° from Chequer House Farm, Ranby.
Drilled 1966 by George Shaw Co. Ltd for Central Notts. Water Board.

DRIFT: Terrace gravel to 4½ ft.

PERMO-TRIASSIC: PB to 308 ft, LMS + UPM to bottom of borehole at 395 ft.

Clark's Dye Works Borehole, East Retford

Ht above O.D. abt 49 ft.   6-in SK 78 SW.
Site [7025 8190], Hallcroft Works, 1030 yd at 234° from Moorgate House, Bolham. Drilled 1930 by C. Isler & Co. Ltd for Clark's Dye Works Ltd.

DRIFT: Terrace deposits to 11½ ft.

PERMO-TRIASSIC: PB to bottom of borehole at 350 ft.

An earlier (1911) borehole at this site was drilled to 250 ft.

Cottam Borehole

Ht above O.D. 12 ft.   6-in SK 87 NW.
Presumed site [8223 7987] 480 yd at 86° from Cottam station.
Drilled 1912. Cores examined by J. Ford. For details of section see Lamplugh and Smith 1914, p. 66.

DRIFT: Alluvium with peat, on sand and gravel to 47¼ ft.

PERMO-TRIASSIC: KM to 630 ft, GB to 700 ft 4 in, PB to bottom of borehole at 1458 ft.

Cowlishaw Plantation Borehole

Ht above O.D. 138 ft.   6-in SK 68 SW.
Site [6072 8253] 430 yd at 26° from Carlton Forest, Worksop.
Drilled 1963 by Foraky Co. for Central Notts. Water Board.
Open holed to 31 ft, cored 31 ft to bottom. Cores examined by C. G. Godwin.

PERMO-TRIASSIC: PB to between 80 and 104 ft, LMS to 170½ ft, UPM to 174 ft 8 in, UML to bottom of borehole at 177½ ft.

This is the first of a series of three boreholes in or near Cowlishaw Plantation. The second [6053 8219] was 201 ft deep, and the third [6124 8236] 200 ft deep.

Crookford Borehole

Ht above O.D. 141 ft.   6-in SK 67 NE.
Site [6749 7565] 420 yd at 61° from Crookford Farm, Elkesley.
Drilled by R. Timmins for N.C.B. Open holed to 1026 ft and from
1048 to 1077 ft, cored 1026–1048 ft and 1077 ft to bottom. Cores
examined by R. A. Eden.

PERMO-TRIASSIC: No details above 1026 ft, LML to between 1089 and 1103 ft, LPM to 1107 ft, BPS to $1107\frac{1}{4}$ ft.

UPPER COAL MEASURES: Ackworth Div. to 1223 ft.

MIDDLE COAL MEASURES: Top M.B. at 1230 ft 10 in; Shafton M.B. at 1279 ft 2 in; *Edmondia* Band at 1391 ft 8 in.

Bottom of borehole 1445 ft.

Dunham Bridge Pumping Station No. 1 Well

Ht above O.D. 17 ft.   6-in SK 87 SW.
Site [8210 7436] 650 yd at 103° from St Oswald's Church, Dunham.
Drilled 1945 for Lincoln City Water Department; now owned by
Lincoln and District Water Board.

DRIFT: Alluvium and Terrace deposits to abt 29 ft.

PERMO-TRIASSIC: KM to abt 538 ft, Waterstones to 717 ft, GB (including 10 ft of conglomerate at base) to 760 ft, PB to bottom of borehole at 1250 ft.

No. 2 Well [8263 7426] and No. 3 Well [8207 7386], respectively 1412 and 1360 ft deep, proved similar sections.

Eaton Borehole

Ht above O.D. 65 ft.   6-in SK 77 NW.
Site [7103 7810] 320 yd at 266° from Eaton Hall. Drilled 1956-7
by Foraky Co. for N.C.B. Cores examined by R. A. Eden,
T. R. W. Hawkins and G. H. Rhys.

	Thickness ft in	Depth ft in		Thickness ft in	Depth ft in
DRIFT			*Horridonia horrida, Lingula credneri, Strophalosia morrisiana,* cf. *Strobeus geinitzianus,* echinoid plates, ?worm burrows	70 0	1275 0
Glacial Sand and Gravel to abt 7 ft on clay with boulders to abt 14 ft..	14 0	14 0			
PERMO-TRIASSIC					
Not cored: Waterstones + GB to 130 ft, PB + LMS to abt 769 ft, UPM to 848 ft, UML to 905 ft, MPM to 1050 ft	1036 0	1050 0	LOWER PERMIAN MARL Marl, grey; carbonate bands in upper part, Marl Slate in lower part (probably below 1288 ft 8 in); cf. *Dielasma elongatum,* cf. *H. horrida, Strophalosia morrisiana,* indet. palaeoniscid remains	24 0	1299 0
LOWER MAGNESIAN LIMESTONE					
Not cored	155 0	1205 0	BASAL PERMIAN BRECCIA Conglomerate ..	0 10	**1299 10**
Dolomite, grey, and grey marl; *Acanthocladia anceps,* indet. cryptostome bryozoa, *Crurithyris clannyana,*					

UPPER COAL MEASURES

HEMSWORTH DIVISION

Stratum	Thickness ft in	Depth ft in
Mudstone breccia, red, grey and green ..	0 9	1300 7
Mudstone, red-mottled	12 5	1313 0
Mudstone, dark, with red mottling	0 4	1313 4
Seatearth, red-mottled	2 10	1316 2
Siltstone and silty mudstone, largely red or pink; seatearth-like band with sphaerosiderite in middle; ironstone in lower part; worm tracks and burrows	17 6	1333 8
Mudstone, grey and red; plants..	0 2	1333 10
Shale, black; plants ..	0 2	1334 0
Dirty coal 6 in ..	**0 6**	**1334 6**
Seatearth, coaly at top; sphaerosiderite ..	4 3	1338 9
Mudstone with pink-mottled bands; rootlets; sphaerosiderite..	12 2	1350 11
Seatearth with sphaerosiderite	5 1	1356 0
Siltstone with sandstone bands; ironstone; plants.. ..	20 2	1376 2

FOURTH CHERRY TREE MARKER

Stratum	Thickness ft in	Depth ft in
Mudstone, dark, with ironstone; *Anthraconaia pruvosti, Euestheria simoni*..	3 3	**1379 5**

BRIERLEY DIVISION

Stratum	Thickness ft in	Depth ft in
Seatearth with sphaerosiderite	2 7	1382 0
Siltstone with sandstone bands	10 8	1392 8
Seatearth	5 4	1398 0
Siltstone and sandstone	8 8	1406 8
Mudstone, largely silty; ironstone; *Anomalonema defretinae, ?'Estheria' sp.* ..	5 0	1411 8
Mudstone, dark; ironstone; shell fragments	0 10	1412 6
Seatearth with sphaerosiderite	10 4	1422 10
Coal 12 in / Dirt and coal 14 in ..	**2 2**	**1425 0**
Seatearth	3 0	1428 0
Mudstone, greenish, largely silty and siltstone; sphaerosiderite; plants..	22 0	1450 0
Seatearth	4 0	1454 0
Mudstone, silty, with ironstone	3 2	1457 2
Mudstone, dark; ironstone; *Euestheria simoni* and fish ..	4 6	1461 8
Coal 8 in (*no recovery*)	**0 8**	**1462 4**
Seatearth with sphaerosiderite	5 8	1468 0
Siltstone	5 6	1473 6
Mudstone, greenish, with dark bands; ironstone	8 6	1482 0

BLYTH

Stratum	Thickness ft in	Depth ft in
Coal 10 in	**0 10**	**1482 10**

ACKWORTH DIVISION

Stratum	Thickness ft in	Depth ft in
Seatearth	3 4	1486 2
Mudstone, partly silty, with sandstone bands	9 8	1495 10
Seatearth	1 4	1497 2
Coal 12 in (9¾ *in recovered*) ..	**1 0**	**1498 2**
Seatearth with sphaerosiderite	5 7	1503 9
Mudstone, greenish at top, dark at base; sphaerosiderite ..	7 6	1511 3
Seatearth with sphaerosiderite	2 8	1513 11
Mudstone, dark ..	0 9	1514 8
Dirty coal 2 in ..	**0 2**	**1514 10**
Seatearth with sphaerosiderite	5 5	1520 3
Mudstone, largely silty, and siltstone; ironstone	33 7	1553 10

	Thickness ft in	Depth ft in
Sandstone; ironstone; coal rafts	19 4	1573 2
Mudstone, largely silty; ironstone	5 3	1578 5
Sandstone with mudstone band	3 11	1582 4
Mudstone, partly silty; ironstone; plants ..	8 9	1591 1
Mudstone, dark ..	1 4	1592 5
Sandstone and siltstone	6 2	1598 7
Mudstone and siltstone; sandstone bands; ironstone; *Anthraconauta* aff. *phillipsii*	7 6	1606 1
Shale, canneloid, with fish debris	0 3	1606 4
SCOFTON **Dirty coal** 8 in ..	**0 8**	**1607 0**
Seatearth with sphaerosiderite	19 4	1626 4
Mudstone, greenish ..	3 2	1629 6
Mudstone with ironstone	1 6	1631 0
Coal 16 in	**1 4**	**1632 4**
Seatearth	2 7	1634 11
Mudstone, dark ..	1 8	1636 7
Seatearth	2 4	1638 11
Mudstone, largely silty, and siltstone, with sandstone bands ..	28 4	1667 3

MIDDLE COAL MEASURES

TOP MARINE BAND

	Thickness ft in	Depth ft in
Mudstone, largely dark; ironstone; *Glomospira* **sp.**, *Planolites ophthalmoides*	2 7	1669 10
Mudstone, dark; ironstone; *Crurithyris* sp., *Lingula mytilloides, Orbiculoidea* cf. *nitida, Coleolus* sp., *Platyconcha* aff. *hindi, Retispira?, Dunbarella* sp., *Palaeoneilo* sp., *Polidevcia* cf. *stilla,*		

	Thickness ft in	Depth ft in
Schizodus ?, Politoceras kitchini, cf. *Tomaculum* sp., ?faecal pellets, 'fucoids', fish remains	6 0	**1675 10**
Mudstone, dark ..	0 5	1676 3
Seatearth	2 9	1679 0
Siltstone and sandstone; plants..	12 3	1691 3
Mudstone, dark at base; *Naiadites* sp., *Lioestheria vinti*	2 3	1693 6
Clay (*fault gouge*) ..	2 1	1695 7
Dirty coal 13 in ..	**1 1**	**1696 8**
Seatearth	2 6	1699 2
Siltstone and sandstone; ironstone	10 2	1709 4
Mudstone with ironstone	2 0	1711 4
Mudstone, black; fish ..	0 1	1711 5
Seatearth with sphaerosiderite	3 5	1714 10
Mudstone, largely dark, and dark siltstone; ironstone; *L. vinti* 1721 ft 4–7 in ..	7 2	1722 0
SHAFTON MARINE BAND Mudstone, dark, with ironstone; foraminifera including *Ammodiscus sp.,Anthraconaia spathulata, L. vinti*	4 4	**1726 4**
Sandstone	30 7	1756 11
Siltstone and dark silty mudstone; plants ..	4 0	1760 11
Coal 10 in	**0 10**	**1761 9**
Seatearth with sphaerosiderite	7 11	1769 8
Siltstone with sphaerosiderite	2 4	1772 0
Seatearth and mudstone	4 6	1776 6
Siltstone with ironstone; coalified plants ..	23 1	1799 7
Coal 0¼ in	**0 0¼**	**1799 7¼**
Seatearth	1 10¾	1801 6

	Thickness ft in	Depth ft in
Mudstone, largely dark, with ironstone; *Naiadites* cf. *hindi*, *Carbonita* cf. *pungens*, fish remains	2 3	1803 9
Dirty coal 2 in ..	**0 2**	**1803 11**
Seaetarth	3 8	1807 7
Siltstone with sandstone bands; plants ..	6 7	1814 2
Mudstone, dark; *Naiadites sp.*, *L. vinti*, fish scales	3 0	1817 2
Seatearth; coal streaks at top..	10 7	1827 9
Striped beds	3 3	1831 0
Mudstone, dark with ironstone; plants and shells	0 9	1831 9
Mudstone	4 4	1836 1
Seatearth	8 11	1845 0
Siltstone and silty mudstone; ironstone ..	8 7	1853 7
Mudstone with ironstone; *Geisina subarcuata* at 1860¾ ft ..	8 2	1861 9
Edmondia BAND		
Mudstone, silty; foraminifera including *Tolypammina sp.*, *Geisina sp.*, *Hollinella sp.*	0 3	1862 0
Mudstone, dark, with ironstone;canneloid at base; foraminifera including *Glomospira sp.*, *Glomospirella sp.*, *Edmondia* cf. *transversa*, *Myalina compressa*, *Geisina sp.* [at base], *Hollinella sp.*, *Paraparchites ?*, *Rhabdoderma sp.* ..	7 5	**1869 5**
Shale, canneloid; *Naiadites melvillei*, fish ..	0 11	1870 4
Mudstone, silty; *N. sp.*, palaeoniscid scale ..	4 8	1875 0

	Thickness ft in	Depth ft in
Core broken; fault ..	0 6	1875 6
Mudstone, dark; *Spirorbis sp.*, *N. melvillei* ..	1 6	1877 0
Siltstone; *N. sp.* ..	6 4	1883 4
Mudstone, dark at base; *N.* cf. *daviesi*, *N. sp.* (cf. *productus*) ..	2 9	1886 1
Seatearth, greenish grey; sphaerosiderite ..	1 11	1888 0
Siltstone with ironstone; sphaerosiderite ..	8 6	1896 6
Mudstone, silty; *N. sp.* cf. *productus* ..	3 5	1899 11
Seatearth with canneloid shale band with shells	4 4	1904 3
Siltstone and sandstone	5 6	1909 9
Mudstone, dark; *Anthraconaia hindi*, *A.* aff. *stobbsi*, *Geisina subarcuata*, *N.* aff. *melvillei*, fish remains	4 3	1914 0
SHARLSTON TOP		
Coal and dirt 18 in (*9 in recovered*) ..	**1 6**	**1915 6**
Seatearth, dark ..	5 0	1920 6
Siltstone	11 1	1931 7
Mudstone, dark, with ironstone	6 11	1938 6
Seatearth; black shale band	6 8	1945 2
Siltstone	2 10	1948 0
Core lost	1 0	1949 0
Mudstone, largely dark; ironstone; *N.* aff. *hindi*, *Carbonita humilis*, *Rhizodopsis sp.*	5 7	1954 7
Core lost	1 0	1955 7
Mudstone, plants ..	0 10	1956 5
SHARLSTON LOW		
Coal and dirt 76 in[1]	**6 4**	**1962 9**
Seatearth	1 0	1963 9
Siltstone and sandstone	27 9	1991 6

[1]Detailed section: coal 10½ in (*2 in recovered*), dirt 26½ in, coal and dirt 11½ in (*2 in recovered*), dirt 26½ in, dirty coal 1 in.

	Thickness ft in	Depth ft in
Mudstone, silty at top; ironstone; cf. *Gyrochorte carbonaria* ..	6 1	1997 7
Mudstone, dark, with ironstone; shells, fish	3 3	2000 10
Seatearth with coal streaks	0 10½	2001 8½
WALES		
Coal and dirt 28½ in[1]	**2 4½**	**2004 1**
Seatearth	3 4	2007 5
Siltstone and sandstone	8 11	2016 4
Mudstone with ironstone	6 8	2023 0
Mudstone, mostly dark, with ironstone; *Planolites sp.*, fish	6 4	2029 4
MANSFIELD MARINE BAND		
Mudstone, dark; *Lingula mytilloides, Orbiculoidea* cf. *nitida*	1 3	2030 7
Mudstone, dark; ironstone; *L. mytilloides, Polidevcia sp., Belinurus* cf. *bellulus*, fish including palaeoniscid scales indet...	14 5	2045 0
Mudstone, dark; *Serpuloides sp., L. mytilloides, O.* cf. *nitida, Coleolus sp., Retispira?, Ianthinopsis?, Platyconcha hindi, Palaeoneilo sp., Polidevcia* cf. *acuta, 'Anthracoceras'?, Politoceras?,* orthocone nautiloid, palaeoniscid scales indet., cf. *Tomaculum sp.,* 'fucoids'	10 0	2055 0
Siltstone; *L. sp.,* chonetoid, *Angyomphalus?, Huanghoceras?*	2 5	2057 5
Mudstone, dark at base; ironstone	2 7	2060 0

	Thickness ft in	Depth ft in
Seatearth with sphaerosiderite	2 6	2062 6
Mudstone, greenish, and siltstone; sphaerosiderite	6 6	2069 0
Coal 2 in	**0 2**	**2069 2**
Seatearth, brownish ..	0 10	2070 0
Siltstone and mudstone, greenish, with sandstone band; sphaerosiderite	11 5	2081 5
Mudstone, dark at base; ironstone; *Anthraconaia?, Anthracosia* cf. *atra, A. rubida, Naiadites sp.* (cf. *productus*), fish? ..	5 2	2086 7
Seatearth, buff, with sphaerosiderite ..	6 8	2093 3
Mudstone, dark ..	3 7	2096 10
SUTTON MARINE BAND		
Mudstone, dark; *Lingula sp., O.* cf. *nitida, Megalichthys sp.*	0 9	**2097 7**
Mudstone, dark, with fish	0 2	2097 9
Coal 8 in		
Dirty coal 1 in ..	**0 9**	**2098 6**
Seatearth	4 10	2103 4
Siltstone and sandstone	5 8	2109 0
Mudstone, dark at base	2 9	2111 9
Coal 4 in	**0 4**	**2112 1**
Seatearth	2 1	2114 2
Mudstone, largely dark, silty, with ironstone; plants..	10 1	2124 3
HAUGHTON MARINE BAND		
Mudstone, dark silty; ironstone; *Serpuloides stubblefieldi, L.* cf. *elongata, O.* cf. *nitida*, fish remains, *Tomaculum?*	15 9	**2140 0**

[1]Detailed section: dirty coal 1½ in, dirt 2 in, dirty coal 11½ in, coal 7 in, dirt 1 in, coal 5½ in.

	Thickness ft in	Depth ft in
Dirty coal 8 in ..	0 8	**2140 8**
Seatearth with sphaerosiderite	8 2	2148 10
Mudstone, silty, with ironstone; plants ..	12 9	2161 7
Shale, black; mussels, fish	1 5	2163 0
Coal 5 in	0 5	**2163 5**
Seatearth	5 1	2168 6
Sandstone and siltstone	1 8	2170 2
Coal 3 in	0 3	**2170 5**
Fragmental Clay Rock	0 1	2170 6
Seatearth, brownish, with sphaerosiderite..	5 6	2176 0
Mudstone, mottled greenish; sphaerosiderite	4 0	2180 0
Mudstone, silty at base; ironstone	2 4	2182 4
Striped beds with plants	6 2	2188 6
Sandstone with coal rafts	5 6	2194 0
Mudstone, silty; cf. *Gyrochorte sp.* ..	2 8	2196 8
Mudstone, dark, with ironstone; *Lepidodendron sp.*, *Anthracosia atra*	2 8	2199 4
Coal 3 in	0 3	**2199 7**
Seatearth with coal streaks	0 6	2200 1
Mudstone, dark at base; shells	1 0	2201 1
CLOWN		
Coal and dirt 49 in[1]	4 1	**2205 2**
Seatearth	1 8	2206 10
Coal 2 in	0 2	**2207 0**
Seatearth	1 1	2208 1
Dirty coal 1 in ..	0 1	**2208 2**
Seatearth, brownish, with sphaerosiderite..	2 10	2211 0
Siltstone and sandstone	5 6	2216 6
Mudstone, silty at top; shells; *small fault* ..	8 0	2224 6
MANTON '*Estheria*' BAND Shale, dark; ironstone; '*Estheria*' sp. ..	2 0	**2226 6**
MAINBRIGHT Coal 30 in (27¼ in recovered)..	2 6	**2229 0**
Seatearth	3 6	2232 6
Siltstone with plants; *fault*	16 9	2249 3
Mudstone	0 9	2250 0
Shale, dark, with ironstone; pyrite and galena; *Naiadites sp.*, fish	1 11	2251 2
Striped beds ..	6 10	2258 0
Mudstone with oolitic ironstone	2 1	2260 1
Ironstone, dark ..	0 4	2260 5
Mudstone, dark silty; *Anthracosia* cf. *lateralis, N. sp.* ..	4 8½	2265 1½
TWO-FOOT MARINE BAND Mudstone, dark silty; *Lingula sp.* ..	0 0½	**2265 2**
TWO-FOOT **Coal** 21 in	1 9	**2266 11**
Seatearth	1 9	2268 8
Siltstone and sandstone	10 1	2278 9
Mudstone, largely dark; ironstone; *Spirorbis sp.*, *N. sp.*	6 3	2285 0
ABDY (OR WINTER) **Coal** 8 in **Dirty coal** 7 in ..	1 3	**2286 3**
Seatearth with coal streaks	2 9	2289 0
Coal and dirt 5 in Seatearth 3 in **Coal** 4 in (*2 in recovered*) Seatearth 15 in **Coal** 2 in	2 5	**2291 5**
Seatearth	3 7	2295 0
Sandstone and siltstone; *Annularia sp.*, *Neuropteris sp.*	41 3	2336 3
Coal 12 in	1 0	**2337 3**
Seatearth; coaly shale band	4 8	2342 0
Siltstone and sandstone	5 0	2347 0

[1]Detailed section: coal 3 in, dirt 14 in, dirty coal 5 in, coal 10 in, dirty coal 2 in, coal 15 in.

	Thickness ft in	Depth ft in
Mudstone and shale, largely dark, with silty bands; ironstone; *Spirorbis sp., Anthraconaia sp., Anthracosia caledonica?, A.* cf. *elliptica, A. sp.* cf. *fulva, A. lateralis, Anthracosphaerium sp.*	21 4	2368 4
HIGH HAZLES (TOP BED) Coal 13 in	1 1	2369 5
Seatearth; coaly shale band	5 9	2375 2
Siltstone and sandstone with ironstone ..	32 4	2407 6
Mudstone, mostly silty, dark at base, and siltstone; ironstone; *Naiadites* cf. *productus,* fish	7 2	2414 8
KENT'S THIN Cannel; mussels, 7 in Black shale; mussels, fish, 12 in Coal and dirt 6 in ..	2 1	2416 9
Seatearth	2 11	2419 8
Striped beds with ironstone	13 10	2433 6
Mudstone, mostly dark; ironstone; *Cochlichnus kochi,* shells	17 5	2450 11
Mudstone, dark, largely silty; ironstone; *Spirorbis sp., Anthraconaia sp., Anthracosia sp.* (intermediate between *caledonica* and *disjuncta*), *A. simulans, A.* cf. *variabilis, Carbonita humilis,* fish remains including *Palaeoxyris* cf. *helicteroides, Cochlichnus kochi* ..	6 6	2457 5

	Thickness ft in	Depth ft in
KENT'S THICK Coal and dirt 47 in[1]	3 11	2461 4
Seatearth	2 8	2464 0
Siltstone and sandstone; ironstone	16 7	2480 7
Mudstone, dark; mussels	6 11	2487 6
Ironstone, oolitic ..	0 1	2487 7
Coal 5 in	0 5	2488 0
Shale, coaly, and seatearth	4 9	2492 9
Siltstone and sandstone; plants..	21 11	2514 8
Mudstone, dark at base; *Planolites sp.,* mussels, sineoid track.. ..	6 4	2521 0
Coal and dirt 17 in[2]	1 5	2522 5
Seatearth	3 7	2526 0
Dirty coal 9½ in Dirt 6 in Dirty coal 6½ in ..	1 10	2527 10
Mudstone with coal streaks; seatearth ..	1 8	2529 6
Siltstone with ironstone Mudstone, dark; *Anthracosia* aff. *nitida,* palaeoniscid scales, vague sineoid tracks..	8 4 / 4 3	2537 10 / 2542 1
Coal 8 in	0 8	2542 9
Seatearth	1 3	2544 0
Siltstone and sandstone	31 6	2575 6
Mudstone, silty; mussels	5 10	2581 4
Shale, dark, with ironstone; mussels ..	0 8	2582 0
TOP HARD + DUNSIL Coal and dirt 97½ in[3]	8 1½	2590 1½
Seatearth	4 1½	2594 3
Mudstone, dark ..	1 7	2595 10
Seatearth	2 4	2598 2
Mudstone, silty; mussels	11 3	2609 5
Shale; coal streaks ..	0 11	2610 4

[1]Detailed section: coal 5 in, dirt ½ in, coal 3½ in, dirty coal ¾ in, coal 9 in, dirty coal 3¼ in, dirt 3 in, dirty coal 2 in, coal 16½ in, dirty coal 3½ in.

[2]Detailed section: dirty coal ¼ in, dirt 1 in, coal and dirt 5¾ in, coal 10 in.

[3]Detailed section: coal 6½ in, dirty coal 11 in, dirt and coal 7½ in, dirty coal 3 in, coal 5 in, dirty coal 6 in, dirt 13½ in, coal 40½ in, dirty coal 4½ in.

	Thickness ft in	Depth ft in
Seatearth	3 0	2613 4
Coal 9 in		
Coal and dirt 5 in ..	**1 2**	**2614 6**
Seatearth	1 0	2615 6
Siltstone and sandstone; ironstone	21 6	2637 0
Shale, dark	0 2	2637 2
1ST WATERLOO		
Cannel 3 in		
Dirt 1 in		
Coal 6 in	**0 10**	**2638 0**
Seatearth	4 6	2642 6
Siltstone and sandstone; ironstone	12 8	2655 2
Mudstone, dark, shaly and canneloid at base; ironstone; *Spirorbis sp., Anthracosia spp., Naiadites quadratus, Carbonita humilis, Rhabdoderma sp., Rhizodopsis sp.* ..	6 11	2662 1
WATERLOO MARKER		
Cannel 6 in	**0 6**	**2662 7**
Siltstone, dark at top; *Anthraconaia salteri, Anthracosia cf. beaniana, A. phrygiana, Anthracosphaerium aff. affine, N. quadratus*	10 5	2673 0
Mudstone with dark bands; ironstone; *Spirorbis sp., Anthraconaia robertsoni, Anthracosia beaniana, N. quadratus, C. humilis, Cochlichnus kochi, Gyrochorte carbonaria*	13 2	2686 2
SWALLOW WOOD +		
HAIGH MOOR		
Coal and dirt 14 in[1]		
Seatearth; canneloid shale at base 25 in		
Coal 25¼ in		
Dirt 5 in		
Coal 20¾ in	**7 6**	**2693 8**
Seatearth	3 4	2697 0
Sandstone, siltstone and silty mudstone ..	16 0	2713 0
Mudstone, mostly silty, with ironstone ..	5 0	2718 0
Mudstone, dark at base; *Spirorbis sp.*, mussels	3 6	2721 6
Coal 1½ in	**0 1½**	**2721 7½**
Shale, black, and seatearth	2 4½	2724 0
Sandstone and siltstone	13 0	2737 0
Mudstone, dark silty, with ironstone ..	10 4	2747 4
Siltstone and sandstone; oolitic ironstone at base	5 9	2753 1
Seatearth '	2 6	2755 7
Sandstone and siltstone	4 5	2760 0
Seatearth	1 7	2761 7
Striped beds; *?Anthracosphaerium affine* ..	26 1	2787 8
Mudstone, dark at base; ironstone; *Calamites sp., ?A. affine, Naiadites sp.*	2 6	2790 2
2ND ELL		
Coal and dirt 21 in[2]	**1 9**	**2791 11**
Seatearth	1 3	2793 2
Sandstone and siltstone; mudstone band ..	16 4	2809 6
Seatearth	0 2	2809 8
Striped beds with ironstone	13 11	2823 7
Siltstone and silty mudstone; ironstone ..	10 5	2834 0
Shale, dark, partly silty; *?Anthraconaia sp., Anthracosia cf. ovum, A. aff. phrygiana* ..	6 8	2840 8
CLAY CROSS MARINE BAND		
Shale, dark, silty at top; *Lingula mytilloides*	6 2	**2846 10**

[1]Detailed section: coal 5 in, dirt 2 in, coal 4 in, dirt 1 in, dirty coal 2 in.

[2]Detailed section: coal 11½ in, dirt 1 in, coal 5½ in, dirty coal 3 in.

	Thickness ft in	Depth ft in
LOWER COAL MEASURES		
Shale, dark, with plants	0 2	2847 0
JOAN		
Coal 16 in		
(*12 in recovered*)		
Seatearth 16 in		
Dirty coal 5 in	**3 1**	**2850 1**
Seatearth	5 6	2855 7
Mudstone, silty; plants, *Naiadites sp.* ..	36 9	2892 4
Bat with ironstone ..	1 2	2893 6
Seatearth	1 9	2895 3
Siltstone and sandstone	6 11	2902 2
Mudstone, dark, silty at top; mussels, ostracods	5 2	2907 4
Ironstone, dark ..	0 1½	2907 5½
DEEP SOFT		
Coal and dirt 60½ in[1]	**5 0½**	**2912 6**
Seatearth	2 11	2915 5
Sandstone and siltstone	108 10	3024 3
PARKGATE (ROOF COAL)		
Coal 8½ in		
(*7½ in recovered*)		
Dirt 5 in		
Dirty coal 2 in	**1 3½**	**3025 6½**
Seatearth with coal streaks at top ..	3 8½	3029 3
Striped beds	13 3	3042 6
Siltstone and sandstone	6 10	3049 4
Mudstone, dark, with shells	1 3	3050 7
PARKGATE (MAIN PART)		
Coal 24½ in	**2 0½**	**3052 7½**
Seatearth	5 6½	3058 2
Siltstone and sandstone	1 2	3059 4
Mudstone, dark, with ironstone; fish including *Rhizodopsis sp.*	1 10	3061 2
Seatearth, sandy ..	1 2	3062 4
Mudstone; plants ..	0 6	3062 10
Sandstone, silty ..	9 7	3072 5
Ironstone	0 1	3072 6
Coal 4 in	**0 4**	**3072 10**
Shale, dark, and seatearth	2 4	3075 2
Sandstone and siltstone	9 6	3084 8
Mudstone; ironstone and shells	1 11	3086 7
Shale, dark; *Naiadites sp.*, *Rhizodopsis sp.* ...	0 8	3087 3
COCKLESHELL		
Coal 4 in		
(*2½ in recovered*) ..	**0 4**	**3087 7**
Seatearth, dark ..	3 5	3091 0
Mudstone, dark silty; fish	6 0	3097 0
Sandstone and siltstone; ironstone	24 0	3121 0
Seatearth	6 6	3127 6
Mudstone, largely silty; ironstone; plants ..	13 9	3141 3
Dirt and coal 23½ in[2]	**1 11½**	**3143 2½**
Seatearth	5 5	3148 7½
THREEQUARTERS		
Coal 37 in		
Dirty coal 1 in	**3 2**	**3151 9½**
Seatearth, coaly at top..	3 8½	3155 6
Mudstone, dark silty; plants..	7 10	3163 4
Mudstone, dark at base; ironstone; plants, *Curvirimula sp.*, fish..	23 2	3186 6
TOP SILKSTONE		
Coal and dirt 46 in[3]	**3 10**	**3190 4**
Mudstone, silty, and seatearth	1 10	3192 2
Siltstone and sandstone	8 2	3200 4
LOW '*Estheria*' BAND		
Mudstone; '*Estheria*' sp. nov.	1 9	**3202 1**

[1]Detailed section: coal 5½ in, dirt 15½ in, coal 15½ in, dirt ½ in, coal 4¼ in, dirt ¼ in, coal 14¼ in, dirty coal 4½ in.

[2]Detailed section: dirt and coal 1 in, dirt 7 in, coal 1 in, dirt 1 in, coal 7 in, dirt 4 in, coal 2½ in.

[3]Detailed section: dirty coal 2½ in, dirt 11 in, dirty coal 12½ in, coal 2 in, dirty cannel 2½ in, coal 3½ in, coal and dirt ½ in, dirt 10 in, inferior cannel 1½ in.

R

	Thickness ft in	Depth ft in		Thickness ft in	Depth ft in
LOW SILKSTONE			Mudstone with siltstone		
Coal 8 in	0 8	3202 9	bands	8 0	3257 9
Seatearth	12 5	3215 2	Shale, dark; plants, fish	0 8	3258 5
Mudstone with ironstone	12 2	3227 4	Dirt and coal 4 in	0 4	3258 9
Seatearth	1 8	3229 0	Seatearth	2 4	3261 1
Mudstone, dark at base; plants..	13 10	3242 10	Coal 5 in	0 5	3261 6
Seatearth	1 1	3243 11	Seatearth	5 4	3266 10
Coal and dirt 22 in[1]	1 10	3245 9	Sandstone and silty		
Seatearth; coal streaks at top..	4 0	3249 9	mudstone	4 6	3271 4

Elkesley Borehole

Ht above O.D. 137 ft.   6-in SK 67 NE.
Site [6788 7603] 1290 yd at 296° from Elkesley Church. Drilled
1962 by Foraky Co. for N.C.B. Open holed to 1820 ft 7 in,
cored 1820 ft 7 in to bottom. Cores examined by C. G. Godwin,
J. G. O. Smart, R. E. Elliott and M. Lock.

PERMO-TRIASSIC: PB + LMS to abt 700 ft, MPM to abt 850 ft, LML to abt 1118 ft,
LPM to abt 1136 ft, ?BPS to abt 1137 ft.

UPPER COAL MEASURES: base not defined, but probably at abt 1200 ft.

MIDDLE COAL MEASURES: ?position of *Edmondia* Band at 1480 ft[2]; ?Sharlston Top
(c. abt 36 in) at 1543 ft[2]; Wales (c. abt 28 in) at 1643 ft[2]; position of Mansfield M.B.
at abt 1712 ft[2]; Mainbright (d.c. 6 in, d. 2 in, c. 8 in, d.c. 4 in, d. 35 in, d.c. 5 in, ca. 4
in on c. 14 in) at $1906\frac{3}{4}$ ft; Two-Foot M.B. at 1924 ft 10 in; Two-Foot (c. 18 in, measures
65 in on c. 8 in) at 1931 ft 5 in; Abdy (d.c. & d. 17 in on c. 1 in) at $1956\frac{3}{4}$ ft; Kilnhurst
(c. 13 in) at 1994 ft 1 in; High Hazles, Top Bed (c. & d.c. 10 in) at 2042 ft 7 in; Kent's
Thin (d.c. 8 in) at 2069 ft 2 in; Kent's Thick (c. 21 in, d. $3\frac{1}{2}$ in, c. $13\frac{1}{2}$ in on d.c. 8 in) at
$2115\frac{1}{4}$ ft; Top Hard + Dunsil (c. & d. 5 in, d. 5 in, c. 48 in, d. 6 in, c. $27\frac{1}{2}$ in on d.c. 7
in) at $2285\frac{1}{2}$ ft; 1st Waterloo (d.c. 6 in, measures 88 in on c. 9 in) at 2317 ft 7 in; 2nd Ell
(c. 22 in on d.c. 1 in) at 2486 ft 8 in; Clay Cross M.B. at $2541\frac{1}{2}$ ft.

LOWER COAL MEASURES: Deep Soft (c. 16 in, d. $2\frac{1}{2}$ in, c. $13\frac{1}{2}$ in, d.c. & d. $9\frac{1}{2}$ in, c.
$2\frac{1}{2}$ in on d.c. 2 in) at 2641 ft 2 in; Deep Hard (c. 8 in) at 2728 ft 2 in; Parkgate (c. 16 in,
d. 1 in, c. $43\frac{1}{2}$ in on d.c. $3\frac{1}{2}$ in) at 2774 ft 10 in; Thorncliff (c. 14 in, measures 26 ft 5 in on
c. 14 in) at 2857 ft 1 in.

Bottom of borehole 2906 ft 5 in.

Elkesley Pumping Station

There are five boreholes at this pumping station, four of which
were over 570 ft deep, and the fifth 330 ft deep. The following
notes apply to the first and deepest borehole.

Ht above O.D. abt 112 ft.   6-in SK 67 NE.
Site [6635 7596] 1050 yd at 302° from Crookford Farm, Elkesley.
Drilled 1909 by A. C. Potter & Co. for Lincoln Corporation;
now owned by Lincoln and District Water Board. For detailed
section see Lamplugh and Smith 1914, p. 69.

DRIFT: Sandy boulder clay, not recorded.

PERMO-TRIASSIC: PB + LMS to 581 ft, UML to bottom of borehole at 587 ft 4 in.

---

[1]Detailed section: coal $11\frac{1}{2}$ in, dirt $2\frac{1}{2}$ in, dirty coal 8 in.

[2]Interpreted from gamma-ray log.

Everton Pumping Station

There are seven boreholes at this pumping station, including that at Rye Hall [6923 8982] described by Lamplugh and Smith 1914, p. 70. The following notes apply to the deepest borehole, now known as No. 3.

Ht above O.D. 50 ft.   6-in SK 69 SE.
Site [6933 9005] 290 yd at 252° from Pusto Hill Farm, Everton. Drilled 1964 by Boldon Drilling and Engineering Co. for Lincoln and District Water Board.

PERMO-TRIASSIC: PB + LMS to abt 678 ft, UPM to abt 793 ft, ?UML to bottom of borehole at 798½ ft.

Firbeck Main Colliery (No. 1 Shaft)

Ht above O.D. 117 ft.   6-in SK 58 NE.
Site [5824 8597] 1380 yd at 290° from Woodhouse Cottages, Carlton in Lindrick. Sunk 1925. For detailed section see Wilcockson 1950, pp. 164–7; Edwards 1951, pp. 171–2.

PERMO-TRIASSIC: UML to 5 ft 4 in, MPM to 123 ft 7 in, LML to 195½ ft, LPM to 231 ft 8 in (BPS absent).

UPPER COAL MEASURES: Badsworth Div. to ?380¾ ft, Hemsworth Div. to ?664 ft 11 in, Brierley Div. to 847 ft 11 in (Blyth, c. 26 in at 847 ft 11 in), Ackworth Div. to abt 1085 ft.

MIDDLE COAL MEASURES: Horizon of Shafton M.B. at 1201 ft 2 in; Sharlston Top (c. 28 in) at 1459¼ ft; Sharlston Low (c. 12 in) at 1540 ft 10 in; Sharlston Yard (c. 13 in) at 1594 ft 4 in; Mansfield M.B. at 1701 ft 2 in; Clown (c. 31 in) at 1924½ ft; Two-Foot M.B. at 2079 ft 2 in; Two-Foot (c. 37 in) at 2082¼ ft; Abdy (c. & d. 35 in) at 2112¾ ft; Kilnhurst (c. 24 in, d. 6 in on c. 8 in) at 2138 ft 4 in; High Hazles, Top Bed (c. 5 in) at 2190¼ ft; Kent's Thin (c. 14 in) at 2218¼ ft; Kent's Thick, top leaf (c. 12 in) at 2290 ft; bottom leaf (c. 13 in, d. 24 in on c. 18 in) at 2312 ft 10 in; Top Hard (c. 50 in, d. 4 in on c. 3 in) at 2484 ft 5 in; Dunsil (c. 20 in, d. 43 in on c. & d. 30 in) at 2556¾ ft.

Bottom of shaft 2569 ft 1 in.

Firbeck No. 1 Underground Borehole

Top in Barnsley workings 2401 ft below O.D.   6-in SK 58 NE.
Site [5848 8706] 1770 yd at 190° from St Mark's Church, Oldcoates. Drilled 1945–6 by Cementation Co. for N.C.B. Open holed to 13 ft, cored from 13 ft to bottom. Cores examined by W. N. Edwards.

MIDDLE COAL MEASURES: Dunsil (c. 24 in, d. & c. 60 in on c. 15 in) at 93¼ ft; Haigh Moor (c. 21 in) at 250¾ ft; Clay Cross M.B. at abt 448 ft.

LOWER COAL MEASURES: Thorncliff (c. 21 in) at 759½ ft. Other seams lost.

Bottom of borehole 940 ft.

Firbeck No. 2 Underground Borehole

Top in Barnsley workings 2472 ft below O.D.   6-in SK 58 NE.
Site [5889 8891] 330 yd at 37° from St Mark's Church, Oldcoates. Drilled 1954 by Craelius Co. for N.C.B. Cored throughout. Cores examined by R. F. Goossens.

MIDDLE COAL MEASURES: Dunsil (c. abt 18 in) at 59 ft 4 in; Swallow Wood (c. 7 in, d. 3 in, c. 12 in, d. 16 in, c. 12 in, measures 33 ft 3 in, ca. 7 in, d. 2 in on c. 14 in) at 205 ft 2 in; Haigh Moor (c. 26 in, d. & d.c. 17 in on c. 17 in) at 220¾ ft; Clay Cross M.B. at 405 ft 11 in.

LOWER COAL MEASURES: Flockton Thick (c. 22 in) at 460 ft 10 in; Deep Soft (c. 9 in,
d.c. 4 in, d. 16 in, d.c. & d. 24 in, measures $6\frac{1}{4}$ ft, d.c. 3 in, c. 7 in, d. & d.c. 22 in, c. 31 in, d.
9 in on c. with 2-in d.c. band, 11 in) at 526 ft 4 in; Parkgate (c. 14 in) at 685 ft 11 in; Thorn-
cliff (c. $35\frac{1}{2}$ in on d.c. with $4\frac{1}{2}$ in d., $29\frac{1}{2}$ in) at 760 ft 11 in; Threequarters (c. 6 in, d. 18 in on
c. 3 in) at $812\frac{3}{4}$ ft; Top Silkstone (ca. $5\frac{1}{2}$ in, d. 2 in, ca. 27 in, c. $16\frac{1}{4}$ in on d.c. $1\frac{1}{4}$ in) at
897 ft 4 in; Low '*Estheria*' Band at 927 ft 8 in; Low Silkstone (ca. 3 in, c. $13\frac{1}{2}$ in on d.c. $1\frac{1}{2}$
in) at 929 ft 2 in.

Bottom of borehole 978 ft 4 in.

Forest Hill Borehole

Ht above O.D. 185 ft.   6-in SK 58 SE.
Site [5988 8138] 390 yd at 40° from Forest Hill, Worksop. Drilled
1955 by Craelius Co. for N.C.B. Open holed to 556 ft, cored 556 ft
to bottom. Cores examined by R. F. Goossens.

PERMO-TRIASSIC: PB to abt 185 ft, LMS to abt 280 ft, UPM to abt 287 ft, UML to abt
290 ft, MPM to abt 384 ft, LML to 559 ft, LPM to 567 ft 8 in, BPS to $568\frac{3}{4}$ ft.

UPPER COAL MEASURES: Hemsworth Div. to $?769\frac{3}{4}$ ft, Brierley Div. (base faulted,
Blyth missing) to $904\frac{1}{4}$ ft, Ackworth Div. to $1099\frac{1}{2}$ ft.

MIDDLE COAL MEASURES: Top M.B. at $1109\frac{3}{4}$ ft; Shafton M.B. at $1204\frac{1}{4}$ ft; Sharlston
Low, ?top leaf (d.c. 6 in) at 1541 ft; Sharlston Yard (d.c. 3 in) at $1606\frac{1}{4}$ ft; ?Houghton
Thin (d.c. & d. 6 in) at 1617 ft 10 in; Mansfield M.B. at $1692\frac{1}{2}$ ft; Clown (c. 32 in) at
1851 ft 7 in; Meltonfield (c. 21 in) at 1916 ft; Two-Foot M.B. at $1964\frac{3}{4}$ ft; Two-Foot
(d.c. $2\frac{1}{2}$ in on c. $28\frac{1}{2}$ in) at 1967 ft 4 in; Abdy (d.c. & d. 30 in) at $1996\frac{1}{2}$ ft; Kent's Thick,
top leaf (d.c. 2 in, d. 3 in on c. 8 in) at 2098 ft 10 in; Top Hard (c. 4 in, d.c. & d. $\frac{1}{2}$ in,
c. $26\frac{1}{2}$ in, d. $1\frac{1}{4}$ in on c. $3\frac{3}{4}$ in) at 2263 ft 4 in; Dunsil (c. 15 in) at 2316 ft 8 in; 1st Waterloo
(d.c. 3 in, d. 2 in on c. 4 in) at $2372\frac{3}{4}$ ft; Swallow Wood, Main Bed (d.c. 6 in, d. & measures
9 ft 3 in on d.c. 4 in) at 2420 ft 5 in; Haigh Moor (c. 10 in, d. 26 in on c. 18 in) at 2460 ft 10
in; Lidget (c. 1 in) at 2528 ft 10 in; Clay Cross M.B. at $2666\frac{1}{2}$ ft.

LOWER COAL MEASURES: Deep Soft (c. with $\frac{3}{4}$ in d., 20 in, d. & measures $55\frac{1}{2}$ in, c.
2 in, d. & d.c. $17\frac{1}{2}$ in, c. & d. 8 in, d. & d.c. 16 in on c. & d. 4 in) at $2787\frac{1}{4}$ ft; ?Deep
Hard Roof Coal (c. 8 in) at 2842 ft 2 in; Deep Hard (d.c. $7\frac{1}{2}$ in, d. $2\frac{1}{2}$ in on c. 15 in) at
2860 ft 4 in; Parkgate (c. 4 in, d. & d.c. 2 in, c. 14 in, measures 11 ft 2 in on c. 6 in) at 2904
ft 4 in.

Bottom of borehole 2965 ft 8 in.

Gainsborough (BP) Oil Bores

There are 53 bores in the Gainsborough Oilfield, of which 50
(all except Nos. 25, 26 and 40) lie within the area of 1-in Sheet 101
(see Fig. 3). Ten (Nos. 45D to 54D) are inclined bores drilled from
the sites of Nos. 29 and 43 bores (Fig. 3). Five bores (including
Nos. 25, 26 and 40) are on the Lias outcrop, six on the Rhaetic
(see Fig. 31) and the rest on the Keuper Marl. All, except Nos. 57
and 60 bores, which ended in Middle and Lower Coal Measures
respectively, and Nos. 1 to 5 and 43 bores, were drilled into the top
400 ft of the Millstone Grit Series. No. 1 Bore proved abt 1075 ft
of Millstone Grit rocks, No. 2 abt 1760 ft, No. 3 abt 1100 ft
(faulted), and Nos. 4, 5 and 43 between 500 and 675 ft. Notes on
Nos. 1 and 2 bores are given below.

No. 1 Bore

R. T. E. above O.D. 99 ft.   6-in SK 89 SW.
Site [8326 9026] 440 yd at 145° from Highfield Grange, Gainsborough. Drilled 1958–9 by BP Exploration Co. Ltd. Largely open

holed, cored 2240–55 ft, 3642–4080 ft, 4250–960 ft (with small gaps) and 5678 ft to bottom of bore. Cores examined by G. D. Gaunt, G. H. Rhys and I. P. Stevenson.

DRIFT: Boulder clay, not recorded.

PERMO-TRIASSIC: Rhaetic to 28 ft, KM to 895 ft (Tea Green Marl to abt 44 ft), PB + LMS to abt 1804 ft, UPM to 1985 ft, UML to 2065 ft, MPM to abt 2210 ft, LML to abt 2510 ft, LPM to abt 2525 ft, BPS to 2555 ft.

MIDDLE COAL MEASURES: ?Kent's Thick at abt 3194 ft; Top Hard at abt 3321 ft; position of Clay Cross M.B. at abt 3635 ft.

	Thickness ft in	Depth ft in
LOWER COAL MEASURES	From	3642 0
Sandstone	2 3	3644 3
Mudstone, dark, silty at top; megaspores, *Spirorbis sp., Anthracosia regularis, Naiadites sp., Carbonita humilis* [rare], *Geisina arcuata* [common] (*2½ ft core lost*)	9 8	3653 11
Seatearth; sphaerosiderite	2 1	3656 0
Mudstone, silty ..	2 6	3658 6
Sandstone and siltstone (*7 ft 4 in core lost*) ..	56 3	3714 9
Core lost	5 3	3720 0
Sandstone; thin bands of breccia (*5 ft 11 in core lost*)	20 8	3740 8
Coal 2 in	0 2	3740 10
Seatearth	0 11	3741 9
Core lost	1 3	3743 0
DEEP SOFT **Coal** 12 in[1]	1 0	3744 0
Mudstone, carbonaceous	0 1	3744 1
Seatearth	2 7	3746 8
Mudstone, silty; sphaerosiderite ..	6 4	3753 0
Mudstone, silty, and siltstone; thin sandstone bands; plants ..	32 0	3785 0
Siltstone and sandstone (*1 ft 8 in core lost*) ..	53 0	3838 0

	Thickness ft in	Depth ft in
Mudstone, silty ..	4 8	3842 8
Sandstone and siltstone; thin bands of breccia in lowest part ..	32 4	3875 0
Core lost	2 0	3877 0
PARKGATE **Coal** 36 in[2]	**3 0**	**3880 0**
Seatearth	4 6	3884 6
Mudstone, dark; *Spirorbis sp., Anthracosphaerium ?sp. nov., Carbonicola oslancis, C. rhomboidalis, Carbonita humilis, G. arcuata,* fish remains, including palaeoniscid scales, *Rhizodopsis sp.*	8 8	3893 2
Sandstone	2 10	3896 0
Siltstone, ?rootlets ..	2 6	3898 6
Mudstone, silty in parts; ironstone; *Spirorbis sp., Carbonicola oslancis, C. rhomboidalis, C.* cf. *robusta, Naiadites flexuosus, G. arcuata*	5 6	3904 0
THORNCLIFF (TOP LEAF) **Coal trace** (*no thickness given, no recovery*) ..		3904 0
Seatearth and mudstone; ?mussels	10 7	3914 7
Mudstone, silty ..	6 5	3921 0
Mudstone, dark; coaly plants..	1 0	3922 0

[1]Thickness recorded by BP Company; 1 in recovered.

[2]Thickness estimated from geophysical logs; 1 in recovered.

THORNCLIFF	Thickness ft in	Depth ft in
(BOTTOM LEAF)		
Coal trace		
(*no thickness given,*		
no recovery) ..		3922 0
Seatearth, silty ..	4 0	3926 0
Mudstone, silty (*2 ft core lost*)	8 0	3934 0
Sandstone	8 0	3942 0
Mudstone, dark, coaly	2 4	3944 4
Seatearth	2 8	3947 0
Mudstone, silty, and siltstone	5 1	3952 1
Sandstone	17 5	3969 6
THREEQUARTERS		
Coal 12 in[1]	**1 0**	**3970 6**
Seatearth	5 6	3976 0
Mudstone, silty ..	3 0	3979 0
Core lost	1 0	3980 0
Mudstone	5 0	3985 0
Mudstone, dark silty, and siltstone; *Carbonicola* aff. *pseudorobusta*, *Curvirimula candela* ..	9 5	3994 5
Seatearth	0 3	3994 8

	Thickness ft in	Depth ft in
Siltstone	8 4	4003 0
Mudstone; ironstone	14 10	4017 10
Mudstone, dark; fish remains including *Rhabdoderma sp.* ..	2 0	4019 10
Seatearth	1 2	4021 0
TOP SILKSTONE		
Coal 12 in[2]	**1 0**	**4022 0**
Seatearth	2 7	4024 7
Sandstone	7 5	4032 0
Core lost, probably sandstone	10 0	4042 0
Sandstone	7 8	4049 8
Mudstone, silty, and siltstone	4 10	4054 6
Mudstone; ironstone ..	5 6	4060 0
LOW '*Estheria*' BAND Mudstone, carbonaceous; '*Estheria*' *sp. nov.* [widely spaced lirae], *Rhabdoderma sp.*	2 4	**4062 4**
Seatearth; thin carbonaceous mudstone band near top	5 5	4067 9
Mainly silty mudstone ..	12 3	4080 0

Kilburn horizon at 4312½ ft; ?Alton M.B. at 4570 ft 7 in; Pot Clay M.B. at 4615 ft 11 in. MILLSTONE GRIT SERIES: ?*G. cancellatum* M.B. at abt 4690 ft; ?*R. superbilingue* M.B. at 4723¼ ft; ?*R. bilingue* M.B. at 4850 ft 4 in. For details of cores 4690–960 ft see pp. 19–20. Bottom of bore 5706 ft.

No. 2 Bore

R.T.E. above O.D. 104 ft.   6-in SK 89 SW.
Site [8177 9079] 1390 yd at 280° from Highfield Grange, Gainsborough. Drilled 1959 by BP Exploration Co. Ltd. Largely open holed, cored 3202–16 ft, 3378–90 ft, 3920–4003 ft, 4530–691 ft and 5294–8 ft. Cores examined by G. H. Rhys and I. P. Stevenson.

PERMO-TRIASSIC: KM to abt 792 ft, PB + LMS to abt 1720 ft, UPM to abt 1870 ft, UML to abt 2000 ft, MPM to abt 2110 ft, LML to abt 2440 ft, LPM to abt 2455 ft, BPS to abt 2484 ft.

MIDDLE COAL MEASURES: ?Kent's Thick at abt 3085 ft; position of Clay Cross M.B. at abt 3580 ft.

LOWER COAL MEASURES: ?Parkgate at abt 3801 ft; Low '*Estheria*' Band at 3994½ ft; Kilburn at abt 4285 ft; position of Pot Clay M.B. at abt 4500 ft.

MILLSTONE GRIT SERIES: ?*G. cancellatum* M.B. at abt 4557 ft; dolerite between 6065 ft and 6170 ft.

Bottom of bore 6259 ft.

---

[1]Thickness recorded by BP Company; 2 in recovered.
[2]Thickness recorded by BP Company; 4 in recovered.

Gainsborough Waterworks (Lea Road Pumping Station)

There are three boreholes at this pumping station. No. 1 was drilled in 1885–7 to 1351 ft and No. 2 in 1894–1900 to 1515 ft—for details see Lamplugh and Smith 1914, p. 72. The following notes apply to No. 3 Borehole.

Ht above O.D. abt 20 ft.   6-in SK 88 NW.
Site [8161 8892] 1060 yd at 199° from Gainsborough station. Drilled 1933–8 by C. Isler & Co. Ltd for Gainsborough U.D.C.; now owned by Lincoln and District Water Board.

DRIFT: Terrace deposits to abt 15 ft.

PERMO-TRIASSIC: KM to abt 710 ft, PB to bottom of borehole at 1634 ft.

Gamston Wells

There are two wells of interest in the village of Gamston [6-in SK 77 NW]. Both were drilled in 1912 by Le Grand, Sutcliffe and Gell Ltd. The northern well [7092 7644], 460 yd at 8° from St Peter's Church and abt 60 ft above O.D. was dug to 18½ ft, sunk through Green Beds to abt 50½ ft and through Pebble Beds to the bottom at 103 ft. The southern well [7092 7633], 340 yd at 10° from St Peter's Church and abt 80 ft above O.D., was dug to 44 ft, sunk through Green Beds to 72 ft and through Pebble Beds to the bottom at 130 ft.

Gringley (Anglo-American) No. 1 Oil Bore

R.T.E. above O.D. 252 ft.   6-in SK 79 SW.
Site [7457 9064] 1020 yd at 92° from St Peter and St Paul's Church, Gringley on the Hill. Drilled 1945 by Anglo-American Oil Co. Ltd. Largely open holed, but cores taken at intervals between 4836 ft and 5441 ft.

DRIFT: Glacial Sand and Gravel to abt 40 ft.

PERMO-TRIASSIC: KM with GB at base to ?630 ft (but see p. 180), PB + LMS to abt 1405 ft, UPM to abt 1455 ft, UML to abt 1525 ft, MPM to abt 1620 or abt 1655 ft, LML to abt 1892 ft (LPM and BPS ? absent).

UPPER COAL MEASURES: base not recorded but probably at abt 2230 ft.

MIDDLE COAL MEASURES: Top Hard at abt 3297 ft; position of Clay Cross M.B. at abt 3700 ft.

LOWER COAL MEASURES: Deep Soft at abt 3795 ft; Parkgate at abt 3895 ft; Top Silkstone at abt 4094 ft; Kilburn at abt 4495 ft; ?Forty-Yards M.B. at 4871 ft; position of Pot Clay M.B. at abt 5000 ft.

MILLSTONE GRIT SERIES to bottom of bore at 5574 ft.

Grove Hall Borehole

Ht above O.D. 275 ft.   6-in SK 77 NW.
Site [7396 7993] 510 yd at 25° from St Helen's Church, Grove. Drilled 1907. For detailed section see Lamplugh and Smith 1914, p. 75.

PERMO-TRIASSIC: KM + Waterstones to 385 ft, GB (see p. 180) to 436 ft, PB to bottom of borehole at 602 ft.

Grove (BP) Oil Bores

There are two oil bores at Grove. Notes on No. 1 are given
below. No. 2 [7508 8157] was drilled to 4730 ft into the Millstone
Grit Series abt 375 ft below the *G. cancellatum* M.B. horizon.

No. 1 Bore

R.T.E. above O.D. 224 ft.   6-in SK 78 SE.
Site [7523 8070] 510 yd at 333° from Grovemoor Farm, Grove.
Drilled 1960 by BP Exploration Co. Ltd. Largely open holed, cored
4066–202 ft and 5154 ft to bottom of bore.

PERMO-TRIASSIC: KM to abt 390 ft, ?Waterstones to 400 ft, GB to 450 ft, PB to abt
1220 ft, LMS + UPM to 1330 ft, UML to 1385 ft, MPM to 1530 ft, LML to abt 1815 ft,
LPM to 1830 ft (no BPS recorded).

UPPER COAL MEASURES to abt 2200 ft.

MIDDLE COAL MEASURES: Mainbright at abt 2765 ft; Top Hard at abt 3120 ft; horizon
of Clay Cross M.B. at abt 3400 ft.

LOWER COAL MEASURES: Low Silkstone at abt 3780 ft; Kilburn at abt 3980 ft; base
at abt 4340 ft.

MILLSTONE GRIT SERIES to abt 5090 ft.

CARBONIFEROUS LIMESTONE SERIES to bottom of bore at 5161 ft.

Grove Pumping Station

There are four boreholes at this pumping station, drilled between
1957 and 1963. They are all 1100 ft deep and they provide similar
sections. The notes below apply to No. 2.

Ht above O.D. abt 295 ft.   6-in SK 78 SW.
Site [7407 8033] 1170 yd at 143° from Little Gringley. Drilled 1958
by F. Smith and Son Ltd for East Retford R.D.C.; now owned by
Lincoln and District Water Board.

PERMO-TRIASSIC: KM to abt 348 ft, ?Waterstones (see p. 183) to 415 ft, GB (for details
see p. 180) to 465 ft, PB to bottom of borehole at 1100 ft.

Harness Grove Borehole

Ht above O.D. abt 210 ft.   6-in SK 57 NE.
Site [5540 7821] 890 yd at 9° from Ratcliffe Grange, Worksop.
Drilled 1933 by Doncaster Well Borers Ltd for Captain Ward-
Jones.

PERMO-TRIASSIC: MPM to 57 ft, LML to 126 ft, LPM to 204 ft, BPS to bottom of borehole
at 235 ft.

Harworth Colliery (No. 1 Shaft)

Ht above O.D. 111 ft.   6-in SK 69 SW.
Site [6254 9125] 1370 yd at 113° from All Saints' Church, Harworth.
Sunk 1920–3. For detailed section see Wilcockson 1950, pp. 201–4;
Edwards 1951, pp. 177–9.

DRIFT: Sandy boulder clay to abt 24 ft.

PERMO-TRIASSIC: PB to 274 ft, LMS to 353 ft 4 in, UPM to 390½ ft, UML to 431 ft 10 in,
MPM to 512 ft 10 in, LML to 698 ft 4 in, LPM to 720 ft 8 in (BPS absent).

UPPER COAL MEASURES: Badsworth Div. to ?1023 ft 10 in, Hemsworth Div. to 1265 ft
11 in, Brierley Div. to 1445 ft (Blyth, c. 36 in at 1445 ft), Ackworth Div. to 1658 ft.

MIDDLE COAL MEASURES: ?Top M.B. at 1670¼ ft; Shafton M.B. at 1748¾ ft; Shafton (c. 10 in) at 1768 ft 11 in; Sharlston Top (d.c. 7 in) at 1913 ft 7 in; Sharlston Yard (c. 12 in) at 2054 ft 5 in; ?Houghton Thin (c. 9 in) at 2069 ft 4 in; Mansfield M.B. at 2115 ft; Clown (c. 15 in) at 2324 ft 2 in; Mainbright (c. 8 in on c. & d. 12 in) at 2374 ft 11 in; Two-Foot M.B. at 2430 ft 2 in; Two-Foot (c. 36 in) at 2433 ft 2 in; Abdy (c. & d. 13 in) at 2457 ft; Kilnhurst (c. 25 in) at 2492 ft 4 in; High Hazles, Top Bed (c. 15 in) at 2551 ft 10 in; Kent's Thin (ca. 1½ in, c. 18½ in on d. & c. 14 in) at 2579 ft 7 in; Kent's Thick (d. & d.c. 104 in, d. & c. 2 in, c. 4 in, d. 14 in on c. 25 in) at 2658¼ ft; Top Hard + Dunsil (c. 34½ in, d. 7½ in, c. 20½ in, d. 1½ in on c. 17 in) at 2787 ft 4 in.

Bottom of shaft 2898 ft 11 in.

Hayton Smeeth Borehole

Ht above O.D. abt 30 ft.   6-in SK 78 SW.
Site [7143 8340] 850 yd at 62° from Bolham Hall, East Retford. Drilled 1918–20 for Butterley Co. Open holed to 783 ft, partly open holed and partly cored 783 ft to 1300 ft, cored 1300 ft to bottom. For detailed section see Wilson 1926, pp. 199–204; Edwards 1951, pp. 185–6.

DRIFT: Terrace deposits to abt 9 ft.

PERMO-TRIASSIC: Waterstones + GB to abt 100 ft, PB to abt 740 ft, LMS to 824½ ft, UPM to 953¼ ft, UML to 977 ft (but see p. 156), MPM to 1143 ft, LML to 1360 ft, LPM to 1376 ft, BPS to 1379 ft 4 in.

UPPER COAL MEASURES: Badsworth Div. to abt 1585 ft, Hemsworth Div. to abt 1685 ft, Brierley Div. ? to 1768½ ft, Ackworth Div. to 1919½ ft.

MIDDLE COAL MEASURES: Top M.B. at 1929 ft; position of Shafton M.B. at 1981 ft 4 in; Wales (c. 4 in) at 2236½ ft; Mansfield M.B. at abt 2300 ft; ?Kent's Thick (c. 15 in, d. 5 in on c. 3 in) at 2656¾ ft; ?Top Hard + Dunsil (c. with ½ in d., 37½ in, d. 7 in, c. 3 in, d. 7 in on c. 3 in) at 2792 ft 2 in; ? position of Clay Cross M.B. at 3071¾ ft.

LOWER COAL MEASURES: ?Deep Soft (c. 5 in, d. 6 in, c. 4 in, d. 4 in, c. 2 in, d. 3 in on c. 29 in) at 3130¼ ft.

Bottom of borehole 3131 ft.

Jockey House Borehole

Ht above O.D. 75 ft.   6-in SK 67 NE.
Site [6897 7684] 1450 yd at 3° from Elkesley church. Drilled 1942 by Foraky Co. for Wigan Coal Corporation. Open holed to 870 ft, cored 870 ft to bottom. Cores examined by W. Edwards.

PERMO-TRIASSIC: PB + LMS to 618¼ ft, UPM to 676 ft 4 in, UML to 706 ft (but see p. 156), MPM to 862 ft, LML to 1100½ ft, LPM to 1112 ft 8 in, BPS to 1113 ft.

UPPER COAL MEASURES: Brierley Div. to ?1185¼ ft (? Blyth, c. 8 in at 1185¼ ft), Ackworth Div. to 1351½ ft.

MIDDLE COAL MEASURES: Top M.B. at 1359 ft; Shafton M.B. at 1410 ft 4 in; *Edmondia* Band at 1552 ft; Sharlston Top (c. 14 in) at 1601 ft 5 in; Sharlston Low (c. 10 in) at 1667 ft 7 in; Wales (c. 1 in, d. 9 in, c. 22 in, d. 3 in on c. 22 in) at 1724 ft 1 in; Mansfield M.B. at 1786 ft; Haughton M.B. at 1881½ ft; Clown (c. 8 in) at 1950 ft; Two-Foot M.B. at 2000½ ft; Two-Foot (c. 8 in) at 2001 ft 2 in; Kilnhurst (c. 12 in) at 2067 ft 2 in; High Hazles, Top Bed (d.c. 2 in) at 2112½ ft; Kent's Thin (ca. 6 in, d. 3 in on c. 2 in) at 2154 ft 4 in; Kent's Thick (c. 48 in) at 2207 ft; Top Hard + Dunsil (c. 43½ in) at 2354½ ft; Haigh Moor (c. 30 in) at 2470 ft; 2nd Ell (c. 20 in) at 2575 ft 8 in; Clay Cross M.B. at 2637 ft 8 in.

LOWER COAL MEASURES: Deep Soft (c. 43 in) at 2720 ft 4 in; Deep Hard (c. 3 in) at 2811 ft 10 in; Parkgate (c. 70 in, d. 3½ in on c. 7½ in) at 2865 ft 11 in; Cockleshell (c. 12 in) at 2913 ft 4 in; Thorncliff, bottom leaf (c. 1 in) at 2936 ft; Threequarters (c. 8 in on d. & c. 20 in) at 2956 ft 4 in; Top Silkstone (d.c. 18 in) at 3023 ft 8 in.

Bottom of borehole 3023 ft 8 in.

Lound Borehole

Ht above O.D. 28 ft.   6-in SK 78 NW.
Site [7045 8585] 1660 yd at 179° from Wild Goose Cottage, Lound. Drilled 1957 by Foraky Co. for N.C.B. Open holed to 1257 ft 4 in, cored 1257 ft 4 in to bottom. Cores examined by E. G. Smith and T. R. W. Hawkins.

DRIFT: Alluvium and Terrace deposits to abt 10½ ft.

PERMO-TRIASSIC: PB + LMS to 680 ft, UPM to 821½ ft, UML to 924 ft, MPM to 1034½ ft, LML to 1296¾ ft, LPM to 1304 ft 11 in, BPS to 1308 ft.

UPPER COAL MEASURES: Badsworth Div. to ?1386 ft 11 in, Hemsworth Div. to ? 1610 ft 4 in, Brierley Div. to 1726½ ft (Blyth c. 16 in, measures 7 ft 9½ in on c. 18 in at 1726½ ft), Ackworth Div. to 1896 ft 2 in.

MIDDLE COAL MEASURES: Top M.B. at 1918 ft 7 in; Shafton M.B. at 1970 ft 8 in; Sharlston Top (c. 10 in, d.c. 2½ in, c. 1½ in on d.c. 8½ in) at 2154 ft 1 in; Sharlston Low (c. 17 in) at 2196 ft 5 in; Wales (d.c. 9 in, d. & measures 82 in on d.c. 8 in) at 2247 ft 7 in; Mansfield M.B. at 2304 ft; Sutton M.B. at 2343 ft 2 in; Haughton M.B. at 2383 ft 4 in; Clown (c. 6 in) at 2432¾ ft; Two-Foot M.B. at abt 2512¼ ft; Two-Foot (c. 7½ in, d.c. 7 in on d. & d.c. 12 in) at 2515 ft 8 in; Abdy (c. with 5¼ in d.c., 19¾ in, d. 1½ in on c. 3¾ in) at 2555 ft 5 in; Kent's Thin (c. 13½ in) at 2664 ft 5 in; Kent's Thick (d.c. 5½ in, c. 6½ in, d. 5½ in, d.c. 6 in on c. 6½ in) at 2705 ft 7 in; Top Hard + Dunsil (d.c. 5 in, c. 12 in, d.c. 2 in, d. 17½ in, d.c. & d. 24 in, c. with 1¼ in d.c., 11¼ in, d. 25½ in on c. 8¾ in) at 2788 ft 5 in; Clay Cross M.B. at 3006 ft 11 in.

LOWER COAL MEASURES: Deep Soft (c. 1 in) at 3106 ft 1 in; Parkgate (d.c. ½ in on c. 29½ in) at 3248 ft 1 in; Thorncliff (c. 16½ in, d. 1½ in, c. 4½ in, d.c. 2 in, c. 2½ in, d. 1½ in, c. with ½ in d.c., 2¾ in, d.c. & d. 12 in, c. 2¾ in on d.c. 1¾ in) at 3310 ft 1 in; Threequarters (c. 6 in) at 3336 ft 2 in; Top Silkstone (d.c. 2 in, d. 7 in, d.c. 6 in, c. 16 in, d. 4 in on c. 3 in) at 3411 ft 2 in; Low 'Estheria' Band at 3445 ft 10 in; Low Silkstone (d.c. 2 in, d. 7 in, c. 11¾ in on d.c. 1¼ in) at 3447 ft 8 in.

Bottom of borehole 3495 ft 5 in.

Loxdale Cottages Borehole, Oldcoates

Ht above O.D. abt 60 ft.   6-in SK 68 NW.
Site [6036 8795] 1780 yd at 15° from Hodsock Lodge, Hodsock.
Drilled 1962 by Doncaster Well Borers Ltd for Mr. S. Heggie.

PERMO-TRIASSIC: LMS to 6 ft, UPM to 50 ft, UML to 70 ft, MPM to 158½ ft, LML to bottom of borehole at 332 ft.

Maltby Main Colliery (No. 2 Shaft)

Ht above O.D. 259 ft.   6-in SK 59 SE.
Site [5512 9246] 1330 yd at 328° from Sandbeck Lodge, Maltby.
Sunk 1910. For detailed section see Wilcockson 1950, pp. 273–7; Edwards 1951, pp. 203–6. Edwards (1951, pp. 91–2) gives an extensive list of plants collected from the Coal Measures.

MADE GROUND: including soil, to 2 ft.

PERMO-TRIASSIC: LML to 157 ft 5 in (LPM and BPS absent).

UPPER COAL MEASURES: Badsworth Div. to 292¾ ft, Hemsworth Div. to 598 ft 1 in, Brierley Div. to 824 ft 5 in (Blyth, ca. 2 in, c. 24 in, d. 2 in on c. 8 in at 824 ft 5 in), Ackworth Div. to 1085 ft 7 in.

MIDDLE COAL MEASURES: Shafton M.B. at 1207 ft 8 in; *Edmondia* Band at abt 1387 ft; Sharlston Top (c. 10 in) at 1466 ft 11 in; Sharlston Low (c. 8 in, d. & measures 10 ft 8 in on c. 7 in) at 1581 ft 7 in; Sharlston Yard (c. 24 in) at 1630 ft 5 in; Houghton Thin (c. 16 in) at 1659 ft 7 in; Mansfield M.B. at 1743 ft 7 in; Haughton M.B. at 1888 ft 1 in; Clown (c. 27 in) at 1987 ft; Meltonfield (c. 12 in, d. 1 in, c. 2 in, d. 1½ in, c. 1½ in, d. 1 in, c. 9 in, d. 1 in on c. 6½ in) at 2059¼ ft; Two-Foot M.B. at 2110 ft 2 in; Two-Foot (d.c. 23 in, ca. 43¼ in on c. 4¾ in) at 2116 ft 1 in; Abdy (c. 29 in) at 2146 ft; Kilnhurst (c. 2 in, d. 24 in on c. 6 in) at 2177 ft; Kent's Thin (c. 17 in on c. & d. 4 in) at 2259 ft 11 in; Kent's Thick, top leaf (c. 14 in, d. ½ in on c. 20½ in) at 2322 ft 11 in; bottom leaf (c. 14 in, d. 1¼ in on c. 5¾ in) at 2352½ ft; Barnsley (c. 102¾ in) at 2460¾ ft; ?Dunsil (c. 14 in) at 2493 ft 2 in.

Bottom of shaft 2578 ft 4 in.

Maltby Main Colliery No. 7 Underground Borehole

Top in Barnsley Workings at 2454 ft below O.D. 6-in SK 59 SE. Site [5805 9200] 800 yd at 167° from Limestone Hill, Tickhill. Drilled 1969–70 by N.C.B. Open holed to 42 ft 11 in, cored 42 ft 11 in to bottom. Cores examined by T. Robbins and H. W. Pearson.

MIDDLE COAL MEASURES: Swallow Wood (c. 21¾ in, d. ¾ in, c. 15½ in, d. 30 in on c. 11 in) at 212 ft 2 in; Haigh Moor (c. 15 in, d. 5 in on c. 13 in) at 228 ft 1 in.

Bottom of borehole 232 ft 8 in.

Manton Borehole

Ht above O.D. abt 105 ft.   6-in SK 67 NW.
Site uncertain but thought to be [6138 7832] 1570 yd at 345° from Manton Forest Farm. Drilled 1896 or 1897 by Vivians Johnson and Sons for Wigan Coal and Iron Co. Ltd.

PERMO-TRIASSIC: PB + LMS + ?UPM to 241 ft 4 in, UML to 246½ ft, MPM to 337¾ ft, LML to 615¼ ft, LPM to 628 ft 5 in, BPS to 629 ft 5 in.

UPPER COAL MEASURES: Ackworth Div. to abt 724 ft 5 in.

MIDDLE COAL MEASURES: Horizon of Top M.B. at abt 733 ft; horizon of Shafton M.B. at 830 ft 4 in; horizon of Mansfield M.B. at abt 1346 ft; ?Clown (c. 25 in) at 1555 ft 4 in; ? Mainbright (c. 34 in) at 1620 ft 7 in.

Bottom of borehole 1707 ft 11 in.

Manton Colliery (No. 4 Shaft)

Ht above O.D. 149 ft.   6-in SK 67 NW.
Site [6085 7816] 1650 yd at 324° from Manton Forest Farm. Sunk 1946–52. A detailed section of No. 2 Shaft is given by Wilcockson 1950, pp. 279–82; Edwards 1951, pp. 208–10.

PERMO-TRIASSIC: PB to ?188½ ft, LMS to 255 ft, UPM absent, UML to 266 ft, MPM to 352½ ft (but see pp. 130 and 149), LML to abt 630 ft, LPM to 653 ft, BPS to 654 ft 2 in.

UPPER COAL MEASURES: Brierley Div. to 685 ft 11 in (Blyth, d.c. 30 in at 685 ft 11 in), Ackworth Div. to 844¾ ft.

MIDDLE COAL MEASURES: Top M.B. at 852¾ ft; Shafton M.B. at 951½ ft; *Edmondia* Band at 1123 ft; Sharlston Low (c. 18 in, d. & measures 52 in on c. 9 in) at 1296½ ft; Sharlston Yard (d.c. 27 in) at 1347 ft; ?Houghton Thin (d.c. 36 in) at 1357¾ ft; Mansfield M.B. at 1443 ft 8 in; Haughton M.B. at 1578 ft; Clown (c. 37½ in) at 1682½ ft; Meltonfield (ca. 6 in, c. 23 in, d. 61 in, c. 9 in, d. 4 in on c. 11 in) at 1761¾ ft; Two-Foot M.B. at 1807 ft 1 in; Two-Foot (ca. 6 in, c. 25 in on d.c. 7 in) at 1810 ft 5 in; Abdy (d.c. 21 in) at 1840¾ ft; High Hazles, Top Bed (d.c. 36 in) at 1953½ ft; Kent's Thin (ca. 3 in on c. 21 in) at 1970½ ft; Kent's Thick, top leaf (c. 20 in, d. 8 in on c. 7 in) at 2031¾ ft; bottom leaf (c. 17 in) at 2061 ft 5 in; Top Hard (c. 37 in, d. 2½ in on c. 2½ in) at 2195¾ ft; Dunsil (c. & d. 30 in) at 2261¾ ft; 1st Waterloo (ca. 6 in, d. 6 in on d.c. 15 in) at 2316½ ft; Swallow Wood (c. 9 in, d. 6 in, c. 6 in, d.c. 4 in, c. 8 in, d. 54 in on c. 21 in) at 2390 ft; Haigh Moor (c. 14 in, d. 63 in on c. 30 in) at 2408 ft 2 in; Lidget (c. & d. 12 in) at 2479 ft; 2nd Ell (c. 9 in) at 2532¼ ft; Clay Cross M.B. at abt 2594 ft.

LOWER COAL MEASURES: Deep Soft (c. 25½ in, d. 12½ in, c. 27½ in, d. ½ in, c. 3 in, d. 12 in, c. 10½ in, d. 4 in, c. 7 in, d. 19 in on c. 5 in) at 2735½ ft; ?Deep Hard Roof Coal (c. 10 in) at 2785 ft 8 in; Deep Hard (c. 21 in, measures 14½ ft on c. 2 in) at 2823 ft 2 in; Parkgate (c. 10 in, d. 8 in, d.c. 3½ in on c. 47 in) at 2870 ft 2 in; Thorncliff (d.c. 3 in, d. ½ in, c. & d. 5½ in, d.c. 5½ in, c. & d. 1½ in, measures 31 ft 5 in, c. 1 in, d. & d.c. 59 in on c. 11 in) at 2960 ft 11 in.

Bottom of shaft 3000 ft.

Manton Colliery Underground Boreholes

There are eight underground boreholes from the Top Hard (Barnsley) workings at Manton Colliery. No. 1 [5791 7736] is an up-bore of 253 ft to the High Hazles; No. 2 [6067 7833] was drilled down 151 ft to 35 ft below the 1st Waterloo; No. 3, on the same site as No. 2, was drilled down to the Dunsil at 66 ft; Nos. 4 [6064 7835], 5 [5850 7756], 6 [5922 7530] and 7 [6379 7633], respectively 1021, 1024, 889 and 804 ft deep, were drilled from the Top Hard to below the Low Silkstone; No. 8 [5949 7706] is an up-bore of 549 ft to the Clown. Notes on Nos. 5 and 8 boreholes are given below.

Manton Colliery No. 5 Underground Borehole

Top in Top Hard workings 2004 ft below O.D.   6-in SK 57 NE. Site [5850 7756] 1530 yd at 287° from Worksop College. Drilled 1952–3 by Craelius Co. for N.C.B. Cored except for top 6 ft. Cores examined by R. F. Goossens.

MIDDLE COAL MEASURES: Dunsil (c. 11 in, d.c. 1 in, d. 6 in, c. 21¼ in on d.c.¶1 in) at 83 ft; 1st Waterloo (d.c. 6 in, d. 18 in, ca. 6 in on d.c. 3 in) at 125¼ ft; Swallow Wood, Main Bed (d.c. 3 in) at 179½ ft; Haigh Moor (c. 9 in, d. & measures 13 ft 7 in, d.c. 1 in, c. 9½ in, d. 1 in on c. 19½ in) at 226 ft 8 in; Lidget (c. 5½ in, d.c. 3 in on c. 9 in) at 311 ft; 2nd Ell (c. 6 in) at 358¾ ft; Clay Cross M.B. at 421¾ ft.

LOWER COAL MEASURES: ?Deep Soft, top leaf (d.c. & d. 6 in on c. 6 in) at 527½ ft; Deep Soft, main leaf (c. & d. 92 in) at 578 ft 2 in; ?Deep Hard Roof Coal (c. 3 in) at 607¾ ft; Deep Hard (c. 20 in) at 644 ft 8 in; Parkgate (c. 16½ in, d. 5 in, c. 7½ in, d. 6 in, d.c. 6 in, c. 35½ in, d. 3½ in, c. 19½ in on d.c. 2½ in) at 712 ft; Thorncliff (c. 17 in, inferior ca. 32 in, measures 20 ft 10 in, d.c. 12½ in on c. 3½ in) at 796 ft 7 in; Top Silkstone (d.c. 6 in, c. 15 in, d. 53 in on c. 1 in) at 913 ft; Low '*Estheria*' Band at 926¾ ft; Low Silkstone (c. 6 in) at 927¼ ft.

Bottom of borehole 1023¾ ft.

Manton Colliery No. 8 Underground Borehole (up-bore)

Start in Top Hard workings 2151 ft below O.D. 6-in SK 57 NE.
Site [5949 7706] 1290 yd at 199° from Worksop College. Drilled
1969 by N.C.B. Open holed to 140 ft above start, cored 140 ft– 318
ft 5 in, open holed 318 ft 5 in to 399 ft, cored 399 ft – 548 ft 8 in.
Cores examined by H. W. Pearson. Measurements from top of
borehole at 1602 ft below O.D.

MIDDLE COAL MEASURES: Clown (c. 40 in) at $5\frac{1}{4}$ ft; Mainbright or Meltonfield (*core lost*
$37\frac{1}{2}$ in on c. $25\frac{1}{2}$ in) at 88 ft 8 in; Two-Foot M.B. at 129 ft 1 in; Two-Foot (d.c. $3\frac{1}{4}$ in, c.
$32\frac{1}{4}$ in, d.c. $1\frac{1}{2}$ in, d. $5\frac{1}{2}$ in on c. $3\frac{1}{2}$ in) at 134 ft 7 in; High Hazles (c. $22\frac{3}{4}$ in, d. $8\frac{1}{4}$ in on
d.c. $\frac{1}{2}$ in) at 314 ft 10 in; Top Hard workings at 548 ft 8 in.

Start of borehole at 548 ft 8 in.

Markham Moor House Borehole

Ht above O.D. 76 ft.   6-in SK 77 SW.
Site [7153 7404] 190 yd at 152° from Old Eel Pie House, Markham
Moor. Drilled prior to 1914 for Newcastle Estate. For detailed
section see Lamplugh and Smith 1914, p. 91.

DRIFT: Terrace deposits (including peat) to 6 ft.

PERMO TRIASSIC: Waterstones to abt 98 ft, GB to 132 ft, PB to bottom of borehole at
222 ft.

Mattersey Borehole

Ht above O.D. 25 ft.   6-in SK 68 NE.
Site [6862 8898] 650 yd at 229° from Mattersey church. Drilled
1954–5 by Foraky Co. for N.C.B. Open holed to 1000 ft, cored
1000 ft to bottom. Cores examined by G. H. Rhys, B. W. Conway
and E. G. Smith.

DRIFT: Alluvium, not recorded.

PERMO-TRIASSIC: PB + LMS + UPM to 685 ft, UML to 748 ft, MPM to 866 ft, LML to
$1117\frac{1}{2}$ ft, LPM to 1125 ft 5 in, BPS to 1125 ft 10 in.

UPPER COAL MEASURES: Badsworth Div. to ?1250 ft 7 in, Hemsworth Div. to $1497\frac{3}{4}$ ft,
Brierley Div. to $1619\frac{3}{4}$ ft (Blyth, c. 36 in at $1619\frac{3}{4}$ ft), Ackworth Div. to 1793 ft 10 in.

MIDDLE COAL MEASURES: Top M.B. at $1797\frac{1}{4}$ ft; Shafton M.B. at 1853 ft 2 in; *Edmondia*
Band at 1991 ft 7 in; Sharlston Low (c. 18 in) at 2141 ft 8 in; Sharlston Yard (c. 7 in, d.
14 in on c. 8 in) at $2176\frac{3}{4}$ ft; ?Houghton Thin (c. 9 in) at $2188\frac{3}{4}$ ft; Mansfield M.B. at
$2248\frac{1}{4}$ ft; Haughton M.B. at 2343 ft 8 in; Swinton Pottery (d.c. 18 in) at 2349 ft 8 in;
Two-Foot (c. 8 in, d.c. 1 in, c. $9\frac{1}{2}$ in, d. $\frac{1}{2}$ in, c. 9 in on d.c. 2 in) at 2536 ft 4 in; Kent's Thin
(c. 13 in, d. with $\frac{1}{2}$ in c., $34\frac{1}{2}$ in on d.c. $9\frac{1}{2}$ in) at 2666 ft 2 in; Kent's Thick, top leaf (c.
22 in, d. 7 in, c. 8 in, d. $\frac{3}{4}$ in on c. $22\frac{1}{4}$ in) at $2735\frac{1}{2}$ ft; Top Hard + Dunsil (d.c. & d.
$1\frac{1}{2}$ in, c. $14\frac{1}{2}$ in, d. 3 in, c. 20 in, d.c. 2 in on c. $18\frac{1}{2}$ in) at 2870 ft 11 in; Swallow Wood
(c. $15\frac{1}{2}$ in, d. 13 in, d.c. 1 in, d. $1\frac{1}{2}$ in, c. 6 in, d. $2\frac{1}{2}$ in on d.c. $5\frac{1}{2}$ in) at $3020\frac{1}{4}$ ft; Haigh
Moor (ca. 5 in, c. 5 in, d.c. 2 in, c. 13 in, d. & measures 17 ft 10 in, c. 25 in, d. 5 in on c.
10 in) at 3062 ft 7 in; Lidget (c. 6 in) at 3122 ft; Clay Cross M.B. at $3238\frac{1}{2}$ ft.

LOWER COAL MEASURES: Deep Soft (c. with $\frac{1}{2}$ in d. and 1 in d.c., 76 in) at $3335\frac{3}{4}$ ft;
Thorncliff (c. 27 in on d.c. 4 in) at 3558 ft 10 in; Threequarters (c. 6 in) at 3591 ft 8 in;
Low '*Estheria*' Band at 3714 ft 2 in; Low Silkstone (c. 8 in) at 3714 ft 10 in.

Bottom of borehole 3762 ft.

Morton (BP) No. 1 Oil Bore

R.T.E. above O.D. 14 ft.   6-in SK 79 SE.
Site [7932 9241] 1020 yd at 329° from Point Farm, Walkeringham.
Drilled 1965 by BP Petroleum Development Ltd. Open holed to
4658 ft and from 4675 ft to bottom, cored 4658–75 ft. Cores examined by C. G. Godwin.

DRIFT: Alluvium to abt 30 ft, gravels to 48 ft.

PERMO-TRIASSIC: KM to 575 ft, PB + LMS to abt 1525 ft, UPM to 1650 ft, UML to
1732 ft, MPM to abt 1865 ft, LML to 2210 ft, LPM to 2218 ft, BPS to 2225 ft.

UPPER COAL MEASURES: Ackworth Div. to abt 2350 ft.

MIDDLE COAL MEASURES: position of Top M.B. at abt 2355 ft; position of Mansfield
M.B. at abt 2715 ft; Top Hard ?+ Dunsil at abt 3160 ft; position of Clay Cross M.B.
at abt 3462 ft.

LOWER COAL MEASURES: Parkgate at abt 3696 ft; Top Silkstone at abt 3855 ft; Kilburn
at abt 4214 ft; position of Pot Clay M.B. at abt 4602 ft.

MILLSTONE GRIT SERIES: *G. cancellatum* M.B. at 4663 ft 3½ in (but see p. 23).

Bottom of well 5494 ft.

Nornay Borehole

Ht above O.D. 40 ft.   6-in SK 68 NW.
Site [6251 8868] 1510 yd at 5° from St Mary and St Martin's
Church, Blyth. Drilled 1954 by Foraky Co. for N.C.B. Cores
examined by R. A. Eden and R. E. Elliott.

	Thickness ft in	Depth ft in		Thickness ft in	Depth ft in
PERMO-TRIASSIC			Siltstone and sandstone, purple	8 4	672 4
Not cored: PB + LMS to 135 ft, UPM to 147 ft, UML to 183 ft, MPM to 274 ft, LML to abt 473 ft (LPM & BPS not recorded) ..	473 0	473 0	Mudstone, silty, purple	0 8	673 0
			Seatearth, purple and grey	4 0	677 0
			Striped beds, largely purple	10 9	687 9
UPPER COAL MEASURES			Sandstone, red and purple in parts, medium–coarse grained towards base	50 1	737 10
BADSWORTH DIVISION					
Not cored	127 6	600 6	Mudstone, purple to red; 'Estheria' ?	1 10	739 8
Core lost	2 6	603 0	Seatearth with red iron- stone and sphaero- siderite	6 0	745 8
Mudstone, brown, purple and greenish; plants..	1 3	604 3	Sandstone, partly purple	7 4	753 0
Sandstone and siltstone, purple	6 9	611 0	Mudstone, silty, partly purple	7 3	760 3
Mudstone, purple and greenish	1 9	612 9	Mudstone, red and green; 'Estheria' ?, *Planolites sp*... ..	1 8	761 11
Sandstone and siltstone, largely purple ..	37 9	650 6	Seatearth, purple and green	7 11	769 10
Mudstone, with some siltstone, multi- coloured	13 6	664 0			

	Thickness ft in	Depth ft in		Thickness ft in	Depth ft in
Siltstone, purple and green; plants ..	9 2	779 0	Siltstone and mudstone, greenish; reddish brown ironstone at top	9 0	856 8
Mudstone, purplish; *Euestheria simoni*, *Cochlichnus kochi* ..	1 11	780 11	Seatearth, greenish and brown	4 0	860 8
Shale, black; plants and mussels	0 1	781 0	Striped beds with ironstone	9 6	870 2
Seatearth with green and red patches; sphaerosiderite	6 0	787 0	Mudstone, pale greenish, with dark bands; 'Estheria'? ..	3 1	873 3
Siltstone, reddish brown in part; sphaerosiderite	4 0	791 0	Seatearth, brown and green	2 5	875 8
Seatearth, reddish brown in part	4 0	795 0	Sandstone and siltstone; sphaerosiderite at top	35 7	911 3
Sandstone and siltstone, mostly greenish; sphaerosiderite ..	3 0	798 0	Mudstone, silty, and siltstone	3 2	914 5
Mudstone, silty, green, purple and brown ..	5 6	803 6	SECOND CHERRY TREE MARKER Mudstone, dark in parts; ironstone; *Anthraconaia pruvosti, Euestheria simoni*	5 7	920 0
Seatearth, greenish; sphaerosiderite ..	1 0	804 6			
Sandstone with siltstone bands, greenish ..	4 0	808 6			
Mudstone, green; sphaerosiderite ..	1 0	809 6	**Coal and dirt** 2 in **Coal** 16 in	**1 6**	**921 6**
Sandstone, pale green ..	5 6	815 0	Seatearth and mudstone, partly red-mottled; sphaerosiderite ..	24 6	946 0
Striped beds, light green	8 0	823 0	Siltstone, greenish; sandy in parts ..	9 0	955 0
Mudstone, silty at top; ironstone	2 0	825 0	Mudstone, greenish; ironstone	0 6	955 6
FIRST CHERRY TREE MARKER Shale, dark; *Anthraconauta phillipsii*, ostracods, fish debris	2 3	**827 3**	Shale, black; sphaerosiderite; fish	0 6	956 0
			Seatearth, greenish in part	5 9	961 9
HEMSWORTH DIVISION Mudstone, dark silty, with plants	0 9	828 0	Sandstone and siltstone; pebble bands; coal inclusions	153 9	1115 6
Seatearth with sphaerosiderite	6 0	834 0	Mudstone, silty, dark grey and green ..	0 1	1115 7
Siltstone, green; sphaerosiderite ..	4 9	838 9	BRIERLEY DIVISION Seatearth, greenish ..	10 9	1126 4
Mudstone and siltstone, greenish	6 6	845 3	Siltstone, sandy, with ironstone	1 4	1127 8
Seatearth, greenish in part	2 5	847 8	Shale, carbonaceous ..	0 1	1127 9
			Seatearth	2 0	1129 9

	Thickness ft in	Depth ft in
Sandstone and siltstone; sphaerosiderite at top	17 3	1147 0
Mudstone, silty; ironstone; *Anthraconaia?, Anthraconauta?,* 'Estheria'?	1 5	1148 5
Shale, dark; 'Estheria'?, fish	0 1	1148 6
Coal 10 in / Dirt 6 in / **Dirty coal** 2 in	1 6	1150 0
Seatearth	4 7	1154 7
Shale, dark; *Anthraconauta phillipsii, Carbonita humilis, Euestheria simoni* ..	0 1	1154 8
Mudstone with ironstone	1 4	1156 0
Dirty coal 3 in	0 3	1156 3
Seatearth with sphaerosiderite	2 10	1159 1
Mudstone, silty, and siltstone	5 4	1164 5
Seatearth	3 2	1167 7
Siltstone and sandstone	6 5	1174 0
Mudstone with shells ..	0 4	1174 4
Shale, black, with shells	0 2	1174 6
Seatearth	5 6	1180 0
Shale, black; plants ..	0 1	1180 1
Seatearth	3 2½	1183 3½
Coal 0½ in	0 0½	1183 4
Core lost	2 2	1185 6
Seatearth	1 11	1187 5
Siltstone and sandstone	24 7	1212 0
Mudstone and shale, mostly dark; ironstone; *Anthraconauta sp., Carbonita* cf. *agnes, C. humilis,* fish remains	5 10	1217 10
Dirty coal 7 in / Dirt 21 in / **Coal** 16 in / (12½ in recovered) ..	3 8	1221 6
Seatearth	7 6	1229 0
Siltstone and sandstone; pebble bands in lower part; plants	43 0	1272 0

	Thickness ft in	Depth ft in
BLYTH **Coal** 38 in (*33¼ in recovered*) ..	3 2	1275 2
ACKWORTH DIVISION Seatearth	2 10	1278 0
Siltstone with sandstone bands; sphaerosiderite	16 0	1294 0
Mudstone, dark at base; ironstone	3 10	1297 10
Mudstone and shale, black at base, silty; shells and fish remains	6 8	1304 6
Seatearth	2 6	1307 0
Striped beds	5 7	1312 7
Shale, dark to black, with ironstone; *A.* aff. *phillipsii, C. humilis, C. pungens*	2 5	1315 0
Coal 4 in	0 4	1315 4
Seatearth	3 10	1319 2
Sandstone	10 7	1329 9
Mudstone, dark at base; ironstone; *A.* aff. *phillipsii* and fish ..	5 6	1335 3
Seatearth with sphaerosiderite	3 0	1338 3
Siltstone and silty mudstone; sphaerosiderite	7 10	1346 1
Seatearth with 1-in coaly shale	6 5	1352 6
Striped beds	15 9	1368 3
Mudstone, dark, with ironstone	2 5	1370 8
Shale, black, silty; fish..	0 1	1370 9
Coal 3 in (*0½ in recovered*) ..	0 3	1371 0
Seatearth	2 6	1373 6
Siltstone and silty mudstone; ironstone ..	8 10	1382 4
Mudstone with ironstone; *Lioestheria vinti*	4 0	1386 4
Shale, black	0 5	1386 9
SCOFTON **Coal** 7 in (*4 in recovered*) ..	0 7	1387 4

	Thickness ft in	Depth ft in		Thickness ft in	Depth ft in
Seatearth with sphaero-siderite	11 8	1399 0	**Coal** 7 in (*3 in recovered*) ..	**0 7**	**1540 9**
ACKWORTH ROCK			Seatearth	2 3	1543 0
Sandstone; ironstone pebbles ..	89 1	1488 1	Sandstone and striped beds; plants	17 0	1560 0
MIDDLE COAL MEASURES			Mudstone, mostly dark, silty at top; ironstone; *Naiadites sp., L. vinti*	6 11	1566 11
TOP MARINE BAND			SHAFTON MARINE BAND		
Mudstone, dark at base; ironstone; *Crurithyris sp., Lingula mytilloides, Dunbarella sp.,* ?coiled nautiloid, 'fucoids'	8 3	**1496 4**	Mudstone, dark; fora-minifera including *Ammodiscus sp., Lingula mytilloides, 'Estheria' sp., Hollinella sp., Anthraconaia sp., Edmondia sp.,* indet. bivalve (*?Myalina*)	1 0	**1567 11**
Siltstone, black; fish ..	0 2	1496 6			
Coal 1 in **Dirt and coal** 26 in	**2 3**	**1498 9**	**Coal and dirt** 13 in **Coal** 6 in **Coal and dirt** 2 in	**1 9**	**1569 8**
Seatearth	1 1	1499 10	Seatearth; greenish band; sphaerosiderite	17 7	1581 3
Striped beds	8 4	1508 2	Sandstone	2 5	1583 8
Mudstone, silty ..	1 4	1509 6	Mudstone, mostly dark, silty, greenish at top..	23 4	1607 0
Mudstone, dark, partly silty; ironstone; mussels, *Lioestheria vinti*	2 1	1511 7	Mudstone, mostly dark; *L. vinti*	3 0	1610 0
Siltstone, dark; mussels, *L. vinti, Carbonita sp.,* fish	1 0	1515 6	Shale, coaly	2 10	1612 10
Dirty coal 8 in **Dirty cannel** 8 in	**1 4**	**1516 10**	Seatearth	1 2	1614 0
Seatearth	2 11½	1519 9½	Sandstone and siltstone	14 3	1628 3
Siltstone, dark; shells and fish	0 0½	1519 10	Mudstone, mostly silty; dark bands; plants, mussels and fish ..	4 9	1633 0
Coal 7 in	**0 7**	**1520 5**	Sandstone and siltstone; pebble bands; coal inclusions	132 4	1765 4
Seatearth	3 1	1523 6	Mudstone, silty, and silt-stone; plants.. ..	3 3	1768 7
Sandstone and siltstone; ironstone	9 6	1533 0	Mudstone, dark at base; ironstone; shells; *Spirorbis sp., Naiadites sp.* cf. *quadratus, Carbonita humilis,* fish scales	4 7	1773 2
Mudstone and shale, dark; ironstone; *Spirorbis sp.,* mussels, *L. vinti, Geisina sub-arcuata,* palaeoniscid scale	3 9	1536 9			
Seatearth with sphaero-siderite	2 10	1539 7	Shale with coal bands ..	0 9½	1773 11½
Shale, black, with iron-stone	0 7	1540 2	Seatearth with sphaero-siderite	8 1½	1782 1

R

Description	Thickness ft in	Depth ft in
Siltstone and sandstone bands; ironstone; plants..	32 5	1814 6
SHARLSTON TOP **Coal** 4 in	**0 4**	**1814 10**
Seatearth	0 8	1815 6
Striped beds	11 11	1827 5
Mudstone, silty at top; mussels, sineoid markings	4 7	1832 0
Mudstone, dark; black shale bands; sphaerosiderite; *Spirorbis sp.*, *Naiadites sp.*, fish, sineoid marking ..	4 4	1836 4
Coal 1 in	**0 1**	**1836 5**
Seatearth with sphaerosiderite at top ..	5 5	1841 10
Striped beds; lower part with mussels and fish	9 11	1851 9
Mudstone, dark; shelly ironstone; *Spirorbis sp.*, *Naiadites sp.* (cf. *triangularis*), indet. bivalve (*?Anthraconaia hindi* group), *C. humilis*, fish ..	3 3	1855 0
Mudstone, silty; plants	0 8	1855 8
Seatearth	1 8	1857 4
Striped beds and silty mudstone; ironstone; *Spirorbis sp.*, mussels	4 3	1861 7
Siltstone and silty mudstone; plants.. ..	16 4	1877 11
Mudstone, silty; *N.* cf. *productus*	9 2	1887 1
Mudstone, dark, with ironstone; *Spirorbis sp.*, *N.* cf. *triangularis*, *C. humilis*, fish including platysomid scales	2 11	1890 0
SHARLSTON LOW **Coal** 13 in	**1 1**	**1891 1**
Seatearth	3 8	1894 9
Siltstone and sandstone; some ironstone ..	19 1	1913 10

Description	Thickness ft in	Depth ft in
Siltstone; *N.* cf. *productus*	3 1	1916 11
Mudstone, silty at top..	2 3	1919 2
Mudstone, dark; mussels	5 2	1924 4
SHARLSTON YARD **Dirty coal** 8 in **Coal** 2 in	**0 10**	**1925 2**
Seatearth	4 10	1930 0
Siltstone; sandstone and mudstone bands ..	16 5	1946 5
Mudstone, partly dark; plants, mussels ..	3 8	1950 1
Seatearth, dark ..	1 1	1951 2
Mudstone, dark, and ironstone; mussels ..	1 8	1952 10
Shale, coaly	0 4	1953 2
Seatearth, sandy ..	1 4	1954 6
Siltstone and sandstone	29 6	1984 0
Mudstone, silty at top; ironstone	3 1	1987 1
Mudstone, dark; ironstone; mussels ..	1 11	1989 0
MANSFIELD MARINE BAND Mudstone, dark, with ironstone; foraminifera including *Ammodiscus sp.*, *Glomospira sp.*, *Hyperammina sp.*, sponge spicules, *L. mytilloides*, turreted gastropod, *Polidevcia acuta*, *Hollinella sp.*, fish scales including *Rhabdoderma sp.* ..	8 2	1997 2
Mudstone, dark, with ironstone; *L. mytilloides*, bellerophontoid gastropod, fish remains including *Strepsodus sp.* ..	12 10	2010 0
Mudstone, dark, with ironstone; sponge spicules, *Lingula sp.*, *Euphemites sp.*, *Platyconcha hindi*, *Aviculopecten?*, nuculoids indet., *Palaeoneilo taffiana*,		

	Thickness ft in	Depth ft in
coiled nautiloid, *Hollinella?*, fish including *Rhizodopsis* scale, 'fucoids' ..	5 3	2015 3
Siltstone, dark; *Dunbarella* cf. *macgregori*, '*Anthracoceras*' sp., orthocone nautiloid	1 5	2016 8
Mudstone, dark ..	0 6	**2017** 2
Seatearth	3 3	2020 5
Sandstone and siltstone; ironstone	9 11	2030 4
Mudstone, silty ..	4 3	2034 7
Mudstone, coaly ..	0 2	2034 9
Coal 8 in (*4 in recovered*) ..	**0 8**	**2035** 5
Seatearth	3 10	2039 3
Siltstone and silty mudstone	11 0	2050 3
Mudstone, dark ..	1 4	2051 7
Seatearth	3 2	2054 9
Siltstone with ironstone	16 7	2071 4
Shale, black sandy ..	0 1	2071 5
Seatearth, sandy at top; sphaerosiderite ..	5 8	2077 1
Sandstone and siltstone; pebble bands; coal streaks	75 2	2152 3
Seatearth and sphaerosiderite	4 0	2156 3
Striped beds and siltstone	6 7	2162 10
Mudstone, dark at base; plants, mussels ..	3 8	2166 6
Shale, dark; ironstone; mussels, fish ..	0 11	2167 5
Siltstone and silty mudstone; ironstone; mussels	6 1	2173 6
Seatearth	3 8	2177 2
Siltstone and sandstone	6 5	2183 7
Mudstone, dark, with ironstone; *Planolites* sp., mussels, fish ..	5 0	2188 7
Seatearth with sphaerosiderite	8 5	2197 0

	Thickness ft in	Depth ft in
Mudstone, silty ..	8 0	2205 0
Mudstone; mussels ..	2 0	2207 0
Mudstone, dark silty; mussels, *Carbonita humilis*, *C. sp.* ..	4 7½	2211 7½
CLOWN		
Coal 16½ in (*14½ in recovered*)		
Dirty coal 4 in	**1 8½**	**2213** 4
Seatearth	3 6	2216 10
Mudstone, silty; *small fault*	6 11	2223 9
Mudstone, largely dark; mussels	2 9	2226 6
Shale, black silty; fish ..	0 1	2226 7
Seatearth with sphaerosiderite	4 6	2231 1
Siltstone and silty mudstone	7 3	2238 4
Mudstone with ironstone; *Planolites sp*...	11 8	2250 0
Shale, black silty ..	1 0	2251 0
?MAINBRIGHT		
Coal 6 in	**0 6**	**2251** 6
Seatearth	8 3	2259 9
Siltstone	6 0	2265 9
Mudstone	1 7	2267 4
Striped beds; ironstone; plants..	15 10	2283 2
Mudstone, silty and siltstone; ironstone; plants..	10 7	2293 9
Mudstone, largely dark; coal streaks	2 3	2296 0
Seatearth; coaly shale band	7 7	2303 7
Striped beds; ironstone; plants, *Naiadites obliquus*	16 9	2320 4
Mudstone, silty, and siltstone, largely dark; *N.* aff. *alatus*, *N. obliquus*, *N. sp.* ..	8 10	2329 2
TWO-FOOT MARINE BAND Mudstone, black silty; *Lingula mytilloides*	1 0	**2330** 2

	Thickness ft in	Depth ft in
TWO FOOT		
Coal and dirt 37 in[1] (*34½ in recovered*) ..	**3 1**	**2333 3**
Seatearth with sphaerosiderite	4 4	2337 7
Siltstone and silty mudstone; ironstone ..	9 2	2346 9
Mudstone, largely dark; ironstone; *Spirorbis sp.*, cf. *Anthracosia simulans*	8 7	2355 4
Shale, black; fish ..	0 8	2356 0
ABDY (OR WINTER)		
Coal and dirt 22 in[2] (*15 in recovered*) ..	**1 10**	**2357 10**
Seatearth	1 4	2359 2
Siltstone and sandstone; ironstone	4 3	2363 5
Mudstone, largely silty; plants..	6 6	2369 11
Sandstone and siltstone	35 7	2405 6
KILNHURST		
Coal 18 in (*poor recovery*) ..	**1 6**	**2407 0**
Shale, coaly, and seatearth	1 11	2408 11
Striped beds; plants, sineoid marking ..	10 4	2419 3
Mudstone, black at top, silty	2 10	2422 1
Seatearth	1 11	2424 0
Siltstone and sandstone	2 0	2426 0
Mudstone, dark at base; ironstone; mussels ..	3 6	2429 6
Shale, black, silty ..	0 3	2429 9
Seatearth	1 3	2431 0
Striped beds	17 6	2448 6
Mudstone, dark at base; *Planolites sp.*, mussels, sineoid marking ..	10 0	2458 6
HIGH HAZLES (TOP BED)		
Coal 10 in (*poor recovery*) ..	**0 10**	**2459 4**
Seatearth	5 4	2464 8
Sandstone and siltstone	25 8	2490 4
KENT'S THIN		
Coal 18 in (*poor recovery*) ..	**1 6**	**2491 10**
Seatearth	2 6	2494 4
Siltstone and sandstone	6 0	2500 4
Shale, coaly, and seatearth	1 6	2501 10
Sandstone and siltstone	41 0	2542 10
Mudstone, largely silty; ironstone; *Spirorbis sp.*, *A.* cf. *phrygiana*, *A. sp.* (*faba/planitumida* group), *Naiadites sp.*, *Carbonita sp.*	19 2	2562 0
Mudstone, dark; *Planolites sp.*, sineoid marking	6 8	2568 8
KENT'S THICK		
Coal and dirt 93 in[3] (*57½ in recovered*) ..	**7 9**	**2576 5**
Seatearth	2 2	2578 7
Mudstone, silty, and siltstone; ironstone ..	10 2	2588 9
Mudstone, silty; mussels	1 9	2590 6
Mudstone, dark; *Spirorbis sp.*, *N.* cf. *productus*	5 9	2596 3
Coal 7 in (*poor recovery*) ..	**0 7**	**2596 10**
Seatearth	2 6	2599 4
Sandstone and siltstone	10 2	2609 6
Siltstone; *Spirorbis sp.*, cf. *Anthraconaia pulchella*, *?Anthracosia simulans*, cf. *Anthracosphaerium propinquum* ..	1 8	2611 2
Mudstone, dark, silty at top; shelly ironstone; *Anthracosia sp.* between *ovum* and *lateralis*	2 5	2613 7

[1]Detailed section: dirty coal 2 in, dirt ¾ in, dirty coal 3 in, coal 31¼ in.

[2]Detailed section: coal and dirty coal 5½ in, dirt 1½ in, dirty coal 6 in, coal 8 in.

[3]Detailed section (estimated): dirty coal and dirt 21 in, dirt 6 in, coal 15 in, dirt 26 in, coal 25 in.

	Thickness ft in	Depth ft in
Siltstone and sandstone	28 5	2642 0
Mudstone, largely dark, silty at top; plants, mussels, fish	13 0	2655 2
Dirt and coal 30½ in[1] (*29½ in recovered*)	**2 6½**	**2657 8½**
Seatearth	8 7½	2666 4
Siltstone	1 7	2667 11
Mudstone with ironstone; mussels	3 5	2671 4
Coal 7 in (*poor recovery*)	**0 7**	**2671 11**
Seatearth	2 1	2674 0
Mudstone and siltstone; plants	9 5	2683 5
Striped beds and siltstone	9 1	2692 6
Mudstone, dark; *Anthracosphaerium?*, *N. productus*	4 6	2697 0
TOP HARD **Coal** 31 in (*22¾ in recovered*)	**2 7**	**2699 7**
Seatearth	6 10	2706 5
Mudstone, silty, and siltstone	5 4	2711 9
Mudstone; plants	0 6	2712 3
Coal 3 in (*core lost*)	**0 3**	**2712 6**
Seatearth	1 3	2713 9
Siltstone and silty mudstone	28 3	2742 0
Mudstone, silty, dark at base; *Anthracosia* cf. *nitida*	7 5	2749 5
Mudstone, dark; sineoid marking	1 7	2751 0
BLIDWORTH + DUNSIL **Coal** 8 in **Dirt** 2 in **Coal** 26 in (*24½ in recovered*)	**3 0**	**2754 0**
Mudstone with coal streaks	3 11	2757 11
Seatearth	3 5	2761 4
Coal 2 in	**0 2**	**2761 6**
Siltstone and silty mudstone	5 6	2767 0
Mudstone, dark, with ironstone	2 6	2769 6
Coal 2 in (*½ in recovered*)	**0 2**	**2769 8**
Seatearth	5 4	2775 0
Sandstone and silty mudstone	3 6	2778 6
Mudstone, dark; mussels	0 6	2779 0
Mudstone, mainly silty, and siltstone; ironstone	7 2	2786 2
Mudstone, dark	0 10	2787 0
1ST WATERLOO **Dirty coal** 10 in **Dirt** 3 in **Coal** 5 in (*3 in recovered*)	**1 6**	**2788 6**
Core lost	1 2	2789 8
Seatearth with coaly band	5 1	2794 9
Mudstone, silty, with ironstone	3 5	2798 2
Sandstone and siltstone	8 10	2807 0
Mudstone, largely dark; shelly ironstone; *A.* cf. *disjuncta*, *A. phrygiana*	6 10	2813 10
WATERLOO MARKER **Coal** 4 in (*2 in recovered*)	**0 4**	**2814 2**
Seatearth	1 8	2815 10
Siltstone and sandstone; *Mariopteris?*	27 0	2842 10
Mudstone, silty, with ironstone	0 9	2843 7
Mudstone, dark, with ironstone; *Planolites sp.*, mussels, ostracods, fish, sineoid markings	8 4	2851 2

[1]Detailed section: dirt and coal 4 in, coal 4 in, dirt and coal 5½ in, dirty coal 6 in, dirt and coal 2½ in, dirty coal 3½ in, dirt and coal 3½ in, dirty coal 1½ in.

	Thickness ft in	Depth ft in
SWALLOW WOOD		
(MAIN AND BOTTOM BEDS)		
Coal abt 13½ in		
Dirt abt 7 in		
Coal abt 13½ in	2 10	2854 0
Seatearth	1 0	2855 0
Core lost	1 10	2856 10
Seatearth	0 5	2857 3
Siltstone and silty mudstone; ironstone ..	11 7	2868 10
Shale, dark; fish ..	1 9	2870 7
SWALLOW WOOD		
(FLOOR COAL)		
Cannel 2 in		
Dirty coal 8 in		
(5 in recovered)		
Coal and dirt 4 in	1 2	2871 9
Seatearth	1 10	2873 7
Mudstone, dark; ironstone; plants.. ..	2 0	2875 7
Seatearth	3 1	2878 8
Siltstone with ironstone	3 2	2881 10
Striped beds with ironstone	37 1	2918 11
Mudstone, dark; plants	0 8	2919 7
HAIGH MOOR		
(LOWER LEAF)		
Coal 6 in		
Dirt 3 in		
Coal 22 in		
(13¾ in recovered) ..	2 7	2922 2
Seatearth	4 6	2926 8
Siltstone and mudstone; ironstone; plants ..	11 4	2938 0
Core lost	0 8	2938 8
Coal 5 in		
Dirty coal 2 in		
Dirt 9 in		
Dirty coal 3 in	1 7	2940 3
Seatearth	2 6	2942 9
Coal 3 in	0 3	2943 0
Seatearth	3 8	2946 8
Mudstone; plants ..	3 9	2950 5
Striped beds	16 7	2967 0
Siltstone	10 0	2977 0
Mudstone, silty; *A.* cf. *nitida, A. phrygiana* ..	6 9	2983 9

	Thickness ft in	Depth ft in
Shale, black, mostly silty; mussels ..	1 9	2985 6
LIDGET		
Dirty coal 31 in		
(8 in recovered) ..	2 7	2988 1
Seatearth	3 5	2991 6
Siltstone with ironstone	3 6	2995 0
Seatearth	2 5	2997 5
Dirty coal 3 in		
(1½ in recovered) ..	0 3	2997 8
Seatearth	1 2	2998 10
Mudstone, silty at top; mussels	3 1	3001 11
Striped beds	15 4	3017 3
Mudstone, dark; mussels	1 2	3018 5
Seatearth; coaly band ..	1 9	3020 2
Mudstone, dark at base; ironstone; mussels, fish	1 4	3021 6
Seatearth	2 0	3023 6
Striped beds and silty mudstone	9 9	3033 3
Seatearth	1 11	3035 2
Sandstone	9 10	3045 0
Mudstone, dark at base, mostly silty; shelly ironstone; mussels ..	2 11	3047 11
Siltstone and sandstone; ironstone	24 11	3072 10
Mudstone, largely dark, silty; ironstone; *Spirorbis sp., A. ovum, A.* cf. *phrygiana, A.* aff. *regularis, N.* aff. *quadratus*	16 2	3089 0
CLAY CROSS MARINE BAND		
Mudstone, dark silty; *Lingula mytilloides,* mussels	3 9	3092 9
Mudstone, dark silty; *L. mytilloides, Dunbarella sp., Anthracoceratites sp., Hollinella (Praehollinella) claycrossensis* ..	0 3	3093 0

	Thickness ft in	Depth ft in
LOWER COAL MEASURES		
Shale, dark; *Anthracosia?*, *N. sp.*, *Geisina arcuata*, fish	0 10	3093 10
Ironstone, oolitic	0 1	3093 11
Seatearth	4 1	3098 0
Striped beds	8 9	3106 9
Mudstone; ironstone	1 0	3107 9
Seatearth	2 3	3110 0
Sandstone and siltstone	14 0	3124 0
Mudstone, silty, with ironstone	4 8	3128 8
Mudstone, dark; shelly ironstone; *Anthracosia sp. nov.* (*?ovum/regularis*), *G. arcuata*	2 0	3130 8
Seatearth	8 10	3139 6
Mudstone, dark; *A. regularis*, *G. arcuata*	3 1	3142 7
?FLOCKTON THICK		
Dirty coal 17 in **Canneloid coal** 32 in (*24 in recovered*)	**4 1**	**3146 8**
Mudstone, black canneloid	0 2	3146 10
Siltstone and sandstone	15 5	3162 3
Mudstone with dark bands, largely silty; ironstone; *Planolites sp.*, mussels, ostracods	15 9	3178 0
DEEP SOFT		
Coal and dirt 101 in[1] (*94¾ in recovered*)	**8 5**	**3186 5**
Seatearth with sphaerosiderite	5 0	3191 5
Siltstone and silty mudstone; ironstone; plants	9 4	3200 9
Mudstone, plants	1 3	3202 0
Seatearth	1 0	3203 0
Siltstone and silty mudstone; ironstone; plants, mussels	13 9	3216 9

	Thickness ft in	Depth ft in
Coal 13 in (*5 in recovered*)	**1 1**	**3217 10**
Seatearth	2 8	3220 6
Sandstone and siltstone; bands of ironstone pellets and siltstone breccia	114 0	3334 6
PARKGATE		
Coal 1 in	**0 1**	**3334 7**
Seatearth	2 5	3337 0
Siltstone and silty mudstone; ironstone	7 0	3344 0
Mudstone, largely dark; silty at base; shelly ironstone; *Carbonicola sp.*, *G. arcuata*	2 0	3346 0
Seatearth	1 0	3347 0
Siltstone and sandstone	10 2	3357 2
Mudstone, largely silty; ironstone; *C. cristagalli*, *C. oslancis*, *C. sp.* (*?rhomboidalis*)	3 6	3360 8
Seatearth	1 7	3362 3
Siltstone and sandstone; plants	36 6	3398 9
Mudstone, silty; mussels	5 3	3404 0
Mudstone, dark at base; ironstone; mussels, fish, sineoid marking	5 1	3409 1
THORNCLIFF		
Coal and dirt 61 in[2] (*52 in recovered*)	**5 1**	**3414 2**
Seatearth	6 10	3421 0
Siltstone and sandstone; ironstone	15 6	3436 6
Mudstone and silty mudstone, largely dark; plants, *C. communis*, *Curvirimula candela*, *N. flexuosus*, fish	16 2	3452 8

[1]Detailed section: coal 1¾ in, dirt ¾ in, coal 25½ in, dirty coal 3½ in, dirt ½ in, dirty coal 6 in, dirt 6 in, dirty coal 2 in, dirt and coal 3 in, dirty coal 7¼ in, dirt 1¼ in, coal 3½ in, dirty coal 3½ in, dirt 1 in, dirty coal 1½ in, coal 31 in, dirty coal 1½ in, coal 1½ in.

[2]Detailed section: *core missing* 9 in, coal 28 in, dirt and coal 1 in, coal 4½ in, dirty coal 1 in, coal 11½ in, dirty coal 1 in, coal 4 in, dirty coal 1 in.

	Thickness ft in	Depth ft in
THREEQUARTERS		
Coal 9 in		
(*poor recovery*)		
Dirt 5 in		
Coal 5 in		
(*poor recovery*) ..	1 7	**3454** 3
Seatearth	1 6	3455 9
Sandstone and siltstone; plants..	7 6	3463 3
Mudstone, silty; ironstone	6 2	3469 5
Core lost	1 10	3471 3
Mudstone, dark; sineoid marking	4 10	3476 1
Mudstone, silty and siltstone; ironstone; *C. candela*	11 5	3487 6
Mudstone with ironstone; mussels, sineoid mark ..	2 4	3489 10
Seatearth	0 7	3490 5
Mudstone and silty mudstone, largely dark; ironstone; *Carbonicola pseudorobusta, Curvirimula candela, N. flexuosus, G. arcuata,* fish ..	36 9	3527 2
TOP SILKSTONE		
Cannel and dirt 47 in[1]		
(*38½ in recovered*) ..	**3 11**	**3531** 1

	Thickness ft in	Depth ft in
Siltstone and sandstone; ironstone; roots at top	11 11	3543 0
Mudstone, silty; ironstone; *Anthraconaia ?sp. nov., Curvirimula subovata*	3 6	3546 6
LOW '*Estheria*' BAND		
Mudstone, silty; '*Estheria*' sp. nov. ...	5 3	3551 9
Mudstone, dark, silty at top; '*Estheria*' sp. nov., fish	6 10	**3558** 7
LOW SILKSTONE		
Coal 13 in		
(*poor recovery*) ..	**1 1**	**3559** 8
Seatearth (*2 ft 2 in core lost*)	4 9	3564 5
Siltstone and sandstone	20 10	3585 3
Mudstone, dark; *Anthracosphaerium sp. nov.* cf. *dawsoni, Carbonicola browni, Curvirimula candela, G. arcuata,* fish ..	4 5	3589 8
Dirty coal 4 in	**0 4**	**3590** 0
Seatearth	1 0	3591 0
Striped beds	1 6	3592 6
Core lost	0 6	3593 0

Ordsall Road Pumping Station, East Retford

There are a number of boreholes at this pumping station, drilled between 1885 and 1949. Some of the boreholes are connected by headings. Details of one of the old boreholes, drilled in 1902 and 350 ft deep, are given by Lamplugh and Smith 1914, p. 107. The following notes apply to the 1949 (Whisker Hill) borehole.

Ht above O.D. abt 102 ft.   6-in SK 68 SE.
Site [6916 8003] 1070 yd at 148° from All Saints' Church, Babworth. Drilled 1949 by C. Isler & Co. Ltd, for East Retford Corporation; now owned by Lincoln and District Water Board.

PERMO-TRIASSIC: PB to bottom of borehole at 600 ft.

---

[1]Detailed section: inferior cannel 8 in, ironstone 2 in, cannel 19 in, dirty coal 7 in, dirt 3 in, dirty coal 8 in.

Ordsall Works (British Ropes) Borehole, East Retford

Ht above O.D. abt 67 ft.   6-in SK 78 SW.
Site [7018 8003] 300 yd at 181° from East Retford station. Drilled
1953 by Doncaster Well Borers Ltd for British Ropes Ltd.

MADE GROUND: to 3 ft or 12 ft.

PERMO-TRIASSIC: PB to bottom of borehole at 400 ft.

Another borehole, 200 ft deep, was drilled at this site in 1939.

Rampton Hospital Borehole

Ht above O.D. abt 110 ft.   6-in SK 77 NE.
Site [7753 7773] 1360 yd at 21° from Hardings Farm, East Drayton.
Drilled 1909 by A. C. Potter & Co. For detailed section see
Lamplugh and Smith 1914, pp. 105–6.

PERMO-TRIASSIC: KM to 344 ft, Waterstones to 509 ft, GB to 538 ft, PB to bottom of
borehole at 1005 ft.

Ranby Camp Borehole

Ht above O.D. 146 ft.   6-in SK 68 SE.
Site [6639 8074] 820 yd at 111° from Ranby House, Ranby.
Drilled 1955–6 by Foraky Co. for N.C.B. Open holed to 898¾ ft,
cored 898¾ ft to bottom. Cores examined by E. G. Smith and
R. A. Eden.

DRIFT: Sandy boulder clay to abt 20 ft.

PERMO-TRIASSIC: PB + LMS to abt 476 ft, UPM to abt 570 ft, UML to abt 610 ft, MPM
to abt 788 ft, LML to 1018 ft 10 in, LPM to 1031½ ft, BPS to 1031 ft 7 in.

UPPER COAL MEASURES: Hemsworth Div. to 1114 ft 1 in, Brierley Div. to 1231¾ ft
(Blyth, c. 25 in at 1231¾ ft), Ackworth Div. to 1442 ft 5 in.

MIDDLE COAL MEASURES: Top M.B. at 1445 ft 5 in; Shafton M.B. at 1508¼ ft; Sharlston Top (c. 10 in, d. & c. 8 in, d.c. 8 in on c. 6 in) at 1697 ft 4 in; Sharlston Low (c. 8½
in, d. 57 in on c. 4 in) at 1768½ ft; Sharlston Yard (d.c. 4½ in on c. 3½ in) at 1800¾ ft;
?Houghton Thin (d.c. 3½ in on c. 4½ in) at 1806 ft 8 in; Mansfield M.B. at 1911¼ ft;
Haughton M.B. at 2001 ft 7 in; Clown (d.c. 4 in, d. 14 in on c. 33 in) at 2062 ft 4 in;
Two-Foot M.B. at 2181 ft 10 in; Two-Foot (c. 30 in) at 2184 ft 4 in; Abdy (c. 12¼ in, d.
1 in, c. 2 in, d. & d.c. 15¼ in on c. 5 in) at 2211 ft 5 in; Kilnhurst (d.c. 3½ in on c. 10¼ in)
at 2236¾ ft; High Hazles, Top Bed (d.c. 4 in) at 2277 ft 4 in; Kent's Thin (ca. & d.c.
21 in) at 2327¾ ft; Kent's Thick (c. 20¼ in, d. & d.c. 8 in, c. 14½ in on d.c. 3¼ in) at 2370
ft; Top Hard + Dunsil (c. 57½ in, d. 1½ in, c. 4½ in on d.c. 2½ in) at 2517 ft 1 in; 1st
Waterloo (d. & c. 1¾ in, c. 12¼ in on c. 4 in) at 2554 ft 7 in; Clay Cross M.B. at 2796 ft
2 in.

LOWER COAL MEASURES: ?Deep Hard Roof Coal (c. 1 in) at 2963 ft 10 in; Deep Hard
(d.c. 3 in, d. 3 in, d.c. 9 in, d. 3 in, c. 1 in, d. 4¼ ft on d.c. 4 in) at 2992 ft; Parkgate (c.
9 in, d. 10 in, d.c. 5 in, c. 33½ in, d. 2 in on d.c. 2½ in) at 3034¾ ft; Thorncliff (c. ¾ in, d.
6 in on c. 5 in) at 3064 ft 2 in; Top Silkstone (d.c. 3½ in, d. 6½ in, d.c. 3 in, c. 4½ in, inferior
cannel 7½ in, d. & measures 7 ft 5 in on d.c. 7 in) at 3207 ft 4 in; Low 'Estheria' Band at
3240¾ ft; Low Silkstone (c. & d.c. 17 in) at 3242 ft 2 in.

Bottom of borehole 3246 ft.

Ranby Hall Borehole

Ht above O.D. 99 ft.   6-in SK 68 SW.

Site [6488 8238] 980 yd at 2° from Chequer House Farm, Ranby.
Drilled 1956 by Foraky Co. for N.C.B. Open holed to 780 ft,
cored 780 ft to bottom. Cores examined by L. S. O. Morris and
R. A. Eden.

PERMO-TRIASSIC: PB + LMS to 372 ft, UPM to 471 ft, UML to 513 ft 8 in, MPM to
654 ft, LML to 871$\frac{3}{4}$ ft, LPM to 885 ft, BPS to 889$\frac{1}{2}$ ft.

UPPER COAL MEASURES: Hemsworth Div. to 1132$\frac{1}{2}$ ft, Brierley Div. to 1299 ft 10 in
(Blyth, c. 30 in at 1299 ft 10 in), Ackworth Div. to 1482 ft.

MIDDLE COAL MEASURES: Top M.B. at 1497 ft 1 in; Shafton M.B. at 1576 ft; Sharlston
Low, main coal (c. 12 in, d.c. 1$\frac{3}{4}$ in, c. 1$\frac{3}{4}$ in, d. 3$\frac{1}{4}$ in on c. 7$\frac{1}{4}$ in) at 1921$\frac{1}{2}$ ft; Sharlston
Yard (d.c. & d. on c., 24 in) at 1969 ft; ?Houghton Thin (d.c. 11 in, d. 15 in on d.c. 2 in)
at 1978 ft; horizons of Mansfield M.B. and Haughton M.B. faulted out; Mainbright
(d.c. & d. 20 in, d. & measures 10 ft 5 in, c. 3 in, d. 5 in, c. 2 in, d. 12 in on d.c. 3 in)
at 2019 ft 11 in; Two-Foot M.B. at 2060 ft 11 in; Two-Foot (c. with 2$\frac{1}{4}$ in d.c., 31 in) at
2063$\frac{1}{2}$ ft; Abdy (d.c. 14$\frac{3}{4}$ in, d. & d.c. 4$\frac{1}{2}$ in, c. 2 in, d. 2 in, c. 2$\frac{1}{4}$ in, d.c. 3 in, d. & measures
4 ft 9$\frac{1}{2}$ in on c. 3 in) at 2096 ft 7 in; High Hazles, Top Bed (c. 5 in, d. 7 in on c. 1 in) at
2185 ft 10 in; Kent's Thin (ca. 3 in, c. 10$\frac{1}{4}$ in on d.c. 1$\frac{1}{4}$ in) at 2231 ft 5 in; Kent's Thick
(c., d.c. & d. 17 in, d. 7 in on c. 25 in) at 2282 ft 10 in; Top Hard (ca. $\frac{3}{4}$ in on c. 30$\frac{1}{4}$ in) at
2442 ft 7 in; Blidworth + Dunsil (c. 15$\frac{1}{4}$ in, d. 1$\frac{3}{4}$ in on c. 20 in) at 2456 ft 11 in; 1st
Waterloo, top leaf (c. 1$\frac{1}{2}$ in, d. $\frac{3}{4}$ in on c. 12$\frac{3}{4}$ in) at 2508 ft 2 in; bottom leaf (c. 1 in, d.
12 in, c. 2 in, d.c. 5 in on c. 1 in) at 2545 ft 11 in; Swallow Wood, Main Bed (d.c. 12$\frac{1}{4}$ in)
at 2598$\frac{1}{2}$ ft; Haigh Moor + ?Swallow Wood, Floor Coal (d.c. 1$\frac{3}{4}$ in, d. 2$\frac{1}{2}$ in, c. 1$\frac{3}{4}$ in,
d.c. 1$\frac{1}{4}$ in, c. 4$\frac{1}{2}$ in, d. 1$\frac{3}{4}$ in, d.c. 2 in, c. 27$\frac{1}{2}$ in, d.c. 5$\frac{1}{4}$ in, d. 24 in, c. 3 in, d.c. 2$\frac{1}{2}$ in on c.
21 in) at 2610 ft 10 in; Lidget (c. 4 in, d. 15 in on c. 1 in) at 2680$\frac{3}{4}$ ft; 2nd Ell (c. 1 in) at
2730 ft; Clay Cross M.B. at 2804 ft 4 in.

LOWER COAL MEASURES: Deep Soft (c. with 1 in d., 20 in, d.c. 1$\frac{1}{2}$ in, d. 1$\frac{3}{4}$ in,
d.c. $\frac{3}{4}$ in, c. 24$\frac{3}{4}$ in, d. 1 in, c. 3$\frac{1}{4}$ in, d.c. 1 in, c. 1$\frac{1}{2}$ in, d. $\frac{3}{4}$ in, c. 20$\frac{3}{4}$ in, d. 2$\frac{3}{4}$ in, d.c. 1$\frac{1}{4}$ in,
c. 4$\frac{1}{4}$ in on d. with c. and d.c., 35$\frac{1}{2}$ in) at 2910 ft 8$\frac{3}{4}$ in; Deep Hard (d.c. & d. 10$\frac{1}{2}$ in, c.
13 in, d.c. & d. 9$\frac{3}{4}$ in on c. 5$\frac{3}{4}$ in) at 2963 ft 10 in; Parkgate (c. 5$\frac{1}{4}$ in, d.c. 1$\frac{3}{4}$ in, d. 20 in,
d.c. & d. 10$\frac{3}{4}$ in, measures 12 ft 11 in, c. 34$\frac{1}{2}$ in, d. 1 in, c. 3 in on d.c. 2$\frac{1}{2}$ in) at 3004 ft
7 in; Thorncliff (c. 9$\frac{1}{2}$ in, measures 30 ft 7 in on d.c. & d. 25 in) at 3080 ft 2 in; Top
Silkstone (d.c. 2$\frac{1}{2}$ in, d. 4 in, d.c. 5$\frac{3}{4}$ in, c. 13$\frac{1}{4}$ in, inferior ca. 2 in, d. $\frac{3}{4}$ in, c. 9$\frac{3}{4}$ in on d. &
d.c. 18 in) at 3193$\frac{1}{2}$ ft; Low *Estheria* Band at 3232 ft 7 in; Low Silkstone (c. & d. abt
11$\frac{1}{2}$ in, c. abt 13 in on d.c. 3$\frac{1}{2}$ in) at 3235 ft 1 in.

Bottom of borehole 3263 ft 2 in.

Ranskill Borehole

Ht above O.D. abt 55 ft.   6-in SK 68 NE.

Site [6535 8910] 1650 yd at 328° from Ranskill station. Drilled
1954 by Foraky Co. for N.C.B. Open holed to 790$\frac{1}{4}$ ft, cored
790$\frac{1}{4}$ ft to bottom. Cores examined by R.A. Eden and R. E. Elliott.

DRIFT: Glacial Sand and Gravel to 32 ft.

PERMO-TRIASSIC: PB to 330 ft, LMS + UPM to 460 ft, UML to 512 ft, MPM + LML
to 823$\frac{3}{4}$ ft, LPM to 845 ft, BPS to 845$\frac{1}{4}$ ft.

UPPER COAL MEASURES: Badsworth Div. to 1035 ft 7 in, Hemsworth Div. to 1313 ft
8 in, Brierley Div. to 1453 ft 10 in (Blyth, c. 33 in at 1453 ft 10 in), Ackworth Div. to
1675 ft.

MIDDLE COAL MEASURES: Top M.B. at 1682 ft 10 in; Shafton M.B. at 1749 ft; Shafton
(c. 26 in) at 1773 ft; *Edmondia* Band at 1881$\frac{1}{2}$ ft; Sharlston Top (d.c. 3$\frac{1}{2}$ in) at 1933 ft 1 in;
Sharlston Low (c. 16 in) at 2023 ft; Sharlston Yard (d.c. 2 in, d. 9 in on c. 14 in) at 2051 ft

2 in; ?Houghton Thin (c. 8 in, d. 6 in on c. 1 in) at 2065 ft 2 in; Mansfield M.B. at 2131 ft 10 in; Haughton M.B. at 2228 ft 8 in; Clown (d.c. 1 in, c. 3 in, d. $\frac{1}{2}$ in on c. $13\frac{1}{2}$ in) at 2298 ft 1 in; Mainbright, top leaf (d.c. 6 in) at 2347 ft 8 in; bottom leaf (d.c. 8 in?) at 2369 ft 4 in; Two-Foot M.B. at 2396 ft 10 in; Two-Foot (d.c. $6\frac{1}{2}$ in on c. $22\frac{1}{2}$ in) at $2399\frac{1}{4}$ ft; Abdy (c. with $1\frac{1}{4}$ in d.c. & d., 22 in, d. 2 in on d.c. 1 in) at $2426\frac{3}{4}$ ft; Kilnhurst (d.c. 6 in) at $2462\frac{1}{4}$ ft; High Hazles, Top Bed (c. 18 in) at 2519 ft; Kent's Thin (c. 12 in) at 2544 ft 10 in; Kent's Thick (d.c. 1 in, c. 17 in, d. 9 in on c. 19 in) at 2608 ft; Top Hard (d.c. & d. $2\frac{1}{2}$ in, c. $17\frac{1}{2}$ in on d.c. & d. 12 in) at 2737 ft 7 in; Blidworth + Dunsil (c. 26 in) at $2758\frac{3}{4}$ ft; 1st Waterloo (c. 3 in, d. & measures 40 in, c. 9 in, d. $1\frac{1}{2}$ in on d.c. $5\frac{1}{2}$ in) at 2827 ft 4 in; Swallow Wood (c. 11 in, d. 13 in, c. 14 in, measures 19 ft 3 in, c. 8 in, d. 4 in on c. 8 in) at 2904 ft 10 in; Haigh Moor (c. 18 in, d. 12 in on c. 10 in) at 2927 ft 10 in; Lidget (c. & d.c. 16 in) at 3012 ft 4 in; 2nd Ell (d.c. 4 in) at $3042\frac{1}{4}$ ft; Clay Cross M.B. at $3109\frac{1}{4}$ ft.

LOWER COAL MEASURES: Deep Soft (c. $33\frac{1}{2}$ in, d. $\frac{1}{2}$ in, c. 6 in, d.c. $2\frac{1}{2}$ in, c. 10 in, d.c. $5\frac{1}{2}$ in on c. 21 in) at 3192 ft 11 in; Parkgate (c. abt 18 in, d. abt 11 in on d.c. abt 27 in) at $3341\frac{1}{4}$ ft[1]; Thorncliff (c. on d.c. abt 36 in, d. 8 in on c. abt 12 in) at 3405 ft 7 in; Three-quarters (c. 1 in) at $3435\frac{1}{4}$ ft; Top Silkstone (ca. 10 in, d.c. 2 in, d. 1 in, c. & ca. 14 in, d. & measures 60 in on d.c. 3 in) at $3516\frac{1}{2}$ ft; Low '*Estheria*' Band at 3545 ft 4 in; Low Silkstone (ca. 4 in, d. 2 in on c. 5 in) at $3546\frac{1}{4}$ ft.

Bottom of borehole 3597 ft.

Ranskill (BP) No. 1 Oil Bore

R.T.E. above O.D. 54 ft.   6-in SK 68 NW.
Site [6423 8814] 1550 yd at 147° from Serlby Hall. Drilled 1965 by BP Petroleum Development Ltd. Largely open holed, cored 4327–56 ft and 5662 ft to bottom. Cores examined by C. G. Godwin.

PERMO-TRIASSIC: PB + LMS + UPM to abt 334 ft, UML to abt 385 ft, MPM to abt 535 ft, LML to abt 728 ft, LPM + BPS to $743\frac{1}{2}$ ft.

UPPER COAL MEASURES: Badsworth, Hemsworth and Brierley Divisions to abt 1365 ft, Ackworth Div. to abt 1565 ft.

MIDDLE COAL MEASURES: position of Mansfield M.B. at abt 2050 ft; position of Two-Foot M.B. at abt 2350 ft; Kent's Thick at abt 2570 ft; Top Hard at abt 2710 ft; position of Clay Cross M.B. at abt 3045 ft.

LOWER COAL MEASURES: Deep Soft at abt 3130 ft; Parkgate at abt 3275 ft; Top Silkstone at abt 3440 ft; Kilburn at abt 3835 ft; Pot Clay M.B. at 4355 ft 2 in.

MILLSTONE GRIT SERIES to bottom of bore at 5674 ft.

Rockley Borehole

Ht above O.D. abt 70 ft.   6-in SK 77 NW.
Site [7125 7509] 1075 yd at 159° from St Peter's Church, Gamston. Drilled 1896 for Wigan Coal and Iron Co. Ltd. For detailed section see Lamplugh and Smith 1914, pp. 72–3.

DRIFT: Terrace deposits (largely clay) to 8 ft.

PERMO-TRIASSIC: Waterstones to 62 ft 5 in, GB to 100 ft 8 in, PB + LMS to $840\frac{1}{2}$ ft, UML to 871 ft 2 in, MPM to bottom of borehole at 883 ft 2 in.

Scaftworth Borehole

Ht above O.D. 62 ft.   6-in SK 69 SE.
Site [6761 9167] 1050 yd at 90° from Scaftworth Hall. Drilled 1956 by Foraky Co. for N.C.B. Cores examined by E. G. Smith.

---

[1]Section and depth as proved in diversion.

Description	Thickness ft	in	Depth ft	in
PERMO-TRIASSIC				
Not cored: PB + LMS to 659¾ ft, UPM to 800 ft, UML to 842 ft, MPM to abt 978 ft ..	978	0	978	0
LOWER MAGNESIAN LIMESTONE				
Not cored	91	2	1069	2
Dolomite, light grey, with marly and marl bands; *Dielasma elongatum, Horridonia horrida, Strophalosia morrisiana*	46	10	1116	0
Interbanded grey dolomite and marl; traces of blende and galena; *Acanthocladia anceps,* cf. *Batostomella crassa, Fenestella retiformis, D. elongatum, H. horrida, Lingula credneri, S. morrisiana* abt	74	6	1190	6
LOWER PERMIAN MARL				
Marl, grey laminated (Marl Slate), with carbonate bands; veinlets of galena abt	10	6	1201	0
BASAL PERMIAN SANDS				
Sand, grey, with rounded grains; sporadic pebbles	0	2	**1201**	**2**
UPPER COAL MEASURES				
HEMSWORTH DIVISION				
Sandstone, partly red ..	4	2	1205	4
Mudstone, greenish and red, largely silty ..	5	7	1210	11
Core lost	6	2	1217	1
Seatearth, greenish brown and red ..	3	9	1220	10
Core lost	9	6	1230	4
Seatearth, partly red ..	3	2	1233	6
Siltstone and mudstone with red patches ..	4	0	1237	6
Seatearth, partly red ..	5	4	1242	10
Sandstone and siltstone, partly red	10	2	1253	0
Mudstone, largely silty, purple, red and greenish; ironstone ..	7	0	1260	0
Siltstone with ironstone	5	2	1265	2
SECOND CHERRY TREE MARKER				
Mudstone, with ironstone; *Anthraconauta* aff. *phillipsii, Euestheria simoni* ..	2	0	**1267**	**2**
Coal 16 in	**1**	**4**	**1268**	**6**
Seatearth	2	5	1270	11
Mudstone, silty, purple	1	0	1271	11
Striped beds	0	10	1272	9
Seatearth, red-mottled; sphaerosiderite at base	13	1	1285	10
Striped beds with sphaerosiderite ..	0	10	1286	8
Seatearth, red-mottled; sphaerosiderite ..	19	5	1306	1
Siltstone with sphaerosiderite	4	10	1310	11
Mudstone, silty in parts, partly red; sphaerosiderite; *Planolites montanus*	6	9	1317	8
Seatearth and siltstone	10	2	1327	10
Sandstone with siltstone bands; conglomeratic at base. *3 ft 2 in of core lost*	102	11	1430	9
BRIERLEY DIVISION				
Seatearth	5	8	1436	5
Siltstone	2	10	1439	3
Seatearth	8	0	1447	3
Sandstone	7	7	1454	10
Striped beds	4	1	1458	11
Mudstone and siltstone; plant debris, *Cochlichnus kochi* ..	3	5½	1462	4½
Coal 9¾ in / **Dirt** 9¾ in / **Coal** 11½ in / **Dirty coal** 1 in	**2**	**8**	**1465**	**1**
Seatearth	2	11	1468	0
Mudstone with sphaerosiderite	1	11	1469	11

	Thickness ft in	Depth ft in
Seatearth	4 7	1474 6
Siltstone	1 1	1475 7
Seatearth	3 0	1478 7
Mudstone, silty, and siltstone	3 0	1481 7
Striped beds	3 5	1485 0
Siltstone and mudstone; *Anthraconauta sp.*, 'Estheria' sp. ..	6 4	1491 4
Mudstone and coaly shale; abundant plants	1 0½	1492 4½
Dirty coal 2½ in	**0 2½**	**1492 7**
Seatearth	6 2	1498 9
Siltstone and mudstone; *C. kochi, Anthraconauta sp.,* 'Estheria'?	7 1	1505 10
Shale, black, with ironstone; plants, *A.* aff. *phillipsii, A.* cf. *wrighti, Carbonita humilis,* fish tooth	0 6	1506 4
Siltstone and silty mudstone; plants.. ..	4 1	1510 5
Cannel 0½ in **Coal** 0½ in	**0 1**	**1510 6**
Shale, coaly, with ironstone; *C. humilis* and fish debris	0 3	1510 9
Seatearth	0 3	1511 0
Coal 1 in Dirt 4 in **Dirty coal** 4 in	**0 9**	**1511 9**
Seatearth with sphaerosiderite	4 5	1516 2
Dirty coal 2 in	**0 2**	**1516 4**
Seatearth with sphaerosiderite	11 9	1528 1
Siltstone with plants ..	15 8	1543 9
Sandstone, conglomeratic at base ..	46 5	1590 2
ACKWORTH DIVISION		
Siltstone; ironstone nodules	14 1	1604 3
Mudstone with dark bands; ironstone; *A.* aff. *phillipsii, C. humilis*	13 1	1617 4

	Thickness ft in	Depth ft in
Shale, black, silty; plants	0 10	1618 2
Seatearth	1 10	1620 0
Siltstone and mudstone with ironstone ..	4 10	1624 10
Mudstone, black, with ironstone	0 5	1625 3
Mudstone, soft.. ..	0 4	1625 7
Dirty coal 12 in	**1 0**	**1626 7**
Seatearth	3 8	1630 3
Siltstone and sandstone	6 2	1636 5
Mudstone; ironstone and fish	0 2	1636 7
Dirty coal 13 in	**1 1**	**1637 8**
Seatearth	7 1	1644 9
Sandstone and silty mudstone	11 9	1656 6
Mudstone	1 7	1658 1
Coal 0¾ in Dirt 10¾ in **Dirt and coal** 2½ in	**1 2**	**1659 3**
Seatearth	2 0	1661 3
Sandstone and siltstone	6 1	1667 4
Mudstone, silty at top..	7 1	1674 5
SCOFTON **Coal** 12 in	**1 0**	**1675 5**
Seatearth; sphaerosiderite at base ..	7 5½	1682 10½
Siltstone, sandy; plants	3 2½	1686 1
Seatearth	1 7	1687 8
Siltstone	4 4	1692 0
ACKWORTH ROCK Sandstone with pebble bands	71 4	1763 4
MIDDLE COAL MEASURES TOP MARINE BAND Mudstone, silty, with ironstone; 'fucoids'	0 10	1764 2
Mudstone, silty, with ironstone; *Spirorbis sp., Lingula mytilloides, Orbiculoidea* cf. *nitida, Dunbarella sp.,* 'Anthracoceras' ?, palaeoniscid scale	2 6	1766 8

Description	Thickness ft in	Depth ft in
Mudstone, silty, and siltstone; plants ..	6 1	1772 9
Mudstone, silty at top, with ironstone; *Crurithyris sp.*, *L. mytilloides*, *O.* cf. *nitida*, *D.* cf. *macgregori*, *Myalina sp.*, *Coelogasteroceras dubium*, *Cypridina 'phillipsii'*	12 5	1785 2
Mudstone with ironstone; *L. mytilloides*, 'fucoids', cf. *Tomaculum*	2 5	1787 7
Mudstone, dark; *L. mytilloides*, *Rhabdoderma sp.*, *Rhizodopsis sp.*, 'fucoids' ..	0 4	**1787 11**
Coal 9 in (*poor recovery*) ..	**0 9**	**1788 8**
Seatearth	3 5	1792 1
Sandstone and siltstone	6 0	1798 1
Mudstone, dark at top; ironstone; *Naiadites sp.*, *Lioestheria vinti*, palaeoniscid scale ..	2 3	1800 4
Seatearth	3 3	1803 7
Sandstone and siltstone	4 5	1808 0
Mudstone, silty in parts, with 1-in coaly shale band	3 5	1811 5
Coal 2 in / **Dirt** 2 in / **Coal** 11 in	**1 3**	**1812 8**
Seatearth	7 6	1820 2
Siltstone	1 0	1821 2
Mudstone, dark at base, silty in parts; *N.* cf. *hindi*, *L. vinti*, *Geisina subarcuata*	5 0	1826 2
Seatearth with sphaerosiderite ..	6 9	1832 11
Shale, black; ironstone; *N.* cf. *hindi*, *L. vinti*, *Carbonita* cf. *claripunctata*, *G. subarcuata*, fish remains including *Rhabdoderma sp.* ..	4 7	1837 6

Description	Thickness ft in	Depth ft in
Seatearth, dark ..	0 5	1837 11
Sandstone and siltstone	6 1	1844 0
Mudstone, silty; at top, *N.* cf. *hindi*, *L. vinti* ..	9 7	1853 7
Striped beds	1 2	1854 9
Mudstone, silty at top; ironstone and *N.* cf. *hindi*, *L. vinti* ..	6 11	1861 8
SHAFTON MARINE BAND Mudstone, dark at base; foraminifera including *Ammodiscus sp.* and *Glomospira sp.*, *Lingula mytilloides*, *Anthraconaia sp.* (?*spathulata*), *Edmondia?*, *Myalina compressa*, *Hollinella sp.*, cf. *Tomaculum sp.*	3 4	**1865 0**
Dirt and coal 31 in[1] (*30 in recovered*) ..	**2 7**	**1867 7**
Seatearth with sphaerosiderite	6 10	1874 5
Siltstone and sandstone	9 11	1884 4
Mudstone with ironstone; *N. sp.* (cf. *hindi*), '*Estheria*' *sp.*, ostracods, *Rhabdoderma sp.*	4 2	1888 6
Shale, dark, canneloid at base; megaspore, *N. sp.* (?*daviesi*), *Lioestheria vinti*, *C. humilis*, *C. pungens*, *C.* cf. *salteriana*, *Geisina subarcuata* and fish remains ..	0 7	1889 1
SHAFTON **Coal and dirt** 46½ in[2]	**3 10½**	**1892 11½**
Seatearth with sphaerosiderite. *4 ft of core lost*	27 1½	1920 1
Siltstone and sandstone	4 8	1924 9
Mudstone	6 1	1930 10
Shale, dark; ironstone; *N.* cf. *hindi*, *C. humilis*	0 6	1931 4

[1] Detailed section: dirty coal 3¾ in, coal 3¼ in, dirty coal 7 in, dirt 13 in, dirty coal 4 in.
[2] Detailed section: dirty coal 4 in, coal 18¾ in, dirt 3 in, coal 6¼ in, dirt 13 in, coal 1½ in.

	Thickness ft in	Depth ft in
Mudstone	0 8	1932 0
Seatearth	1 1	1933 1
Sandstone	1 5	1934 6
Mudstone and siltstone with ironstone; plants	9 6	1944 0
Seatearth	7 9	1951 9
Siltstone and sandstone	67 1½	2018 10½
Dirty coal 2½ in	0 2½	2019 1
Seatearth, brown at top	9 4	2028 5
Siltstone, sandy ..	4 2	2032 7
Mudstone, dark; plants, fish	0 7	2033 2
Siltstone, dark; plants, Naiadites sp. ..	0 5	2033 7
Seatearth	4 7	2038 2
Mudstone, dark; Naiadites sp.	0 7½	2038 9½
Seatearth	2 1½	2040 11
SHARLSTON TOP		
Coal and dirt 29 in[1] (no recovery) ..	2 5	2043 4
Seatearth; coal streaks at top..	2 7	2045 11
Sandstone and siltstone	17 7	2063 6
Mudstone with ironstone; megaspores, Naiadites sp., C. humilis, fish remains including Elonichthys sp.	6 9	2069 11
Shale, dark, cannaloid at base; ironstone; N. cf. hindi, fish ..	1 9	2071 8
Seatearth, greenish at top	2 2	2073 10
Mudstone, hard greenish	0 1½	2073 11½
Mudstone with ironstone	2 5½	2076 5
Shale, dark, canneloid in part; ironstone; Anthraconaia hindi, Naiadites sp., fish including Rhadinichthys sp., Rhizodopsis sp. ..	1 1	2077 6
Mudstone, mostly silty; Naiadites sp., fish ..	5 6	2083 0
Sandstone; plants ..	2 6	2085 6
Striped beds	13 8½	2099 2½
Mudstone, dark at base	0 3½	2099 6
SHARLSTON LOW		
Coal and dirt 13 in[2] (poor recovery) ..	1 1	2100 7
Seatearth with sphaerosiderite	18 11	2119 6
Mudstone, silty in parts, with ironstone; plants	45 0	2164 6
Sandstone and siltstone	2 8	2167 2
Mudstone, silty at top; plants..	5 3	2172 5
Ironstone	0 3	2172 8
Shale, dark silty; Lepidodendron sp., fish including platysomid, Rhabdoderma sp., and Rhizodopsis sp. ..	1 4	2174 0
SHARLSTON YARD		
Dirty coal 1 in		
Coal 2 in	0 3	2174 3
Seatearth	9 5	2183 8
Sandstone and siltstone	3 3	2186 11
Shale, dark, silty in part	1 0	2187 11
?HOUGHTON THIN		
Coal 21 in (poor recovery) ..	1 9	2189 8
Shale, coaly and seatearth	2 2	2191 10
Sandstone and siltstone	15 11	2207 9
Mudstone, silty at top; ironstone	5 7	2213 4
Mudstone, dark silty; N. cf. productus, cf. Geisina sp., Rhizodopsis sp.	5 0	2218 4
MANSFIELD MARINE BAND		
Mudstone, dark, with ironstone; foraminifera including Ammodiscus sp.,		

[1]Detailed section according to penetration recorder: coal 3 in, dirt with coal bands 12 in, coal 14 in.

[2]Detailed section according to driller: coal 7 in, dirt 2 in, coal 4 in.

	Thickness ft in	Depth ft in
Glomospira sp. and *Tolypammina sp.*, hexactinellid sponge spicules, *Lingula mytilloides, Donaldina ?, Myalina* cf. *compressa, Polidevcia acuta, Hollinella ?*, fish remains including *Megalichthys sp.*, ?'fucoids'	15 8	2234 0
Mudstone, dark, with ironstone; *L. mytilloides, Orbiculoidea* cf. *nitida, Coelogasteroceras sp.*, fish remains including *Listracanthus sp.* ..	6 5	2240 5
Mudstone, dark, with ironstone; ?plant strands, *Serpuloides* cf. *stubblefieldi, Rugosochonetes skipseyi, Euphemites ?, 'Anthracoceras' hindi, Coelogasteroceras sp.*, 'fucoids' ..	4 1	2244 6
Mudstone, dark, 'canky'; *Lingula sp., Orbiculoidea sp., 'A'. hindi, Metacoceras* cf. *cornutum*, indet. goniatite fragments, *Hollinella sp.* ..	1 8	**2246 2**
Seatearth; sphaerosiderite at base ..	9 10	2256 0
Mudstone, mostly dark and silty; ironstone; plants, *N.* aff. *alatus* ..	16 6	2272 6
Shale with plants ..	1 5	2273 11
Ironstone with traces of galena and siltstone..	1 7	2275 6
Mudstone with ironstone	4 4	2279 10
Shale, dark; plants ..	0 2	2280 0
Coal and dirt 13 in (*6 in recovered*) ..	**1 1**	**2281 1**
Seatearth	7 3	2288 4

	Thickness ft in	Depth ft in
SUTTON MARINE BAND Mudstone, dark; foraminifera, *Anthraconaia* (cf. *spathulata*), fish scales, 'fucoids'	1 4	**2289 8**
Seatearth	2 1	2291 9
Siltstone and sandstone	7 0	2298 9
Mudstone, silty at top ..	4 0	2302 9
Shale, dark; *Anthracosia atra ?, Naiadites sp., Carbonita humilis, C. scalpellus*, fish ..	0 11	2303 8
Siltstone and mudstone with ironstone ..	6 11	2310 7
Coal 8 in (*poor recovery*) ..	**0 8**	**2311 3**
Seatearth with sphaerosiderite	5 0	2316 3
Ironstone	0 6	2316 9
HAUGHTON MARINE BAND Mudstone, dark, silty at top; ironstone; *Serpuloides stubblefieldi, Lingula* cf. *elongata, Orbiculoidea* cf. *nitida*, cf. *Tomaculum sp.*, fish, 'fucoids'	19 6	**2336 3**
Mudstone, dark; fish including *Rhizodopsis sp.*	0 5	2336 8
Siltstone, dark, and sandstone	0 11	2337 7
Seatearth	1 8	2339 3
Striped beds	3 0	2342 3
Siltstone	4 2	2346 5
Mudstone, dark at base	0 5	2346 10
SWINTON POTTERY **Coal and dirt** 32 in[1] (*27½ in recovered*) ..	**2 8**	**2349 6**
Seatearth	7 1	2356 7
Shale, dark; *Anthracosia* aff. *atra* ..	0 1	2356 8
Seatearth	2 6	2359 2
Mudstone; silty near base; *A. sp.*	1 2	2360 4

[1]Detailed section: coal 8 in, dirty coal 12 in, coal 12 in.

	Thickness ft in	Depth ft in
Seatearth	2 1	2362 5
Sandstone and siltstone	5 5	2367 10
Mudstone with ironstone	2 4	2370 2
Seatearth with sphaerosiderite	3 11	2374 1
Mudstone, silty at top; ironstone	4 10	2378 11
Mudstone, dark; *A. atra*	0 1	2379 0
Seatearth	4 0	2383 0
Siltstone and mudstone; *Anthracosia?, Naiadites sp., Rhizodopsis sp.*	13 0	2396 0
Coal 1½ in	**0 1½**	**2396 1½**
Shale, coaly	0 3½	2396 5
Seatearth	6 1	2402 6
Mudstone, silty and ironstone; plants ..	2 6	2405 0
Shale, dark; *Anthracosia sp.*	0 5	2405 5
Mudstone, silty ..	1 10	2407 3
Ironstone, oolitic ..	0 3½	2407 6½
?CLOWN		
Coal 0½ in	**0 0½**	**2407 7**
Seatearth with sphaerosiderite	3 0	2410 7
Siltstone and sandstone	12 8	2423 3
Mudstone, silty at top; ironstone	16 7	2439 10
Mudstone, dark slickensided	0 9	2440 7
Mudstone with ironstone; *Anthracosia sp.*	1 7	2442 2
Shale, dark, partly silty; *A.* cf. *atra, Rhadinichthys sp.*	1 10	2444 0
MAINBRIGHT (OR MELTONFIELD)		
Dirt and coal 43 in[1]	**3 7**	**2447 7**
Seatearth; coal bands at top	3 6	2451 1

	Thickness ft in	Depth ft in
Siltstone, sandy; ironstone	1 11	2453 0
Mudstone, silty ..	0 7	2453 7
Seatearth, dark; coal bands	0 3	2453 10
Coal and dirt 3 in		
Dirt 34 in		
Dirty coal 1 in	**3 2**	**2457 0**
Seatearth	0 3½	2457 3½
Shale, coaly	0 3½	2457 7
Siltstone and sandstone	9 2	2466 9
Mudstone, partly silty, and siltstone; *Anthraconaia librata, Anthracosia aquilinoides, A.* cf. *aquilina, A. concinna, Naiadites sp.* ...	6 5	2473 2
Dirty coal 2 in		
Coal 3 in	**0 5**	**2473 7**
Seatearth; coal bands at top	7 5	2481 0
Siltstone and silty mudstone with ironstone; plants..	13 1	2494 1
Siltstone and sandstone	48 1	2542 2
Seatearth	2 1	2544 3
Siltstone and sandstone	11 1	2555 4
Striped beds with plants	26 6	2581 10
Sandstone; conglomeratic at base	46 0	2627 10
Siltstone and sandstone	2 4	2630 2
Mudstone, silty at top; *A.* cf. *aquilina* ..	3 8	2633 10
Mudstone, dark; ironstone; *A. sp.* cf. *fulva, Naiadites sp.* ..	1 4½	2635 2½
HIGH HAZLES (TOP BED)		
Coal 3½ in	**0 3½**	**2635 6**
Seatearth, coaly at top..	1 0	2636 6
Sandstone	5 5	2641 11
Seatearth	2 1	2644 0
Sandstone and siltstone	4 3	2648 3
Mudstone, silty; ironstone; mussels ..	9 11	2658 2

[1]Detailed section: coal 5 in, coal and dirt 6½ in, coal 1¼ in, dirt 7 in, coal and dirt 2 in, dirt and coal 5½ in, coal and dirt 1¼ in, dirt 7¾ in, coal 6¾ in. This section may represent only the top leaf of the Mainbright.

	Thickness ft in	Depth ft in
Shale, dark; ironstone; *Anthracosia sp.* ..	6 10	2665 0
KENT'S THIN		
Coal 14½ in		
Dirt and coal 11 in		
Coal 15½ in	**3 5**	**2668 5**
Seatearth	14 10	2683 3
Striped beds; ironstone and plants	33 3	2716 6
Seatearth	5 4	2721 10
Sandstone and siltstone	21 6	2743 4
Mudstone, dark, silty; shelly ironstone; *Anthraconaia* aff. *librata, Anthracosia sp.* cf. *fulva, A.* cf. *planitumida*	18 2	2761 6
Mudstone; mussels ..	4 5	2765 11
KENT'S THICK		
Coal and dirt 59 in[1] (*43 in recovered*) ..	**4 11**	**2770 10**
Seatearth	3 8	2774 6
Mudstone, silty; ironstone; plants.. ..	5 0	2779 6
Mudstone; *Naiadites sp.*	2 1	2781 7
Mudstone, dark silty ..	0 11	2782 6
Coal 5 in (*core lost*) ..	**0 5**	**2782 11**
Seatearth	3 0	2785 11
Sandstone and siltstone	9 4	2795 3
Seatearth	2 5	2797 8
Siltstone	13 7	2811 3
Mudstone, silty at top..	10 6	2821 9
Siltstone, dark; fish ..	0 11	2822 8
Coal and dirt 25 in[2] (*24¼ in recovered*) ..	**2 1**	**2824 9**
Seatearth	9 3	2834 0
Mudstone, silty at base; ironstone; mussels ..	4 3	2838 3
Ironstone and silty mudstone	2 7½	2840 10½
Siltstone and sandstone	21 10½	2862 9

	Thickness ft in	Depth ft in
Mudstone with ironstone	4 5	2867 2
Coal and dirt 26 in[3] (*25¼ in recovered*) ..	**2 2**	**2869 4**
Seatearth	3 3	2872 7
Siltstone with ironstone; plants..	10 11	2883 6
Mudstone, silty at top; ironstone	11 4½	2894 10½
Shale, dark, canneloid at base; fish remains including platysomid, palaeoniscid, *Rhabdoderma sp.* and *Rhizodopsis sp.*, sineoid marking	0 5	2895 3½
Cannel 1½ in Coal 2 in	**0 3½**	**2895 7**
Seatearth, dark at top ..	4 2	2899 9
Siltstone and sandstone	5 3	2905 0
Mudstone, silty, with ironstone; mussels ..	3 1	2908 1
Mudstone, dark; *Naiadites sp.*, fish (*1¾ ft core lost*)	6 3	2914 4
TOP HARD		
Coal 46 in (*42¾ in recovered*) ..	**3 10**	**2918 2**
Seatearth	2 10	2921 0
Mudstone with ironstone; slickensided; plants.. ..	4 7	2925 7
BLIDWORTH + DUNSIL		
Coal 34 in (*27¼ in recovered*) ..	**2 10**	**2928 5**
Seatearth	29 4	2957 9
Striped beds	3 6	2961 3
Mudstone, dark silty ..	2 4	2963 7
Dirty coal 1 in	**0 1**	**2963 8**
Shale, coaly, and seatearth	2 8	2966 4
Striped beds	0 10	2967 2

[1]Detailed section: coal 20 in, dirt 5 in, coal 34 in.

[2]Detailed section: dirt and coal 4¾ in, dirty coal 3¾ in, coal 16½ in.

[3]Detailed section: coal 3½ in, dirt and coal 3 in, coal 13¼ in, dirt 2¾ in, dirty coal 3½ in.

	Thickness ft in	Depth ft in
Mudstone, dark silty; plants, *Anthracosia phrygiana*	0 8	2967 10
Sandstone and siltstone	0 10	2968 8
Mudstone, silty at top; ironstone; *N. triangularis?*	8 4	2977 0
1ST WATERLOO		
Coal 5 in		
Seatearth 70 in		
Dirt and coal 49 in[1]		
(*47½ in recovered*) ..	10 4	2987 4
Seatearth	1 2	2988 6
Striped beds	7 1	2995 7
Mudstone, silty, and siltstone	6 1	3001 8
Mudstone with ironstone; *A. ovum, A. phrygiana*	4 5	3006 1
Mudstone, dark silty; *Spirorbis sp.*, mussels, fish	0 4	3006 5
WATERLOO MARKER (TOP LEAF)		
Coal 7 in	0 7	3007 0
Seatearth	10 4	3017 4
Siltstone and sandstone; plants..	8 11	3026 3
Mudstone with ironstone; plants.. ..	1 4	3027 7
WATERLOO MARKER (BOTTOM LEAF)		
Coal 1 in	0 1	3027 8
Siltstone and sandstone; ironstone; plants, *Anthracosia?*.. ..	14 2	3041 10
Mudstone with ironstone; *A.* aff. *phrygiana*	5 10	3047 8
Shale, dark	0 9	3048 5
SWALLOW WOOD (MAIN AND BOTTOM BEDS)		
Coal and dirt 52 in[2]	4 4	3052 9
Seatearth, dark ..	0 2½	3052 11½
Shale, dark, with ironstone; mussels, fish ..	0 3½	3053 3
Inferior cannel; *Anthraconaia salteri, Anthracosia* cf. *aquilina*	0 3	3053 6
Seatearth	3 11½	3057 5½
Mudstone, silty at top..	0 11	3058 4½
SWALLOW WOOD (FLOOR COAL)		
Coal and dirty coal 28 in[3]	2 4	3060 8½
Seatearth	3 6½	3064 3
Siltstone and sandstone; ironstone	25 0	3089 3
Mudstone with ironstone; plants.. ..	2 9	3092 0
HAIGH MOOR		
Coal and dirt 49 in[4]		
(*34½ in recovered*) ..	4 1	3096 1
Seatearth	5 2	3101 3
Shale; *A. aquilina?* ..	2 1	3103 4
Coal 4 in		
Dirty coal 2 in	0 6	3103 10
Shale, coaly, and seatearth	4 7	3108 5
Sandstone and siltstone	1 3	3109 8
Seatearth	5 7	3115 3
Siltstone and sandstone	5 4	3120 7
Shale; coal streaks at base	0 7½	3121 2½
Seatearth	1 11½	3123 2
Siltstone and sandstone	18 1	3141 3
Mudstone, mostly silty; shelly ironstone; *A. beaniana, A. phrygiana*, ?arthropod	11 0	3152 3
Shale, black, silty at top; fish	1 8	3153 11
Seatearth	0 4	3154 3

[1]Detailed section: dirty coal 1 in, coal 8½ in, coal and dirt 2 in, coal 4 in, dirt 16½ in, dirty coal 1½ in, coal 5½ in, dirt 7¾ in, dirty coal 2¼ in.

[2]Detailed section: coal 18 in, dirt 18 in, dirty coal 4 in, coal 12 in.

[3]Detailed section: coal 8½ in, dirty coal ½ in, coal 7 in, dirty coal 1½ in, coal 8½ in, dirty coal 2 in.

[4]Detailed section: coal 30 in, dirt 2 in, coal 17 in.

	Thickness ft in	Depth ft in
LIDGET		
Coal 6 in		
Dirty coal 1½ in	0 7½	3154 10½
Seatearth	4 0½	3158 11
Sandstone and striped beds	27 5	3186 4
Dirt and coal 42 in[1]	3 6	**3189 10**
Seatearth	2 1	3191 11
Sandstone and siltstone with ironstone ..	6 5	3198 4
Mudstone, silty at top; ironstone; mussels ..	3 3	3201 7
Seatearth	1 6	3203 1
Sandstone and siltstone; ironstone	9 0	3212 1
Mudstone, silty ..	1 7	3213 8
Sandstone and siltstone; coal inclusions ..	2 5	3216 1
Seatearth	4 9	3220 10
Striped beds	9 0	3229 10
Siltstone	3 10	3233 8
Seatearth	1 1	3234 9
Coal 1 in	0 1	**3234 10**
Seatearth	1 10	3236 8
Striped beds and sandstone	9 7	3246 3
Mudstone, silty; *A.* aff. *phrygiana, Anthracosphaerium exiguum*	1 0	3247 3
Striped beds and silty mudstone	24 3	3271 6
Mudstone, silty at top; ironstone; mussels at base	12 5	3283 11
Mudstone, dark silty; *Spirorbis sp., Anthracosia* cf. *beaniana, A. ovum, A. sp. phrygiana /ovum, A.* aff. *subrecta, Naiadites quadratus,* fish debris including *Megalichthys sp.* and palaeoniscid scales ..	10 9½	3294 8½

	Thickness ft in	Depth ft in
CLAY CROSS MARINE BAND		
Mudstone, dark silty; *Lingula mytilloides, Anthracosia* cf. '*carissima*', *A.* cf. *phrygiana*	4 3½	3299 0
Mudstone, dark silty; *L. mytilloides, Dunbarella papyracea* mut. δ, *Posidonia* cf. *sulcata,* orthocone nautiloid, *Anthracoceratites vanderbeckei, Metacoceras* cf. *cornutum*	5 3	3304 3
Shale, dark; *L. mytilloides, Anthracosia sp., Myalina sp., Carbonita spp., Geisina arcuata, Hollinella (Praehollinella) claycrossensis, Paraparchites sp.,* fish including *Elonichthys sp.*	3 1	**3307 4**
LOWER COAL MEASURES		
Shale, dark; *A. regularis, Naiadites sp., 'Estheria' sp., G. arcuata,* fish including acanthodian spine, *Elonichthys sp.,* platysomid scales, *Rhabdoderma sp.,* and *Rhizodopsis sp.* ..	3 8	3311 0
Shale, dark; *Spirorbis sp., A. regularis, Naiadites sp. nov., G. arcuata,* palaeoniscid scale	2 6	3313 6
Ironstone; *N. sp. nov., G. arcuata*	0 5	3313 11
Seatearth	1 2	3315 1
Mudstone, dark; *Spirorbis sp., A. regularis*	1 1	3316 2
Mudstone, dark silty; ironstone	1 0	3317 2

[1]Detailed section: coal 10½ in, dirt 8½ in, coal 6 in, dirt 15 in, coal 2 in.

	Thickness ft	Thickness in	Depth ft	Depth in
Shale, coaly	0	6	3317	8
Seatearth	3	1½	3320	9½
Coal 1½ in	**0**	**1½**	**3320**	**11**
Seatearth; *13 in core lost*	9	9	3330	8
Siltstone and sandstone	14	10	3345	6
Mudstone, silty, and siltstone; ironstone; plants, *A. regularis* (*minor core losses*)	25	0	3370	6
DEEP SOFT				
Coal and dirt 83 in[1] (*81 in recovered*)	**6**	**11**	**3377**	**5**
Seatearth	6	2½	3383	7½
Coal 1½ in	**0**	**1½**	**3383**	**9**
Seatearth	18	7	3402	4
Dirt and coal 21 in[2]	**1**	**9**	**3404**	**1**
Seatearth	3	2	3407	3
Coal 3½ in Dirt 5 in **Coal** 1½ in	**0**	**10**	**3408**	**1**
Seatearth	6	3	3414	4
Siltstone and sandstone; ironstone at base	5	0	3419	4
Mudstone, silty, with ironstone; plants, *Naiadites sp.*	10	5	3429	9
Dirty coal 2 in	**0**	**2**	**3429**	**11**
Seatearth	3	8	3433	7
Coal 4 in	**0**	**4**	**3433**	**11**
Seatearth	3	4	3437	3
Striped beds and sandstone	63	11	3501	2
Sandstone with pebbles; coal inclusions at top	41	11	3543	1
PARKGATE				
Coal 20¾ in **Dirty coal** 1¾ in **Coal** 19½ in	**3**	**6**	**3546**	**7**
Seatearth	5	9	3552	4
Mudstone, silty	3	0	3555	4
Mudstone, dark; shelly ironstone; *Spirorbis sp., Carbonicola oslancis, C. sp.* (*oslancis/cristagalli*), *G. arcuata*, fish	3	0	3558	4
Seatearth	8	10	3567	2
Siltstone and sandstone; ironstone at base; plants	12	6	3579	8
Mudstone, dark at base; *Anthracosphaerium* cf. *dawsoni*, *C. cristagalli*	4	10½	3584	6½
THORNCLIFF (TOP LEAF)				
Coal 13¾ in Dirt 2½ in **Inferior cannel** 8¼ in	**2**	**0½**	**3586**	**7**
Seatearth	4	4	3590	11
Siltstone	13	8	3604	7
Mudstone, silty at top	1	5	3606	0
Shale, dark; plants	0	7	3606	7
THORNCLIFF (BOTTOM LEAF)				
Coal and dirt 8 in	**0**	**8**	**3607**	**3**
Seatearth	2	7	3609	10
Siltstone and sandstone; plants	12	11	3622	9
Mudstone, dark; *C. cristagalli, C. pseudorobusta, C. rhomboidalis, C. sp. cristagalli/rhomboidalis, G. arcuata*, fish	1	0	3623	9
Dirty coal 8 in (*poor recovery*)	**0**	**8**	**3624**	**5**
Seatearth with sphaerosiderite	6	6	3630	11
Siltstone and sandstone	1	11	3632	10
Ironstone	0	8	3633	6
Mudstone, silty; mussels	2	0	3635	6
Mudstone, dark; shelly ironstone; mussels	4	10	3640	4
Mudstone	3	0	3643	4

[1]Detailed section: coal ½ in, dirt ½ in, coal 7¾ in, dirty coal 1 in, coal 38 in, dirt 8 in, dirty coal 3¾ in, dirt 1 in, coal 22½ in.

[2]Detailed section: coal 3 in, dirt 7 in, coal 1 in, dirt and coal 3 in, coal 1 in, dirt 3 in, coal 3 in.

	Thickness ft in	Depth ft in		Thickness ft in	Depth ft in
THREEQUARTERS			TOP SILKSTONE		
Coal 3 in			Coal and dirt 30½ in[1]		
Dirt 17 in			(29¾ in recovered) ..	2 6½	3744 6
Coal 7 in	2 3	3645 7	Seatearth	3 4	3747 10
Seatearth	14 11	3660 6	Sandstone and siltstone	13 11	3761 9
Siltstone and silty mudstone; ironstone; cf.			Mudstone, silty ..	1 10	3763 7
Planolites montanus ..	16 3	3676 9	Mudstone, dark at top; Curvirimula sp., G.		
Core lost	3 0	3679 9	arcuata, fish	1 10	3765 5
Mudstone, dark ..	1 1	3680 10	Siltstone and sandstone; 4 ft 3 in core lost ..	12 9	3778 2
Mudstone, dark silty; ironstone	8 10	3689 8	Mudstone, silty with ironstone; Carboni-		
Seatearth	1 5	3691 1	cola circinata ..	18 11	3797 1
Core lost	7 11	3699 0	LOW 'Estheria' BAND		
Siltstone and sandstone	10 7	3709 7	Mudstone; 'Estheria' sp. nov.	2 2	3799 3
Core lost	2 6	3712 1	Shale, dark; 'E.' sp. nov., fish	1 7	3800 10
Mudstone, silty, with ironstone	14 5	3726 6	LOW SILKSTONE		
Mudstone, dark; N.			Coal 1 in		
flexuosus	9 3	3735 9	Coal and dirt 22 in[2]	1 11	3802 9
Mudstone, dark; plants	6 2½	3741 11½	Seatearth	3 6	3806 3

Scofton Borehole

Ht above O.D. 91 ft.   6-in SK 68 SW.
Site [6278 8053] 230 yd at 349° from Scofton church. Drilled 1954 by Craelius Co. for N.C.B. Cores examined by R. F. Goossens.

	Thickness ft in	Depth ft in		Thickness ft in	Depth ft in
PERMO-TRIASSIC			LOWER PERMIAN MARL		
Not cored: PB to 185 ft,			Marl grey (partly or wholly Marl Slate);		
LMS to 302 ft ..	302 0	302 0	fish debris	13 3	651 6
MIDDLE PERMIAN MARL					
Not cored	23 0	325 0	UPPER COAL MEASURES		
Marl, red, with grey bands; sandstone			HEMSWORTH DIVISION		
bands up to 9 ft thick			Seatearth, with vertical cracks containing fine		
totalling 15¾ ft ..	80 6	405 6	conglomerate ..	3 3	654 9
LOWER MAGNESIAN LIMESTONE			Seatearth, partly red ..	7 9	662 6
Dolomite, grey, with numerous bands of			Siltstone, red and green	6 6	669 0
grey marl below 479½ ft	232 9	638 3	Mudstone, partly purple and brown	1 0	670 0

[1]Detailed section: cannel 9 in, coal 19 in, dirt 1¼ in, dirty coal 1¼ in.

[2]*Core lost*; thickness according to penetration recorder and Schlumberger temperature log.

	Thickness ft in	Depth ft in
Seatearth, partly purple	4 3	674 3
Siltstone and sandstone, partly red	6 3	680 6
FOURTH CHERRY TREE MARKER		
Mudstone, partly brown; *Spirorbis sp.*, *Anthraconauta phillipsii*, *Euestheria simoni*	6 0	686 6
Shale, black; *Spirorbis sp.*, *A. phillipsii*	0 6	687 0
Siltstone, black, with plants	0 6	687 6
BRIERLEY DIVISION		
Seatearth	2 0	689 6
Sandstone	2 0	691 6
Siltstone, partly red	3 3	694 9
Mudstone	1 0	695 9
Mudstone, dark, with shell fragments	1 3	697 0
Seatearth	2 6	699 6
Ironstone	0 4	699 10
Mudstone, black, with mussels	2 2	702 0
Seatearth, black at top, coaly at base	3 9	705 9
Dirty coal 6 in	0 6	706 3
Seatearth	2 6	708 9
Siltstone	5 3	714 0
Mudstone, dark at base; ironstone	2 0	716 0
Seatearth	1 9	717 9
Siltstone and sandstone	6 3	724 0
Mudstone, dark at base; ironstone	6 6	730 6
Mudstone, black; *A. phillipsii*, *Carbonita sp.*	0 9	731 3
Seatearth, greenish	3 0	734 3
Siltstone and sandstone	7 9	742 0
Seatearth	1 6	743 6
Siltstone	2 6	746 0
Seatearth	4 0	750 0
Siltstone	10 6	760 6
Mudstone, black silty	0 3	760 9

	Thickness ft in	Depth ft in
Shale, canneloid, with ironstone; *Anthraconauta sp.*, *C. humilis*, *Rhizodopsis sp.*	0 6	761 3
Coal 3 in	**0 3**	**761 6**
Seatearth and mudstone with roots	6 9	768 3
Shale, black	0 1	768 4
Coal 3 in	**0 3**	**768 7**
Seatearth	2 8	771 3
Siltstone and sandstone; sphaerosiderite	3 6	774 9
Seatearth, greenish	2 6	777 3
Siltstone with ironstone	7 0	784 3
Seatearth; sphaerosiderite	3 3	787 6
Dirty coal 9 in / Seatearth 21 in / **Dirty coal** 6 in	3 0	790 6
Seatearth	5 0	795 6
Siltstone and sandstone	12 6	808 0
Mudstone with ironstone	1 9	809 9
Mudstone, dark, with shell fragments	0 6	810 3
Mudstone and siltstone with ironstone	4 10	815 1
Shale, black	0 2	815 3
Seatearth	3 6	818 9
Sandstone and siltstone; sphaerosiderite at top	29 0	847 9
Seatearth	4 3	852 0
Siltstone with ironstone; plants	30 0	882 0
BLYTH		
Dirty coal 12 in	**1 0**	**883 0**
ACKWORTH DIVISION		
Seatearth	6 6	889 6
Siltstone with ironstone	10 0	899 6
Mudstone with ironstone	3 3	902 9
Seatearth	3 0	905 9
Siltstone and sandstone; conglomerate at 979 ft. *Fault between 918½ and 925 ft*	122 3	1028 0

	Thickness ft in	Depth ft in
Mudstone with ironstone	1 6	1029 6
Mudstone, dark, with *Anthraconauta sp.* ..	3 6	1033 0
Shale, black, with *Lioestheria vinti*, acanthodian spine, *Rhabdoderma sp.* and *Rhizodopsis sp.* ..	0 3	1033 3
SCOFTON		
Coal　　　　9 in		
(*Core lost*) ..	**0 9**	**1034 0**
Seatearth	3 9	1037 9
Siltstone with sphaerosiderite	9 3	1047 0
Mudstone	0 9	1047 9
Seatearth, dark at base	1 0	1048 9
Coal　　　　3 in		
Seatearth　　　9 in		
Dirty coal　　9 in	**1 9**	**1050 6**
Seatearth, largely black	0 9	1051 3
ACKWORTH ROCK		
Siltstone and sandstone	35 3	1086 6
Mudstone, silty ..	4 0	1090 6
MIDDLE COAL MEASURES		
TOP MARINE BAND		
Mudstone with ironstone; foraminifera, '*Anthracoceras*' ?, 'fucoids'	5 9	1096 3
Mudstone, dark with ironstone; *Serpuloides sp.*, *Spirorbis sp.*, *Crurithyris sp.*, *Lingula mytilloides*, *Dunbarella sp.*, fish remains, '*fucoids*' ..	3 9	**1100 0**
Seatearth	10 9	1110 9
Mudstone with ironstone; *Naiadites sp.*, *Carbonita sp.*, *Lioestheria vinti* ..	2 9	1113 6
Siltstone with ironstone	2 6	1116 0
Mudstone, dark, with ironstone; *N. sp.*, *Geisina subarcuata*, *L. vinti*	2 0	1118 0

	Thickness ft in	Depth ft in
Mudstone, dark silty ..	1 0	1119 0
Seatearth and siltstone	5 3	1124 3
Mudstone, dark, silty at base; *N. sp.*, '*Guilielmites*'	1 1	1125 4
Ironstone; white ooliths	0 3	1125 7
Seatearth	7 1	1132 8
Mudstone, dark, silty; fish debris including platysomid scale ..	0 1	1132 9
Seatearth, partly greenish	4 0	1136 9
Siltstone and silty mudstone	17 0	1153 9
Mudstone, dark at base; *N.* cf. *daviesi*, *Geisina subarcuata*	3 8	1157 5
Ironstone; *N.* cf. *daviesi*, *G. subarcuata*, *Hemicycloleaia* cf. *minima*, *L. vinti*, fish including acanthodian spine, palaeoniscid scales, platysomid scales, *Rhizodopsis sp.* ..	0 2	1157 7
Mudstone, dark with ironstone; *N.* cf. *daviesi*, *G. subarcuata*, *L. vinti*, fish including platysomid scales	1 5	1159 0
Seatearth	5 0	1164 0
Siltstone and sandstone	16 0	1180 0
Mudstone, silty at top, with ironstone; *L. vinti*	4 9	1184 9
SHAFTON MARINE BAND		
Mudstone with ironstone; *Ammodiscus sp.*, *Anthraconaia spathulata* ?, *Myalina compressa*, 'fucoids'	1 9	1186 6
Mudstone, dark with ironstone; *Ammodiscus sp.*, *Lingula mytilloides*, *Edmondia sp.*, *Myalina sp.*, *Lioestheria vinti* ..	1 3	1187 9

	Thickness ft in	Depth ft in
Mudstone, dark silty; *Lingula mytilloides*, fish remains including *Rhabdoderma sp.*	2 0	**1189 9**
Mudstone, coaly ...	0 1	1189 10
Dirty coal 15½ in / Dirt 1 in / **Dirty coal** 4½ in	**1 9**	**1191 7**
Seatearth with coal streaks	4 11	1196 6
MEXBOROUGH ROCK		
Sandstone and siltstone	13 3	1209 9
Sandstone, mostly coarse, with occasional pebbles ...	148 5	1358 2
Coal 22 in	**1 10**	**1360 0**
Seatearth	4 0	1364 0
Sandstone and siltstone	14 0	1378 0
Mudstone, dark, with ironstone; *N.* aff. *daviesi*	4 8	1382 8
Seatearth; some sphaerosiderite ...	11 10	1394 6
Siltstone and silty mudstone	4 0	1398 6
Mudstone with ironstone	3 6	1402 0
Mudstone, dark; *Anthraconaia* cf. *adamsi* group, fish including palaeoniscid scales, *Rhabdoderma sp.*, *Rhizodopsis sp.* ...	0 6	1402 6
Mudstone, dark; some ironstone	1 6	1404 0
Core lost	1 10	1405 10
Seatearth	3 2	1409 0
Siltstone	15 0	1424 0
Mudstone, dark at base; ironstone	5 9	1429 9
Mudstone, dark; *Naiadites sp.*, *Carbonita sp.* (*humilis?*) ...	0 6	1430 3
Ironstone, dark ...	0 1	1430 4
Coal 5 in	**0 5**	**1430 9**

	Thickness ft in	Depth ft in
Seatearth	2 6	1433 3
Siltstone and silty mudstone	5 3	1438 6
Mudstone, dark ...	0 6	1439 0
Dirty coal 1 in / Seatearth 1 in / Coal 7 in	**0 9**	**1439 9**
Seatearth	5 0	1444 9
Sandstone and siltstone	15 6	1460 3
Mudstone, silty at top; *Naiadites sp.* ...	0 11	1461 2
Mudstone, dark, with ironstone; mussels ..	1 0	1462 2
Seatearth	1 10	1464 0
Sandstone and siltstone	14 3	1478 3
Mudstone, silty ...	2 11	1481 2
Mudstone, dark, with ironstone; *Spirorbis sp.*, *N. melvillei*, *Carbonita sp.*, *Geisina subarcuata*, fish ...	0 5	1481 7
SHARLSTON LOW		
Coal 13 in	**1 1**	**1482 8**
Seatearth	6 7	1489 3
Sandstone and siltstone	25 3	1514 6
Mudstone, dark at base, with ironstone; *N. sp.*, fish including palaeoniscid scale ...	9 8	1524 2
Mudstone, dark silty ..	1 2	1525 4
SHARLSTON YARD		
Dirty coal 3 in	**0 3**	**1525 7**
Seatearth, partly greenish; some sphaerosiderite ...	17 2	1542 9
Siltstone, greenish, with sphaerosiderite ...	4 3	1547 0
Seatearth	6 6	1553 6
?HOUGHTON THIN		
Dirty coal 5 in	**0 5**	**1553 11**
Seatearth and siltstone	4 4	1558 3
Mudstone, mainly dark; *Naiadites sp.*, bivalve *Gen. et sp. nov.*, *C. humilis*	1 4	1559 7
Sandstone and siltstone	28 5	1588 0

	Thickness ft in	Depth ft in
Mudstone with iron-stone	5 5	1593 5
Mudstone, dark, with ironstone; *N. sp.*, fish remains	2 5	1595 10

MANSFIELD MARINE BAND

	Thickness ft in	Depth ft in
Mudstone with iron-stone; *Planolites ophthalmoides*, 'fucoids'	2 5	1598 3
Mudstone, dark; foraminifera including *Glomospira sp.*, *Tolypammina sp.*, sponge spicules, *Lingula mytilloides*, 'fucoids'	8 9	1607 0
Mudstone, dark, with ironstone; *Serpuloides sp., Crurithyris sp., Lingula sp., Tornquistia diminuta, Euphemites sp., Platyconcha sp., Palaeoneilo taffiana, 'Anthracoceras'?,* orthocone nautiloid, *Cypridina sp.*, crinoid columnals, fish including *Rhabdoderma sp.*, cf. *Tomaculum sp.*, pyrite-filled burrows	16 8	1623 8
Siltstone, dark; marine fossils	0 11	1624 7
Mudstone, dark; mar-ine fossils	1 3	**1625 10**
Mudstone, dark; plant debris..	0 9	1626 7
Seatearth with sphaero-siderite	2 5	1629 0
Siltstone, greenish, with ironstone	5 6	1634 6
Seatearth	2 0	1636 6
Siltstone, part greenish, with sphaerosiderite..	14 3	1650 9
Mudstone, dark at base, with ironstone ..	8 3	1659 0
Seatearth with sphaero-siderite	3 6	1662 6

	Thickness ft in	Depth ft in
Mudstone, mostly silty, dark at base ..	3 6	1666 0
Seatearth	9 6	1675 6

SUTTON MARINE BAND

	Thickness ft in	Depth ft in
Mudstone, dark silty, with ironstone; *Lingula sp., Orbiculoidea* cf. *nitida* ..	4 0	**1679 6**
Mudstone, dark, with ironstone	1 10	1681 4
Seatearth	4 2	1685 6
Siltstone	14 0	1699 6
Mudstone with iron-stone	2 9	1702 3
Seatearth, dark at base	7 9	1710 0

HAUGHTON MARINE BAND

	Thickness ft in	Depth ft in
Mudstone, dark silty, with ironstone; *L. mytilloides, Rhizodopsis sp.*	13 9	1723 9
Mudstone, dark silty, with ironstone; *L. mytilloides,* bellerophontoid indet., cf. *Tomaculum sp.,* pyritized burrow ..	7 9	1731 6
Mudstone, dark silty, with ironstone; sponge spicules [pyritized], *Serpuloides stubblefieldi*..	1 3	**1732 9**
Mudstone, dark silty; fish	0 3	1733 0
Mudstone, dark, silty at top, with ironstone; *Naiadites sp.* ..	5 3	1738 3

SWINTON POTTERY

		Thickness ft in	Depth ft in
Coal	3 in	0 3	1738 6
Seatearth with sphaero-siderite		3 6	1742 0
Siltstone		20 0	1762 0
Mudstone, silty; *N. angustus*		3 0	1765 0
Mudstone, mostly dark and silty, with iron-stone; *N. angustus,* fish including platyso-mid		4 3	1769 3

	Thickness ft in	Depth ft in
Seatearth	3 6	1772 9
Sandstone and siltstone	8 9	1781 6
Mudstone, silty ..	4 6	1786 0
Mudstone with iron-stone	3 6	1789 6
Mudstone, dark, mostly silty; shell fragments, fish remains ..	1 9	1791 3
Seatearth, greenish, with sphaerosiderite ..	6 9	1798 0
Mudstone, greenish ..	3 5	1801 5
Mudstone, dark, with ironstone	1 4	1802 9

CLOWN

		Thickness ft in	Depth ft in
Dirty coal	5 in		
Seatearth	17 in		
Dirty coal	8 in		
Coal	30 in	5 0	1807 9

	Thickness ft in	Depth ft in
Seatearth	3 6	1811 3
Mudstone, dark, with ironstone; *N. angustus*	4 3	1815 6

MANTON 'Estheria' BAND

	Thickness ft in	Depth ft in
Mudstone, dark; 'Estheria' sp. ..	0 9	1816 3
Seatearth with sphaero-siderite	5 0	1821 3
Siltstone	1 9	1823 0
Mudstone, partly dark; plants..	1 3	1824 3
Siltstone and sandstone	25 6	1849 9
Mudstone, silty at top, with ironstone ..	13 3	1863 0
Mudstone, dark, with ironstone; *Anthracosphaerium* aff. *propinquum* ..	4 6	1867 6
Mudstone, dark ..	1 0	1868 6

MAINBRIGHT
(OR MELTONFIELD)

		Thickness ft in	Depth ft in
Coal	19¼ in		
Dirt	0¾ in		
Coal	7 in	2 3	1870 9

	Thickness ft in	Depth ft in
Seatearth	2 3	1873 0

		Thickness ft in	Depth ft in
Coal	3 in		
Seatearth	3 in		
Coal	3 in		
Seatearth	5 in		
Dirty coal	2 in	1 4	1874 4

	Thickness ft in	Depth ft in
Seatearth	2 8	1877 0
Siltstone and sandstone; ironstone ..	11 9	1888 9
Mudstone, silty at top, with ironstone ..	4 7	1893 4
Mudstone, dark silty ..	1 0	1894 4

TWO-FOOT MARINE BAND

	Thickness ft in	Depth ft in
Mudstone, dark silty; *L. mytilloides* ..	2 8	1897 0
Mudstone, dark silty, foraminifera including *Glomospira sp.*, *L. mytilloides, Myalina compressa, Hollinella sp., Paraparchites sp.*	1 3	1898 3
Mudstone, dark silty; foraminifera, *L. mytilloides* ..	2 11	1901 2
Mudstone, coaly ..	0 2	1901 4

TWO-FOOT

		Thickness ft in	Depth ft in
Coal and dirt	3 in		
Coal	24 in	2 3	1903 7

	Thickness ft in	Depth ft in
Seatearth	4 2	1907 9
Sandstone and siltstone	11 3	1919 0
Mudstone, dark at base; ironstone	9 9	1928 9

ABDY (OR WINTER)

		Thickness ft in	Depth ft in
Coal and dirt	23 in[1]	1 11	1930 8

	Thickness ft in	Depth ft in
Seatearth	2 4	1933 0
Siltstone with ironstone; plants..	5 9	1938 9
Mudstone with ironstone; plants.. ..	5 6	1944 3
Seatearth	7 6	1952 0
Mudstone, silty, with ironstone; plants ..	9 3	1961 3
Siltstone and sandstone	62 3	2023 6
Mudstone with ironstone; plants.. ..	11 0	2034 6

[1]Detailed section: coal and dirt 10 in, dirt 8 in, coal 1 in, dirt 3 in, coal 1 in.

	Thickness ft in	Depth ft in
Ironstone with coal streaks	2 3	2036 9
Siltstone with ironstone	10 9	2047 6
Mudstone, silty at top, with ironstone; *Anthracosia caledonica, N. productus*, fish debris including *Rhizodopsis sp.*	8 6	2056 0
Siltstone with ironstone	12 9	2068 9
Mudstone, silty at top, with ironstone	10 7	2079 4
Mudstone, dark; fish	0 1	2079 5
KENT'S THIN **Cannel** 2½ in **Coal** 16½ in	**1 7**	**2081 0**
Seatearth with coaly layer	2 3	2083 3
Sandstone and siltstone	9 3	2092 6
Mudstone with ironstone; *Anthraconaia sp.* cf. *confusa, Naiadites sp.*	1 6	2094 0
Siltstone with ironstone	8 9	2102 9
Mudstone, dark at base, with shelly ironstone; *Anthraconaia?, Anthracosia* cf. *faba, A.* cf. *aquilinoides, Anthracosphaerium* cf. *exiguum, N. productus, Carbonita humilis*	11 10	2114 7
Mudstone, dark silty; plants	3 5	2118 0
Siltstone	9 6	2127 6
Mudstone, silty at top; ironstone	7 8	2135 2
Mudstone, dark	1 6	2136 8
KENT'S THICK **Coal and dirt** 67 in[1]	**5 7**	**2142 3**
Seatearth, coal streaks at top	8 0	2150 3
Siltstone and sandstone	2 9	2153 0
Mudstone, silty with ironstone	3 0	2156 0
Mudstone, dark at base; *Anthracosia* aff. *caledonica, N. productus*	5 10	2161 10
Dirty coal 2 in	**0 2**	**2162 0**
Seatearth	2 4	2164 4
Siltstone and sandstone	10 6	2174 10
Mudstone, coaly	0 1	2174 11
Seatearth	7 7	2182 6
Mudstone	5 3	2187 9
Siltstone and silty mudstone	9 3	2197 0
Mudstone, dark and silty at base; ironstone; *Planolites sp., Naiadites sp.*, fish including *Rhizodopsis sp.*	8 0	2205 0
Siltstone, dark	0 3	2205 3
Dirty coal 3 in	**0 3**	**2205 6**
Seatearth	4 3	2209 9
Siltstone	34 3	2244 0
Sandstone with pebbles in lower part	42 0	2286 0
TOP HARD **Coal** 31¾ in **Dirt** 0¾ in **Coal** 5½ in	**3 2**	**2289 2**
Seatearth	7 7	2296 9
Mudstone	0 3	2297 0
BLIDWORTH **Coal** 9 in	**0 9**	**2297 9**
Seatearth	6 6	2304 3
Sandstone and siltstone	23 0½	2327 3½
DUNSIL **Coal** 18½ in **Dirt** 3½ in **Coal** 7½ in	**2 5½**	**2329 9**
Seatearth	9 3	2339 0
Siltstone, dark, with ironstone; plants	4 9	2343 9
Siltstone and sandstone	11 9	2355 6
Mudstone, silty, with ironstone (*fault*)	6 6	2362 0

[1]Detailed section: coal 11½ in, dirty coal 2½ in, coal 14½ in, mudstone 19 in, dirty coal 4 in, coal 15½ in.

	Thickness ft in	Depth ft in
Mudstone, dark, mostly silty	0 10	2362 10
Dirty coal 5 in	**0 5**	**2363 3**
Seatearth	1 9	2365 0
Sandstone and siltstone	3 6	2368 6
Mudstone, dark at base; ironstone; mussels ..	8 6½	2377 0½
1ST WATERLOO		
Coal 18½ in		
Seatearth 1 in		
Coal 3 in	**1 10½**	**2378 11**
Seatearth	5 1	2384 0
Sandstone and siltstone	7 6	2391 6
Mudstone, dark at base; shelly ironstone; *Spirorbis sp., Anthraconaia?, Anthracosia nitida, A.* aff. *ovum, A. phrygiana, N. quadratus,* fish including palaeoniscid scales	9 8	2401 2
WATERLOO MARKER (TOP LEAF)		
Cannel[1] 5 in		
Shale, black; mussels 5 in		
Dirty coal 1 in	**0 11**	**2402 1**
Seatearth	3 2	2405 3
Siltstone and sandstone; some ironstone ..	8 6	2413 9
Mudstone, dark, with ironstone	0 11	2414 8
WATERLOO MARKER (BOTTOM LEAF)		
Coal 1 in	**0 1**	**2414 9**
Seatearth	0 6	2415 3
Siltstone	12 9	2428 0
Mudstone, dark at base; ironstone; *Spirorbis sp., Naiadites sp...*	11 0	2439 0
SWALLOW WOOD + HAIGH MOOR		
Coal and dirt 122 in[2]	**10 2**	**2449 2**
Seatearth	3 7	2452 9

	Thickness ft in	Depth ft in
Sandstone and siltstone	5 9	2458 6
Mudstone, mostly silty; ironstone	5 0	2463 6
Mudstone, dark; mussels	0 9	2464 3
Seatearth	3 6	2467 9
Siltstone; plants ..	11 3	2479 0
Seatearth, greenish ..	11 6	2490 6
Siltstone with ironstone	11 9	2502 3
Sandstone	0 9	2503 0
Mudstone, dark at base; ironstone; mussels ..	0 5¾	2503 5¾
Coal 0¼ in	**0 0¼**	**2503 6**
Seatearth	1 9	2505 3
Siltstone with some ironstone	15 0	2520 3
Mudstone, dark at base; shelly ironstone; *Spirorbis sp., A. ovum, N. quadratus* ..	9 9	2530 0
LIDGET		
Coal 3 in	**0 3**	**2530 3**
Seatearth, coaly at top..	8 9	2539 0
Mudstone with ironstone; *Spirorbis sp.,* mussels	5 6	2544 6
Coal 2 in	**0 2**	**2544 8**
Seatearth, partly greenish, and siltstone	15 10	2560 6
Siltstone	47 0	2607 6
Sandstone with ironstone pebbles at two horizons	41 3	2648 9
Siltstone	2 0	2650 9
Mudstone, mainly dark and silty; ironstone; *Spirorbis sp., A. spp., N. sp.* (*?quadratus*), fish remains ..	7 0	2657 9
CLAY CROSS MARINE BAND		
Mudstone, dark silty; *Lingula mytilloides, Anthracosia sp.* ..	1 9	2659 6

[1]This cannel yielded: *Anthracosia beaniana, A* aff. *phrygiana, Naiadites sp.*

[2]Detailed section: dirty coal 3 in, seatearth 1 in, dirty coal 3 in, seatearth 32 in, coal 4 in, coaly dirt 5 in, coal 4 in, coal and dirt 2 in, coal 32 in, seatearth 17 in, coal 19 in.

	Thickness ft in	Depth ft in
Mudstone, dark, silty; *L. mytilloides, Paraconularia sp.,* palaeoniscid scales	5 0	**2664 6**

LOWER COAL MEASURES

	Thickness ft in	Depth ft in
Seatearth	2 9	2667 3
Siltstone and silty mudstone with ironstone	8 3	2675 6
Mudstone; *Spirorbis sp.,* *A. regularis*	1 0	2676 6
Ironstone	0 3	2676 9
Seatearth	3 6	2680 3
Siltstone with ironstone; plants..	67 0	2747 3
Siltstone and sandstone	27 9	2775 0
Siltstone, dark, with ironstone; plants ..	8 3	2783 3
Mudstone, dark silty ..	0 3	2783 6

DEEP SOFT

	Thickness ft in	Depth ft in
Coal and dirt 63 in[1]	5 3	**2788 9**
Seatearth	5 4	2794 1
Dirty coal 2 in **Seatearth** 7 in **Coal** 5 in	1 2	**2795 3**
Seatearth	0 3	2795 6
Core lost	2 3	2797 9
Seatearth	4 9	2802 6
Siltstone with ironstone; plants..	8 0	2810 6
Sandstone	2 3	2812 9
Seatearth	1 6	2814 3
Siltstone and silty mudstone; ironstone ..	10 3	2824 6
Mudstone with ironstone; mussels ..	1 0	2825 6

?DEEP HARD ROOF COAL

	Thickness ft in	Depth ft in
Coal 8 in	0 8	**2826 2**
Seatearth	4 4	2830 6
Sandstone and siltstone	12 3	2842 9
Mudstone with ironstone	7 0	2849 9

	Thickness ft in	Depth ft in
Mudstone, dark; *Spirorbis sp., Naiadites sp.*	0 9	2850 6

DEEP HARD (MAIN LEAF)

	Thickness ft in	Depth ft in
Dirty coal 3 in **Seatearth** 3 in **Coal** 10 in	1 4	**2851 10**
Seatearth	6 2	2858 0
Mudstone, dark and silty at base	0 9	2858 9

DEEP HARD (BOTTOM LEAF)

	Thickness ft in	Depth ft in
Coal 5 in	0 5	**2859 2**
Seatearth	1 1	2860 3
Siltstone and sandstone	8 0	2868 3
Mudstone, silty, with ironstone ..	11 0	2879 3
Mudstone, dark at base; ironstone	11 1	2890 4

PARKGATE

	Thickness ft in	Depth ft in
Coal and dirt 90 in[2]	7 6	**2897 10**
Seatearth, coaly at top..	3 5	2901 3
Siltstone	3 0	2904 3
Mudstone with ironstone	2 6	2906 9
Mudstone, dark, silty in lower part; fish debris including acanthodian spine, palaeoniscid scales, *Rhizodopsis sp.*	0 9	2907 6
Ironstone	0 3	2907 9
Siltstone; some ironstone	9 9	2917 6
Mudstone with ironstone; *Spirorbis sp., Carbonicola sp.* ..	2 0	2919 6
Seatearth	4 9	2924 3
Sandstone and siltstone	7 9	2932 0
Mudstone, silty at top, dark at base; ironstone; *Carbonicola sp., Naiadites sp., Geisina arcuata* ..	7 3	2939 3

[1]Detailed section: coal 23 in, dirt 5½ in, dirty coal 4¼ in, dirt ¾ in, coal 7 in, dirt 2½ in, dirty coal 9½ in, dirt 5 in, dirty coal 5½ in.

[2]Detailed section: dirty coal 9 in, seatearth 19 in, dirty coal 6 in, coal 22½ in, dirt 24 in, coal 9½ in.

		Thickness ft in	Depth ft in
THORNCLIFF			
Coal	5 in	0 5	2939 8
Shale, coaly		0 8	2940 4
Seatearth		3 8	2944 0
Siltstone; below 2980 ft with *Anthraconaia potoriba, Naiadites sp.*		54 9	2998 9
Mudstone, silty, with ironstone		4 3	3003 0
Sandstone and siltstone; few pebbles		66 4	3069 4
Mudstone, dark, partly silty; shelly ironstone		3 6	3072 10

		Thickness ft in	Depth ft in
Mudstone, dark and silty at base; shells ..		5 2	3078 0
Siltstone		9 0	3087 0
Mudstone, silty, with ironstone		5 6	3092 6
Mudstone, dark; ironstone and shells ..		4 4	3096 10
TOP SILKSTONE			
Coal and dirt	46 in[1]	3 10	3100 8
Seatearth and coaly shale		3 10	3104 6
Siltstone and sandstone		16 4	3120 10

Shireoaks Colliery (No. 1 Shaft)

Ht above O.D. 195 ft.   6-in SK 58 SE.
Site [5592 8093] 600 yd at 90° from St Luke's Church, Shireoaks. Sunk 1854–9. For detailed section see Lancaster and Wright 1860, pp. 137–45; Wilcockson 1950, pp. 440–3; Edwards 1951, pp. 241–3, but depths of seams do not agree with those given in the notes below, which are taken from the original record.

MADE GROUND: Pit bank to 17¼ ft.

PERMO-TRIASSIC: MPM to 64¾ ft, LML to 188½ ft, LPM to 222½ ft, BPS to 223 ft 11 in.

MIDDLE COAL MEASURES: ?Position of Shafton M.B. at abt 264 ft; Sharlston Top (c. 9 in) at 550 ft 2 in; Sharlston Low (d.c. 12 in, d. 16 in on d.c. 6 in) at 631 ft 1 in; ?Sharlston Yard (c. 14 in) at 685 ft 10 in; ?Houghton Thin (c. 4 in, d. & measures 10 ft 1 in on c. & d. 6 in) at 713 ft; position of Mansfield M.B. at abt 801 ft; position of Haughton M.B. at abt 966 ft; Clown (c. 52 in) at 1039 ft 1 in; Mainbright (d. & c. 37 in) at 1092 ft 2 in; Two-Foot (c. 32 in) at 1155 ft 8 in; Abdy (c. 12 in, d. 20 in on d. & c. 8 in) at 1178 ft 4 in; Kilnhurst (c. 24 in) at 1227½ ft; High Hazles (c. 6 in, d. 7 in on c. 34 in) at 1300¼ ft; Top Hard (c. 46 in) at 1547 ft 7 in.

Bottom of shaft 1548 ft (bored to 1661¾ ft according to Edwards 1951).

South Carr Pumping Station

There are three boreholes, all 301 ft deep, and with reputedly identical sections, at this pumping station. They were drilled between 1920 and 1935 by C. Isler & Co. Ltd for Barber, Walker and Co. Ltd, owners of Harworth Colliery; they are now owned by the N.C.B. The notes below apply to No. 1 Borehole.

Ht above O.D. 31 ft.   6-in SK 69 SW.
Site [6398 9122] 2970 yd at 102° from All Saints' Church, Harworth. Drilled 1920–2.

PERMO-TRIASSIC: PB (? with LMS at base) to bottom of borehole at 301 ft.

---

[1]Detailed section: coal 1½ in, seatearth 5½ in, dirty coal 7 in, cannel 14 in, coaly shale 3 in, cannel 9 in, seatearth 3 in, dirty coal 3 in.

South Leverton Borehole

Ht above O.D. abt 25 ft.   6-in SK 78 SE.
Site [7945 8103] 1250 yd at 95° from All Saints' Church, South
Leverton. Drilled 1913–4. For detailed section see Lamplugh
and Smith 1914, p. 87; Ford 1920, pp. 99–100; Wilcockson 1950, pp.
462–3; Edwards 1951, p. 244; Lamplugh and Smith provide most
details of the Permo-Triassic, but these are at variance with the
other accounts.

DRIFT: Alluvium to abt 5 ft.

PERMO-TRIASSIC: KM to 352¾ ft, Waterstones to 504½ ft, GB to 547 ft, PB to 1325 ft,
  LMS to 1437 ft 7 in (but see p. 168), UPM to abt 1464¼ ft, UML to 1518 ft 5 in, MPM
  to 1664 ft 2 in, LML to 1914¼ ft, LPM to 1958¾ ft, BPS to 1959½ ft.

COAL MEASURES to bottom of borehole at 2592 ft 7 in.

South Leverton (BP) Oil Bores

There are fifteen bores in the South Leverton Oilfield, fourteen
of which are grouped between South Leverton and Treswell
(see Fig. 3). No. 18 Bore [7619 7881] is abt 1¼ miles WSW of
Treswell. Notes on No. 1 Bore are given below. Nos. 6 to 13 and
15 bores were drilled to the approximate base of the Coal Measures,
and Nos. 2 to 5 and 18 penetrated a short distance into the Millstone Grit Series, but in no case was the Pot Clay M.B. proved.

No. 1 Bore

R.T.E. above O.D. 37 ft.   6-in SK 78 SE.
Site [7933 8040] 1320 yd at 126° from All Saints' Church, South
Leverton. Drilled 1960 by BP Exploration Co. Ltd. Largely open
holed, cored 3417–41 ft, at intervals between 4130 ft and 4710 ft,
and from 5105 ft to bottom of bore. Cores examined by C. G.
Godwin.

DRIFT: Head, not recorded.

PERMO-TRIASSIC: KM to abt 330 ft, Waterstones to abt 490 ft, GB to 520 ft, PB to 1350 ft,
  LMS + UPM to 1460 ft, UML to 1510 ft, MPM to 1660 ft, LML to abt 1935 ft, LPM
  to abt 1947 ft, BPS to 1951 ft.

UPPER COAL MEASURES to abt 2090 ft.

MIDDLE COAL MEASURES: Position of Top M.B. at abt 2093 ft; position of Mansfield
  M.B. at abt 2545 ft; Top Hard at abt 3010 ft; position of Clay Cross M.B. at abt 3310 ft.

LOWER COAL MEASURES: Parkgate at abt 3557 ft; Kilburn at abt 3858 ft; fault cutting
  out abt 80 to 100 ft of measures at abt 3920 ft; base at abt 4125 ft.

MILLSTONE GRIT SERIES: M.B. with *R.* cf. *reticulatum* at 4575½ ft; base at abt 4970 ft·

CARBONIFEROUS LIMESTONE SERIES to bottom of bore at 5125 ft.

Steetley Colliery

Ht above O.D. 217 ft.   6-in SK 57 NE.
Site [5520 7848] 1180 yd at 356° from Ratcliffe Grange, Worksop.
Sinking commenced in 1873. For detailed section see Lamplugh
and Smith 1914, pp. 117–8; Wilcockson 1950, pp. 472–3; Edwards
1951, pp. 245–7.

MADE GROUND: Pit bank to 28¾ ft.

PERMO-TRIASSIC: MPM to 36 ft 1 in, LML to 115 ft 11 in, LPM to 176 ft 4 in, BPS to
  223¾ ft.

MIDDLE COAL MEASURES: ?position of Shafton M.B. at abt 328 ft; position of *Edmondia* Band at abt 506 ft; Sharlston Low (c. 18 in, d. 4 in on c. 14 in) at 736 ft; ?Sharlston Yard (c. 6 in) at 776 ft 7 in; ?Houghton Thin (c. 18 in, d. 55 in, c. 7 in, d. 84 in on c. 7 in) at 810 ft 4 in; position of Mansfield M.B. at abt 907 ft; position of Haughton M.B. at abt 1086 ft; Clown (c. 39 in) at 1169 ft 4 in; Mainbright (c. 20 in) at 1238 ft 11 in; Two-Foot (c. 33 in) at 1307 ft 10 in; Abdy (c. 17 in) at 1337 ft 10 in; Kilnhurst (c. 13 in) at 1398 ft 7 in; High Hazles (c. 36 in) at 1494 ft 10 in; Kent's Thick (d. & c. 38 in) at 1551 ft 5 in; Top Hard (c. 30 in on d.c. 10 in) at 1768¾ ft.

Bottom of shaft 1825 ft 11 in.

Steetley Colliery No. 1 Underground Borehole

Top in Top Hard workings 1561 ft below O.D. 6-in SK 57 NE.
Site [5520 7831] 1000 yd at 355° from Ratcliffe Grange, Worksop.
Drilled 1954 by Craelius Co. for N.C.B. Cored throughout.
Cores examined by R. F. Goossens.

MIDDLE COAL MEASURES: Dunsil (c. 16 in, d. & measures 8 ft 10 in on c. 16 in) at 80½ ft; 1st Waterloo (ca. 3 in on c. 24 in) at 111¼ ft; Swallow Wood (c. 2 in, d. 19 in, d.c. 2½ in, c. 15½ in, d. 35 in on c. 15 in) at 181 ft 2 in; Haigh Moor (c. 32 in) at 225 ft 7 in; 2nd Ell (d.c. 3 in) at 349 ft 7 in; Clay Cross M.B. at 413¾ ft.

LOWER COAL MEASURES: Flockton Thick (d.c. & d. 57 in on c. 18 in) at 486 ft 4 in; ?Deep Hard Roof Coal (c. 14 in) at 634 ft 8 in; Deep Hard (d.c. & d. 19 in, c. 9 in, d. 8 in on d.c. 3 in) at 668 ft; Parkgate (d.c. 2½ in, c. 21 in, d. 6 in, c. 19 in on d.c. 3½ in) at 744 ft 10 in; Thorncliff (c. & d.c. 30¾ in, d.c. 16¼ in, d. 4 in, d.c. 9 in on c. 15 in) at 808 ft 10 in; Top Silkstone (c. 1 in, d. 11 in, d.c. 12 in, c. & d. 13 in, d.c. 2 in, d. 20 in on c. 4 in) at 939 ft 8 in; Low Silkstone (c. 9 in) at 960 ft 5 in.

Bottom of borehole 981¾ ft.

Steetley Colliery (Water) Boreholes

There are two boreholes near Steetley Colliery shafts. No geological details are available for No. 1 [5516 7851]. The following notes apply to No. 2 Borehole.

Ht above O.D. 186 ft.   6-in SK 57 NE.
Site [5512 7844] 1180 yd at 352° from Ratcliffe Grange, Worksop.
Drilled 1948 by C. Isler & Co. Ltd for N.C.B.

MADE GROUND to 5 ft.

PERMO-TRIASSIC: MPM to 6½ ft, LML to abt 84 ft, LPM to 150 ft, BPS to 191 ft.

MIDDLE COAL MEASURES to bottom of borehole at 250 or 251 ft.

Steetley Wood Borehole

This is one of two boreholes in Steetley Quarry. No geological details are available for the other, which was 200 ft deep.

Ht above O.D. abt 150 ft.   6-in SK 57 NE.
Site [5510 7898] 1490 yd at 265° from Lodge Farm, Worksop.
Drilled 1944 by J. T. Hymas for Steetley Co. Ltd. Cores below 50 ft examined by W. G. Fearnsides, W. H. Wilcockson and W. Edwards.

MADE GROUND to 8 ft.

PERMO-TRIASSIC: LML to 94 ft, LPM to 129½ ft, BPS to bottom of borehole at 160 ft.

T

Stone Borehole

Ht above O.D. 215 ft.   6-in SK 58 NE.
Site [5556 8992] 1610 yd at 317° from Firbeck Hall. Drilled 1965 by Foraky Co. for N.C.B. Open holed to 100 ft, cored 100 ft to bottom. Cores examined by R. F. Goossens.

PERMO-TRIASSIC: LML to 109 ft, LPM to 123 ft, BPS to $126\frac{1}{4}$ ft.

UPPER COAL MEASURES: Badsworth Div. to 416 ft 8 in, Hemsworth Div. to $729\frac{3}{4}$ ft, Brierley Div. to 953 ft 11 in (Blyth, ca. $7\frac{1}{2}$ in, c. $13\frac{3}{4}$ in, d. 1 in, d.c. $\frac{1}{2}$ in on c. $10\frac{1}{2}$ in,) at 953 ft 11 in, Ackworth Div. to 1192 ft.

MIDDLE COAL MEASURES: Top M.B. at $1197\frac{1}{2}$ ft; Shafton M.B. at 1342 ft; *Edmondia* Band at 1559 ft; Sharlston Top (c. $11\frac{3}{4}$ in, d. 2 in on d.c. $2\frac{1}{4}$ in) at 1619 ft 10 in; Sharlston Low (c. 12 in, d. 21 in, d.c. 3 in, d. 19 in on d.c. 3 in) at $1735\frac{1}{4}$ ft; Sharlston Yard (c. with $\frac{1}{4}$ in d., 8 in, d.c. 1 in on c. $14\frac{1}{2}$ in) at 1807 ft 4 in; Houghton Thin (c. 8 in) at $1829\frac{1}{4}$ ft; Mansfield M.B. at 1907 ft 1 in; Meltonfield (c. 10 in, d. with $5\frac{3}{4}$ in d.c., $27\frac{1}{4}$ in, d.c. $3\frac{3}{4}$ in, d. 1 in on d.c. 10 in) at 2044 ft 1 in; Two-Foot M.B. at 2096 ft 7 in; Two-Foot (d.c. 1 in on c. 20 in) at 2098 ft 4 in; Abdy (d.c. $3\frac{1}{2}$ in, c. with $\frac{1}{2}$ in d., $10\frac{1}{4}$ in, d. $2\frac{1}{4}$ in, d.c. $\frac{1}{2}$ in on c. $6\frac{1}{4}$ in) at $2123\frac{1}{4}$ ft; Kilnhurst (c. 1 in, d. 6 in, c. 2 in, d. 4 in on c. 22 in) at 2173 ft 7 in; Kent's Thin (c. 19 in, d. 1 in, d.c. $2\frac{1}{4}$ in, d. $1\frac{1}{2}$ in on d.c. $1\frac{1}{4}$ in) at 2254 ft 5 in; Kent's Thick, top leaf (c. $10\frac{1}{4}$ in, d. $1\frac{1}{4}$ in, c. $2\frac{3}{4}$ in, d. $7\frac{3}{4}$ in on d.c. 2 in) at 2309 ft 5 in; bottom leaf (d.c. $5\frac{3}{4}$ in, d. $3\frac{3}{4}$ in on d.c. 10 in) at $2336\frac{3}{4}$ ft; Barnsley (d.c. $\frac{3}{4}$ in, d. $1\frac{3}{4}$ in, c. & d.c. 6 in, d. $5\frac{1}{2}$ in, c. 14 in, d. $\frac{3}{4}$ in, c. $1\frac{1}{4}$ in, d. & d.c. $9\frac{1}{2}$ in, c. 40 in, d. 17 ft 4 in on c. 3 in) at 2476 ft 8 in; Dunsil (c. $9\frac{1}{2}$ in) at 2508 ft 10 in; 1st Waterloo (c. 4 in, d. 1 in on c. 2 in) at 2591 ft 10 in; Swallow Wood (c. $2\frac{1}{4}$ in, d.c. $1\frac{1}{4}$ in, c. $\frac{1}{2}$ in, d. $1\frac{1}{4}$ in, c. with $\frac{1}{2}$ in d., $22\frac{3}{4}$ in, d. $10\frac{1}{4}$ in, d.c. 2 in, c. $2\frac{1}{2}$ in, d. 2 in, d.c. $\frac{1}{2}$ in, c. $17\frac{3}{4}$ in, d.c. 1 in, d. 2 in on c. 7 in) at $2667\frac{1}{4}$ ft; Haigh Moor (c. 5 in, d. $\frac{3}{4}$ in, c. $1\frac{3}{4}$ in, d.c. $1\frac{1}{4}$ in, d. 1 in, c. $15\frac{1}{4}$ in, d. & d.c. $10\frac{1}{2}$ in on c. $14\frac{1}{2}$ in) at 2710 ft; Lidget (d.c. 4 in, c. $1\frac{1}{4}$ in, d.c. 1 in on c. $16\frac{3}{4}$ in) at 2796 ft 11 in; Clay Cross M.B. at $2915\frac{1}{4}$ ft.

LOWER COAL MEASURES: Flockton Thick (d.c. & d. $2\frac{1}{4}$ in, c. $14\frac{3}{4}$ in, d. $3\frac{1}{4}$ in on c. $6\frac{1}{4}$ in) at 2963 ft 8 in; Deep Soft, top leaf (inferior ca. $2\frac{1}{4}$ in, c. $1\frac{3}{4}$ in, d. 2 in on c. 12 in) at 2987 ft 8 in; Deep Soft, main bed (d. & d.c. $33\frac{1}{2}$ in, c. 6 in, d. 47 in, c. 24 in, d.c. $4\frac{1}{2}$ in on c. $2\frac{1}{2}$ in) at 3026 ft 4 in; Deep Hard (c. $4\frac{3}{4}$ in, d. $4\frac{3}{4}$ in, d.c. $2\frac{1}{2}$ in, d. 13 in, c. $2\frac{3}{4}$ in, d. $13\frac{1}{2}$ in, c. $7\frac{1}{4}$ in, d. 49 in on c. 19 in) at $3084\frac{1}{4}$ ft; Parkgate (c. & d.c. $3\frac{1}{4}$ in, d. 1 in on c. $52\frac{1}{4}$ in) at 3162 ft 8 in; Thorncliff (c. 16 in, d.c. with $\frac{1}{2}$ in d., $5\frac{1}{2}$ in, d. $9\frac{1}{2}$ in on d.c. & d. $27\frac{1}{2}$ in) at 3227 ft 5 in.

Bottom of borehole 3233 ft 5 in.

Sutton Borehole

Ht above O.D. abt 45 ft.   6-in SK 68 SE.
Site [6817 8385] 1220 yd at 177° from Sutton church. Drilled 1925–6 by Foraky Co. for J. Nimmo & Co. Ltd. Cored except for part of LML. For detailed section above 2738 ft see Wilson 1927, pp. 138–40, and above $2890\frac{1}{2}$ ft see Edwards 1951, pp. 247–9.

DRIFT: Terrace deposits, largely sand and gravel, to 24 ft 4 in.

PERMO-TRIASSIC: PB to $501\frac{1}{4}$ ft, LMS to $597\frac{3}{4}$ ft, UPM to 632 ft 8 in, UML to 677 ft 8 in, MPM to 838 ft, LML to 1083 ft 11 in, LPM to 1091 ft 7 in, BPS to 1096 ft 10 in.

UPPER COAL MEASURES: Badsworth Div. to ?1158 ft, Hemsworth Div. to ?1361 ft 2 in, Brierley Div. to abt 1475 ft, Ackworth Div. to abt 1645 ft.

MIDDLE COAL MEASURES: Top M.B. at $1659\frac{3}{4}$ ft; Shafton M.B. at $1715\frac{1}{2}$ ft; Wales (c. 2 in, d. 5 in on c. 3 in) at 2032 ft 4 in; Mansfield M.B. at $2099\frac{1}{2}$ ft; Sutton M.B. at 2128 ft 5 in; Haughton M.B. at 2174 ft 5 in; Clown (c. 31 in) at $2231\frac{1}{4}$ ft; Two-Foot (c. $22\frac{1}{2}$ in) at 2320 ft 4 in; Top Hard + Dunsil (c. $57\frac{1}{2}$ in) at $2682\frac{1}{4}$ ft; 1st Waterloo (c. $13\frac{1}{2}$ in) at 2710 ft 11 in; Swallow Wood (c. $21\frac{1}{2}$ in) at 2770 ft 5 in; Clay Cross M.B. at $2908\frac{1}{2}$ ft.

LOWER COAL MEASURES: ?Deep Hard (c. 2 in) at 3099 ft 8 in; Parkgate (d.c. 8 in, d. 43 in on c. 33 in) at 3163¾ ft; Low '*Estheria*' Band at 3362 ft 8 in; Low Silkstone (c. 7½ in) at 3363 ft 4 in.

Bottom of borehole 3552 ft 8 in.

Tickhill (BP) No. 1 Oil Bore

R.T.E. above O.D. 95 ft.   6-in SK 59 SE.
Site [5773 9297] 310 yd at 332° from Limestone Hill, Tickhill. Drilled 1958 by BP Exploration Co. Ltd. Largely open holed to 4250 ft with sporadic cores between 2570 ft and 3285 ft, largely cored 4250–5050 ft, open holed 5050 ft to bottom. Some cores examined by E. G. Smith.

PERMO-TRIASSIC: MPM to 50 ft, LML to abt 225 ft, LPM to abt 228 ft (BPS not recorded but may be present to abt 245 ft).

UPPER COAL MEASURES: Badsworth, Hemsworth and Brierley Divisions to abt 879 ft, Ackworth Div. to abt 1135 ft.

MIDDLE COAL MEASURES: Top M.B. horizon at abt 1140 ft; ?Two-Foot at abt 2030 ft; Barnsley goaf at 2390 ft; position of Clay Cross M.B. at 2792 ft.

LOWER COAL MEASURES: Deep Soft at abt 2990 ft; Parkgate at abt 3050 ft; ?Top Silkstone at abt 3265 ft; Kilburn horizon at abt 3760 ft; Pot Clay M.B. at 4428¼ ft.

MILLSTONE GRIT SERIES: *G. cumbriense* M.B. at 4471 ft, *G. cancellatum* M.B. at 4533 ft 8 in.

Bottom of borehole 5612 ft.

Torworth Borehole

Ht above O.D. 81 ft.   6-in SK 68 NW.
Site [6495 8559] 640 yd at 120° from Jubilee Farm, Torworth. Drilled 1953 by Foraky Co. for N.C.B. Open holed to 767 ft, cored 767 ft to bottom. Core examined by R. A. Eden and R. E. Elliott.

PERMO-TRIASSIC: PB + LMS to 390 ft, UPM to 431¾ ft, UML to 470 ft, MPM to 590 ft, LML to 781 ft 10 in, LPM to 791 ft 10 in, BPS to 864 ft.

UPPER COAL MEASURES: Badsworth Div. to 998½ ft, Hemsworth Div. to 1255 ft 7 in, Brierley Div. to 1404 ft 8 in (Blyth, c. 1½ in, pyrite 1 in on c. 33½ in at 1404 ft 8 in), Ackworth Div. to 1625 ft 1 in.

MIDDLE COAL MEASURES: Shafton M.B. at 1664 ft 7 in; Shafton (c. 18 in) at 1710 ft 10 in; Sharlston Low (c. 10 in, d. & measures 10 ft 1 in on c. 2 in) at 1959 ft 2 in; Sharlston Yard (d.c. 1 in, d. 3 in, d.c. 3 in on c. 11 in) at 1992 ft; ?Houghton Thin (c. 7 in, d. 3 in on d.c. 10 in) at 2004 ft; Mansfield M.B. at 2062 ft 11 in; Clown (c. 30 in) at 2231½ ft; Two-Foot M.B. at 2342 ft 4 in; Two-Foot (d.c. 10 in, c. 11½ in, d.c. 4 in on c. 6½ in) at 2345 ft; Abdy (d.c. & c. with 1 in d., 27 in) at 2378 ft; Kilnhurst (c. 11 in) at 2401 ft; High Hazles, Top Bed (c. 14 in) at 1475¼ ft; Kent's Thin (c. 21 in) at 2508 ft 1 in; Kent's Thick, top leaf (c. 18 in) at 2551 ft 4 in; bottom leaf (c. 8 in, d. 2 in on c. 3 in) at 2577 ft 7 in; Top Hard (ca. 6 in, c. 7 in, ca. 2 in, d. 20½ in, d.c. 2 in, c. 6½ in, d. 1 in, c. 2 in, ca. 14 in on d.c. 2 in) at 2725 ft 5 in; Blidworth + Dunsil[1] (c. with 1¼ in d., 37 in) at 2764 ft 5 in; 1st Waterloo (c. & d.c. 13 in) at 2793 ft 10 in; Swallow Wood (c. 8 in) at 2848 ft 8 in; Haigh Moor (c. 26 in) at 2872 ft 4 in; Lidget (c. 8 in) at 2946½ ft; Clay Cross M.B. at 3080 ft 8 in.

---

[1]Section and thickness as proved in diversion.

LOWER COAL MEASURES: Deep Soft (c. 15 in, d.c. 4 in, c. $9\frac{1}{2}$ in, d.c. 3 in, c. 2 in, d.c. $2\frac{1}{2}$ in, c. 12 in, d.c. 3 in, d. 1 in on c. 32 in) at 3194 ft 4 in; Deep Hard (c. 8 in, d. 59 in on d.c. & c. 27 in) at 3258 ft 8 in; Parkgate[1] (c. 10 in, d. 73 in, d.c. $10\frac{1}{2}$ in, c. $16\frac{1}{2}$ in, d. 4 in, c. $1\frac{1}{2}$ in, d.c. 4 in on c. 24 in) at 3321 ft $1\frac{1}{2}$ in; Thorncliff (c. 17 in, d. 2 in, d.c. & c. $5\frac{1}{2}$ in, d. 8 in, d.c. 6 in, c. $3\frac{1}{2}$ in on d.c. $5\frac{1}{2}$ in) at $3373\frac{3}{4}$ ft; Threequarters (c. 7 in) at 3407 ft 7 in; Low '*Estheria*' Band at $3498\frac{3}{4}$ ft; Low Silkstone (c. $7\frac{1}{2}$ in) at 3499 ft 4 in.

Bottom of borehole $3555\frac{1}{2}$ ft.

Twyford Bridge Borehole

Ht above O.D. 75 ft.   6-in SK 67 NE.
Site [6980 7545] 980 yd at 95° from Elkesley church. Drilled 1962 by Foraky Co. for N.C.B. Open holed to 1950 ft, cored 1950 ft to bottom. Cores examined by C. G. Godwin, M. Lock and R. E. Elliott.

PERMO-TRIASSIC: GB to abt 19 ft, PB to abt 600 ft, LMS to abt 625 ft, UPM to 740 ft, UML to 759 ft, MPM + LML to 1160 ft, LPM to 1170 ft, BPS to 1172 ft.

UPPER COAL MEASURES: base not defined, may be about 1300 ft.

MIDDLE COAL MEASURES: ?position of *Edmondia* Band at abt 1575 ft[2]; ?Sharlston Top (c. & d. 26 in) at abt 1623 ft[2]; ?Wales (d.c. on c. 42 in) at abt 1728 ft[2]; position of Mansfield M.B. at 1781 ft[2]; Clown (c. 27 in) at abt 1927 ft[2]; Mainbright (d.c. 2 in, c. 21 in, d. 14 in on c. 7 in) at 1964 ft 11 in; Two-Foot M.B. at 1980 ft 7 in; Two-Foot (c. 13 in) at 1981 ft 8 in; Abdy (c. 13 in on d. & c. 8 in) at 2007 ft 4 in; Kilnhurst (c. 11 in) at 2050 ft 7 in; High Hazles, Top Bed (d.c. $5\frac{1}{2}$ in on c. 5 in) at $2084\frac{1}{2}$ ft; Kent's Thin (d.c. 2 in, c. 5 in, d. 5 in on d.c. 2 in) at 2118 ft 10 in; Kent's Thick (d.c. $3\frac{1}{2}$ in, c. 6 in, d.c. 3 in, c. 6 in, d.c. 1 in on c. 4 in) at 2164 ft 8 in; Top Hard + Dunsil (c. 5 in, d. 1 in, c. $13\frac{1}{2}$ in, d.c. $7\frac{1}{2}$ in, c. 13 in, d.c. 2 in, d. 1 in, c. $14\frac{1}{2}$ in, d. $2\frac{1}{2}$ in, c. 15 in on d.c. 16 in) at 2334 ft; Swallow Wood + Haigh Moor (c. 14 in, d. $18\frac{1}{2}$ in, c. $31\frac{1}{4}$ in, d. $2\frac{1}{4}$ in, c. $21\frac{1}{2}$ in on d.c. 2 in) at $2452\frac{1}{2}$ ft; Lidget (c. 5 in) at 2523 ft 5 in; 2nd Ell (c. $7\frac{1}{4}$ in, d.c. $7\frac{1}{4}$ in, c. $3\frac{1}{2}$ in, d.c. 3 in on c. 5 in) at $2572\frac{1}{2}$ ft; Clay Cross M.B. at 2636 ft 10 in.

LOWER COAL MEASURES: Deep Soft (c. $14\frac{1}{2}$ in, d. 4 in, d.c. 3 in, c. $30\frac{1}{2}$ in, d. $3\frac{1}{2}$ in on c. $2\frac{1}{2}$ in) at 2729 ft 4 in; ?Deep Hard Roof Coal (c. 3 in) at $2774\frac{1}{4}$ ft; Deep Hard (c. 5 in) at 2793 ft 5 in; Parkgate (c. $12\frac{1}{2}$ in, d. 2 in, c. $2\frac{1}{2}$ in, d. $\frac{1}{2}$ in, d.c. 1 in, c. 35 in, d. $2\frac{1}{2}$ in on d.c. $4\frac{1}{2}$ in) at 2838 ft; Cockleshell (c. $\frac{1}{2}$ in) at 2877 ft $1\frac{1}{2}$ in; Threequarters (c. & d. 7 in, measures 7 ft, d.c. $3\frac{1}{2}$ in on c. $29\frac{1}{2}$ in) at 2953 ft 1 in.

Bottom of borehole 2987 ft.

Walkeringham (BP) Oil Bores

There are two oil bores at Walkeringham. Notes on No. 1 are given below. No. 2 [7583 9091] was drilled to 5595 ft, abt 900 ft into the Millstone Grit Series.

No. 1 Bore

R.T.E. above O.D. 125 ft.   6-in SK 79 SE.
Site [7555 9190] 210 yd at 164° from Highfield Farm, Walkeringham. Drilled 1959 by BP Exploration Co. Ltd. Largely open holed, cored at intervals between 3700 ft and 5460 ft and from 6342 ft to bottom of bore. Some of cores examined by G. D. Gaunt and G. H. Rhys.

---

[1]Section of seam below 73-in dirt is that proved in diversion. The section of the upper leaf and the depth of the base of the whole seam are as proved in the main hole.

[2]Interpreted from gamma-ray log.

PERMO-TRIASSIC: KM to 430 ft, PB to 1220 ft, LMS to abt 1290 ft, UPM to 1430 ft, UML to abt 1520 ft, MPM to 1645 ft, LML to 1945 ft, LPM to abt 1952 ft, BPS to 1961 ft.

UPPER COAL MEASURES to abt 2100 ft.

MIDDLE COAL MEASURES: Shafton at 2181 ft; Top Hard at 3043 ft; position of Clay Cross M.B. at abt 3445 ft.

LOWER COAL MEASURES: Parkgate at 3619 ft; Low '*Estheria*' Band at 3840½ ft; Kilburn at abt 4172 ft; Alton M.B. at 4552 ft; Pot Clay M.B. at 4607 ft 2 in.

MILLSTONE GRIT SERIES: *G. cancellatum* M.B. at abt 4693 ft.

Bottom of borehole 6390 ft.

Wallingbrook Wood Borehole

Ht above O.D. 174 ft.   6-in SK 57 NE.
Site [5536 7609] 710 yd at 229° from New Farm, Worksop. Drilled 1955–6 by Boyles Bros. Ltd for N.C.B. Open holed to 20¾ ft, cored 20¾ ft to bottom. Cores examined by E. G. Smith, L. S. O. Morris and G. H. Rhys.

PERMO-TRIASSIC: MPM to abt 42 ft 4 in, LML to 170 ft, LPM to below 196 ft 4 in, BPS at abt 200 ft.

MIDDLE COAL MEASURES: *Edmondia* Band at 371¼ ft; Sharlston Top (c. 6 in, d. 12 in, c. 4 in, d. 9 in, d.c. 5 in, d. 20 in on d.c. 4 in) at 438¾ ft; Sharlston Low (c. 7 in, d. & d.c. 6 in on c. 6 in) at 529 ft 7 in; ?Houghton Thin (c. 9 in, d. 51 in on d.c. 22 in) at 608¾ ft; Mansfield M.B. at abt 699 ft; Haughton M.B. at 845½ ft; Clown (c. 38 in) at 950 ft 2 in; Mainbright (c. with d.c. & a 6-in dirt band, 40 in on d. with three thin coal bands, abt 5½ ft) at 1048¾ ft.

Bottom of borehole 1052 ft 1 in.

Wallingwells Borehole

Ht above O.D. abt 105 ft.   6-in SK 58 NE.
Site [5789 8629] 820 yd at 105° from Langold Farm, Letwell. Drilled 1912. Open holed through Permo-Triassic and probably through upper part of Coal Measures. For detailed section see Wilson 1926, pp. 262–4; Wilcockson 1950, pp. 512–3.

PERMO-TRIASSIC: MPM to 115 ft, LML to abt 198 ft, LPM to 232 ft (BPS absent).

UPPER COAL MEASURES: Badsworth Div. to abt 292 ft, Hemsworth Div. to abt 670 ft, Brierley Div. to 856 ft, Ackworth Div. to abt 1095 ft (base not defined).

MIDDLE COAL MEASURES: position of Top M.B. abt 1100 ft; Shafton M.B. at abt 1206 ft; Sharlston Low (c. 5 in) at 1560 ft 5 in; Mansfield M.B. at 1715 ft; Clown (c. 36 in) at 1930 ft; Two-Foot M.B. at 2069 ft; Two-Foot (c. 33 in) at 2071¼ ft; Abdy (c. 16 in) at 2098 ft 4 in; Kilnhurst (c. 16 in) at 2134 ft 4 in; Kent's Thin (c. 14 in) at 2209 ft 7 in; Kent's Thick (c. 12 in, d. 24 in on c. 12 in) at 2312 ft; Barnsley (c. 49 in) at 2497 ft 1 in; Dunsil (c. 23 in, d. 47 in, c. 8 in, d. 5 in, c. 5 in, d. 9 in on c. 12 in) at 2580 ft 7 in; 1st Waterloo (c. 12 in) at 2607 ft.

Bottom of borehole 2642½ ft.

Welbeck Abbey Borehole

Ht above O.D. 171 ft.   6-in SK 57 SE.
Site [5629 7477] 1240 yd at 265° from Welbeck Woodhouse. Drilled 1956 by Boyles Bros. Ltd for N.C.B. Open holed to 10 ft, cored 10 ft to bottom. Cores examined by G. H. Rhys.

PERMO-TRIASSIC: MPM to abt 99¾ ft, LML to 264¾ ft, LPM to 291 ft 5 in, BPS to 291¾ ft.

MIDDLE COAL MEASURES: ?Sharlston Top (d.c. 2 in, d. 19 in, d.c. 3 in, d. 11 in on c. 2 in) at 332 ft 4 in; Sharlston Low (d.c. 12 in, d. 21 in on c. 9 in) at 423½ ft; Sharlston Yard (c. & d.c. 8 in) at 450 ft; ?Houghton Thin (c. 8 in, d. 9 in on c. 17 in) at 478 ft 8 in; Mansfield M.B. at 565 ft; Sutton M.B. at 629 ft 4 in; Haughton M.B. at 696¾ ft; Clown (c. 25 in) at 781½ ft; Mainbright (c. & d.c. abt 30½ in, d. 3½ in, c. 1¼ in, d.c. abt 1¾ in, d. abt 3½ in, d.c. 4 in, d. ½ in on c. 4½ in) at 872 ft 7 in.

Bottom of borehole 881½ ft.

West Drayton No. 1 Borehole (and BP Oil Bore)

Ht above O.D. 80 ft.   6-in SK 77 SW.
Site [7024 7448] 1000 yd at 254° from St Paul's Church, West Drayton. Drilled 1940 by Foraky Co. for Wigan Coal Corporation to 3358 ft and by D'Arcy Exploration Co. Ltd to 3901 ft. Open holed to 979¾ ft, cored 979¾ ft to bottom. Cores examined by W. Edwards.

DRIFT: Terrace deposits to abt 10 ft.

PERMO-TRIASSIC: Waterstones + GB to 50 ft +, PB + LMS to 790 ft, UML to 810 ft, MPM to 964¾ ft, LML to 1205 ft, LPM to 1220 ft 5 in, BPS to 1223¼ ft.

UPPER COAL MEASURES: Brierley Div. to ?1287 ft 11 in (?Blyth, c. 6 in at 1287 ft 11 in), Ackworth Div. to 1454 ft 4 in.

MIDDLE COAL MEASURES: Top M.B. at 1466 ft 8 in; Shafton M.B. at 1500 ft; *Edmondia* Band at 1639 ft; Sharlston Top (c. 29 in) at 1685 ft 10 in; Wales (d. & d.c. 34 in on c. 2 in) at 1781 ft; Mansfield M.B. at 1830 ft 4 in; Sutton M.B. at 1870½ ft; Haughton M.B. at 1912½ ft; Clown (c. 12 in) at 1965 ft; Mainbright (c. 14 in) at 2003 ft 2 in; Two-Foot (c. 20 in) at 2026 ft 1 in; Abdy (c. 10 in) at 2046 ft; Kilnhurst (c. 7 in) at 2094 ft 11 in; Kent's Thin (c. & ca. 2 in, d. 3 in on c. with 2 in ca., 22 in) at 2175 ft 4 in; Kent's Thick (c. & d. 14 in) at 2206½ ft; Top Hard + Dunsil (c. 34 in, d. 10 in, c. 16 in, d. 5 in, c. 15 in, d. 5 in on c. 26 in) at 2369¾ ft; Swallow Wood + Haigh Moor (c. 9 in, d. 3 in, c. 17 in, d. 6 in on c. 23 in) at 2483 ft 7 in; Lidget (c. & d. 8 in on c. 2 in) at 2557 ft 1 in; 2nd Ell (c. 2 in, d. 4 in on c. 12 in) at 2603 ft 5 in; Clay Cross M.B. at 2668 ft.

LOWER COAL MEASURES: Deep Soft (d. & c. 2 in) at 2765 ft 4 in; Parkgate (c. 72 in) at 2877 ft 1 in; Thorncliff, lower leaf (c. 3 in) at 2940¼ ft; Threequarters (c. 1 in, measures 13 ft 8 in on c. 42 in) at 2993 ft; Top Silkstone (c. 10 in) at 3046 ft; Low Silkstone (c. 6 in, d. 8 in on c. 1 in) at 3077 ft; position of Kilburn at abt 3427 ft; ?Upper Band (c. 9 in) at 3620 ft; Alton M.B. at abt 3708 ft; position of Pot Clay M.B. at abt 3827 ft.

MILLSTONE GRIT SERIES to bottom of borehole at 3901 ft.

West Drayton (BP) No. 2 Oil Bore

R.T.E. above O.D. 103 ft.   6-in SK 67 SE.
Site [6986 7404] 1650 yd at 85° from Haughton Park House, Bothamsall. Drilled 1953-4 by D'Arcy Exploration Co. Ltd. Largely open holed, cored 2260-300 ft, 3815-25 ft.

DRIFT: Terrace deposits, not recorded.

PERMO-TRIASSIC: Waterstones + GB to abt 55 ft, PB + LMS to 775 ft, UML to 800 ft, MPM to 960 ft, LML to 1190 ft, LPM to abt 1210 ft, BPS to 1212 ft.

UPPER COAL MEASURES: Ackworth Div. to abt 1390 ft.

MIDDLE COAL MEASURES: position of Top M.B. at abt 1395 ft; position of Mansfield M.B. at abt 1760 ft; Kent's Thick at abt 2135 ft; Top Hard at abt 2290 ft; position of Clay Cross M.B. at abt 2585 ft.

LOWER COAL MEASURES: Deep Soft at abt 2660 ft; Parkgate at abt 2790 ft; ?Top Silkstone at abt 2945 ft; base at abt 3700 ft.

MILLSTONE GRIT SERIES: Igneous rocks 3719-94 ft.

Bottom of borehole 3825 ft.

Wigthorpe Borehole

Ht above O.D. 90 ft. 6-in SK 58 SE.
Site [5891 8339] 570 yd at 170° from St John's Church, Carlton in Lindrick. Drilled 1953 by Craelius Co. Ltd for N.C.B. Open holed to 22 ft, cored 22 ft to bottom. Cores examined by R. F. Goossens.

MADE GROUND: Ash infilling old quarry 5 ft.

PERMO-TRIASSIC: UML to 10 ft, MPM to 143 ft, LML to 273 ft, LPM to 284½ ft, BPS to 287 ft.

UPPER COAL MEASURES: Hemsworth Div. to 435 ft 4 in, Brierley Div. to 609½ ft (Blyth, c. 21 in at 609½ ft), Ackworth Div. to 845 ft.

MIDDLE COAL MEASURES: Top M.B. at 850¾ ft; Shafton M.B. at 943¼ ft; Sharlston Top (d.c. 7 in) at 1193 ft; Sharlston Low (c. 4 in, d. 32 in on c. 7 in) at 1276¼ ft; Sharlston Yard + ?Houghton Thin (c. 15 in, d. 97 in on d.c. 2 in) at 1352 ft; Mansfield M.B. at 1441 ft; Clown (d.c. 5½ in on c. 23½ in) at 1660 ft 2 in; Meltonfield (c. 6¾ in, d.c. ¼ in on c. with thin d., 14 in) at 1720¾ ft; Two-Foot M.B. at 1790 ft 8 in; Two-Foot (d.c. 1½ in, c. 1 in, d.c. 7 in on c. 24½ in) at 1793½ ft; Abdy (c. 13 in, c. & d. 2 in, d. 4½ in, d.c. & d. 5 in on c. 3½ in) at 1823 ft 4 in; Kilnhurst (d.c. 1½ in, d. & d.c. 13½ in, c. with ¾ in d., 21½ in on d.c. & d. 3½ in) at 1847¾ ft; High Hazles, Top Bed (c. 2 in) at 1904¾ ft; Kent's Thin (c. 13 in, d. 1 in, c. 1 in, d. 11 in on d.c. 5 in) at 1939 ft 1 in; Kent's Thick, top leaf (d.c. 15½ in, d. 14½ in on c. 8½ in) at 1995 ft 1 in; bottom leaf (d.c. 3 in, d. 26½ in on c. 18½ in) at 2009¼ ft; ?Top Hard (c. 1 in) at 2189¾ ft (most of coal washed out); Dunsil (c. 15½ in, c. & d. 1½ in, d. 63 in, d.c. 3 in, d. 6 in on d.c. 6 in) at 2247¾ ft; 1st Waterloo (d.c. 3 in, d. 3 in, c. 2 in, d. 2 in, c. 5 in, d. 2 in, c. 1 in, d. 69 in on d.c. 1 in) at 2307 ft 10 in; Swallow Wood, Main Bed (d.c. 2¼ in, d. 3 in, c. 8 in, d.c. 2 in on c. 16½ in) at 2358¼ ft; Haigh Moor (c. 7¼ in, d. 10¾ in, c. 6 in, d.c. 1 in, c. 11 in, d. 36 in, c. 8½ in, d. 1 in on c. 10½ in) at 2396 ft; Lidget (c. 17 in) at 2492 ft 10 in; 2nd Ell (c. 4 in, d. 2 in on d.c. 4 in) at 2552 ft; Clay Cross M.B. at 2618½ ft.

LOWER COAL MEASURES: Flockton Thick (d.c. 15 in, d. 10 in, c. 1 in, d. 14 in on c. 3 in) at 2665¼ ft; Deep Soft, top leaf (c. 5½ in, d. ½ in on c. 10 in) at 2702¼ ft; Deep Soft, main bed (c. & d. 5¼ in, c. 9¼ in, d. 1 in, c. 6¼ in on d.c. & d. 1¼ in) at 2760 ft 2 in; ?Deep Hard Roof Coal (c. 12 in) at 2804½ ft; Deep Hard (d.c., ca. & d. 26 in, c. 10½ in, d. 1½ in on c. 3 in) at 2830 ft 2 in; Parkgate (c. 19½ in, d.c. 2 in, d. & measures 17 ft 5 in on c. 8 in) at 2902¼ ft; Thorncliff (c. 22 in, d.c. 4 in on d. & coal partings 39 in) at 2971 ft 1 in; ?Threequarters (d.c. & d. 19 in) at 2999 ft 10 in; Top Silkstone (ca. 24½ in, c. 6 in, d. 5½ in on c. & d.c. 7 in) at 3091 ft 4 in.

Bottom of borehole 3122 ft.

Worksop Brewery No. 2 Borehole

Ht above O.D. abt 106 ft. 6-in SK 57 NE.
Site [5893 7922] 720 yd at 145° from Worksop station. Drilled 1956–7 by C. Isler & Co. Ltd for Worksop and Retford Brewery Co. Ltd. Open holed to 24 ft, cored 24 ft to bottom. Cores examined by E. G. Smith.

MADE GROUND to 4 ft.

DRIFT: Alluvial clay and gravel to 15 ft.

PERMO-TRIASSIC: LMS to 33 ft 11 in (see p. 168), UML to 34¾ ft (but see p. 156), MPM to 178 ft (see pp. 150–1), LML to bottom of borehole at 300 ft (see pp. 132–3).

Worksop Brewery No. 1 Borehole [5900 7919], drilled in 1901, showed a similar succession to 201 ft. For detailed section see Lamplugh and Smith 1914, p. 127.

Worksop Waterworks (Sunnyside Pumping Station)

There are several boreholes at this pumping station, drilled between abt 1878 and 1948. Details of one borehole, drilled prior to 1908 and referred to in this memoir as No. 1 or the old borehole, are given by Lamplugh and Smith 1914, p. 126. The following notes apply to the 1948 borehole, referred to in the memoir as No. 2 or the new borehole.

Ht above O.D. abt 240 ft.   6-in SK 58 SE.
Site [5901 8064] 500 yd at 326° from Worksop Infirmary. Drilled 1947–8 by C. Isler & Co. Ltd for Worksop U.D.C.; now owned by Central Notts. Water Board. Open holed to 150 ft, cored 150 ft to bottom. Cores examined by R. F. Goossens.

PERMO-TRIASSIC: PB to abt 70 ft, LMS to $207\frac{1}{4}$ ft (see pp. 168–9), UPM absent, UML to 215 ft (see p. 156), MPM to bottom of borehole at 315 ft (see p. 150).

REFERENCES

EDWARDS, W. 1951. The concealed coalfield of Yorkshire and Nottinghamshire. 3rd edit. *Mem. geol. Surv. Gt Br.*

—— 1967. Geology of the country around Ollerton. 2nd edit. *Mem. geol. Surv. Gt Br.*

FORD, J. 1920. Record of the deep borings at Kelham and South Leverton. *Trans. Instn min. Engrs*, **58**, 94–107.

LAMPLUGH, G. W. and SMITH, B. 1914. The water supply of Nottinghamshire from underground sources. *Mem. geol. Surv. Gt Br.*

LANCASTER, J. and WRIGHT, C. C. 1860. On the sinking at Shireoak Colliery, Worksop, to the 'Top Hard Coal' or 'Barnsley Coal'. *Q. Jnl geol. Soc. Lond.*, **16**, 137–45.

WILCOCKSON, W. H. 1950. Sections of strata of the Coal Measures of Yorkshire. 3rd edit. *Midld Inst. min. Engrs.*

WILSON, G. V. 1926. The concealed coalfield of Yorkshire and Nottinghamshire. 2nd edit. *Mem. geol. Surv. Gt Br.*

—— 1927. The eastern boundary of the concealed coalfield of Yorkshire and Nottinghamshire. *Summ. Prog. geol. Surv. Gt Br. for 1926*, App. VII, 138–46.

LIST OF GEOLOGICAL SURVEY PHOTOGRAPHS
(One-inch Sheet 101)

Copies of these photographs are deposited for reference in the libraries of the Institute of Geological Sciences at Exhibition Road, South Kensington, London, SW7 2DE and Ring Road Halton, Leeds LS15 8TQ. Prints (those marked with an asterisk are available in colour) and lantern slides may be supplied at a fixed tariff. National Grid References, all in 100-km square SK, are given in square brackets; those of general views are of the viewpoints. Dates of photographs are also given.

PLEISTOCENE AND RECENT

A 8249–51	Glacial sand and gravel: Toulson's sand and gravel pit, Crossley Hill, $3\frac{1}{2}$ miles NE of Worksop [610 832]: 1946.
A 8257–62	Sand and gravel (Head) on Bunter Pebble Beds (contorted junction): old quarry, $\frac{1}{2}$ mile WSW of Retford station [6960 8013]: 1946.
L 529, 530	Sandy boulder clay on Bunter Pebble Beds: sand-pit at Lords Wood, Harworth [626 907] (L 529 = Plate XIXA): 1962.
L 531	Glacial sand and gravel showing festoon bedding: quarry north-west of Carlton Forest [6013 8218]: 1962.
L 532	Gravelly Head on Bunter Pebble Beds: sand-pit near Whisker Hill Junction, East Retford [6960 8013]: 1962.
L 533	Head at Walkeringham Brick Works [7548 9240]: 1962.
L 534–6	Terrace gravel of River Idle, at North Notts. Gravel Company's pit, near Sutton [700 841] (L 535 = Plate XIXB): 1962.
L 537	Terrace gravel of River Idle: gravel pit between Torworth and Danes Hill [666 866]: 1962.
L 538	Glacial sand and gravel: Scrooby Top Quarry [6525 8890]: 1962.
L 539	False-bedded glacial sands: another locality in Scrooby Top Quarry [652 890]: 1962.
L 540	Glacial sand and gravel: pit at Hollins Holt, near Scrooby Top [6554 8912]: 1962.

PERMO-TRIASSIC

A 5106–7	Weathering of dolomite masonry (Lower Magnesian Limestone): Welbeck Abbey [564 742]: 1930.
A 5108–9	Weathering of dolomitic sandstone masonry (Mansfield Sandstone, Lower Magnesian Limestone, Sheet 112): Welbeck Abbey: 1930.
A 5110–2	As 5106–7: 1930.
A 5113–4	Lower Magnesian Limestone: Steetley Quarries [549 791]: 1930.

A 5116   Weathering of dolomite masonry (Lower Magnesian Limestone): St Luke's Church, Shireoaks [5537 8094]: 1930.

A 5117   Ditto: Shireoaks Church School [5537 8089]: 1930.

A 5118–9 Lower Magnesian Limestone in quarry near Lindrick Common [556 826]: 1930.

A 5120–1 Weathering of dolomite masonry (Lower Magnesian Limestone): St Mary's Church, Tickhill [5918 9308]: 1930.

A 8242–5 Sandstone in Middle Permian Marl: sand-pit at Gateford [576 813]: 1946.

A 8246   Sandstone in Middle Permian Marl: sand-pit north-east of Fox Covert, near Gateford [564 823]: 1946.

A 8247   Basal beds of Upper Magnesian Limestone: railway cutting west of Bagley Farm, Tickhill [5870 9164]: 1946.

A 8248   Upper Magnesian Limestone: railway cutting 1 mile NNE of Old-coates [5958 9000]: 1946.

A 8252   Lower Mottled Sandstone at Wigthorpe [5914 8315]: 1946.

A 8253–4 Bunter Pebble Beds: old quarry ⅓ mile ESE of Worksop station [591 796]: 1946.

A 8255   Upper Magnesian Limestone on Middle Permian Marl: railway cutting at Langold [5874 8694]: 1946.

L 541   False-bedded Bunter Pebble Beds: old quarry at Drakeholes, Wiseton [7056 9036]: 1962.

L 542–4   Green Beds (mainly grassed over) on Bunter Pebble Beds: railway cutting at Thrumpton Goods Yard, East Retford [710 805]: 1962.

L 545   Keuper Marl with thin skerry bands: 400 yd SW of All Saints' Church, South Leverton [7808 8081]: 1962.

L 546   Keuper Marl with gypsum: east bank of River Trent, ½ mile S of Dunham Bridge [8213 7377]: 1962.

L 547–8   Lower Rhaetic shales: railway cutting at Lea [8342 8658]: 1962.

L 549   Gypsum near top of Keuper Marl: Gainsborough station [8181 9008]: 1962.

L 550   Keuper Marl: temporary excavation at West Burton Power Station site [795 853]: 1962.

L 551–2   False-bedded Bunter Pebble Beds: old quarry at Ellis Plantation, Mattersey [6869 8775]: 1962.

L 553–4   Bunter Pebble Beds with mudstone lens: Mattersey Sand Quarry, Ellis Plantation [6861 8778]: 1962.

L 555, 556*, Clarborough Beds in Keuper Marl: gypsum pit near Clarborough
L 557, 558*   [7396 8336]: 1962 and 1970.

L 559*   Bunter Pebble Beds with overlying Green Beds (overgrown): Bolham Lane, East Retford [7073 8207]: 1970.

L 561–2* As A 5118–9 (L 561 = Plate XII): 1971.

L 563–5* Lower Mottled Sandstone on sandstone in Middle Permian Marl: sand-pit at Gateford [576 813] (L 564 = Plate I): 1971.

L 566*   Working plant in sand-pit at Gateford [576 813]: 1971.

L 567*   As A 8242–5 (Plate XVA): 1971.

L 594–5*   Bunter Pebble Beds: Carlton Forest Quarry, 2 miles NNE of Worksop
             [6013 8218] (L 595 = Plate XVB): 1971.

L 596–7*   As L 555–8 (L 596 = Plate XVIA): 1971.

L 598*   As L 546 (Plate XVIB): 1971.

GENERAL

A 5105   Welbeck Abbey, on outcrop of Middle Permian Marl [564 742] : 1930.

A 5115   Limekilns at Steetley Quarry [549 791]: 1930.

A 8256   Topography of Permian and Bunter outcrops: view eastwards from
             Firbeck Colliery [586 861]: 1946.

A 8263–4   Anglo-American Oil Company's drilling rig: Gringley No. 1 Oil Bore
             [7457 9064]: 1946.

L 560   Alluvium of River Trent and scarp of Keuper Marl, general view looking
             towards High Marnham Power Station, from east bank of River Trent,
             ½ mile S of Dunham Bridge [8213 7377]: 1962.

MLD 5943   British Petroleum Company's drilling rig: Gainsborough No. 31 Oil
             Bore [8320 9112]: 1962.

INDEX OF FOSSILS

No distinction is made here between a positively determined species and variants of the species or examples doubtfully referred to it (i.e. with the qualifications aff., cf. or ?).

Fossils identifiable at generic level only (e.g. *Calamites sp.*) are listed after the named species.

A. PLANTS

Acanthotriletes castanea Butterworth & Williams, 26
Alethopteris serli Brongniart, 60
—— *valida* Boulay, 95
Algae, reef-forming calcareous, 120, 123, 127, 132, 154
——, others, 125; see also *Tubulites permianus*
Algites virgatus (Münster), Pl. X
Alisporites aequalis Mädler 1964, 196
—— *grauvogeli* Klaus 1964, 195–6
—— *microreticulatus* Reinhardt 1964, 195–6
—— *nuthallensis* Clarke 1965, 136–7
—— *parvus* de Jersey 1962, 196
Anapiculatisporites telephorus (Pautsch 1958) Klaus 1960, 196
Anaplanisporites protumulosus (Reinhardt 1964) Schulz 1965, 195
Angustisulcites gorpii Visscher 1966, 195–6
—— *grandis* (Freudenthal 1964) Visscher 1966, 195–6
—— *klausii* Freudenthal 1964, 195–6
Annularia microphylla Sauveur, 81
—— *sp.*, 60, 258
Apiculatasporites plicatus Visscher 1966, 195–6
Aratrisporites quadriiuga (Visscher 1966) Visscher & Commissaris 1968, 196
—— *saturni* (Thiergart 1949) Mädler 1964, 196
Asterophyllites longifolium var. *striata* Weiss, 81

Baltisphaeridium debilispinum Wall & Downie 1963, 137, 139, Pl. XIII
Bellispores nitidus (Horst) Sullivan, 26
Bothrodendron minutifolium Boulay, 81
Brachysaccus neomundanus (Leschik 1955) Mädler 1964, 196
—— *sp.*, 196

Calamites cisti Brongniart, 98
—— *suckowi* Brongniart, 96, 98–9
—— *sp.*, 260

Calamospora sp., 137, 195–6
Calathella dictyonemoides Stoneley, Pl. X
Callipteris martinsi Zeiller, Pl. X
Camptotriletes cristatus Sullivan & Marshall, 28
Chordasporites magnus Klaus 1964, 196
Cirratriradites rarus (Ibrahim) Schopf, Wilson & Bentall, 26, 28
Colpectopollis ellipsoideus Visscher 1966, 195–6
Convolutispora tesselata Hoffmeister, Staplin & Malloy, 26
—— *wicheri* (Thiergart 1949) Schulz 1965, 195
Cordaites principalis (Germar), 95, 98
Crassispora kosankei (Potonié & Kremp) Bharadwaj, 26–8
Crustaesporites globosus Leschik 1956, 136–7, Pl. XIII
Cycadopites accerrimus (Leschik 1955) Clarke 1965, 195–6
—— *subgranulosus* (Couper 1958) Clarke 1965, 195
cf. —— *sp.* R. Jansonius 1962, 136–7
—— *sp.*, 196
Cyclogranisporites arenosus Mädler 1964, 196
Cyclotriletes margaritatus Mädler 1964, 196
—— *microgranifer* Mädler 1964, 196
—— *oligogranifer* Mädler 1964, 195
cf. —— *pustulatus* Mädler 1964, 196
? —— *triassicus* Mädler 1964, 195
—— *sp.*, 196
Cycloverrutriletes presselensis Schulz 1964, 195

Densoisporites caretteae Visscher 1966, 195
—— *sp.*, 196
Densosporites anulatus (Loose) Smith & Butterworth, 27
Dictyotriletes tuberosus Neves, 28

321

B. ANIMALS

GENERAL INDEX

Printed in England for Her Majesty's Stationery Office
by Hull Printers Ltd., Willerby, Hull HU10 6DH

Dd 503880   K12